Texts in Applied Mathematics 22

Editors
J.E. Marsden
L. Sirovich
M. Golubitsky
W. Jäger

Advisor
G. Iooss

T0184399

Springer
New York
Berlin
Heidelberg
Barcelona
Hong Kong
London
Milan
Paris
Singapore
Tokyo

Texts in Applied Mathematics

J. W. Thomas

Numerical Partial Differential Equations: Finite Difference Methods

With 70 Illustrations

 Springer

J.W. Thomas
Department of Mathematics
Colorado State University
Fort Collins, CO 80543
USA

Series Editors

J.E. Marsden
Control and Dynamical Systems, 104–44
California Institute of Technology
Pasadena, CA 91125
USA

L. Sirovich
Division of Applied Mathematics
Brown University
Providence, RI 02912
USA

M. Golubitsky
Department of Mathematics
University of Houston
Houston, TX 77204-3476
USA

W. Jäger
Department of
   Applied Mathematics
Universität Heidelberg
Im Neuenheimer Feld 294
69120 Heidelberg, Germany

Mathematics Subject Classification: 35xx, 39xx

Library of Congress Cataloging-in-Publication Data
Thomas J.W. (James William), 1941–
    Numerical partial differential equations : finite difference
methods / J.W. Thomas.
       p.   cm. − (Texts in applied mathematics ; 22)
    Includes bibliographical references and index.
    ISBN 978-1-4419-3105-4
    1. Differential equations, Partial−Numerical solutions.
    2. Finite differences.   I. Title.   II. Series.
    QA377.T495   1995
    315'.353−dc20   95-17143

Printed on acid-free paper.

Production managed by Natalie Johnson; manufacturing supervised by Joseph Quatela.
Photocomposed copy prepared from the author's LAT$_E$X files.

Printed in the United States of America.

9 8 7 6 5 4 3

This book is dedicated to Ann,
David, Michael,
Carrie, Susan
and Jane.

Series Preface

Mathematics is playing an ever more important role in the physical and biological sciences, provoking a blurring of boundaries between scientific disciplines and a resurgence of interest in the modern as well as the classical techniques of applied mathematics. This renewal of interest, both in research and teaching, has led to the establishment of the series: *Texts in Applied Mathematics* (*TAM*).

The development of new courses is a natural consequence of a high level of excitement on the research frontier as newer techniques, such as numerical and symbolic computer systems, dynamical systems, and chaos, mix with and reinforce the traditional methods of applied mathematics. Thus, the purpose of this textbook series is to meet the current and future needs of these advances and encourage the teaching of new courses.

TAM will publish textbooks suitable for use in advanced undergraduate and beginning graduate courses and will complement the *Applied Mathematical Sciences* (*AMS*) series, which will focus on advanced textbooks and research level monographs.

Preface

This textbook is in two parts. The first part contains Chapters 1–7 and is subtitled *Finite Difference Methods*. The second part contains Chapters 8–11 and is subtitled *Conservation Laws and Elliptic Equations*. This text was developed from material presented in a year long, graduate course on using difference methods for the numerical solution of partial differential equations. Writing this text has been an iterative process, and much like the Jacobi iteration scheme presented in Chapter 10, convergence has been slow. The course at Colorado State University is designed for graduate students in both applied mathematics and engineering. The students are required to have at least one semester of partial differential equations and some programming capability. Generally, the classes include a broad spectrum of types of students, ranging from first year mathematics graduate students with almost no physical intuition into the types of problems we might solve, to third year engineering graduate students who have a lot of physical intuition and know what types of problems they personally want to solve and why they want to solve them. Since the students definitely help shape the class that is taught, they probably have also helped to shape this text.

There are several distinct goals of the courses. One definite goal is to prepare the students to be competent practiioners, capable of solving a large range of problems, evaluating numerical results and understanding how and why results might be bad. Another goal is to prepare the applied mathematics Ph.D. students to take additional courses (the third course in our sequence is a course in computational fluid dynamics which requires both semesters taught out of this text) and to write theses in applied mathematics.

One of the premises on which this text is based is that in order to understand the numerical solution of partial differential equations the student must solve partial differential equations. The text includes homework problems that implement different aspects of most of the schemes discussed. As a part of the implementation phase of the text, discussions are included on how to implement the various schemes. In later parts of the text, we return to earlier problems to discuss the results obtained (or that should have been obtained) and to explain why the students got the results they did. Throughout the text, the problems sometimes lead to bad numerical results. As I explain to my students, since these types of results are very common in the area of numerical solutions of partial differential equations, they must learn how to recognize them and deal with them. A point of emphasis in my course, which I hope that I convey also in the text, is teaching the students to become experimentalists. I explain that before one runs an experiment, one should know as much as possible about the problem. A complete problem usually includes the physical problem, the mathematical problem, the numerical scheme and the computer. (In this text, the physical problem is often slighted.) I then try to show how to run numerical experiments. As part of the training to be a numerical experimentalist, I include in the Prelude four nonlinear problems. I assume that the students do not generally know much about these problems initially. As we proceed in the text, I suggest that they try generalizations of some of our linear methods on these nonlinear problems. Of course, these methods are not always successful and in these cases I try to explain why we get the results that we get.

The implementation aspect of the text obviously includes a large amount of computing. Another aspect of computing included in the text is symbolic computing. When we introduce the concept of consistency, we show the calculations as being done on paper. However, after we have seen a few of these, we emphasize that a computer with a symbolic manipulator should be doing these computations. When we give algorithms for symbolic computations, we have tried to give it in a pseudo code that can be used by any of the symbolic manipulators. Another aspect of the new technologies that we use extensively is graphics. Of course, we provide plots of our numerical results and ask the students to provide plots of their results. We also use graphics for analysis. For example, for the analyses of dissipation and dispersion, where much of this has traditionally been done analytically (where one obtains only asymptotic results), we emphasize how easy it is to plot these results and interpret the dissipativity and dispersivity properties from the plots.

Though there is a strong emphasis in the text on implementing the schemes, there is also a strong emphasis on theory. Because of the audience, the theory is usually set in what might be called computational space (where the computations are or might be done) and the convergence is done in $\ell_{2,\Delta x}$ spaces in terms of the energy norm. Though at times these spaces might not be as nice mathematically as some other spaces that might be used, it seems that working in spaces that mimic the computational space is easier for the students to grasp. Throughout the text, we emphasize the meaning of consistency, stability and convergence. In my classes I emphasize that it is dangerous for a person who is using difference methods not to understand what it means for a scheme to converge. In my class and in the text, I emphasize that we sometimes get necessary and sufficient conditions for convergence and sometimes get only necessary conditions (then we must learn to accept that we have only necessary conditions and proceed with caution and numerical experimentation). In the text, not only do we prove the Lax Theorem, but we return to the proof to see how to choose an initialization scheme for multilevel schemes and how we can change the definition of stability when we consider higher order partial differential equations. For several topics (specifically for many results in Chapters 8, 9 and 10) we do not include all of the theory (specifically not all of the proofs) but discuss and present the material in a theoretically logical order. When theorems are used without proof, references are included.

Lastly, it is hoped that the text will become a reference book for the students. In the preparation of the text, I have tried to include as many aspects of the numerical solution of partial differential equations as possible. I do not have time to include some of these topics in my course and might not want to include them even if I had time. I feel that these topics must be available to the students so that they have a reference point when they are confronted with them. One such topic is the derivation of numerical schemes. I personally do not have a preference on whether a given numerical scheme is derived mathematically or based on some physical principles. I feel that it is important for the student to know that they can be derived both ways and that both ways can lead to good schemes and bad schemes. In Chapter 1, we begin by first deriving the basic difference mathematically, and then show how the same difference scheme can be derived by using the integral form of the conservation law. We emphasize in this section that the errors using the latter approach are errors in numerical integration. This is a topic that I discuss and that I want the students to know is there and that it is a possible approach. It is also a topic that I do not develop fully in my class. Throughout the text, we return to this approach to show how it differs when we have two dimensional problems, hyperbolic problems, etc. Also, throughout the text we derive difference schemes purely mathematically (heuristically, by the method of undetermined coefficients or by other methods). It is hoped the readers will understand that if they have to derive their own schemes for a partial differential equation not previously considered, they will know where to find some tools that they can use.

Because of the length of the text, as was stated earlier, the material is being given in two parts. The first part includes most of the basic material on time dependent equations including parabolic and hyperbolic problems, multi-dimensional problems, systems and dissipation and dispersion. The second part includes chapters on stability theory for initial–boundary value problems (the GKSO theory), numerical schemes for conservation laws, numerical solution of elliptic problems and an introduction to irregular regions and irregular grids. When I teach the course, I usually cover most of the first five chapters during the first semester. During the second semester I usually cover Chapters 6 and 7 (systems and dissipation and dispersion), Chapter 10 (elliptic equations) and selected topics from Chapters 8, 9 and 11. In other instances, I have covered Chapters 8 and 9 during the second semester, and on one occasion, I used a full semester to teach Chapter 9. Other people who have used the notes have covered parts of Chapters 1–7 and Chapter 10 in one semester. In either case, there seems to be sufficient material for at least two semesters of course work.

At the end of most of the chapters of the text and in the middle of several, we include sections which we refer to as "Computational Interludes." The original idea of these sections was to stop working on new methods, take a break from theory and compute for a while. These sections do include this aspect of the material, but as they developed, they also began to include more than just computational material. It is in these sections that we discuss results from previous homework problems. It is also in these sections that we suggest it is time for the students to try one of their new methods on one of the problems HW0.0.1–HW0.0.4 from the Prelude. There are also some topics included in these sections that did not find a home elsewhere. At times a more appropriate title for these sections might have been "etc.".

At this time I would like to acknowledge some people who have helped me with various aspects of this text. I thank Drs. Michael Kirby, Steve McKay, K. McArthur and K. Bowers for teaching parts of the text and providing me with feedback. I also thank Drs. Kirby, McArthur, Jay Bourland, Paul DuChateau and David Zachmann for many discussions about various aspects of the text. Finally, I thank the many students who over the years put up with the dreadfully slow convergence of this material from notes to text. Whatever the result, without their input the result would not be as good. And, finally, though all of the people mentioned above and others have tried to help me, there are surely still some typos and errors of thought (though, hopefully, many mistakes have been corrected for the Second Printing). Though I do so sadly, I take the blame for all of these mistakes. I would appreciate it if you would send any mistakes that you find to thomas@math.colostate.edu. Thank you.

J.W. Thomas

Contents

Contents of Part 2: Conservation Laws and Elliptic Equations

0
Prelude

Numerical partial differential equations is a large area of study. The subject includes components in the areas of applications, mathematics and computers. These three aspects of a problem are so strongly tied together that it is virtually impossible to consider one in isolation from the other two. It is impossible (or at least fool-hardy) to consider the applied aspect of a problem without considering at least some of the mathematical and computing aspects of that problem. Often, the mathematical aspects of numerical partial differential equations can be developed without considering applications or computing, but experience shows that this approach does not generally yield useful results.

Of the many different approaches to solving partial differential equations numerically (finite differences, finite elements, spectral methods, collocation methods, etc.), we shall study **difference methods**. We will provide a review of a large range of methods. A certain amount of theory and rigor will be included, but at all times implementation of the methods will be stressed. The goal is to come out of this course with a large number of methods about which you have **theoretical knowledge** and with which you have **numerical experience**.

Often, when numerical techniques are going to be used to solve a physical problem, it is not possible to thoroughly analyze the methods that are used. Any time we use methods that have not been thoroughly analyzed, we must resort to methods that become a part of **numerical experimentation**. As we shall see, often such experimentation will also become necessary for linear problems. In fact, we often do not even know what to try to prove analytically until we have run a well-designed series of experiments.

As in other areas of experimental work, the experiments must be carefully designed. The total problem may involve the physics that describes the original problem; the mathematics involved in the partial differential equation, the difference equation and the solution algorithm for solving the difference equation; and the computer on which the numerical work will be done. Ideally, before experiments are designed, one should know as much as possible about each of the above aspects of the problem.

Several times throughout the text we will assign small experiments that will hopefully clarify some aspects of the schemes we will be studying. It is hoped that these experiments will begin to teach some useful experimental techniques that can be used in the area of numerical solution of partial differential equations. In other parts of the text, it will be necessary or best for the reader to decide that a small experiment is needed, and then design and run that experiment. It is hoped that one thing the reader will learn from this text is that *numerical experimentation is a part of the subject of numerical solution of partial differential equations.*

Numerical methods for solving partial differential equations, like the analytic methods, often depend on the type of equation, i.e. elliptic, parabolic or hyperbolic. A slight background in the methods and theory of partial differential equations will be assumed. An elementary course in partial differential equations is generally sufficient. In addition, since part of the emphasis of the text is on the implementation of the difference schemes, it will be assumed that the reader has sufficient programming skills to implement the schemes and appreciate the implementations that are discussed.

The methods that we develop, analyze and implement will usually be illustrated using common model equations. Most often these will be the heat equation, the one way wave equation or the Poisson equation. We will generally separate the methods to correspond to the different equation types by first doing methods for parabolic equations, then methods for hyperbolic equations and finally methods for elliptic equations. However, we reserve the right to introduce an equation of a different type and a method for equations of different types at any time.

Much of the useful work in numerical partial differential equations that is being done today involves nonlinear equations. The one class of nonlinear equations that we will study is in Chapter 9, Part 2, ref. [13], where we consider the numerical solution of conservation laws. Often an analysis of the method developed to solve a nonlinear partial differential equation is impossible. In addition, the discrete version of the nonlinear problem may not itself be solvable. (Of course, methods like Newton's method can be used to solve the nonlinear problems. But Newton's method is really an iteration of a series of linear problems.) A common technique is to develop and test the schemes based on model linear problems that have similarities to the desired nonlinear problem. We will try to illustrate some methods that can be used to solve nonlinear problems. Below we include four model nonlinear problems. The reader should become familiar with these problems. When we develop linear methods that might be applicable to these nonlinear problems, a numerical experiment should be conducted using the linearized method to try to solve the nonlinear problem. Hints and discussions of methods will be given when we think we should be trying to solve problems HW0.0.1–0.0.4.

HW 0.0.1 (Viscous Burgers' Equation)

$$v_t + vv_x = \nu v_{xx}, \ x \in (0,1), \ t > 0 \qquad (0.0.1)$$
$$v(0,t) = v(1,t) = 0, \ t > 0 \qquad (0.0.2)$$
$$v(x,0) = \sin 2\pi x, \ x \in [0,1]. \qquad (0.0.3)$$

We wish to find solutions for $\nu = 1.0, \ 0.1, \ 0.01, \ 0.001, \ 0.0001$ and 0.00001.

The reader should be aware that there has been a large amount of work done on the above problem. There is an exact solution to the problem which is not especially useful to us. It is probably better to try to work this problem without knowing the exact solution since that is the situation in which a working numerical analyst must function.

HW 0.0.2 (Inviscid Burgers' Equation)

$$v_t + vv_x = 0, \ x \in (0,1), \ t > 0 \qquad (0.0.4)$$
$$v(0,t) = v(1,t) = 0, \ t > 0 \qquad (0.0.5)$$
$$v(x,0) = \sin 2\pi x, \ x \in [0,1]. \qquad (0.0.6)$$

We note that the only difference between this problem and HW0.0.1 is that the v_{xx} term is not included. As we shall see later, this difference theoretically changes the whole character of the problem. We also note that this problem is well-posed with a boundary condition at both $x = 0$ and $x = 1$.

HW 0.0.3 (Shock Tube Problem)

Consider a tube filled with gas where a membrane separates the tube into two sections. For the moment, suppose that the tube is infinitely long, the membrane is situated at $x = 0$, for $x < 0$ the density and pressure are equal to 2, for $x > 0$ the density and pressure are equal to 1 and the velocity is zero everywhere. At time $t = 0$, the membrane is removed and the problem is to determine the resulting flow of gas in the tube. We assume that the gas is inviscid, the flow is one dimensional and write the conservation laws for the flow (mass, momentum and energy) as

$$\rho_t + (\rho v)_x = 0 \qquad (0.0.7)$$
$$(\rho v)_t + (\rho v^2 + p)_x = 0 \qquad (0.0.8)$$
$$E_t + [v(E + p)]_x = 0 \qquad (0.0.9)$$

where ρ, v, p and E are the density, velocity, pressure and total energy, respectively. In addition, we assume that the gas is polytropic and write the equation of state as

$$p = (\gamma - 1) \left[E - \frac{1}{2}\rho v^2 \right] \qquad (0.0.10)$$

where γ is the ratio of specific heats which we take to be equal to 1.4. Clearly, both as a physical problem and as a practical numerical problem, a finite domain and boundary conditions are necessary. These conditions will be discussed in Chapter 6 when we prod you into beginning to solve this problem.

HW 0.0.4 (Thin Disturbance Transonic Flow Equation)

Consider potential flow over a symmetric, thin, circular arc airfoil. Expansions based on the thickness of the airfoil reduce the problem to one involving the thin disturbance transonic flow equation. For example, we consider the following problem describing the transonic flow past a 5% half circular arc airfoil.

$$[1 - M_\infty^2 - (\gamma + 1)M_\infty^2 \phi_x] \phi_{xx} + \phi_{yy} = 0,$$
$$(x, y) \in R = (-1, 2) \times (0, 1) \qquad (0.0.11)$$
$$\phi_n = 0 \quad \text{on} \quad \partial R - \{(x, 0) : 0 \le x \le 1\} \qquad (0.0.12)$$
$$\phi_y = \frac{-(x - \frac{1}{2})}{\sqrt{6.375625 - (x - \frac{1}{2})^2}} \quad \text{on} \ \{(x, 0) : 0 \le x \le 1\} \qquad (0.0.13)$$

(i) Solve this problem for $\gamma = 1.4$ and $M_\infty = 0.7$ and
(ii) $\gamma = 1.4$ and $M_\infty = 0.78$.
Though nonlinear, the partial differential equation above will look like either an elliptic equation or a hyperbolic equation, depending on the sign of the term $[1 - M_\infty^2 - (\gamma + 1)M_\infty^2 \phi_x]$.

As you will see as you attempt to solve problems HW0.0.1–0.0.4, when we apply schemes developed for model linear problems to nonlinear problems, the results are often not what we anticipated. For this reason, much care must be taken when we work the above problems.

1

Introduction to Finite Differences

1.1 Introduction

This chapter serves as an introduction to the subject of finite difference methods for solving partial differential equations. Some of the goals of the chapter include

- introducing finite difference grids and notation for functions defined on grids,

- introducing a finite difference approximation of a partial differential equation

  - with an explanation of why this *might* be a logical approximation,

  - along with an introduction to the treatment of several different types of boundary conditions and forcing functions, and

- getting the reader involved in the implementation of the finite difference scheme (if the text is being used for a class, hopefully as early as the first day of class).

In addition, the chapter includes a finite volume derivation of difference equations including grid centered equations, cell centered equations and cell averaged equations.

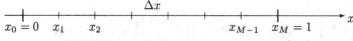

FIGURE 1.2.1. Uniform grid on the interval $[0, 1]$.

1.2 Getting Started

We begin our discussion of numerical methods for partial differential equations by getting started solving a problem numerically. We consider the following initial–boundary–value problem

$$v_t = \nu v_{xx}, \ x \in (0, 1), \ t > 0 \tag{1.2.1}$$

$$v(x, 0) = f(x), \ x \in [0, 1] \tag{1.2.2}$$

$$v(0, t) = a(t), \ v(1, t) = b(t), \ t \geq 0 \tag{1.2.3}$$

where $f(0) = a(0)$, and $f(1) = b(0)$.

Of course, we know how to solve this problem analytically. For purposes of illustration, we shall solve it numerically. (Note that if numerical values for the solution are desired, one can generally produce them faster by the numerical method than by solving the problem analytically and evaluating the solution.) Our method is to reduce the problem above to a discrete problem that we are able to solve. We begin by discretizing the spatial domain by placing a grid over the domain. For convenience, we will use a uniform grid, with grid spacing $\Delta x = 1/M$, as is shown in Figure 1.2.1.

If we wish to refer to one of the points in the grid, we shall call the points x_k, $k = 0, \cdots, M$ where $x_k = k\Delta x$, $k = 0 \cdots, M$. Likewise, we discretize the time domain similarly by place a grid on the temporal axis with grid spacing Δt. The resulting grid in the time-space domain is illustrated in Figure 1.2.2.

The space-time domain of our problem will then be approximated by the lattice of points in Figure 1.2.2. We will attempt to approximate the solution to our problem at the points on this lattice. Notationally, we will define u_k^n to be a function defined at the point $(k\Delta x, n\Delta t)$ or the lattice point (k, n). The function u_k^n will be our approximation to the solution of problem (1.2.1)–(1.2.3) at the point $(k\Delta x, n\Delta t)$.

We now have a grid that approximates our domain. The next step is to approximate problem (1.2.1)–(1.2.3) on this grid. We begin by noting that since

$$v_t(x, t) = \lim_{\Delta t \to 0} \frac{v(x, t + \Delta t) - v(x, t)}{\Delta t},$$

a reasonable approximation of $v_t(k\Delta x, n\Delta t)$ can be given by

$$\frac{u_k^{n+1} - u_k^n}{\Delta t}. \tag{1.2.4}$$

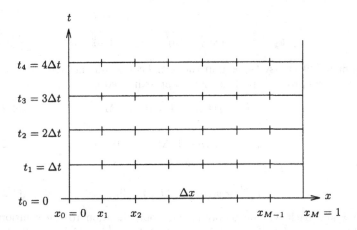

FIGURE 1.2.2. Grid on time-space domain.

Can the above expression be also used to approximate v_t at $(k\Delta x, (n + 1)\Delta t)$? Of course it can. It was only claimed that the former approximation was reasonable.

In a similar fashion we approximate v_{xx} at $(k\Delta x, n\Delta t)$ by

$$\frac{u_{k+1}^n - 2u_k^n + u_{k-1}^n}{\Delta x^2}. \tag{1.2.5}$$

To see that this is a reasonable approximation, consider that

$$\frac{(v_x)_{k+\frac{1}{2}}^n - (v_x)_{k-\frac{1}{2}}^n}{\Delta x}$$

approximates $(v_{xx})_k^n$, and that $(v_x)_{k+\frac{1}{2}}^n$ and $(v_x)_{k-\frac{1}{2}}^n$ can be approximated by

$$\frac{u_{k+1}^n - u_k^n}{\Delta x} \quad \text{and} \quad \frac{u_k^n - u_{k-1}^n}{\Delta x},$$

respectively. Then

$$(v_{xx})_k^n \approx \frac{(v_x)_{k+\frac{1}{2}}^n - (v_x)_{k-\frac{1}{2}}^n}{\Delta x} \tag{1.2.6}$$

$$\approx \frac{\frac{u_{k+1}^n - u_k^n}{\Delta x} - \frac{u_k^n - u_{k-1}^n}{\Delta x}}{\Delta x}. \tag{1.2.7}$$

Thus the expressions given in (1.2.4) and (1.2.5) can be used to approximate partial differential equation (1.2.1) at the point $(k\Delta x, n\Delta t)$ by

$$\frac{u_k^{n+1} - u_k^n}{\Delta t} = \nu \frac{u_{k+1}^n - 2u_k^n + u_{k-1}^n}{\Delta x^2} \tag{1.2.8}$$

or

$$u_k^{n+1} = u_k^n + \nu \frac{\Delta t}{\Delta x^2}(u_{k+1}^n - 2u_k^n + u_{k-1}^n). \tag{1.2.9}$$

And finally, it is easy to see that the initial condition and boundary conditions for the problem are reasonably approximated by

$$u_k^0 = f(k\Delta x), \ \ k = 0, \cdots, M \tag{1.2.10}$$

$$u_0^{n+1} = a((n+1)\Delta t), \ n = 0, \cdots \tag{1.2.11}$$

and

$$u_M^{n+1} = b((n+1)\Delta t), n = 0, \cdots. \tag{1.2.12}$$

Our approach will be to obtain an approximation to the solution to problem (1.2.1)–(1.2.3) by solving the discrete problem (1.2.9)–(1.2.12). Of course we must still choose Δx (or M) and Δt, which will determine much about the accuracy and behavior of our solution. If we temporarily ignore these details, we see that equation (1.2.10) gives u_k^0 for $k = 0, \cdots, M$; equation (1.2.9) with $n = 0$ can then be used to determine u_k^1 for $k = 1, \cdots, M - 1$; and, finally, equations (1.2.11) and (1.2.12) can be used to determine u_0^1 and u_M^1. Thus, equations (1.2.9), (1.2.11) and (1.2.12) use the information at $n = 0$ to determine u at the first time step. Now that u_k^1, $k = 0, \cdots, M$ is known, equations (1.2.9), (1.2.11) and (1.2.12) can be used to determine u for $n = 2$. And, of course, this process can be continued to determine u_k^n, $k = 0, \cdots, M$ to any desired time level n.

It should be noted that it was not possible to determine u_0^1 and u_M^1 by using equation (1.2.9), since one of the subscripts $k + 1$ and $k - 1$ would reach out of bounds (less than 0 or greater than 1) for either of the calculations. Hence it will almost always be necessary to have some sort of boundary treatment such as equations (1.2.11) and (1.2.12). In this problem the treatment was obvious and very easy. That will not always be the case.

We now have a numerical scheme to approximate the solution of initial–boundary–value problem (1.2.1)–(1.2.3). We call this scheme an **explicit scheme** because we are able to solve for the variable at the $(n + 1)st$ time level explicitly.

We might ask if this is the only finite difference scheme for solving problem (1.2.1)–(1.2.3), but, hopefully, we are not that naive. We should ask whether it is a good scheme, and the answer is that we shall see as we proceed in the text. And finally, we should ask whether the solution is useful physically. The general claim is that if it is a sufficiently good approximation to the solution of problem (1.2.1)–(1.2.3), it will be a good physical solution. This is another question that hopefully will be answered as we proceed through the material.

HW 1.2.1 Write a code to approximately solve problem (1.2.1)–(1.2.3). Use $f(x) = \sin 2\pi x$, $a = b = 0$, $M = 10$ and $\nu = 1/6$. Find solutions at $t = 0.06$, $t = 0.1$, $t = 0.9$ and $t = 50.0$. For the first three values of t, use $\Delta t = 0.02$. To speed solution to the last value of t, you might use a larger value for Δt. Determine how large you can choose Δt and still get results that might be correct. Compare and contrast your solutions to the exact solutions.

1.2.1 Implementation

The plan in this text is to provide numerical experience as a part of the learning process. To aid in obtaining numerical experience, many computational problems will be given. For this reason, and because it is best to develop good habits early, it is suggested that the codes (beginning with the code assigned in HW1.2.1) be developed carefully. If modular codes are written, many pieces of the codes can be used often. A basic code for numerically solving a time-dependent partial differential equation should look approximately as follows.

Call Initial

    For time $\leq$ Final time

        Call solution scheme

        If desired, Call output

        Set uold=unew

    Next time

Of course, the above code is given is some sort of pseudo code, but the analogous codes in Fortran, c, Matlab, Basic or Pascal would all look similar to the above pseudo code.

In the initialization subroutine, parameters are set (Final time, Δx, Δt, etc.), uold (the value of u at the old time step) is initialized (from the function f) and maybe the boundary conditions are set (at least in the case of constant boundary conditions). This subroutine, or a slight variation of this subroutine, will be used in all of the one-dimensional problems that we do.

Usually the For-Next loop for the time step would be determined by some integer, the final time step being set in Initial. The subroutine Solution Scheme will be the part of code that changes the most from problem to problem. This section may be more than one routine for certain schemes, or if it is sufficiently simple it may be written *in line*. For HW1.2.1 this section will consist of looping from $k = 1, \cdots, M - 1$ using equation (1.2.9) to determine u at the new time level. (So of course, this is a good candidate for just including the Solution Scheme *in line.*) We will return to this part of the code in later problems where more interesting Solution Scheme routines are necessary.

The next step is to output information. Since it may lead to more numbers than you care to contend with (and use up too many trees in the process), you may not want to print out every value at every time step. The output routine is another routine that will be used in all of our one dimensional codes and should be constructed with care. If you want numerical values outputted at certain times, the output routine should determine whether or not it is time to output the solution, outputting the solution (and other related parameters) if it is time and proceeding if it is not.

Another topic that might be considered at this point is graphics. Often it is extremely convenient to be able to examine graphs of the solutions rather than a large array of numbers. So one option in the output routine is to consider the possibility of including some sort of graphical output. Some of the possibilities here include using True Basic, TurboC, TurboPascal, Matlab or any language in conjunction with an independent graphics routine.

And finally, before we return to perform the next time step, we must adjust our u values in that the unew of this time step will be the uold of the next time step. This generally consists of an easy loop through the data.

1.3 Consistency

We next begin to investigate how good scheme (1.2.9)–(1.2.12) is for approximating the solution to problem (1.2.1)–(1.2.3). One piece of information towards this end is to see how well difference equation (1.2.9) approximates partial differential equation (1.2.1). The tool that we shall use is the Taylor series expansion.

To arrive at the finite difference equation used to approximate partial differential equation (1.2.1), we used the approximation

$$v_t(n\Delta t, k\Delta x) \approx \frac{u_k^{n+1} - u_k^n}{\Delta t}.$$

We see, using a Taylor series expansion, that

$$v_k^{n+1} = v(k\Delta x, (n+1)\Delta t) = v(k\Delta x, n\Delta t) + \frac{\partial v}{\partial t}(k\Delta x, n\Delta t)\frac{\Delta t}{1!}$$
$$+ \frac{\partial^2 v}{\partial t^2}(k\Delta x, n\Delta t)\frac{\Delta t^2}{2!} + \cdots. \tag{1.3.1}$$

So

$$\frac{v_k^{n+1} - v_k^n}{\Delta t} = \frac{\partial v}{\partial t}(k\Delta x, n\Delta t) + \frac{\Delta t}{2}\left(\frac{\partial^2 v}{\partial t^2}\right)_k^n + \cdots. \tag{1.3.2}$$

We generally write this in terms of the "big $\mathcal{O}$" notation as

$$\frac{v_k^{n+1} - v_k^n}{\Delta t} = \frac{\partial v}{\partial t}(k\Delta x, n\Delta t) + \mathcal{O}(\Delta t), \qquad (1.3.3)$$

where the above expression assumes that the higher order derivatives of v at $(k\Delta x, n\Delta t)$ are bounded. The definition of the "big $\mathcal{O}$" notation is that $f(s) = \mathcal{O}(\phi(s))$ for $s \in S$ if there exists a constant A such that $|f(s)| \le A|\phi(s)|$ for all $s \in S$. We say that $f(x)$ is "big $\mathcal{O}$" of $\phi(x)$ or that $f(s)$ is of order $\phi(s)$.

The above Taylor series expansion of $v(k\Delta x, (n+1)\Delta t)$ shows what we are *throwing away* when we replace v_t in the partial differential equation by $\frac{u_k^{n+1} - u_k^n}{\Delta t}$.

It should be noted that the constant associated with the big $\mathcal{O}$ notation can be very large. In fact when solving problems where the solution has sharp changes with respect to time, this constant will be large. But generally, if Δt is sufficiently small, we are throwing away things that are order Δt, so $\frac{u_k^{n+1} - u_k^n}{\Delta t}$ is a good approximation of v_t.

The same approach can be used to show that

$$\frac{v_{k+1}^n - v_k^n}{\Delta x} = \frac{\partial v}{\partial x}(k\Delta x, n\Delta t) + \mathcal{O}(\Delta x), \qquad (1.3.4)$$

$$\frac{v_k^n - v_{k-1}^n}{\Delta x} = \frac{\partial v}{\partial x}(k\Delta x, n\Delta t) + \mathcal{O}(\Delta x), \qquad (1.3.5)$$

and

$$\frac{v_{k+1}^n - v_{k-1}^n}{2\Delta x} = \frac{\partial v}{\partial x}(k\Delta x, n\Delta t) + \mathcal{O}(\Delta x^2). \qquad (1.3.6)$$

Thus we note that the **centered difference** approximates the first derivative with respect to x more accurately than either of the *one sided differences*, $\mathcal{O}(\Delta x^2)$ verses $\mathcal{O}(\Delta x)$. Again using the Taylor series expansion we can show that

$$\frac{v_{k+1}^n - 2v_k^n + v_{k-1}^n}{\Delta x^2} = \frac{\partial^2 v}{\partial x^2}(k\Delta x, n\Delta t) + \mathcal{O}(\Delta x^2). \qquad (1.3.7)$$

The order of approximation found in equation (1.3.7) is logical since in the derivation of $\frac{u_{k+1}^n - 2u_k^n + u_{k-1}^n}{\Delta x^2}$ as an approximation to $\frac{\partial^2 v}{\partial x^2}$, all differences used were centered differences.

If we now return to partial differential equation (1.2.1), we see that

$$v_t(k\Delta x, n\Delta t) - \nu v_{xx}(k\Delta x, n\Delta t) = \frac{v_k^{n+1} - v_k^n}{\Delta t}$$

$$- \frac{\nu}{\Delta x^2}(v_{k+1}^n - 2v_k^n + v_{k-1}^n) + \mathcal{O}(\Delta t) + \mathcal{O}(\Delta x^2). \quad (1.3.8)$$

Thus we see that difference equation (1.2.8) approximates partial differential equation (1.2.1) to the first order in Δt and the second order in Δx. Another way to view equation (1.3.8) is that *if $v = v(x, t)$ is a solution to the partial differential equation (so that the left hand side of equation (1.3.8) is zero), then v satisfies the difference equation to the first order in Δt and the second order in Δx.*

Either way, we see that equation (1.3.8) shows how well the difference equation approximates the partial differential equation. We should emphasize that *this in no way tells us how well the solution to the difference equation will approximate the solution to the partial differential equation.* This is still a very important topic that we must investigate. We will claim at this time that the solution to the difference scheme will generally approximate the solution to the partial differential equation to the same order that the difference scheme approximates the partial differential equation.

We might ask the question whether we can do a better job of approximating our partial differential equation. It is easy to show that if we use $\frac{u_k^{n+1} - u_k^{n-1}}{2\Delta t}$ as our approximation of v_t, we obtain the difference equation

$$u_k^{n+1} = u_k^{n-1} + \frac{2\nu\Delta t}{\Delta x^2}(u_{k+1}^n - 2u_k^n + u_{k-1}^n) \qquad (1.3.9)$$

which is second order in both space and time. Difference scheme (1.3.9) is called the **leapfrog scheme** and is referred to as a three-level scheme. It should be noted that special consideration must be given to starting the leapfrog scheme, and multilevel schemes in general. It seems logical to claim that if one step of a first order scheme is used to start the leapfrog scheme, the scheme will be first order (since the *contaminated* values from that first step may contaminate the entire solution). We shall see later that this is not the case.

HW 1.3.1 Solve the problem given in HW1.2.1 using the leapfrog scheme, (1.3.9). Do only the part with $\Delta t = 0.02$. (For convenience, use values from the exact solution at Δt to get the leapfrog scheme started. It is also suggested that if the results for this problem are not nice, do not spend the rest of your life on it.)

Before we move on, we include some notation concerning the differences discussed above. We define

$$\delta_+ u_k = u_{k+1} - u_k, \qquad (1.3.10)$$

$$\delta_- u_k = u_k - u_{k-1}, \qquad (1.3.11)$$

$$\delta_0 u_k = u_{k+1} - u_{k-1}, \qquad (1.3.12)$$

and

$$\delta^2 u_k = u_{k+1} - 2u_k + u_{k-1} \tag{1.3.13}$$

to be the forward, backward, centered and centered second difference operators, respectively. When the the variable to which the difference is to be applied is not obvious, we will use the notation $\delta_{t+}u_k^n$. Using this notation, the difference equation (1.2.9) can be written as

$$u_k^{n+1} = u_k^n + \frac{\nu \Delta t}{\Delta x^2} \delta^2 u_k^n, \tag{1.3.14}$$

and will be referred to as the forward time, centered space (FTCS) scheme for solving partial differential equation (1.2.1).

1.3.1 Special Choice of Δx and Δt

It is fairly clear, and we will discuss it later, that different difference operators can be used to give higher order approximations of the partial differential equation. It is also possible to do better than one might expect for a particular scheme, say the one given in equation (1.2.9). We now show that by a particular choice of Δx and Δt, the order of the scheme given in equation (1.2.9) goes from first order in time and second order in space to second order in time and fourth order in space. *We do this, not so much to convince you that you should try to choose Δx and Δt in this manner, but to demonstrate why at times the results you obtain from a scheme are more accurate than you would expect them to be.* We begin by expanding the terms in equation (1.3.8) more carefully to see that

$$(v_t - \nu v_{xx}))_k^n - [\tfrac{\delta_{t+}}{\Delta t} - \tfrac{\nu}{\Delta x^2}\delta^2]v_k^n = -\tfrac{\Delta t}{2}\left(v_{tt}\right)_k^n - \tfrac{\Delta t^2}{3!}\left(v_{ttt}\right)_k^n - \cdots$$
$$+ \nu[2\tfrac{\Delta x^2}{4!}\left(\tfrac{\partial^4 v}{\partial x^4}\right)_k^n + 2\tfrac{\Delta x^4}{6!}\left(\tfrac{\partial^6 v}{\partial x^6}\right)_k^n + \cdots]. \tag{1.3.15}$$

Then since $v_t = \nu v_{xx}$,

$$v_{tt} = \nu v_{xxt} = \nu(v_t)_{xx} = \nu(\nu v_{xx})_{xx} = \nu^2 v_{xxxx}.$$

Using this expression on the right hand side of the expression given in (1.3.15) gives

$$(v_t - \nu v_{xx})_k^n - [\tfrac{\delta_{t+}}{\Delta t} - \tfrac{\nu}{\Delta x^2}\delta^2]v_k^n = -\tfrac{\Delta t}{2}\nu^2\left(v_{xxxx}\right)_k^n - \tfrac{\Delta t^2}{3!}\left(v_{ttt}\right)_k^n - \cdots$$
$$+ \nu 2\tfrac{\Delta x^2}{4!}\left(v_{xxxx}\right)_k^n + \nu[2\tfrac{\Delta x^4}{6!}\left(\tfrac{\partial^6 v}{\partial x^6}\right)_k^n + \cdots]$$
$$= \nu(-\nu\tfrac{\Delta t}{2} + 2\tfrac{\Delta x^2}{4!})\left(v_{xxxx}\right)_k^n + \mathcal{O}(\Delta t^2) + \mathcal{O}(\Delta x^4).$$

Hence if we choose $\Delta t = \frac{\Delta x^2}{6\nu}$, we have a scheme that is second order in time and fourth order in space.

HW 1.3.2 Use the above choice of Δt in your solution to the problem in HW1.2.1 Compare and contrast the error in these results to the error in the results of HW1.2.1.

1.4   Neumann Boundary Conditions

In the only problem that we have solved, we had Dirichlet boundary conditions. We next consider what changes are necessary to consider Neumann boundary conditions. We introduce a boundary condition treatment by considering the following problem.

$$v_t = \nu v_{xx}, \ x \in (0,1), \ t > 0 \tag{1.4.1}$$

$$v(x,0) = \cos \frac{\pi x}{2}, \quad x \in [0,1] \tag{1.4.2}$$

$$v(1,t) = 0, \quad t \geq 0 \tag{1.4.3}$$

$$v_x(0,t) = 0, \quad t \geq 0 \tag{1.4.4}$$

We approximate the partial differential equation, initial condition and Dirichlet boundary condition as we did before by

$$u_k^{n+1} = u_k^n + \frac{\nu \Delta t}{\Delta x^2} \delta^2 u_k^n, \ k = 1, \cdots, M-1, \tag{1.4.5}$$

$$u_k^0 = \cos \frac{\pi k \Delta x}{2}, \ k = 0, \cdots, M, \tag{1.4.6}$$

and

$$u_M^{n+1} = 0. \tag{1.4.7}$$

We treat the Neumann boundary condition two different ways. The first is by using a one-sided difference approximation of $v_x(0,t) = 0$ to get

$$\frac{u_1^{n+1} - u_0^{n+1}}{\Delta x} = 0,$$

or

$$u_0^{n+1} = u_1^{n+1}. \tag{1.4.8}$$

We note that at the new time step, we can use difference equation (1.4.5) to determine u_k^{n+1}, $k = 1, \cdots, M-1$, and then use equation (1.4.8) to determine u_0^{n+1}. This form of boundary treatment for a Neumann boundary condition is common and often sufficient. The inconsistency with using approximation (1.4.8) along with the rest of our formulation is that we have chosen to use scheme (1.4.5) to approximate our partial differential equation. This approximation is second order in space. Equation (1.4.8) is an approximation of our Neumann boundary condition that is first order in space. It is possible that our final results might be only first order in space, completely due to the first order approximation used in the boundary condition. In general, the order approximation of a problem is the lowest order used in any of the parts of the problem.

To approximate the Neumann boundary condition in a manner consistent with our difference scheme, we derive a second order approximation. If we apply a centered difference at the boundary, the difference operator reaches out of the region. For this reason, we place a *ghost point*, $x_{-1} = -\Delta x$, outside of the region, and approximate the boundary condition by

$$\frac{u_1^{n+1} - u_{-1}^{n+1}}{2\Delta x} = 0. \tag{1.4.9}$$

Since we now have introduced another point to our domain, the ghost point really doesn't solve our problem. Even though the system of equations that we solve are trivial to solve, we are always solving a system of equations. For example, when we use the first order approximation for the Neumann boundary condition, equation (1.4.8) is the equation associated with the point x_0. If we do a count, we realize that difference equation (1.4.5) provides us with $M - 1$ equations, the Dirichlet boundary condition provides us with one equation and equation (1.4.8) brings the total number of equations up to $M + 1$. And since we need a solution for $k = 0, \cdots, M$, we have the correct number of equations.

If we now use the second order Neumann boundary condition approximation (1.4.9) with the ghost point, we still have $M + 1$ equations, but we now have $M + 2$ unknowns. The approach we use can be considered as adding another equation or eliminating one of the unknowns. We apply the difference equation (1.4.5) at the point $(0, n + 1)$ to get

$$u_0^{n+1} = u_0^n + \frac{\nu \Delta t}{\Delta x^2}(u_1^n - 2u_0^n + u_{-1}^n). \tag{1.4.10}$$

We then use the second order approximation (1.4.9), in the form

$$u_{-1}^n = u_1^n, \tag{1.4.11}$$

to eliminate the u_{-1}^n term from equation (1.4.10). We are left with

$$u_0^{n+1} = u_0^n + 2\frac{\nu \Delta t}{\Delta x^2}(u_1^n - u_0^n). \tag{1.4.12}$$

We are now able to use difference equation (1.4.5), initial condition (1.4.6), and boundary conditions (1.4.7) and (1.4.12) to solve our problem. Again we have $M + 1$ equations and $M + 1$ unknowns. In this situation, all of the equations are second order in space.

HW 1.4.1 Solve problem (1.4.1)–(1.4.4) using both the first order and the second order treatments of the Neumann boundary condition. Use $M = 10$, $\Delta t = 0.004$, $\nu = 1.0$ and find solutions at $t = 0.06$, $t = 0.1$ and $t = 0.9$. Compare and contrast the results for both of these problems with each other and with the exact solution.

HW 1.4.2 Resolve the problem given in HW1.4.1 using $M = 20$ and $\Delta t = 0.001$, and using $M = 40$ and $\Delta t = 0.0003$. Compare and contrast all of the first order solutions with each other and with the exact solution. Also compare and contrast all of the second order solutions with each other and with the exact solution.

1.5   Some Variations

1.5.1   Lower Order Terms

In this section we shall lump together several other problems and algorithms that we want to consider at this time. For example, an obvious variation of partial differential equation (1.2.1) is to include lower order terms. We consider the problem

$$v_t + av_x = \nu v_{xx}, \quad x \in (0,1), \ t > 0 \tag{1.5.1}$$

$$v(x,0) = f(x), \quad x \in [0,1] \tag{1.5.2}$$

$$v(0,t) = v(1,t) = 0, \quad t \geq 0 \tag{1.5.3}$$

The obvious first choice for a solution scheme for solving problem (1.5.1)–(1.5.3) is to use a centered difference approximation for the v_x term and arrive at the following difference equation.

$$u_k^{n+1} = u_k^n - \frac{a\Delta t}{2\Delta x}(u_{k+1}^n - u_{k-1}^n) + \frac{\nu \Delta t}{\Delta x^2}\delta^2 u_k^n, \ k = 1, \cdots, M-1 \tag{1.5.4}$$

Of course, we treat the initial conditions and the Dirichlet boundary conditions as we did before. To illustrate how well difference scheme (1.5.4) can work, we include the following problems.

HW 1.5.1 Solve problem (1.5.1)–(1.5.3) using difference scheme (1.5.4) with $\nu = 1.0$, $a = 2$, $M = 20$ and $f(x) = \sin 4\pi x$. Find solutions at $t = 0.06$, $t = 0.1$ and $t = 0.9$ using time steps of $\Delta t = 0.001$.

HW 1.5.2 Repeat HW1.5.1 with $\nu = 0.01$.

HW 1.5.3 Check your results obtained in HW1.5.1 and HW1.5.2 by repeating the calculation using $M = 40$.

1.5.2 Nonhomogeneous Equations and Boundary Conditions

Another slight variation of partial differential equation (1.2.1) that we must consider is the nonhomogeneous partial differential equation

$$v_t = \nu v_{xx} + F(x,t), \ x \in (0,1), \ t > 0. \tag{1.5.5}$$

It is not difficult to devise a logical difference scheme for solving partial differential equation (1.5.5). We let $F_k^n = F(k\Delta x, n\Delta t)$ and use the difference scheme

$$u_k^{n+1} = u_k^n + \frac{\nu \Delta t}{\Delta x^2} \delta^2 u_k^n + \Delta t F_k^n. \tag{1.5.6}$$

We then see, that with just a slight variation of the code used in HW1.2.1, we can solve problems involving nonhomogeneous partial differential equations.

Below we include several problems involving nonhomogeneous partial differential equations and nonhomogeneous boundary conditions. Recall that when you have boundary conditions that depend on time, the implementation of these boundary conditions must be done during your time loop (instead of during your initialization, which was possible for the problems that we have done until this time). Also, if you have nonhomogeneous Neumann boundary conditions, say

$$v_x(0,t) = g(t), \ t > 0, \tag{1.5.7}$$

the first and second order approximations of this Neumann boundary condition become

$$u_0^{n+1} = u_1^{n+1} - \Delta x g^{n+1} \tag{1.5.8}$$

and

$$u_{-1}^n = u_1^n - 2\Delta x g^n, \tag{1.5.9}$$

respectively. We should note that in the case of the second order approximation, the analog to equation (1.4.12) is

$$u_0^{n+1} = u_0^n + \frac{2\nu \Delta t}{\Delta x^2}(u_1^n - u_0^n) - \frac{2\nu \Delta t}{\Delta x} g^n. \tag{1.5.10}$$

HW 1.5.4 Solve the following initial–boundary–value problem

$$v_t = \nu v_{xx}, \ x \in (0,1), \ t > 0$$
$$v(x,0) = f(x), \ x \in [0,1]$$
$$v(0,t) = a(t), \ v(1,t) = b(t), \ t \geq 0$$

where $\nu = 0.1, f = 0, b = 0$, and $a(t) = \sin 4\pi t$. Use $M = 10$, and $\Delta t = 0.05$ and find solutions at $t = 0.1$, $t = 0.9$ and $t = 2.0$.

HW 1.5.5 Solve the initial–boundary–value problem given in HW1.5.4 with $f(x) = \sin 2\pi x$ and everything else as given in HW1.5.4.

HW 1.5.6 (a) Repeat HW1.5.5 with a given by $a(t) = \sin 10\pi t$.
(b) Repeat part (a) using $M = 20$ and $\Delta t = 0.01$.

HW 1.5.7 (a) Solve the following initial–boundary–value problem

$$v_t = \nu v_{xx} + F(x,t), \ x \in (0,1), \ t > 0$$
$$v(x,0) = f(x), \ x \in [0,1]$$
$$v(0,t) = a(t), \ v(1,t) = b(t), \ t \geq 0$$

where $\nu = 0.1, f \equiv 0, a \equiv 0, b \equiv 0$ and $F(x,t) = \sin 2\pi x \sin 4\pi t$. Use $M = 10$ and $\Delta t = 0.05$ and find solutions at $t = 0.1$, $t = 0.9$ and $t = 2.0$.
(b) Resolve the problem given in part (a) using $M = 20$ and $\Delta t = 0.01$.

HW 1.5.8 Solve the following initial–boundary–value problem

$$v_t = \nu v_{xx} + F(x,t), \ x \in (0,1), \ t > 0$$
$$v(x,0) = f(x), \ x \in [0,1]$$
$$v(0,t) = a(t), \ v(1,t) = b(t), \ t \geq 0$$

where $\nu = 0.1, f(x) = x(1-x), a(t) = 10\sin t, b(t) = 4\sin 6t$ and $F(x,t) = \sin 2\pi x \sin 4\pi t$.

HW 1.5.9 Use both the first and second order treatment of the Neumann boundary condition to solve the following initial–boundary–value problem

$$v_t = \nu v_{xx} + F(x,t), \ x \in (0,1), \ t > 0$$
$$v(x,0) = f(x), \ x \in [0,1]$$
$$v_x(0,t) = a(t), \ v(1,t) = b(t), \ t \geq 0$$

where $\nu = 0.1, f(x) = x(1-x), a(t) = 10\sin t, b(t) = 4\sin 6t$ and $F(x,t) = \sin 2\pi x \sin 4\pi t$. Use $M = 10$ and $\Delta t = 0.05$ and find solutions at $t = 0.1$, $t = 0.9$ and $t = 2.0$.

HW 1.5.10 Repeat the solution of HW 1.5.9 using $M = 20$, $\Delta t = 0.01$ and $M = 40$, $\Delta t = 0.002$. Compare and contrast your results with the results of HW 1.5.9.

1.5.3 A Higher Order Scheme

In Section 1.3 we saw that difference equation (1.2.8) approximates partial differential equation (1.2.1) to the first order in Δt and the second order in Δx. Also in Section 1.3, we devised a scheme (difference equation (1.3.9)) that was second order in Δt (though unless we have already solved

HW1.3.1, we do not know whether this scheme is any good). In this section we shall show that it is possible to approximate v_{xx} to an order higher than two. As a part of the derivation of the scheme, we shall illustrate a method that will enable us to find approximations to derivatives to any order that we desire.

The method that we will use is sometimes referred to the **method of undetermined coefficients**. We must first decide which points we want to include in the approximation. In HW1.5.11, we will show that it is not possible to approximate v_{xx} to the fourth order using only the points $x = k\Delta x$ and $x = (k \pm 1)\Delta x$. As we develop the difference approximation of v_{xx}, we shall see that it is possible to obtain a fourth order approximation of v_{xx} if we use the points $x = k\Delta x, x = (k \pm 1)\Delta x$ and $x = (k \pm 2)\Delta x$. Hence, we begin by writing our approximation of v_{xx} as

$$\Delta^4 u_k = c_1 u_{k-2} + c_2 u_{k-1} + c_3 u_k + c_4 u_{k+1} + c_5 u_{k+2} \qquad (1.5.11)$$

where $c_1, \cdots, c_5$ are yet to be determined. Expanding $\Delta^4 u_k$ in a Taylor series expansion about $x = k\Delta x$, we get

$$
\begin{aligned}
\Delta^4 u_k = c_1 & \Big[u_k + (u_x)_k(-2\Delta x) + (u_{xx})_k \frac{(-2\Delta x)^2}{2} + (u_{xxx})_k \frac{(-2\Delta x)^3}{6} \\
& + (u_{xxxx})_k \frac{(-2\Delta x)^4}{24} + \left(u^{(5)}\right)_k \frac{(-2\Delta x)^5}{120} + \mathcal{O}(\Delta x^6) \Big] \\
+ c_2 & \Big[u_k + (u_x)(-\Delta x) + (u_{xx})_k \frac{(-\Delta x)^2}{2} + (u_{xxx})_k \frac{(-\Delta x)^3}{6} \\
& + (u_{xxxx})_k \frac{(-\Delta x)^4}{24} + \left(u^{(5)}\right)_k \frac{(-\Delta x)^5}{120} + \mathcal{O}(\Delta x^6) \Big] \\
+ c_3 & \Big[u_k \Big] \\
+ c_4 & \Big[u_k + (u_x)_k \Delta x + (u_{xx})_k \frac{\Delta x^2}{2} + (u_{xxx})_k \frac{\Delta x^3}{6} \\
& + (u_{xxxx})_k \frac{\Delta x^4}{24} + \left(u^{(5)}\right)_k \frac{\Delta x^5}{120} + \mathcal{O}(\Delta x^6) \Big] \\
+ c_5 & \Big[u_k + (u_x)_k(2\Delta x) + (u_{xx})_k \frac{(2\Delta x)^2}{2} + (u_{xxx})_k \frac{(2\Delta x)^3}{6} \\
& + (u_{xxxx})_k \frac{(2\Delta x)^4}{24} + \left(u^{(5)}\right)_k \frac{(2\Delta x)^5}{120} + \mathcal{O}(\Delta x^6) \Big].
\end{aligned}
$$

Regrouping the terms in the above expansion gives

$$\Delta^4 u_k = [c_1 + c_2 + c_3 + c_4 + c_5]\, u_k \qquad (1.5.12)$$

$$+ \left[-2\Delta x c_1 - \Delta x c_2 + \Delta x c_4 + 2\Delta x c_5 \right] (u_x)_k \qquad (1.5.13)$$

$$+ \left[2\Delta x^2 c_1 + \frac{1}{2}\Delta x^2 c_2 + \frac{1}{2}\Delta x^2 c_4 + 2\Delta x^2 c_5 \right] (u_{xx})_k \qquad (1.5.14)$$

$$+ \left[-\frac{4}{3}\Delta x^3 c_1 - \frac{1}{6}\Delta x^3 c_2 + \frac{1}{6}\Delta x^3 c_4 + \frac{4}{3}\Delta x^3 c_5 \right] (u_{xxx})_k \qquad (1.5.15)$$

$$+ \left[\frac{2}{3}\Delta x^4 c_1 + \frac{1}{24}\Delta x^4 c_2 + \frac{1}{24}\Delta x^4 c_4 + \frac{2}{3}\Delta x^4 c_5 \right] (u_{xxxx})_k \quad (1.5.16)$$

$$+ \left[-\frac{4}{15}\Delta x^5 c_1 - \frac{1}{120}\Delta x^5 c_2 + \frac{1}{120}\Delta x^5 c_4 + \frac{4}{15}\Delta x^5 c_5 \right] \left(u^{(5)} \right)_k$$

$$(1.5.17)$$

$$+ \mathcal{O}(\Delta x^6).$$

Thus, since we have five degrees of freedom (five undetermined coefficients), we can plan on setting the coefficients of u_k, $(u_x)_k$, $(u_{xxx})_k$ and $(u_{xxxx})_k$ (the coefficients in lines (1.5.12), (1.5.13), (1.5.15) and (1.5.16)) equal to zero and the coefficient of $(u_{xx})_k$ (line (1.5.14)) equal to one. We are left with the following system of equations.

$$c_1 + c_2 + c_3 + c_4 + c_5 = 0$$
$$-2c_1 - c_2 + c_4 + 2c_5 = 0$$
$$2c_1 + \frac{1}{2}c_2 + \frac{1}{2}c_4 + 2c_5 = 1/\Delta x^2 \quad (1.5.18)$$
$$-\frac{4}{3}c_1 - \frac{1}{6}c_2 + \frac{1}{6}c_4 + \frac{4}{3}c_5 = 0$$
$$\frac{2}{3}c_1 + \frac{1}{24}c_2 + \frac{1}{24}c_4 + \frac{2}{3}c_5 = 0$$

Solving the above system of equations (using an algebraic manipulator such as Maple or Mathematica makes such a task much easier) gives

$$c_1 = -\frac{1}{12\Delta x^2}, \quad c_2 = \frac{4}{3\Delta x^2}, \quad c_3 = -\frac{5}{2\Delta x^2}, \quad c_4 = \frac{4}{3\Delta x^2} \text{ and } c_5 = -\frac{1}{12\Delta x^2}.$$

Since we note that

$$-\frac{4}{15}\Delta x^5 c_1 - \frac{1}{120}\Delta x^5 c_2 + \frac{1}{120}\Delta x^5 c_4 + \frac{4}{15}\Delta x^5 c_5 = 0$$

(the coefficient of $\left(u^{(5)} \right)_k$ in line (1.5.17)) and since the coefficient of the $\mathcal{O}(\Delta x^6)$ term will include a $1/\Delta x^2$ fact inherited from the $c_1, \cdots, c_5$ terms, $\Delta^4 u_k$ is an order Δx^4 approximation of v_{xx}. Hence, the difference equation

$$u_k^{n+1} = u_k^n + r\Delta^4 u_k^n$$

$$= u_k^n + r \left(-\frac{1}{12}u_{k-2}^n + \frac{4}{3}u_{k-1}^n - \frac{5}{2}u_k^n + \frac{4}{3}u_{k+1}^n - \frac{1}{12}u_{k+2}^n \right) \quad (1.5.19)$$

where $r = \nu \Delta t/\Delta x^2$ is an $\mathcal{O}(\Delta t) + \mathcal{O}(\Delta x^4)$ approximation of partial differential equation (1.2.1).

Since the major step in the implementation of difference scheme (1.5.19) is to replace equation (1.2.8) in the program by equation (1.5.19), higher order schemes might appear very advantageous. However, there is one other problem. Suppose we consider solving the problem given in HW1.2.1 When difference equation (1.5.19) is computing at $k = 1$ (or $k = M - 1$), the difference scheme reaches to $k = -1$ and $k = 0$ (as well as $k = 2$ and $k = 3$, which cause no problem).

Reaching to $k = 0$ is no problem since we have been given a boundary condition. Difference scheme (1.2.8) reached to $k = 0$ also. However, reaching to $k = -1$ does cause a problem because we have no value given there. The point $x = -\Delta x$ is not even in the domain of our problem.

The solution to this problem is that we must prescribe some **numerical boundary conditions**. Numerical boundary conditions are "extra" boundary conditions that are prescribed for the numerical problem that are not analytic boundary conditions. In fact, if we analyzed the partial differential equation with an "extra" boundary condition at each side, we would find that the problem is ill-posed. We must prescribe a numerical boundary condition at each end that does not make the numerical problem ill-posed. The good news is that this is possible. The bad news is that this is most often difficult. For the moment, as in HW1.5.12, we provide some "suggested" numerical boundary conditions (while not promising that they are good ones). We shall return to the topic of numerical boundary conditions often throughout the text.

Hopefully, through the work in this section, we see that it is possible to obtain higher order approximations. In fact, the method of undetermined coefficients will allow you to obtain a method that is essentially as high an order as you want. Though the work to obtain these methods is messy, it is very easy to have an algebraic manipulator do just about the entire calculation for you. We will return to consider higher order methods periodically. There are times when a higher order method is exactly what is needed. However, most often when additional accuracy is required, practitioners generally use finer grids instead of higher order methods. For this reason, we will not spend much time on higher order methods.

HW 1.5.11 Use the method of undetermined coefficients to show that it is impossible to approximate v_{xx} at point $(k\Delta x, n\Delta t)$ to the third or higher order using only the points $(k\Delta x, n\Delta t)$ and $(\pm k\Delta x, n\Delta t)$.

HW 1.5.12 Solve the problem given in HW1.2.1 using difference scheme, (1.5.19). Find solutions at $t = 0.06$, $t = 0.1$ and $t = 0.9$.
(a) Use $u_{-1}^n = 0.0$ and $u_{M+1}^n = 0.0$ as your numerical boundary conditions.
(b) Use $u_{-1}^n = 2u_0^n - u_1^n$ and $u_{M+1}^n = 2u_M^n - u_{M-1}^n$ as your numerical boundary conditions.

In both cases, compare your results with the results found in HW1.2.1.

1.6   Derivation of Difference Equations

In Section 1.2 we developed the explicit difference schemes (1.2.9) and
(1.3.9) by approximating the derivatives in the partial differential equation
by various differences. Likewise, when we developed the approximation to
the Neumann boundary condition, we replaced the derivative by an approximation in terms of differences. By now, hopefully, you have worked
Problem HW1.3.1 and found that all schemes that are developed by replacing derivatives with logical difference approximations need not be good
schemes. Many good and bad schemes can be developed by the Taylor series
expansion approach.

In this section we will demonstrate another approach to developing difference schemes. The method might be called the **conservation law approach** in that it develops the difference scheme using a physical conservation law. The method is a basic application of the **method of finite
volumes**. We might warn you that *though schemes developed using the
conservation approach have some nice properties, it is also possible to develop bad schemes using the conservation approach*. We should also warn
you that in the literature you will find an assortment of methods that are
using a conservation law and/or finite volume approach. Many of them will
be equivalent and most of them will be approximately equivalent. We will
try to prepare you for some of them by introducing certain variations of
the conservation law approach throughout the text.

We begin by demonstrating how one might derive partial differential
equation (1.5.5),

$$v_t = \nu v_{xx} + F(x,t), \tag{1.6.1}$$

as a model for heat flow. We consider a bounded interval $D = (0,1)$ filled
with a thermally conducting solid (really we assume that we have a uniform rod where the properties of the rod, initial conditions and boundary
conditions are such that there will be variations in only the x direction).
Let $v = v(x,t)$ denote the temperature at the point $x \in \overline{D}$ (D closure) and
time $t \geq 0$. Let $R = (a,b)$ be an arbitrary but fixed (with respect to time)
subinterval of D so that the points $x = a$ and $x = b$ represent the boundary
of R. The total heat content of R is given by

$$Q = \int_R c\rho v(x,t)dx,$$

where the constants ρ and c are the **density** and **specific heat** (the amount
of heat required to raise the temperature of one unit of mass one unit of
temperature) of the solid. Then the time rate of change of the total heat
content of the interval R is

$$\frac{dQ}{dt} = \int_R \frac{\partial}{\partial t}(c\rho v)dx.$$

The rate of change of total heat content in R must be due to heat supplied in the interval by internal sources and/or heat that flows across the boundary of R. If we let $\mathcal{F} = \mathcal{F}(x, t)$ (called the **source density**) denote the amount of heat per unit volume per unit time generated at the point $x \in R$ and time $t > 0$, the rate of change of heat content in R due to $\mathcal{F}$ is

$$\int_R \mathcal{F}(x, t)dx.$$

The flux of heat, q, across the boundary of R (the amount of heat flowing across ∂R per unit area per unit time) is assumed by Fourier's law to be proportional to the normal component of the temperature gradient, i.e. in our case $q(a) = -\kappa(a, v)\frac{\partial v}{\partial x}(a, t)$ and $q(b) = \kappa(b, v)\frac{\partial v}{\partial x}(b, t)$, where $\kappa = \kappa(x, v)$ is the **thermal conductivity**. (The differences in signs of the two fluxes is due to the fact that the flux was to be proportional to the *normal* derivative.)

If we then equate the time rate of change of total heat content to the sum of the heat supplied by external sources and the heat that flows across the boundary of R, we get

$$\int_R c\rho \frac{\partial v}{\partial t}dx = \int_R \mathcal{F}(x, t)dx + \left(\kappa \frac{\partial v}{\partial x}\right)(b, t) - \left(\kappa \frac{\partial v}{\partial x}\right)(a, t). \quad (1.6.2)$$

If we then notice that

$$\left(\kappa \frac{\partial v}{\partial x}\right)(b) - \left(\kappa \frac{\partial v}{\partial x}\right)(a) = \int_R (\kappa v_x)_x dx,$$

we can write (1.6.2) as

$$\int_R \left[c\rho v_t - (\kappa v_x)_x - \mathcal{F}\right]dx = 0. \quad (1.6.3)$$

Then since the interval R is arbitrary and the integrand is continuous, we get that

$$c\rho v_t - (\kappa v_x)_x - \mathcal{F} = 0. \quad (1.6.4)$$

And, of course, if κ is constant, we get equation (1.6.1) with $\nu = \kappa/c\rho$ and $F = \mathcal{F}/c\rho$.

The reason that we review this derivation is that we will base our difference scheme on this derivation. The hope is that by doing this, our difference scheme may contain more of the physics of the problem than if we naively approximate the resulting partial differential equation by differences. To apply these ideas to difference schemes, we consider the grid placed on the interval (0, 1) given in Figure 1.6.1. We notice that the grid is the same type used throughout Chapter 1, except that we have marked off

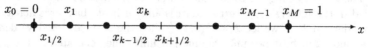

FIGURE 1.6.1. Grid placed on the interval $(0,1)$ with blocks centered at the grid points.

intervals about the grid points, centered at the grid points. We shall refer to the endpoints of the interval about the point x_k as $x_{k\pm 1/2}$. This interval is referred to as the **control volume** associated with the kth grid point.

To derive a difference equation associated with the kth grid point above, we want to apply equation (1.6.2), which we call the **integral form of the conservation law**, to the interval $(x_{k-1/2}, x_{k+1/2})$ (which we refer to as the kth **cell**). Assuming that c, ρ and κ are constants, that we have divided by $c\rho$, set $\nu = \kappa/c\rho$ and $F = \mathcal{F}/c\rho$, we get

$$\int_{x_{k-1/2}}^{x_{k+1/2}} v_t dx = \int_{x_{k-1/2}}^{x_{k+1/2}} F dx + \nu \left[\left(v_x \right)_{k+1/2} - \left(v_x \right)_{k-1/2} \right]. \qquad (1.6.5)$$

We note that both sides of equation (1.6.5) are still functions of t. If we integrate from $t_n = n\Delta t$ to $t_{n+1} = (n+1)\Delta t$, we get

$$\int_{t_n}^{t_{n+1}} \int_{x_{k-1/2}}^{x_{k+1/2}} v_t dx dt = \int_{x_{k-1/2}}^{x_{k+1/2}} \int_{t_n}^{t_{n+1}} v_t dt dx \qquad (1.6.6)$$

$$= \int_{x_{k-1/2}}^{x_{k+1/2}} (v^{n+1} - v^n) dx \qquad (1.6.7)$$

$$= \int_{t_n}^{t_{n+1}} \int_{x_{k-1/2}}^{x_{k+1/2}} F(x,t) dx dt$$

$$+ \nu \int_{t_n}^{t_{n+1}} \left[\left(v_x \right)_{k+1/2} - \left(v_x \right)_{k-1/2} \right] dt \qquad (1.6.8)$$

From the above calculation (lines (1.6.7) and (1.6.8)), we get

$$\int_{x_{k-1/2}}^{x_{k+1/2}} (v^{n+1} - v^n) dx = \int_{t_n}^{t_{n+1}} \int_{x_{k-1/2}}^{x_{k+1/2}} F(x,t) dx dt$$

$$+ \nu \int_{t_n}^{t_{n+1}} \left[\left(v_x \right)_{k+1/2} - \left(v_x \right)_{k-1/2} \right] dt, \qquad (1.6.9)$$

which is equivalent to equation (1.6.2) (which was true for all intervals). Hence, *equation (1.6.9) is an exact equation.*

We now proceed to obtain a finite difference approximation of partial differential equation (1.6.1). We do this by approximating the integral form of the conservation law given in equation (1.6.9). We begin by approximating the integral associated with the nonhomogeneous term by the midpoint rule with respect to x and the rectangular rule with respect to t (evaluating the function at t_n) and the integral on the left by the midpoint rule. We get

$$\int_{t_n}^{t_{n+1}} \int_{x_{k-1/2}}^{x_{k+1/2}} F(x,t)dxdt = \Delta x \Delta t F_k^n + \mathcal{O}(\Delta x^3 \Delta t) + \mathcal{O}(\Delta t^2) \qquad (1.6.10)$$

and

$$\int_{x_{k-1/2}}^{x_{k+1/2}} (v^{n+1} - v^n)dx = \Delta x(v_k^{n+1} - v_k^n) + \mathcal{O}(\Delta t \Delta x^3). \qquad (1.6.11)$$

We might notice that the order of approximation obtained in equation (1.6.11) is not obvious. It is clear that the Δx^3 term comes from using the midpoint rule to approximate integration. The Δt term is due to the fact that the function we are integrating is in the form $v^{n+1} - v^n$. For this reason, when the usual error analysis is done for the integration rule, the appropriate derivatives (with respect to x) can be further expanded in terms of n to introduce a Δt term.

If we next approximate the flux term in equation (1.6.9) (last integral on the right) by the rectangular rule (evaluating the integrand at t_n), we get

$$\int_{t_n}^{t_{n+1}} \left[(v_x)_{k+1/2} - (v_x)_{k-1/2} \right] dt = \Delta t \left[(v_x)_{k+1/2}^n - (v_x)_{k-1/2}^n \right]$$
$$+ \mathcal{O}(\Delta t^2 \Delta x) \qquad (1.6.12)$$

where the Δx term in the order of approximation is due to the fact that the integrand is a difference in k. We are then left to approximate the term $(v_x)_{k+1/2}^n - (v_x)_{k-1/2}^n$. If we proceed as we did in Section 1.3 and expand in Taylor series about x_k, we get

$$(v_x)_{k+1/2}^n - (v_x)_{k-1/2}^n = \Delta x (v_{xx})_k^n + \mathcal{O}(\Delta x^3)$$
$$= \Delta x \left(\frac{\delta^2 v_k^n}{\Delta x^2} + \mathcal{O}(\Delta x^2) \right) + \mathcal{O}(\Delta x^3)$$
$$= \frac{1}{\Delta x} \delta^2 v_k^n + \mathcal{O}(\Delta x^3). \qquad (1.6.13)$$

(We should add that the above approximation is the most common one for the $(v_x)_{k+1/2} - (v_x)_{k-1/2}$ term. When higher order schemes are needed, then this term is approximated to a higher degree of accuracy.)

We then combine equations (1.6.10)–(1.6.13) with equation (1.6.9) to get

$$\Delta x(v_k^{n+1} - v_k^n) = \Delta x \Delta t F_k^n + \frac{\nu \Delta t}{\Delta x} \delta^2 v_k^n$$
$$+ \mathcal{O}(\Delta t \Delta x^3) + \mathcal{O}(\Delta t^2 \Delta x) + \mathcal{O}(\Delta t \Delta x^3). \qquad (1.6.14)$$

And, finally, we note that if we replace the function evaluation v_k^n by the approximation u_k^n and approximate equation (1.6.14) by dropping the $\mathcal{O}$ terms, we obtain difference equation

$$\Delta x(u_k^{n+1} - u_k^n) = \Delta x \Delta t F_k^n + \frac{\nu \Delta t}{\Delta x}\delta^2 u_k^n. \qquad (1.6.15)$$

We should be aware that the truncation analysis performed above is different from those done in Sections 1.2 and 1.3 in that here we are approximating an integral equation instead of the partial differential equation. This is sufficient because our goal was to approximate the conservation law (which happens to be an integral equation). It is very difficult (and sometimes seems impossible) to be careful enough to obtain the same orders of approximation for both approaches. *We recommend that if you want to use the conservation law approach but still want a truncation analysis to the partial differential equation, first derive your difference scheme via the conservation law approach and then perform a separate truncation analysis for the scheme with respect to the partial differential equation.*

Of course, if we approximate equation (1.6.9) in different ways, we obtain other difference schemes. We used three different approximations on equation (1.6.9) (we approximated the derivatives v_x at $k \pm 1/2$ and we evaluated three integrals numerically). We will return to this derivation often in later chapters as we introduce new schemes. Before we leave this section, we use the method to derive another difference scheme for solving partial differential equation (1.6.1).

We return to the integral form for the conservation law, (1.6.5). We integrate equation (1.6.5) from t_{n-1} to t_{n+1} to get

$$\int_{t_{n-1}}^{t_{n+1}} \int_{x_{k-1/2}}^{x_{k+1/2}} v_t\, dx\, dt = \int_{x_{k-1/2}}^{x_{k+1/2}} \int_{t_{n-1}}^{t_{n+1}} v_t\, dt\, dx \qquad (1.6.16)$$

$$= \int_{x_{k-1/2}}^{x_{k+1/2}} (v^{n+1} - v^{n-1})\, dx \qquad (1.6.17)$$

$$= \int_{t_{n-1}}^{t_{n+1}} \int_{x_{k-1/2}}^{x_{k+1/2}} F(x,t)\, dx\, dt$$

$$+ \int_{t_{n-1}}^{t_{n+1}} \left[\left(v_x\right)_{k+1/2} - \left(v_x\right)_{k-1/2}\right] dt. \qquad (1.6.18)$$

If we (1) approximate the integral in (1.6.17) by the midpoint rule, (2) the integral for the source term in (1.6.18) by the midpoint rule with respect to both variables and (3) approximate the flux term in (1.6.18) by the midpoint rule with respect to t and the differences of fluxes as we did in the previous derivation, we get

$$\Delta x(v_k^{n+1} - v_k^{n-1}) + \mathcal{O}(\Delta t \Delta x^3) = 2\Delta x \Delta t F_k^n + \mathcal{O}(\Delta t \Delta x^3) + \mathcal{O}(\Delta t^3 \Delta x)$$

$$+ 2\frac{\nu \Delta t}{\Delta x}\delta^2 v_k^n + \mathcal{O}(\Delta t^3 \Delta x) + \mathcal{O}(\Delta t \Delta x^3). \qquad (1.6.19)$$

The difference scheme associated with equation (1.6.19) is

$$\Delta x(u_k^{n+1} - u_k^{n-1}) = 2\Delta x\Delta t F_k^n + 2\frac{\nu\Delta t}{\Delta x}\delta^2 u_k^n. \qquad (1.6.20)$$

So finally, we see that the leapfrog scheme for the heat equation (which, if you have worked HW 1.3.1, you know may not be a very good scheme) can also be derived using the conservation law approach. Thus, even though schemes that are developed by the conservation law approach have many nice properties, *using the conservation law approach does not guarantee that the scheme is a good one.*

Remark: We notice that our treatment of the nonhomogeneous term in this section was to integrate the nonhomogeneous term at least as accurately as the other terms. This is the most common approach and will be our usual approach throughout the text. However, since the function F is assumed to be known, it is possible (and introduces no error) to treat the integral of the nonhomogeneous term exactly. For example, if we were to replace the $\Delta t F_k^n$ term in difference scheme (1.5.6) by the exact value

$$\Delta t \int_{t_n}^{t_{n+1}} \int_{x_{k-1/2}}^{x_{k+1/2}} F(x,t)dxdt,$$

the derivation given earlier of equation (1.6.15) should make it clear that the resulting equation would be at least as accurate as equation (1.6.15). Because it would only be "as accurate" as equation (1.6.15), this approach is not used often.

HW 1.6.1 Use the appropriate approximations of equation (1.6.9) to derive difference equation (1.5.19).

1.6.1   Neumann Boundary Conditions

We next consider how the conservation law approach will affect our treatment of Neumann boundary conditions. Specifically, we again consider partial differential equation (1.6.1), but now consider the nonhomogeneous Neumann boundary condition (1.5.7) at $x = 0$. We return to Figure 1.6.1 where we apply the integral form of the conservation law (1.6.2) to the half interval marked off to the right of the zero boundary point. We should note that the boundary point has only a half interval because we cannot conserve any quantities outside of the domain. If we apply the integral form of the conservation law to the interval $(0, x_{1/2})$, we get

$$\int_{x_0}^{x_{1/2}} v_t dx = \int_{x_0}^{x_{1/2}} F dx + \nu\left[\left(v_x\right)_{1/2} - \left(v_x\right)_0\right]. \qquad (1.6.21)$$

If, as before, we integrate from t_n to t_{n+1}, we can easily derive the following equation

$$\frac{\Delta x}{2}(v_0^{n+1} - v_0^n) + \mathcal{O}(\Delta t \Delta x^2) = \frac{\Delta x}{2}\Delta t F_0^n + \mathcal{O}(\Delta t \Delta x^2) + \mathcal{O}(\Delta t^2 \Delta x)$$

$$+ \nu \int_{t_n}^{t_{n+1}} \left[(v_x)_{1/2} - (v_x)_0 \right] dt \qquad (1.6.22)$$

where in the first two terms, we now have order Δx^2 instead of Δx^3, because we can no longer approximate the integral using the midpoint rule with respect to x (we use the rectangular rule).

Our approach will now differ slightly from what we did before. Because we have a Neumann boundary condition at $x = 0$, $(v_x)_0$ is known. Hence, equation (1.6.22) can be rewritten as

$$\frac{\Delta x}{2}(v_0^{n+1} - v_0^n) + \mathcal{O}(\Delta t \Delta x^2) = \frac{\Delta x}{2}\Delta t F_0^n + \mathcal{O}(\Delta t \Delta x^2) + \mathcal{O}(\Delta t^2 \Delta x)$$

$$- \nu \Delta t g^n + \mathcal{O}(\Delta t^2) + \nu \int_{t_n}^{t_{n+1}} (v_x)_{1/2} dt, \qquad (1.6.23)$$

where g^n represents the function g evaluated at t_n. If we then approximate the derivative $(v_x)_{1/2}$ by the second order approximation $(v_1 - v_0)/\Delta x + \mathcal{O}(\Delta x^2)$ and approximate the time integral using the rectangular rule, we are left with the following approximation to the integral form of the conservation law.

$$\frac{\Delta x}{2}(v_0^{n+1} - v_0^n) = \frac{\Delta x}{2}\Delta t F_0^n - \nu \Delta t g^n + \nu \frac{v_1^n - v_0^n}{\Delta x}\Delta t$$

$$+ \mathcal{O}(\Delta t \Delta x^2) + \mathcal{O}(\Delta t^2 \Delta x) + \mathcal{O}(\Delta t^2) \qquad (1.6.24)$$

We make special note of the $\mathcal{O}(\Delta t^2)$ term in equation (1.6.24). If we try to divide by $\Delta t \Delta x$ to make equation (1.6.24) look like a partial differential equation, the $\mathcal{O}(\Delta t^2)$ term causes problems. However, since this equation includes boundary condition information (and that is where the $\mathcal{O}(\Delta t^2)$ term came from), there is no special reason that it should look like the partial differential equation. As we shall see later, this will cause us some problems.

If we approximate equation (1.6.24) by eliminating the "big $\mathcal{O}$" terms, we are left with the following difference equation for $k = 0$.

$$u_0^{n+1} = u_0^n + \Delta t F_0^n - \frac{2\nu \Delta t}{\Delta x}g^n + \frac{2\nu \Delta t}{\Delta x^2}(u_1^n - u_0^n) \qquad (1.6.25)$$

If we do as we did in the derivation of the second order boundary condition in Section 1.4, use equation (1.4.10) along with a nonhomogeneous term and eliminate the u^n_{-1} using the second order approximation to the Neumann boundary condition (1.5.7),

$$u^n_{-1} = u^n_1 - 2\Delta x g^n,$$

we obtain equation (1.6.25). Hence, we see by the derivation above in (1.6.24), equation (1.6.25) appears to be order Δx. We also note that the derivation uses the second order approximation of the Neumann boundary condition and that the result is the same as what we have been referring to as the second order approximation to the Neumann boundary condition. Obviously, we must be careful about which order we declare approximations (1.4.12) and (1.6.25) to be. At the moment we will still refer to these approximations as the second order approximation of the Neumann boundary condition and explain the difference later in Section 2.3.2, Example 2.3.3.

Of course, as in the case with the difference equations in the interior, different approximations will yield different difference equations. Some of these approximations will give good difference schemes and some will not. No matter how you arrive at your difference equations, care must be taken to analyze them before you use them (which is what we will learn how to do in the following chapters).

1.6.2 Cell Averaged Equations

We next introduce another derivation of difference equations for the integral form of the conservation law where instead of using the difference variables u^n_k to approximate the function values at the grid points, we let u^n_k approximate the average of the function over the kth cell. This approach is used often in theoretical work-especially in the derivation of difference schemes for conservation laws (see Chapter 9, Part 2, [13]). As we shall see later, most often it will not be necessary to realize which of the two approaches we are using.

To derive a difference equation for cell averages, we return to equation (1.6.5). If we integrate from $t_n = n\Delta t$ to $t_{n+1} = (n+1)\Delta t$, we get

$$\int_{t_n}^{t_{n+1}} \int_{x_{k-1/2}}^{x_{k+1/2}} v_t \, dx dt = \int_{x_{k-1/2}}^{x_{k+1/2}} \int_{t_n}^{t_{n+1}} v_t \, dt dx \tag{1.6.26}$$

$$= \int_{x_{k-1/2}}^{x_{k+1/2}} (v^{n+1} - v^n) dx \tag{1.6.27}$$

$$= \Delta x (\overline{v}^{n+1}_k - \overline{v}^n_k) \tag{1.6.28}$$

$$= \int_{t_n}^{t_{n+1}} \int_{x_{k-1/2}}^{x_{k+1/2}} F(x,t) dx dt$$

$$+ \nu \int_{t_n}^{t_{n+1}} \left[(v_x)_{k+1/2} - (v_x)_{k-1/2} \right] dt \tag{1.6.29}$$

$$= \Delta x \int_{t_n}^{t_{n+1}} F_k + \nu \int_{t_n}^{t_{n+1}} \left[(v_x)_{k+1/2} - (v_x)_{k-1/2} \right] dt \qquad (1.6.30)$$

where $\bar{v}_k^n$ represents the *cell average* of $v(x, t_n)$ over the *k*th cell,

$$\bar{v}_k^n = \frac{1}{\Delta x} \int_{x_{k-1/2}}^{x_{k+1/2}} v^n dx,$$

and F_k is the average of F over the *k*th cell (which, since we know F, we assume is known). From the above calculation, lines (1.6.28) and (1.6.30) give us the equation

$$\Delta x (\bar{v}_k^{n+1} - \bar{v}_k^n) = \Delta x \int_{t_n}^{t_{n+1}} F_k dt + \nu \int_{t_n}^{t_{n+1}} \left[(v_x)_{k+1/2} - (v_x)_{k-1/2} \right] dt,$$

$$(1.6.31)$$

which is equivalent to equation (1.6.2) (which was true for all intervals). Hence, *equation (1.6.31) is an exact equation.*

We now consider how we might approximate equation (1.6.31). The term that must be approximated in the equation is the integral

$$\int_{t_n}^{t_{n+1}} \left[(v_x)_{k+1/2} - (v_x)_{k-1/2} \right] dt. \qquad (1.6.32)$$

Since we are using cell averages as our unknown variables, we must approximate the above integral in terms of cell averages. If we consider our approximation as a piecewise constant function that is equal to the cell average on the appropriate cell, it is easy to see that these flux terms should be given in terms of the same differences that we have previously used. As we shall see, deriving approximations in terms of cell averages is somewhat more difficult than the approach used in the previous section. We note that

$$\bar{v}_{k+1} - \bar{v}_k = \frac{1}{\Delta x} \int_{x_{k+1/2}}^{x_{k+3/2}} v(\xi, t) d\xi - \frac{1}{\Delta x} \int_{x_{k-1/2}}^{x_{k+1/2}} v(x, t) dx \qquad (1.6.33)$$

$$= \frac{1}{\Delta x} \int_{x_{k-1/2}}^{x_{k+1/2}} \left[v(x + \Delta x, t) - v(x, t) \right] dx \qquad (1.6.34)$$

$$= \frac{1}{\Delta x} \int_{x_{k-1/2}}^{x_{k+1/2}} \left[\Delta x v_x + \frac{\Delta x^2}{2} v_{xx} + \frac{\Delta x^3}{6} v_{xxx} \right.$$

$$\left. + \frac{\Delta x^4}{24} v_{xxxx} + \mathcal{O}(\Delta x^5) \right] dx \qquad (1.6.35)$$

$$= \frac{1}{\Delta x} \left\{ \Delta x \left[v_{k+1/2} - v_{k-1/2} \right] + \frac{\Delta x^2}{2} \left[(v_x)_{k+1/2} - (v_x)_{k-1/2} \right] \right.$$

$$\left. + \frac{\Delta x^3}{6} \left[(v_{xx})_{k+1/2} - (v_{xx})_{k-1/2} \right] \right.$$

$$+\frac{\Delta x^4}{24}\Big[(v_{xxx})_{k+1/2}-(v_{xxx})_{k-1/2}\Big]+\mathcal{O}(\Delta x^6)\Big\} \tag{1.6.36}$$

$$=\frac{1}{\Delta x}\Big\{\Delta x\Big[v_{k+1/2}-v_{k-1/2}\Big]+\frac{\Delta x^2}{2}\Big[(v_x)_{k+1/2}-(v_x)_{k-1/2}\Big]$$

$$+\frac{\Delta x^3}{6}\Big[(v_{xx})_{k+1/2}-(v_{xx})_{k-1/2}\Big]$$

$$+\frac{\Delta x^5}{24}(v_{xxxx})_k+\mathcal{O}(\Delta x^6)\Big\} \tag{1.6.37}$$

Performing the analogous calculation for $\bar{v}_k-\bar{v}_{k-1}$ gives

$$\bar{v}_k-\bar{v}_{k-1}=\frac{1}{\Delta x}\Big\{\Delta x\Big[v_{k+1/2}-v_{k-1/2}\Big]-\frac{\Delta x^2}{2}\Big[(v_x)_{k+1/2}-(v_x)_{k-1/2}\Big]$$

$$+\frac{\Delta x^3}{6}\Big[(v_{xx})_{k+1/2}-(v_{xx})_{k-1/2}\Big]-\frac{\Delta x^5}{24}(v_{xxxx})_k+\mathcal{O}(\Delta x^6)\Big\}. \tag{1.6.38}$$

Subtracting equation (1.6.38) from equation (1.6.37) yields

$$\bar{v}_{k+1}-2\bar{v}_k+\bar{v}_{k-1}=\Delta x\Big[(v_x)_{k+1/2}-(v_x)_{k-1/2}\Big]+\mathcal{O}(\Delta x^4). \tag{1.6.39}$$

Inserting this approximation into equation (1.6.31) yields

$$\Delta x(\bar{v}_k^{n+1}-\bar{v}_k^n)=\Delta x\int_{t_n}^{t_{n+1}}F_k+\nu\Delta x\int_{t_n}^{t_{n+1}}\frac{\delta^2\bar{v}_k}{\Delta x^2}dt+\mathcal{O}(\Delta t\Delta x^3). \tag{1.6.40}$$

So finally, we are left with approximating the integrals on the right hand side of equation (1.6.40). Using a rectangular rule approximate integration scheme (evaluating the function at t_n) yields

$$\nu\Delta x\int_{t_n}^{t_{n+1}}\frac{\delta^2\bar{v}_k}{\Delta x^2}dt=\frac{\nu\Delta t}{\Delta x}\delta^2\bar{v}_k^n+\mathcal{O}(\Delta t^2\Delta x) \tag{1.6.41}$$

and

$$\int_{t_n}^{t_{n+1}}F_kdt=\Delta tF_k^n+\mathcal{O}(\Delta t^2). \tag{1.6.42}$$

Hence, equation (1.6.40) becomes

$$\Delta x(\bar{v}_k^{n+1}-\bar{v}_k^n)=\Delta x\Delta tF_k^n+\frac{\nu\Delta t}{\Delta x}\delta^2\bar{v}_k^n+\mathcal{O}(\Delta t^2\Delta x)+\mathcal{O}(\Delta x^3\Delta t). \tag{1.6.43}$$

We note that if we replace the function average $\bar{v}_k^n$ by the approximation u_k^{n+1} and approximate the equation by dropping the $\mathcal{O}(\Delta t^2\Delta x)+\mathcal{O}(\Delta x^3\Delta t)$ term, we obtain difference equation

$$\Delta x(u_k^{n+1}-u_k^n)=\Delta x\Delta tF_k^n+\frac{\nu\Delta t}{\Delta x}\delta^2u_k^n, \tag{1.6.44}$$

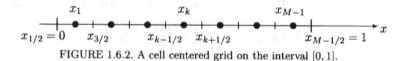

FIGURE 1.6.2. A cell centered grid on the interval $[0, 1]$.

which is the same scheme derived in Sections 1.2 (1.2.9) and 1.6 (1.6.15).

We notice that it is more difficult to rigorously derive and/or perform a truncation analysis for difference equations in terms of cell averages. Of course, it is also possible—but more difficult yet—to derive schemes for Neumann boundary conditions in terms of cell averages. Another approach, which is much easier and works equally well if you are clever enough to do it right, is to perform the approximation of the flux terms with less rigor (and, hence skip all of the work involved) and do a separate consistency analysis on the resulting equation (which we would hope would still be equation (1.6.44)) when you are done. This is the most common approach.

1.6.3 Cell Centered Grids

After observing the derivations in the last three sections, we might get the idea that the grids that we have constructed, which we call **vertex centered grids**, are not the only or the best grids to consider. We might decide that the cells are the important entities, and we may as well just consider the approximate function value or the approximate average value on the cell as being at the center of the cell. To do this, we can consider a **cell centered grid** on the interval $[0, 1]$ as is shown in Figure 1.6.2. We see that the interval is broken up into a series of cells. The grid points are marked off at the center of the cells. The main difference between the cell centered grid shown in Figure 1.6.2 and the vertex centered grid shown in Figure 1.6.1 is at the endpoints. Hence, as long as the conservation law approach is used, we will get the same difference schemes using vertex centered and cell centered schemes.

Depending on how we treat the boundaries, we may get different results due to the boundary conditions. We note that *with the cell centered grid, the endpoint of the interval is not a grid point*. It is an endpoint of a cell. This makes it very nice for treating Neumann boundary conditions. Just as we did in Section 1.6.1 using a half cell, we treat the 1-st cell as we do all of the rest, getting the flux across the left boundary of the cell from the Neumann boundary condition. Hence, if we again treat partial differential equation (1.6.1) and Neumann boundary condition (1.5.7), we see that by applying equation (1.6.9), we get the following equation on the 1-st cell.

$$\int_{x_{1/2}}^{x_{3/2}} (v^{n+1} - v^n)dx = \int_{t_n}^{t_{n+1}} \int_{x_{1/2}}^{x_{3/2}} F(x,t)dxdt$$

$$+ \nu \int_{t_n}^{t_{n+1}} [(v_x)_{3/2} - (v_x)_{1/2}]dt \qquad (1.6.45)$$

If we approximate first integral by the midpoint rule, the integral of the source term as we did in Section 1.6 and use the fact that $(v_x)_{1/2}$ is known, we are left with the equation

$$\Delta x(v_1^{n+1} - v_1^n) + \mathcal{O}(\Delta t \Delta x^3) = \Delta x \Delta t F_1^n + \mathcal{O}(\Delta t \Delta x^3) + \mathcal{O}(\Delta t^2 \Delta x)$$

$$+ \nu \int_{t_n}^{t_{n+1}} [(v_x)_{3/2} - g]dt. \qquad (1.6.46)$$

We approximate $(v_x)_{3/2}$ by $(v_2 - v_1)/\Delta x + \mathcal{O}(\Delta x^2)$ and integrate with respect to time by the rectangular rule. We get

$$\Delta x(v_1^{n+1} - v_1^n) + \mathcal{O}(\Delta t \Delta x^3) = \Delta x \Delta t F_1^n + \mathcal{O}(\Delta t \Delta x^3) + \mathcal{O}(\Delta t^2 \Delta x)$$

$$+ \frac{\nu \Delta t}{\Delta x}(v_2^n - v_1^n) + \mathcal{O}(\Delta t \Delta x^2) - \nu \Delta t g^n + \mathcal{O}(\Delta t^2). \qquad (1.6.47)$$

We obtain the following difference equation approximation to equation (1.6.47).

$$\Delta x(u_1^{n+1} - u_1^n) = \Delta x \Delta t F_1^n + \frac{\nu \Delta t}{\Delta x}(v_2^n - v_1^n) - \nu \Delta t g^n \qquad (1.6.48)$$

We note that though this difference equation appears to be the same as that found in Section 1.6.1, equation (1.6.25), it is different. The difference is *where the functions are evaluated* with respect to the spatial variables. Equation (1.6.25) is centered at $k = 0$ (the boundary) whereas equation (1.6.48) is centered at $k = 1$ (the center of the first cell). We shall see later where the difference of which grid is being used is very important (and, it is not obvious from the form of the difference scheme).

HW 1.6.2 Find an approximation to the solution of the initial–boundary–value problem given in HW1.5.9 using a cell centered approach for the difference equation and Neumann boundary condition treatment, (1.6.48).

It is not as clear how we should treat Dirichlet boundary conditions on a cell centered grid. Often, what is done is to include a ghost cell as shown in Figure 1.6.3. It is then assumed that the Dirichlet boundary condition is prescribed at the center of the ghost cell and the usual approach is followed

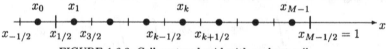

FIGURE 1.6.3. Cell centered grid with a ghost cell.

FIGURE 1.6.4. Cell centered grid with an adjusted half cell near the left boundary of the interval.

to derive the difference equation on the 1-st cell. But, if the solution is strongly dependent on the distance away from the boundary condition, the results using this approach will not be adequate. By applying a Taylor series expansion, it is easy to see that prescribing the Dirichlet boundary condition at the center of the ghost cell instead of at the boundary of the 1-st cell is a first order approximation. And, as stated earlier, if the solution is strongly dependent on this boundary condition, the accuracy will suffer.

Another approach is to include a half cell at the end of the interval as is done in the cell centered grid in Figure 1.6.4. In this case, the difference equation is derived on the 2-nd cell is done as usual (except for the fact that when it reaches to the 1-st cell, it gets the Dirichlet boundary condition). No equation is ever derived on the 1-st cell.

We should realize that both vertex centered and cell centered grids are common. Often, which approach is used depends on the area of application. To be a competent numerical partial differential equation practitioner, you must be familiar with both approaches—and understand that they are not really that different.

1.6.4   Nonuniform Grids

We close this chapter with a description of what needs to be done in the case when we use a nonuniform grid and some of the troubles that can be caused by a naive approach to the nonunform grid treatment. We begin by considering the *kth* cell of a nonuniform, cell centered grid shown in Figure 1.6.5. For convenience, we have chosen the lengths of the three cells involved to be $\Delta x/2$, Δx and $2\Delta x$, respectively. As we do most often, we consider the heat equation $v_t = \nu v_{xx}$ and want to derive a difference equation to approximate the heat equation on the nonuniform grid. Below we give two derivations of difference equations for approximately solving the heat equation on the grid pictured in Figure 1.6.5.

Taylor Series: Our first approach will be to derive a difference equation approximation using a Taylor series approach. The approximation of the

FIGURE 1.6.5. Cell centered nonuniform grid.

time derivative is the same as for uniform grids. To approximate the v_{xx} term on the nonuniform grid, we consider the following Taylor series expansions

$$v_{k+1}^n = v_k^n + (v_x)_k^n \left(\frac{3}{2}\Delta x\right) + \frac{1}{2}(v_{xx})_k^n \left(\frac{3}{2}\Delta x\right)^2 + \mathcal{O}(\Delta x^3) \qquad (1.6.49)$$

and

$$v_{k-1}^n = v_k^n + (v_x)_k^n \left(-\frac{3}{4}\Delta x\right) + \frac{1}{2}(v_{xx})_k^n \left(-\frac{3}{4}\Delta x\right)^2 + \mathcal{O}(\Delta x^3). \qquad (1.6.50)$$

Then using the method of undetermined coefficients, we get

$$av_{k+1}^n + bv_k^n + cv_{k-1}^n = (a+b+c)v_k^n + \left(\frac{3}{2}a\Delta x - \frac{3}{4}c\Delta x\right)(v_x)_k^n$$

$$+ \frac{1}{2}\left[a\left(\frac{3}{2}\Delta x\right)^2 + c\left(\frac{3}{4}\Delta x\right)^2\right](v_{xx})_k^n + \mathcal{O}(\Delta x^3). \qquad (1.6.51)$$

We see that we can choose a, b and c to satisfy

$$a + b + c = 0 \qquad (1.6.52)$$

$$\frac{3}{2}a\Delta x - \frac{3}{4}c\Delta x = 0 \qquad (1.6.53)$$

$$\frac{1}{2}\left[a\left(\frac{3}{2}\Delta x\right)^2 + c\left(\frac{3}{4}\Delta x\right)^2\right] = 1, \qquad (1.6.54)$$

and arrive at the following difference equation which is a $\mathcal{O}(\Delta t) + \mathcal{O}(\Delta x)$ approximation to the heat equation

$$u_k^{n+1} = u_k^n + \nu \frac{16}{27}\frac{\Delta t}{\Delta x^2}(u_{k+1}^n - 3u_k^n + 2u_{k-1}^n). \qquad (1.6.55)$$

Conservation Law: We next proceed to follow the conservation approach used in Sections 1.6–1.6.3. We return to the integral form of the conservation law, equation (1.6.5) with $F = 0$,

$$\int_{x_{k-1/2}}^{x_{k+1/2}} v_t dx = \nu\big[(v_x)_{k+1/2} - (v_x)_{k-1/2}\big]. \qquad (1.6.56)$$

Noting that

$$v_{k+1} = v_{k+1/2} + (v_x)_{k+1/2}\Delta x + \mathcal{O}(\Delta x^2) \qquad (1.6.57)$$

$$v_k = v_{k+1/2} + (v_x)_{k+1/2}\left(-\frac{\Delta x}{2}\right) + \mathcal{O}(\Delta x^2) \qquad (1.6.58)$$

and

$$v_k = v_{k-1/2} + (v_x)_{k-1/2}\frac{\Delta x}{2} + \mathcal{O}(\Delta x^2) \qquad (1.6.59)$$

$$v_{k-1} = v_{k-1/2} + (v_x)_{k+1/2}\left(-\frac{\Delta x}{4}\right) + \mathcal{O}(\Delta x^2), \qquad (1.6.60)$$

we see that we can approximate $(v_x)_{k\pm 1/2}$ as

$$(v_x)_{k+1/2} = \frac{2}{3\Delta x}(v_{k+1} - v_k) + \mathcal{O}(\Delta x) \qquad (1.6.61)$$

$$(v_x)_{k-1/2} = \frac{4}{3\Delta x}(v_k - v_{k-1}) + \mathcal{O}(\Delta x). \qquad (1.6.62)$$

If we were to carry more terms in the expansions above, we would see that no other terms would subtract out. So the order shown above is as well as we can do. The reason that no other terms subtract out is that we have lost all of the symmetry that we had in other derivations.

We now proceed to approximate the flux terms in equation (1.6.56) using equations (1.6.61) and (1.6.62), integrate with respect to t from t_n to t_{n+1} and approximate the term on the left hand side of the equation as we have before to get

$$\Delta x(v_k^{n+1} - v_k^n) + \mathcal{O}(\Delta t \Delta x^3) = \nu\frac{2\Delta t}{3\Delta x}\left[v_{k+1}^n - 3v_k^n + 2v_{k-1}^n\right]$$

$$+ \mathcal{O}(\Delta t \Delta x) + \mathcal{O}(\Delta t^2 \Delta x). \qquad (1.6.63)$$

Thus, we approximate the integral form of the conservation law, (1.6.63), by

$$\Delta x(u_k^{n+1} - u_k^n) = \nu\frac{2\Delta t}{3\Delta x}\left[u_{k+1}^n - 3u_k^n + 2u_{k-1}^n\right]. \qquad (1.6.64)$$

The approximate equation (1.6.64) approximates equation (1.6.63) (and, hence, (1.6.56)) to order $\mathcal{O}(\Delta t \Delta x^3) + \mathcal{O}(\Delta t \Delta x) + \mathcal{O}(\Delta t^2 \Delta x)$. We also comment that difference equation (1.6.64) or derivations equivalent to equation (1.6.64) are commonly used.

We now proceed to do a consistency analysis for difference scheme (1.6.64). We begin by writing difference equation (1.6.64) as

$$\frac{u_k^{n+1} - u_k^n}{\Delta t} = \nu\frac{\frac{2}{3}u_{k+1}^n - 2u_k^n + \frac{4}{3}u_{k-1}^n}{\Delta x^2}. \qquad (1.6.65)$$

If we then use the Taylor series expansions

$$v_k^{n+1} = v_k^n + (v_t)_k^n \Delta t + \mathcal{O}(\Delta t^2) \tag{1.6.66}$$

$$v_{k+1}^n = v_k^n + (v_x)_k^n \left(\frac{3}{2}\Delta x\right) + (v_{xx})_k^n \frac{1}{2}\left(\frac{3}{2}\Delta x\right)^2 + \mathcal{O}(\Delta x^3) \tag{1.6.67}$$

$$v_{k-1}^n = v_k^n - (v_x)_k^n \left(\frac{3}{4}\Delta x\right) + (v_{xx})_k^n \frac{1}{2}\left(\frac{3}{4}\Delta x\right)^2 + \mathcal{O}(\Delta x^3), \tag{1.6.68}$$

we can proceed as we did in Section 1.3 and see that

$$\frac{u_k^{n+1} - u_k^n}{\Delta t} - \nu \frac{\frac{2}{3}u_{k+1}^n - 2u_k^n + \frac{4}{3}u_{k-1}^n}{\Delta x^2} = v_t(k\Delta x, n\Delta t) - \nu \frac{\partial^2 v}{\partial x^2}(k\Delta x, n\Delta t)$$

$$- \frac{\nu}{8}\frac{\partial^2 v}{\partial x^2}(k\Delta x, n\Delta t) + \mathcal{O}(\Delta t) + \mathcal{O}(\Delta x). \tag{1.6.69}$$

Thus we see that *difference scheme (1.6.64) is a zero-th order approximation to the partial differential equation*. Another way to interpret the results given in equation (1.6.69) is that *difference scheme (1.6.64) is a $\mathcal{O}(\Delta x) + \mathcal{O}(\Delta t)$ approximation to the partial differential equation*

$$v_t = \frac{9}{8}\nu v_{xx}. \tag{1.6.70}$$

Thus we see that *difference scheme (1.6.64) is not consistent with the heat equation*.

Obviously, this should be very depressing to us. We have to know why and to be able to predict when we can get a consistent scheme using the conservation law approach. This is not as difficult as it seems. We derived difference equation (1.6.64) to illustrate that it is an easy mistake to make and because there are many people who make this mistake (and, hopefully, no one that reads this section will ever make this mistake). The order arguments used to obtain equation (1.6.63) were done very carefully. After the fact, it is obvious that the $\mathcal{O}(\Delta t \Delta x)$ term is the one that causes the problem. When this term is divided by both a Δt to get the time derivative approximation and a Δx to get rid of the Δx on the left hand side of the equation, *we are left with a $\mathcal{O}(1)$ term, or a 0th order approximation*.

Now that we know where the problem is coming from, it is not hard to correct the problem. The $\mathcal{O}(\Delta t \Delta x)$ term came from the first order approximation to $(v_x)_{k\pm 1/2}$ given in equations (1.6.61) and (1.6.62). *These approximations are just not good enough approximations of the flux.* The approach worked on uniform grids because, when we took the difference of the two terms, the symmetry caused the next highest terms to cancel. This does not happen here because of the nonuniform grids.

Hence, the approach is to find a higher order approximation of $(v_x)_{k\pm 1/2}^n$. The first order approximation for $(v_x)_{k+1/2}^n$ involved the points v_{k+1}^n and

v_k^n. To obtain a second order approximation, we must include a third point. So that we do not increase the size of the stencil, we will use v_{k-1}^n as the third point. The approach for obtaining a second order approximation for $(v_x)_{k-1/2}^n$ is analogous. We begin by considering the following Taylor series expansions.

$$v_{k+1} = v_{k+1/2} + (v_x)_{k+1/2}\Delta x + (v_{xx})_{k+1/2}^n \frac{1}{2}\Delta x^2$$
$$+ \mathcal{O}(\Delta x^3) \tag{1.6.71}$$

$$v_k = v_{k+1/2} + (v_x)_{k+1/2}\left(-\frac{\Delta x}{2}\right) + (v_{xx})_k^n \frac{1}{2}\left(-\frac{\Delta x}{2}\right)^2$$
$$+ \mathcal{O}(\Delta x^3) \tag{1.6.72}$$

$$v_{k-1} = v_{k+1/2} + (v_x)_{k+1/2}\left(-\frac{5\Delta x}{4}\right) + (v_{xx})_{k+1/2}^n \frac{1}{2}\left(-\frac{5\Delta x}{4}\right)^2$$
$$+ \mathcal{O}(\Delta x^3). \tag{1.6.73}$$

Then we can write

$$av_{k+1} + bv_k + cv_{k-1} = (a+b+c)v_{k+1/2} + \Delta x\left(a - \frac{b}{2} - \frac{5c}{4}\right)(v_x)_{k+1/2}$$
$$+ \frac{\Delta x^2}{2}\left(a + \frac{b}{4} + \frac{25c}{16}\right)(v_{xx})_{k+1/2} + \mathcal{O}(\Delta x^3). \tag{1.6.74}$$

If we solve the system of equations

$$a + b + c = 0$$
$$a - \frac{b}{2} - \frac{5c}{4} = \frac{1}{\Delta x}$$
$$a + \frac{b}{4} + \frac{25c}{16} = 0,$$

we see that $a = \frac{14}{27\Delta x}$, $b = \frac{-2}{9\Delta x}$ and $c = \frac{-8}{27\Delta x}$ and we have the following approximation for $(v_x)_{k+1/2}$.

$$(v_x)_{k+1/2} = \frac{2}{27\Delta x}(7v_{k+1} - 3v_k - 4v_{k-1}) + \mathcal{O}(\Delta x^2) \tag{1.6.75}$$

Using the analogous expansions about $x_{k-1/2}$ and the same approach used to develop the approximation (1.6.75), we get

$$(v_x)_{k-1/2} = \frac{2}{27\Delta x}(-v_{k+1} + 21v_k - 20v_{k-1}) + \mathcal{O}(\Delta x^2). \tag{1.6.76}$$

If we then use the approximations for $(v_x)_{k\pm1/2}$ given in equations (1.6.75) and (1.6.76) in equation (1.6.56) and proceed to approximate the rest of equation (1.6.56) as we did before, we are left with the following equation.

$$\Delta x(v_k^{n+1} - v_k^n) + \mathcal{O}(\Delta t \Delta x^3) = \nu \frac{16}{27} \frac{\Delta t}{\Delta x} \left[v_{k+1}^n - 3v_k^n + 2v_{k-1}^n \right]$$
$$+ \mathcal{O}(\Delta t \Delta x^2) + \mathcal{O}(\Delta t^2 \Delta x) \tag{1.6.77}$$

We see that we now have a higher order approximation to the integral form of the conservation equation. The order is high enough so that if we divide through by Δx and Δt, we will not get a zero-th order term. Also, it is easy to see that when we approximate the above equation, we are left with the same equation that we derived via Taylor series expansions (1.6.55). We should emphasize here that this example does not show that anything is wrong with the conservation law or finite volume approaches. This example does show that when using these approaches, or any other approach, care must be taken in how we approximate the terms in our exact equations. In this case, originally we tried to get by using a nice, but naive approximation for the fluxes.

As we see below, the nonuniform grids also cause problems when we derive equations for Neumann boundary conditions.

HW 1.6.3 (a) Consider a grid as given in Figure 1.6.5 where the $k - 1$-st cell is the boundary cell (the cell of length $\Delta x/2$ butts against the left boundary of the domain in question). Use a conservation law approach to derive an approximation of the Neumann boundary condition (1.5.7) approximating the flux across the $x_{k-1/2}$ boundary by $(v_x)_{k-1/2} = (v_k - v_{k-1})/(3\Delta x/4) + \mathcal{O}(\Delta x)$.
(b) Repeat part (a) using three points, x_{k-1}, x_k and x_{k+1} to approximate $(v_x)_{k-1/2}$. Compare and discuss the accuracy of the approximations found in (a) and (b).

2

Some Theoretical Considerations

2.1 Introduction

This chapter begins with the basic definitions of what it means for the solution of the finite difference equations to converge to the solution of the partial differential equation. It is extremely important for a user of finite difference techniques to understand precisely what type of convergence their scheme has, what kinds of assumptions are made to get this convergence and how this convergence affects their accuracy. The most common approach to convergence of difference equations is through the concepts of consistency and stability and the Lax Theorem. The Lax Theorem allows us to prove convergence of a difference scheme by showing that the scheme is both consistent and stable (which are generally easier to show). The goal of this chapter is to include these definitions and the Lax Theorem in a rigorous setting that is intuitive and understandable.

2.2 Convergence

2.2.1 Initial–Value Problems

Finite difference schemes such as those that we have been discussing are used because their solutions approximate the solutions to certain partial differential equations. What is really needed is that the solution of the difference equations can be made to approximate the solution of the partial differential equation to any desired accuracy. Thus we want some sort of convergence of the solution of the finite difference equation to the solution of the partial differential equation. The convergence that is needed is not easy to discuss. We begin by considering initial–value problems. Hence we are considering a partial differential equation, say

$$\mathcal{L}v = F, \tag{2.2.1}$$

for functions v and F that are defined on the entire real line and an initial condition $v(x, 0) = f(x)$. We assume that we have obtained an approximate solution (from a finite difference scheme that we shall refer to as L_k^n where, as in the previous chapter, n corresponds to the time step and k to the spatial mesh point) to this problem, u_k^n, which is defined on a grid (with grid spacings Δx and Δt) and satisfies the initial condition $u_k^0 = f(k\Delta x)$, $k = -\infty, \cdots, \infty$. Let v denote the exact solution to our initial–value problem. We then state the following definition.

Definition 2.2.1 *A difference scheme $L_k^n u_k^n = G_k^n$ approximating the partial differential equation $\mathcal{L}v = F$ is a pointwise convergent scheme if for any x and t, as $(k\Delta x, (n+1)\Delta t)$ converges to (x, t), u_k^n converges to $v(x, t)$ as Δx and Δt converge to 0.*

To demonstrate this definition, we prove convergence for the explicit scheme we used in Chapter 1 (equation (1.2.9)).

Example 2.2.1 Show that the solution of the difference scheme

$$u_k^{n+1} = (1 - 2r)u_k^n + r(u_{k+1}^n + u_{k-1}^n), \tag{2.2.2}$$

$$u_k^0 = f(k\Delta x), \tag{2.2.3}$$

where $r = \nu\Delta t/\Delta x^2$, $0 \le r \le 1/2$, converges pointwise to the solution of the initial–value problem

$$v_t = \nu v_{xx}, \quad x \in \mathbb{R}, \quad t > 0 \tag{2.2.4}$$

$$v(x, 0) = f(x), \quad x \in \mathbb{R}. \tag{2.2.5}$$

Solution: We must realize that since we are considering an initial–value problem on all of $\mathbb{R}$, the k index on u_k^n will range over all of the integers, $-\infty < k < \infty$.

We denote the exact solution to the initial–value problem (2.2.4)–(2.2.5) by $v = v(x, t)$ and set

$$z_k^n = u_k^n - v(k\Delta x, n\Delta t). \tag{2.2.6}$$

If we insert v into equation (1.3.8) (so that it makes the left hand side of equation (1.3.8) zero) and multiply through by Δt, we see that $v_k^n = v(k\Delta x, n\Delta t)$ satisfies

$$v_k^{n+1} = (1 - 2r)v_k^n + r(v_{k+1}^n + v_{k-1}^n) + \mathcal{O}(\Delta t^2) + \mathcal{O}(\Delta t\Delta x^2). \tag{2.2.7}$$

Then by subtracting equation (2.2.7) from equation (2.2.2), we see that z_k^n satisfies

$$z_k^{n+1} = (1 - 2r)z_k^n + r(z_{k+1}^n + z_{k-1}^n) + \mathcal{O}(\Delta t^2) + \mathcal{O}(\Delta t\Delta x^2). \tag{2.2.8}$$

If $0 < r \leq 1/2$, the coefficients on the right hand side of equation (2.2.8) are non-negative and

$$
\begin{aligned}
|z_k^{n+1}| &\leq (1-2r)|z_k^n| + r|z_{k+1}^n| + r|z_{k-1}^n| + A(\Delta t^2 + \Delta t \Delta x^2) \\
&\leq Z^n + A(\Delta t^2 + \Delta t \Delta x^2),
\end{aligned}
$$

where A is the constant associated with the "big $\mathcal{O}$" terms and depends on the assumed bounds of the higher order derivatives of v, and $Z^n = \sup_k\{|z_k^n|\}$. Then taking the sup over k on the left side of the above equation yields

$$
Z^{n+1} \leq Z^n + A(\Delta t^2 + \Delta t \Delta x^2). \tag{2.2.9}
$$

(We should note that taking the supremum over the right hand side included the terms containing Δt and Δx. In this case we are assuming that the constant A is a bound of the second derivative with respect to time and the fourth derivative with respect to space on the entire real line. Thus we have assumed that these appropriate derivatives of the solution are uniformly bounded on the real line.) Applying (2.2.9) repeatedly yields

$$
\begin{aligned}
Z^{n+1} &\leq Z^n + A(\Delta t^2 + \Delta t \Delta x^2) \\
&\leq Z^{n-1} + 2A(\Delta t^2 + \Delta t \Delta x^2) \\
&\;\;\vdots \\
&\leq Z^0 + (n+1)A(\Delta t^2 + \Delta t \Delta x^2).
\end{aligned}
$$

Since $Z^0 = 0$, $|u_k^{n+1} - v(k\Delta x, (n+1)\Delta t)| \leq Z^{n+1}$ and $(n+1)\Delta t \to t$,

$$
\begin{aligned}
|u_k^{n+1} - v(k\Delta x, (n+1)\Delta t)| &\leq (n+1)\Delta t A(\Delta t + \Delta x^2) \\
&\to 0 \text{ as } \Delta t, \Delta x \to 0. \tag{2.2.10}
\end{aligned}
$$

Thus we see that for any x and t, as Δt and Δx approach 0 in such a way that $(k\Delta x, (n+1)\Delta t) \to (x,t)$, u_k^n approaches $v(x,t)$. The reader should note that since the $n+1$ in expression (2.2.10) could potentially mess up our convergence (it is not good to have terms that go to infinity in an expression that we want to go to zero), it was important in the last step that $(n+1)\Delta t \to t$.

Remark 1: One of the things about the above proof that we should notice is that we assume that $0 < r \leq 1/2$ and this assumption was very important in this proof. Of course, this doesn't say that convergence could not be proved without this assumption. This condition that we have placed on r puts a condition on the time step, Δt. Thus we see that for this proof we must have $\Delta t \leq \Delta x^2/2\nu$. You might wish to consider this constraint in light of your results on problem HW 1.2.1.

Remark 2: Another thing that we should realize about the above proof is that we have assumed that the same constant given to us from the "big $\mathcal{O}$" notation works at all of the grid points k and at all of the time levels, i.e. we used A as the constant instead of an A_k^n. Thus to obtain the above result, we have assumed that the appropriate derivatives of the solution(v_{tt} and v_{xxxx}) are uniformly bounded on $\mathbb{R} \times [0,t]$.

We note that we have really proved a much stronger result than we originally set out to prove. In arriving at expression (2.2.10) above, we proved that $Z^{n+1} \to 0$. But since we cared about pointwise convergence, we stated our convergence result as in (2.2.10).

Denote the sup-norm on the space of all bounded sequences, ℓ_∞, by

$$\| \{\alpha_k\} \|_\infty = \sup_{-\infty < k < \infty} |\alpha_k|.$$

If we let $\mathbf{u}^n = (\cdots, u^n_{-1}, u^n_0, u^n_1, \ldots)^T$ and $\mathbf{v}^n = (\cdots, v^n_{-1}, v^n_0, v^n_1, \ldots)^T$, then we have proved that for t such that $(n+1)\Delta t$ approaches t, $\mathbf{u}^{n+1}$ approaches $v(\cdot, t)$ where our concept of "approaches" is that the sup-norm of $\mathbf{u}^{n+1} - \mathbf{v}^{n+1}$ approaches zero as Δt and Δx approach zero.

Since, in general, the pointwise convergence is difficult to prove and is not generally as useful as a more uniform sort of convergence, we shall instead use a definition of convergence in terms of a norm of the difference between the solution to the partial differential equation and the solution of the difference equation. This will be an extension of the convergence that we proved above in terms of the sup-norm. For convenience, we denote the vector of difference equation solution values $\{u^n_k\}$ by $\mathbf{u}^n$ and the vector of solution values of the partial differential equation evaluated at the grid points, $v(k\Delta x, n\Delta t)$, by $\mathbf{v}^n$.

Definition 2.2.2 *A difference scheme $L^n_k u^n_k = G^n_k$ approximating the partial differential equation $\mathcal{L}v = F$ is a convergent scheme at time t if, as $(n+1)\Delta t \to t$,*

$$\| \mathbf{u}^{n+1} - \mathbf{v}^{n+1} \| \to 0 \qquad (2.2.11)$$

as $\Delta x \to 0$ and $\Delta t \to 0$.

Remark 1: It should be specially noted that the norm used above was not specified. This was done in this manner because in different situations, different norms will be used. Above we proved that *the explicit scheme (2.2.2) was convergent according to Definition 2.2.2 in the sup-norm.* At other times we will see that the natural norm to choose will be a variation of the ℓ_2 norm, the $\ell_{2,\Delta x}$ norm. Thus whenever convergence is being discussed, the norm that is being used must be specified.

Remark 2: It must be made clear how Definition 2.2.2 differs from Definition 2.2.1. Using Definition 2.2.1, the rate at which u^n_k converges to $v(x, t)$ (rate in terms of $\Delta x \to 0$ and $\Delta t \to 0$) can vary greatly for different values of x. In fact, Definition 2.2.1 would allow for a scheme to converge for some values of (x, t) and not others. If $\mathbf{u}^n$ is close to $\mathbf{v}^n$ via Definition 2.2.2 (i.e. $\| \mathbf{u}^n - \mathbf{v}^n \|$ is small), then we know that for the given Δt and Δx, u^n_k is close to the associated $(v)^n_k$ for all k. We get no such information from Definition 2.2.1.

At certain times we want to discuss convergence in terms of how fast the solution of the difference equation converges to the solution of the partial differential equation. For this purpose we define *convergence of order (p, q)* as follows.

Definition 2.2.3 *A difference scheme $L_k^n u_k^n = G_k^n$ approximating the partial differential equation $\mathcal{L}v = F$ is a convergent scheme of order (p, q) if for any t, as $(n+1)\Delta t$ converges to t,*

$$\| \mathbf{u}^{n+1} - \mathbf{v}^{n+1} \| = \mathcal{O}(\Delta x^p) + \mathcal{O}(\Delta t^q) \tag{2.2.12}$$

as Δx and Δt converge to 0.

Remark 1: When we use the "$\mathcal{O}$" notation, we must always remember that there is a constant involved, i.e. equation (2.2.12) is really a short hand notation for "there exists a constant C such that $\| \mathbf{u}^{n+1} - \mathbf{v}^{n+1} \| \leq C(\Delta x^p + \Delta t^q)$. In this case, *the constant C will depend on t.*

Remark 2: Using Definition 2.2.3, we are able to discuss the rate of convergence of $\mathbf{u}^n$ to $\mathbf{v}$. For example, if we return to Example 2.2.1 (using the sup-norm convergence discussed after the example rather than the pointwise convergence), we see that difference scheme (2.2.2) is convergent of order $(2, 1)$.

And finally, before we move to initial–boundary–value problems, we point out that for initial–value problems, convergence (and later consistency and stability) will be discussed and proved in infinite dimensional linear spaces. This text is not designed to be at a level of requiring functional analysis as a prerequisite. However, using spaces such as ℓ_∞ and $\ell_{2,\Delta x}$ is the only way that this can be done correctly. Hence, we will include this material. Those not already familiar with infinite dimensional linear spaces and norms should continue, and we are sure you will gain information from the approach.

HW 2.2.1 Use the technique used in Example 2.2.1 to prove that the solution to difference equation

$$u_k^{n+1} = \frac{1}{2}(u_{k+1}^n + u_{k-1}^n) - \frac{R}{2}\delta_0 u_k^n$$

(the Lax-Friedrichs scheme) converges in the sup-norm to the solution of the partial differential equation

$$v_t + a v_x = 0$$

for $|R| \leq 1$ where $R = a\Delta t/\Delta x$.

2.2.2 Initial–Boundary–Value Problems

The difference between convergence for schemes for initial–boundary–value problems and convergence for schemes for initial–value problems lies in the spaces. As we made either Δx or Δt smaller in the scheme that we considered in the last section, the *space* in which we were working continued to be a space of sequences for $-\infty < k < \infty$. If we consider the difference scheme on the interval $[0, 1]$, with zero Dirichlet boundary conditions, the problem is a finite dimensional problem. We are given $u_k^0 = f(k\Delta x)$, $k = 0, \cdots, M$, $u_0^n = 0$, $n = 1, \cdots$, and $u_M^n = 0$, $n = 1, \cdots$, and we find u_k^n, $k = 1, \cdots, M - 1$, $n = 1, \cdots$ by using a difference scheme. As Δx gets small (or as M gets larger), the vector of unknowns gets larger. It is not that we cannot measure the difference between the solution to the difference equation and the solution to the partial differential equation. The point is that there is not "a nice space" in which this convergence takes place.

There are several approaches on how we might confront this problem. The approach that we shall take is to require that Δx approach zero in an orderly manner and *define the convergence in terms of the norms in the appropriate spaces.*

We begin by defining a partition of an interval $[0, 1]$ (or of course, any other interval) to be a uniform grid described by an increment Δx. We then consider any sequence of such partitions with increments $\{\Delta x_j\}, j = 1, \ldots$ such that $\Delta x_j \to 0$ as $j \to \infty$. Let X_j denoted a finite dimensional normed linear space containing the solution associated with increment Δx_j. If we denote a norm on X_j by $\| \cdot \|_j$, we define convergence of the initial–boundary–value problem as follows.

Definition 2.2.4 *A difference scheme approximating a continuous initial–boundary–value problem is a convergent scheme at time t if for any sequence of partitions $\{\Delta x_j\}$, as $(n + 1)\Delta t \to t$,*

$$\| \mathbf{u}^n - \mathbf{v}^n \|_j \to 0 \tag{2.2.13}$$

as $j \to \infty$ and $\Delta t \to 0$.

Of course, in like manner, we can also define convergence of order (p, q).

Remark 1: Though to satisfy the above definition we must consider all partitions that converge to zero, the model for the use of Definition 2.2.4 is to use the domain $[0, 1]$ and $\Delta x = 1/M$ as we have done in all of our examples and problems. Then the space is either an $M - 1$, M or $M + 1$ dimensional space, depending on what type of boundary conditions are being used at each end of the interval. The norms used in the definition of convergence will be the appropriate norms on the finite dimensional spaces on which we work. However, as we will see later, proving convergence for initial–boundary–value problems will be much more difficult than doing so for initial–value problems.

Remark 2: Though not explicitly stated, the implied partition in Remark 1 was the set of points $\{x_k : x_k = k\Delta x, \ k = 0, \cdots, M\}$ where $\Delta x = 1/M$. In this text, this is the grid that is used most often. There are several other partitions of the interval $[0, 1]$ that we will use and will be satisfactory partitions for Definition 2.2.4. Most often, these "other partitions" will be associated with other grid alignments that we consider. For example, when we have a cell centered grid, the uniform partition on the interval $[0, 1]$ will consist of the grid points $\{x_k : x_k = (k - 1)\Delta x + \Delta x/2, \; k = 1, \cdots, M\}$ where $\Delta x = 1/M$. We will be most interested in a partition like this one when we have Neumann boundary conditions on both ends, and often in this situation will use the grid points $x_0 = -\Delta x/2$ and $x_{M+1} = 1 + \Delta x/2$ even though they are not in our partition. When we have a Neumann boundary condition at $x = 0$ and a Dirichlet boundary condition at $x = 1$, we could use partitions of the form $\{x_k : x_k = (k - 1)\Delta x + \Delta x/2, \; k = 1, \cdots, M\}$ where $\Delta x = 2/(2M - 1)$. Some of these grids can be seen in Figures 1.6.1–1.6.5.

To illustrate Definition 2.2.4, we again prove convergence for the explicit scheme (2.2.2), this time for an initial–boundary–value problem.

Example 2.2.2 Show that for $0 \le r \le 1/2$, the solution of the difference scheme

$$u_k^{n+1} = (1 - 2r)u_k^n + r(u_{k-1}^n + u_{k+1}^n), \quad n \ge 0, \; k = 1, \cdots, M - 1 \qquad (2.2.14)$$

$$u_0^{n+1} = u_M^{n+1} = 0, \quad n \ge 0 \qquad (2.2.15)$$

$$u_k^0 = f(k\Delta x), \quad k = 0, \cdots, M \qquad (2.2.16)$$

converges in the sup-norm to the solution of the initial–boundary–value problem

$$v_t = \nu v_{xx}, \; x \in (0,1), \; t > 0 \qquad (2.2.17)$$

$$v(x,0) = f(x), \; x \in [0,1], \qquad (2.2.18)$$

$$v(0,t) = v(1,t) = 0, \; t > 0. \qquad (2.2.19)$$

Solution: We prove the convergence of difference scheme (2.2.14)–(2.2.16) the same way we proved the convergence for the initial–value scheme in Example 2.2.1, except in this case, our vectors will have a finite length and the size will keep changing. Also, because we do it the same way, we will eventually have to make several assumptions to make the proof work. We begin by letting Δx_j be the increment in a partition with $M_j + 1$ points. We let X_j denote the space of $(M_j - 1)$-vectors with the finite dimensional sup-norm,

$$\| (u_1, \cdots, u_{M_j-1})^T \|_{M_j-1,\infty} = \sup_{1 \le k \le M_j-1} |u_k|. \qquad (2.2.20)$$

We again set

$$z_k^n = u_k^n - v(k\Delta x_j, n\Delta t). \qquad (2.2.21)$$

As in Example 2.2.1, z_k^n satisfies

$$z_k^{n+1} = (1 - 2r)z_k^n + r(z_{k+1}^n + z_{k-1}^n) + \mathcal{O}(\Delta t^2) + \mathcal{O}(\Delta t \Delta x_j^2), \quad k = 1, \cdots, M_j - 1. \qquad (2.2.22)$$

If $0 < r \le 1/2$, the coefficients on the right hand side of equation (2.2.22) are non-negative (where we realize that z_0^n and $z_{M_j}^n$ are both zero) and

$$|z_k^{n+1}| \le (1 - 2r)|z_k^n| + r|z_{k+1}^n| + r|z_{k-1}^n| + A(\Delta t^2 + \Delta t(\Delta x_j)^2)$$

$$\le \| \mathbf{z}^n \|_{M_j-1,\infty} + A(\Delta t^2 + \Delta t \Delta x_j^2),$$

where

$$\mathbf{z}^n = [z_1^n, \cdots, z_{M_j-1}^n]^T.$$

If we take the supremum over the left hand side of the above inequality and apply this inequality repeatedly, we are left with

$$\| z^{n+1} \|_{M_j-1,\infty} \leq (n+1)\Delta t A(\Delta t + \Delta x_j^2). \qquad (2.2.23)$$

Hence, as $\Delta t \to 0$, $(n+1)\Delta t \to t$, and $j \to \infty$,

$$\| u^{n+1} - v^{n+1} \|_{M_j-1,\infty} \to 0,$$

and the scheme converges.

Remark 1: In Example 2.2.2, we used a sequence of Δx_j's while we did not subscript Δt. Of course, we took $\Delta t \to 0$ and Δt must approach zero in such a way that Δt and Δx_j satisfies $r \leq 1/2$. We could have just as easily and logically taken a sequence of Δt_j's such that $\nu \Delta t_j / \Delta x_j^2 = r \leq 1/2$ for all j.

Remark 2: We might also mention that it is not necessary to use uniform grids in Definition 2.2.4. Often, a solution procedure will be defined in terms of a non-uniform grid. But for our purposes, the approach using uniform grids will already be difficult enough for theoretical results. Though we shall at times use non-uniform grids in computational problems, our convergence results using Definition 2.2.4 will be done using uniform grids.

HW 2.2.2 Use the explicit scheme (2.2.2) to solve the initial–boundary–value problem

$$v_t = \nu v_{xx}, \ x \in (0,1), \ t > 0$$
$$v(x,0) = \sin 4\pi x, \ x \in [0,1]$$
$$v(0,t) = v(1,t) = 0, \ t > 0$$

with $\nu = 0.1$. Report results at times $t = 0.05$ and $t = 0.1$. Use (i) $\Delta x = 0.1$, $\Delta t = 0.05$; (ii) $\Delta x = 0.05$, $\Delta t = 0.0125$; and (iii) $\Delta x = 0.01$, $\Delta t = 0.0005$. Compare and contrast your results for the three cases and with the solution of the initial–boundary-value problem.

2.2.3   A Review of Linear Algebra

In Sections 2.2.1 and 2.2.2 we see that the definitions of convergence are given in terms of norms defined on finite and infinite dimensional linear spaces. To make the text reasonably self contained, we include here a brief review of some spaces and norms from linear algebra and functional analysis. For more information on matrix theory (finite dimensional spaces), see ref. [3], and for more information on functional analysis (infinite dimensional spaces), see ref. [7].

2.2.3.1 Spaces and Norms

When we work with finite dimensional spaces, we will be using either the real Euclidean space, $\mathbb{R}^N$, or complex Euclidean space, $\mathbb{C}^N$, for some N. The different finite dimensional norms we will use include the usual Euclidean norm

$$\| \mathbf{u} \|_2 = \sqrt{\sum_{k=1}^{N} |u_k|^2} \tag{2.2.24}$$

(where sometimes the subscript 2 will not be included), the $\ell_{2,\Delta x}$ norm

$$\| \mathbf{u} \|_{2,\Delta x} = \sqrt{\sum_{k=1}^{N} |u_k|^2 \Delta x} \tag{2.2.25}$$

(where we will try not to confuse this with the infinite dimensional version), and the sup-norm

$$\| \mathbf{u} \|_{\infty} = \sup_{1 \leq k \leq N} |u_k|. \tag{2.2.26}$$

We should emphasize that when we are in $\mathbb{R}^N$, the notation $|\cdot|$ means the absolute value, while in $\mathbb{C}^N$ the notation means the magnitude of the complex number. And, of course, all of these norms are equivalent on either $\mathbb{R}^N$ or $\mathbb{C}^N$. At different times each of these different norms will be either necessary or convenient.

The infinite dimensional sequence spaces that we shall use include both the real and complex ℓ_2 space,

$$\ell_2 = \{\mathbf{u} = (\cdots, u_{-1}, u_0, u_1, \cdots)^T : \sum_{k=-\infty}^{\infty} |u_k|^2 < \infty\} \tag{2.2.27}$$

with norm

$$\| \mathbf{u} \|_2 = \sqrt{\sum_{k=-\infty}^{\infty} |u_k|^2}, \tag{2.2.28}$$

both the real and complex ℓ_2 space with the energy norm, $\ell_{2,\Delta x}$, with norm

$$\| \mathbf{u} \|_{2,\Delta x} = \sqrt{\sum_{k=-\infty}^{\infty} |u_k|^2 \Delta x}, \tag{2.2.29}$$

and the space of all bounded sequences, ℓ_∞,

$$\ell_\infty = \{\mathbf{u} = (\cdots, u_{-1}, u_0, u_1, \cdots)^T : \sup_{-\infty < k < \infty} |u_k| < \infty\}, \tag{2.2.30}$$

with norm

$$\| \mathbf{u} \|_\infty = \sup_{-\infty < k < \infty} |u_k|. \tag{2.2.31}$$

And finally, when we take transforms, we will get the infinite dimensional linear space of complex valued, Lebesgue square integrable functions, $L_2(R)$,

$$L_2(R) = \{v : R \to \mathbb{C} : \int_R |v(x)|^2 dx < \infty\} \tag{2.2.32}$$

with norm

$$\| v \|_2 = \sqrt{\int_R |v(x)|^2 dx}. \tag{2.2.33}$$

It should be noted that $\int_R$ used above means the integral over R. In this case, the integral is the Lebesgue integral (which the reader need not know much about). Also, the domain of the functions R will generally be the real line, $\mathbb{R}$, or the interval $[-\pi, \pi]$. And, as we stated before, when the functions are complex valued, the notation $|v(x)|$ refers to the magnitude of the complex values.

There are several variations of the above spaces and norms that we will be using. For example, when we consider two dimensional problems, the vectors will then have two subscripts. Though it is often advantageous to sum over each index separately, the approach is to make these two dimensional arrays into a large one dimensional array. Then, for example, the $\ell_{2,\Delta x}$ norm of a vector in the two dimensional, finite dimensional sequence space would be

$$\| \{u_{j\,k}\} \|_{2,\Delta x} = \sqrt{\sum_{j=1}^{N_x} \sum_{k=1}^{N_y} |u_{j\,k}|^2 \Delta x \Delta y}. \tag{2.2.34}$$

(We should notice that the "two dimensional" $\ell_{2,\Delta x}$ norm has both a Δx and a Δy multiplying the ℓ_2 sum. Though we often think of the "vector" as a one dimensional array, this reminds us that it is really two dimensional.) In a similar fashion, we define the infinite dimensional $\ell_{2,\Delta x}$ norm in two dimensions as

$$\| \{u_{j\,k}\} \|_{2,\Delta x} = \sqrt{\sum_{j=-\infty}^{\infty} \sum_{k=-\infty}^{\infty} |u_{j\,k}|^2 \Delta x \Delta y}, \tag{2.2.35}$$

and the two-dimensional L_2 norm as

$$\| v \|_2 = \sqrt{\int_{R \times R} |v(x, y)|^2 dx dy}. \tag{2.2.36}$$

Another variation of the above spaces with which we will have to deal is having either vector valued sequence spaces or L_2 space. For example, if we have the vector $\mathbf{u} = (\cdots, \mathbf{u}_{-1}, \mathbf{u}_0, \mathbf{u}_1, \cdots)^T$ where each vector $\mathbf{u}_k$ is a K-vector, we define

$$\| \mathbf{u} \|_{2,\Delta x} = \sqrt{\sum_{j=-\infty}^{\infty} \| \mathbf{u}_j \|_2 \, \Delta x}, \qquad (2.2.37)$$

where the norm in the summation is the Euclidean norm. And, if we have a function $\mathbf{v} : R \to \mathbb{C}^N$, we define the norm of this $\mathbb{C}^N$-valued, Lebesgue square integrable function as

$$\| \mathbf{v} \|_2 = \sqrt{\int_R \| \mathbf{v}(x) \|_2^2 \, dx}. \qquad (2.2.38)$$

2.2.3.2   Operators

When we consider our problems in the linear spaces described above, the difference operators and transforms become operators on these spaces. We consider an operator Q, $Q : X \to X$ where X is the linear space with norm $\| \cdot \|$. The operators that we consider on sequence spaces will be defined by a formula of how the kth element is defined. For example, the difference operator $u_k^{n+1} = u_k^n + r\delta^2 u_k^n$ can be written as $\mathbf{u}^{n+1} = Q\mathbf{u}^n$ where we have defined the kth element of $Q\mathbf{u}^n$ as $(Q\mathbf{u}^n)_k = u_k^n + r\delta^2 u_k^n$. The operator on L_2 spaces will generally be multiplication by some function, i.e. $\hat{Q}\hat{u}(\xi) = \rho(\xi)\hat{u}(\xi)$. And, operators (including difference operators) defined on the finite dimensional spaces can be and most often will be written as matrices.

We define the induced norm of Q (induced from the norm $\| \cdot \|$) as

$$\| Q \| = \sup_{\|\mathbf{u}\| \leq 1} \| Q\mathbf{u} \| . \qquad (2.2.39)$$

We will discuss norms of operators often. We shall see that we will hardly ever be able to compute the norm of an operator. The two results pertaining to computation of norms that we shall use both involve finite dimensional operators, matrices. Both results assume that we are using the Euclidean norm, and hence, the induced Euclidean operator norm. We first state a result that will enable us to compute the norm of Hermitian matrices (A is a Hermitian matrix if $A^* = A$ where A^* is the conjugate transpose of A, $\bar{A}^T$).

Proposition 2.2.5 *If A is an $N \times N$ Hermitian matrix, then*

$$\| A \| = \sigma(A),$$

where $\sigma(A) = \max\{|\lambda| : \lambda$ is an eigenvalue of $A\}$ is the spectral radius of A.

The result for a general matrix is similar.

Proposition 2.2.6 *If A is an $N \times N$ matrix, then*

$$\| A \| = \sqrt{\sigma(A^*A)}.$$

2.2.3.3   Tridiagonal Matrices

As we shall see, most of the matrices with which we will work will be tridiagonal matrices. Often, (for example when we want to compute the norm of a matrix), we will want to know the eigenvalues of a tridiagonal matrix. There are several interesting tridiagonal matrices for which the eigenvalues and eigenvectors are known. For example, if we define

$$T = Tr(a,b,c) = \begin{pmatrix} b & c & \cdots & & \\ a & b & c & \cdots & \\ & \cdots & \cdots & & \\ \cdots & a & b & c \\ & \cdots & & a & b \end{pmatrix}_{N \times N} \tag{2.2.40}$$

then the eigenvalues and associated eigenvectors are given by

$$\lambda_j = b + 2c\sqrt{\frac{a}{c}} \cos \frac{j\pi}{N+1}, \quad j = 1, \cdots, N, \tag{2.2.41}$$

and

$$\mathbf{u}_j = \begin{bmatrix} u_1 \\ \cdots \\ u_k \\ \cdots \\ u_N \end{bmatrix}, \quad u_k = 2\left(\sqrt{\frac{a}{c}}\right)^k \sin \frac{kj\pi}{N+1}, \quad k = 1, \cdots, N, \tag{2.2.42}$$

for $j = 1, \cdots, N$.

There are several ways to arrive at the above result. The easiest way to see that the above eigenvalues and eigenvectors are correct is to compute $T\mathbf{u}_j$ and see that you get $\lambda_j \mathbf{u}_j$.

Later, we will need the eigenvalues and eigenvectors for several other tridiagonal matrices that are not in the form of (2.2.40). Below we list several tridiagonal matrices along with their eigenvalues and eigenvectors.

In each case we list the matrix, the eigenvalues and the k-th component of the j-th eigenvector. For more information on these matrices, eigenvalues and eigenvectors (including the derivation), see ref. [1].

$$T_{N_1 D} = \begin{pmatrix} 1 & -1 & 0 & \cdots & & \\ -1 & 2 & -1 & 0 & \cdots & \\ & & \cdots & & & \\ \cdots & 0 & -1 & 2 & -1 \\ & \cdots & 0 & -1 & 2 \end{pmatrix}_{N \times N} \tag{2.2.43}$$

$$\lambda_j = 2 - 2\cos\frac{(2j-1)\pi}{2N+1}, \ j = 1, \cdots, N \tag{2.2.44}$$

$$u_k = \cos\frac{(2j-1)\pi x_k}{2}, \ k = 1, \cdots, N; j = 1, \cdots, N \tag{2.2.45}$$

where $x_k = (2k-1)\Delta x/2, \ k = 1, \cdots, N, \ \Delta x = 2/(2N+1)$.

$$T_{N_2 D} = \begin{pmatrix} 2 & -2 & 0 & \cdots & & \\ -1 & 2 & -1 & 0 & \cdots & \\ & & \cdots & & & \\ \cdots & 0 & -1 & 2 & -1 \\ & \cdots & 0 & -1 & 2 \end{pmatrix}_{N \times N} \tag{2.2.46}$$

$$\lambda_j = 2 - 2\cos\frac{(2j-1)\pi}{2N}, \ j = 1, \cdots, N \tag{2.2.47}$$

$$u_k = \cos\frac{(2j-1)\pi x_k}{2}, \ k = 1, \cdots, N; j = 1, \cdots, N \tag{2.2.48}$$

where $x_k = (k-1)\Delta x, \ k = 1, \cdots, N, \ \Delta x = 1/N$.

$$T_{N_2 N_2} = \begin{pmatrix} 2 & -2 & 0 & \cdots & & \\ -1 & 2 & -1 & 0 & \cdots & \\ & & \cdots & & & \\ \cdots & 0 & -1 & 2 & -1 \\ & \cdots & 0 & -2 & 2 \end{pmatrix}_{N \times N} \tag{2.2.49}$$

$$\lambda_j = 2 - 2\cos\frac{(j-1)\pi}{N-1}, \ j = 1, \cdots, N \tag{2.2.50}$$

$$u_k = \cos(j-1)\pi x_k, \ k = 1, \cdots, N; j = 1, \cdots, N \tag{2.2.51}$$

where $x_k = (k-1)\Delta x, \ k = 1, \cdots, N, \ \Delta x = 1/(N-1)$.

2.2.4 Some Additional Convergence Topics

Before proceeding we have several topics about convergence that we should discuss. We should emphasize that as a part of our discussion of convergence we assume that $u_k^0 = f(k\Delta x) = v(k\Delta x, 0)$. In some instances an approximation to the actual initial condition is used. In this case, that approximation must be such that $u_k^0 \rightarrow f(x)$ as $k\Delta x \rightarrow x$, and $\Delta x \rightarrow 0$. This additional level of approximation can be treated in much the same manner as we treat the exact initial conditions. In our discussions, we will not generally include this additional level of approximation.

We also mention that in all of our discussions related to the definitions of convergence, we have only referred to **two level schemes** (a scheme involving only the nth and the $(n+1)st$ time levels.) It should be noted that the definitions given above apply for **multilevel schemes** as well as two level schemes. With the implementation of multilevel schemes, care must be taken to insure that the additional initial conditions that are required to start the scheme are sufficiently well behaved to give us convergence to the continuous problem. This problem is similar to that of approximating the initial conditions that we discussed in the previous paragraph. This is why we cheated in HW 1.3.1 and used the exact solution values at time Δt plus the given initial conditions to get the scheme started.

If we were to try to consider how we might prove convergence for a three or more level scheme, when in Example 2.2.1 we arrived at

$$Z^{n+1} \le Z^0 + (n+1)\Delta t A(\Delta t + \Delta x^2),$$

in a three-level scheme we would instead have a Z^0 and a Z^1 term. It is at this time that the choice of u_k^1 becomes very important. If you had the freedom that we had and chose the exact solution for u_k^1, then both Z^0 and Z^1 would be zero, and we would have convergence. Most often, something other than the exact solution must be chosen and that choice now becomes very important.

At this time, the only three-level scheme that we know is the leapfrog scheme for the heat equation, and, by now, you have worked HW1.2 and believe that the scheme does not converge. We will not illustrate a convergence proof for a multilevel scheme at this time. We will delay any discussion of convergence of multilevel schemes until Chapter 6, where we transform the multilevel schemes into a two level system of equations. For the moment, let us assure you that not all multilevel schemes will be divergent.

Also, since later in this text we will consider multi-dimensional (in space) problems, we must have a concept of convergence of multi-dimensional finite difference schemes. But this is not difficult because Definitions 2.2.2–2.2.4 are easily adapted to include higher dimensional problems. When more spatial variables are included in our problems, the spaces (vectors) will be expanded to include the extra indices. For example, if we consider a two dimensional problem, the solution vector $\mathbf{u}^n = \{u_{j\,k}^n\}$ can be considered as a two-dimensional vector or be realigned into a long one dimensional vector. In either case, the appropriate norm (some of these norms were mentioned in Section 2.2.3) is used to measure the difference between different vectors and the definition of convergence will be the same as Definitions 2.2.2–2.2.4.

And finally, we must consider how to prove convergence. As we stated earlier, it is often difficult to prove convergence directly. A common method is to apply a theorem known as the Lax Equivalence Theorem which states that for consistent schemes for linear problems, convergence is equivalent to stability. Thus it is clear that we must now study consistency and stability.

2.3   Consistency

2.3.1   Initial–Value Problems

As before, denote the partial differential equation under consideration by $\mathcal{L}v = F$ and the corresponding finite difference approximation by $L_k^n u_k^n = G_k^n$ where G_k^n denotes whatever approximation has been made of the source term. We make the following definition.

Definition 2.3.1 *The finite difference scheme $L_k^n u_k^n = G_k^n$ is pointwise consistent with the partial differential equation $\mathcal{L}v = F$ at point (x, t) if for any smooth function $\phi = \phi(x, t)$,*

$$(\mathcal{L}\phi - F)|_k^n - \left[L_k^n \phi(k\Delta x, n\Delta t) - G_k^n \right] \to 0 \qquad (2.3.1)$$

as $\Delta x,\ \Delta t \to 0$ and $(k\Delta x, (n+1)\Delta t) \to (x, t)$.

Remark 1: We should notice that earlier, in equation (1.3.8), we were actually proving that the explicit scheme (1.2.9) was pointwise consistent with the partial differential equation

$$v_t = \nu v_{xx}. \qquad (2.3.2)$$

Remark 2: We should also realize that when using expression (2.3.1) as the consistency condition, we have included the approximation of the source term F as a part of our scheme. Often, the source terms are not included in the definition of consistency and the nonhomogeneous equation is considered separately. Also, it is often the case that the approximation G_k^n is chosen so that the contributions to the source terms in (2.3.1) add out.

Remark 3: As we did when examining consistency in Section 1.3, we can choose ϕ to be the solution, v, to the partial differential equation. Then the expression in Definition 2.3.1 becomes

$$L_k^n v_k^n - G_k^n \to 0 \text{ as } \Delta x,\ \Delta t \to 0. \qquad (2.3.3)$$

If we write the difference scheme as (assuming now that we are working with a two level scheme and a partial differential equation that is first order with respect to t)

$$\mathbf{u}^{n+1} = Q\mathbf{u}^n + \Delta t \mathbf{G}^n \qquad (2.3.4)$$

where

$$\mathbf{u}^n = (\cdots, u_{-1}^n, u_0^n, u_1^n, \cdots)^T,$$
$$\mathbf{G}^n = (\cdots, G_{-1}^n, G_0^n, G_1^n, \cdots)^T$$

and Q is an operator acting on the appropriate space, then a stronger definition of consistency can be given as follows.

Definition 2.3.2 *The difference scheme (2.3.4) is consistent with the partial differential equation in a norm $\|\cdot\|$ if the solution of the partial differential equation, v, satisfies*

$$\mathbf{v}^{n+1} = Q\mathbf{v}^n + \Delta t \mathbf{G}^n + \Delta t \boldsymbol{\tau}^n, \qquad (2.3.5)$$

and

$$\| \boldsymbol{\tau}^n \| \rightarrow 0$$

as Δx, $\Delta t \rightarrow 0$, where $\mathbf{v}^n$ denotes the vector whose kth component is $v(k\Delta x, n\Delta t)$.

Remark 1: It should be noted that the difference between the pointwise consistency and the norm consistency is that the norm consistency forces all of the components of the vector $\boldsymbol{\tau}^n$ to converge to zero in somewhat of a uniform fashion. If the definition of pointwise consistency (using expression (2.3.3)) is applied to difference scheme (2.3.4), pointwise consistency is equivalent to $\tau_k^n \rightarrow 0$ as Δx, $\Delta t \rightarrow 0$. Hence, the difference between the two definitions is whether we want component wise or vector wise convergence of $\boldsymbol{\tau}^n$ to zero. Most often when consistency is discussed in the literature, pointwise consistency is discussed while norm consistency is needed. We will see later in Example 2.3.2 that for implicit schemes, there is more of a difference between the two definitions.

Remark 2: We should also emphasize that the truncation term, $\boldsymbol{\tau}^n$, contains both the truncation due to the approximation of $\mathcal{L}$ by L_k^n and the approximation of F. We shall see later, in Section 2.6.5, that *the approximation of the function F will contribute to the truncation term and a bad approximation of F can lower the order of the scheme.*

Remark 3: A special note should be taken of the Δt that multiplies the $\boldsymbol{\tau}^n$ and $\mathbf{G}^n$ terms. This is due to the fact that from going from the L_k^n form of the difference equation to the Q form of the difference equation, we have multiplied through by the Δt associated with the temporal derivative (assumed to be first order). At times, it will be necessary for us to divide our schemes that have been put in the form of (2.3.4) by Δt to return them to a form that mimics the partial differential equation.

A slight variation of the definition of consistency is given below where the order in which τ^n goes to zero (which is the same order to which the finite difference scheme approximates the partial differential equation) is included.

Definition 2.3.3 *The difference scheme (2.3.4) is said to be accurate of order (p, q) to the given partial differential equation if*

$$\| \tau^n \| = \mathcal{O}(\Delta x^p) + \mathcal{O}(\Delta t^q). \tag{2.3.6}$$

We refer to τ^n or $\| \tau^n \|$ as the **truncation error**.

Remark 1: Of course, it is easy to see that if a scheme is accurate of order (p, q), p, $q \geq 1$, then it is a consistent scheme. Also, it is easy to see that if a scheme is either consistent or accurate of order (p, q), the scheme is pointwise consistent.

Remark 2: It should also be noted that the order condition given in Definition 2.3.3 involves a constant, C (again, constant with respect to Δx and Δt.) As was the case with Definition 2.2.3, *the constant C will generally depend on t.* This is important later in application of this definition. However, C will not depend on k (as the convergence does in Definition 2.3.1).

A comment should be made concerning the calculation that gives the order of accuracy or consistency. As we did earlier, the method is to expand the finite difference equations in Taylor series. The easiest approach is to do the expansion with the solution to the partial differential equation, and then use the fact that the function does satisfy the partial differential equation when it becomes a part of the expansion. To illustrate the differences between how we considered consistency in Section 1.3 and how we apply Definitions 2.3.1–2.3.3, we return to the explicit scheme considered in Section 1.3.

Example 2.3.1 Discuss the consistency of the explicit difference scheme

$$\frac{u_k^{n+1} - u_k^n}{\Delta t} = \nu \frac{u_{k+1}^n - 2u_k^n + u_{k-1}^n}{\Delta x^2} \tag{2.3.7}$$

with partial differential equation

$$v_t = \nu v_{xx}, \quad -\infty < x < \infty, \ t > 0. \tag{2.3.8}$$

Solution: To begin our discussion, we return to the Section 1.3. If we denote a solution of partial differential equation (2.3.8) by v, and insert v into expression (2.3.7), we see that the left hand side of expression (1.3.8) is zero and we are left with

$$\frac{v_k^{n+1} - v_k^n}{\Delta t} - \frac{\nu}{\Delta x^2}(v_{k+1}^n - 2v_k^n + v_{k-1}^n) = \mathcal{O}(\Delta t) + \mathcal{O}(\Delta x^2) \tag{2.3.9}$$

where $r = \nu\Delta t/\Delta x^2$. To apply Definitions 2.3.1–2.3.3, we must be careful that we know what is contained in the $\mathcal{O}(\Delta t) + \mathcal{O}(\Delta x^2)$ term. In Section 1.3 we vaguely replaced the tail ends of the Taylor series expansions by the "big O" notation and quit.

If we are just a little bit more careful and replace the series used in Section 1.3 by Taylor series with a remainder, it is then clear that the "big $\mathcal{O}$" terms contain a second derivative with respect to t evaluated for $x = k\Delta x$ and some t in a neighborhood of $n\Delta t$ and a fourth derivative with respect to x evaluated at $t = n\Delta t$ and for some x in a neighborhood of $k\Delta x$. If we then assume that the second derivative of v with respect to t and the fourth derivative of v with respect to x exist and are bounded in some neighborhood of the point (x, t), expression (2.3.9) will imply that difference scheme (2.3.7) is pointwise consistent with partial differential equation (2.3.8).

To show that difference scheme (2.3.7) is consistent, or accurate of order $(2, 1)$, we must first write the scheme in the form of equation (2.3.4). To write difference scheme (2.3.7) in the form of equation (2.3.4), we multiply through by Δt, solve for u_k^{n+1} and get

$$u_k^{n+1} = u_k^n + \nu \frac{\Delta t}{\Delta x^2}(u_{k+1}^n - 2u_k^n + u_{k-1}^n). \tag{2.3.10}$$

Equation 2.3.10 gives each component of an equation in the form of (2.3.4). Hence, to apply either Definitions 2.3.2 or 2.3.3, we let v be a solution to partial differential equation (2.3.8) and write

$$
\begin{aligned}
\Delta t \tau_k^n &= v_k^{n+1} - \{v_k^n + r[v_{k+1}^n - 2v_k^n + v_{k-1}^n]\} \\
&= v_k^n + (v_t)_k^n \Delta t + v_{tt}(k\Delta x, t_1)\frac{\Delta t^2}{2} \\
&\quad - \left\{ v_k^n + r\left[v_k^n + (v_x)_k^n \Delta x + (v_{xx})_k^n \frac{\Delta x^2}{2} + (v_{xxx})_k^n \frac{\Delta x^3}{6} \right.\right. \\
&\quad + v_{xxxx}(x_1, n\Delta t)\frac{\Delta x^4}{24} - 2v_k^n + v_k^n - (v_x)_k^n \Delta x \\
&\quad \left.\left. + (v_{xx})_k^n \frac{\Delta x^2}{2} - (v_{xxx})_k^n \frac{\Delta x^3}{6} + v_{xxxx}(x_2, n\Delta t)\frac{\Delta x^4}{24} \right]\right\} \\
&= (v_t)_k^n \Delta t - r\Delta x^2 (v_{xx})_k^n + v_{tt}(k\Delta x, t_1)\frac{\Delta t^2}{2} \\
&\quad - r v_{xxxx}(x_1, n\Delta t)\frac{\Delta x^4}{24} - r v_{xxxx}(x_2, n\Delta t)\frac{\Delta x^4}{24} \tag{2.3.11} \\
&= (v_t - \nu v_{xx})_k^n \Delta t + v_{tt}(k\Delta x, t_1)\frac{\Delta t^2}{2} \\
&\quad - \nu v_{xxxx}(x_1, n\Delta t)\frac{\Delta x^2}{24}\Delta t - \nu v_{xxxx}(x_2, n\Delta t)\frac{\Delta x^2}{24}\Delta t \tag{2.3.12}
\end{aligned}
$$

where t_1, x_1 and x_2 are the appropriate points given to us from the Taylor series remainder term and (2.3.12) follows from (2.3.11) because $r = \nu\Delta t/\Delta x^2$. Since $v_t - \nu v_{xx} = 0$, we see that

$$\tau_k^n = v_{tt}(k\Delta x, t_1)\frac{\Delta t}{2} - \nu(v_{xxxx}(x_1, n\Delta t) + v_{xxxx}(x_2, n\Delta t))\frac{\Delta x^2}{24}. \tag{2.3.13}$$

To apply the above result, we must now decide which norm we are using. If we assume that v_{tt} and v_{xxxx} are uniformly bounded on $\mathbb{R} \times [0, t_0]$, (for some $t_0 > t$), we can then use these bounds along with the sup-norm to get that *the scheme is accurate of order* $(2, 1)$ *with respect to the sup-norm*. If we assume that v_{tt} and v_{xxxx} satisfy

$$\sum_{k=-\infty}^{\infty} [(v_{tt})_k^n]^2 \, \Delta x < A < \infty$$

and

$$\sum_{k=-\infty}^{\infty} [(v_{xxxx})_k^n]^2 \, \Delta x < B < \infty$$

for any Δx and Δt, then we see that *the difference scheme is accurate order* $(2,1)$ *with respect to the* $\ell_{2,\Delta x}$ *norm.*

Remark 1: Both the $\ell_{2,\Delta x}$ and ℓ_2 norms were introduced in Section 2.2.3. We shall see in the next chapter that if we could use an ℓ_2 norm, $\| \cdot \|_2$, we would have many tools at our disposal. But, it is impractical to use the ℓ_2 norm to measure the difference between discretizations of functions as $\Delta x \to 0$. To see why this is true, we consider a function $v = v(x)$ defined on the real line, set $\mathbf{v}_{\Delta x} = (\cdots, v(-\Delta x), v(0), v(\Delta x), \cdots)^T$ and calculate $\| \mathbf{v}_{\Delta x} \|_2$. We repeat the above calculation with $\Delta x/2$ in place of Δx. Since the vector $\mathbf{v}_{\Delta x/2}$ is twice as long as the vector $\mathbf{v}_{\Delta x}$ (and, of course, they are both infinitely long) and if Δx is sufficiently small (so that we can approximate two adjacent values in $\mathbf{v}_{\Delta x/2}$ by one of the values in $\mathbf{v}_{\Delta x}$), we see that $\| \mathbf{v}_{\Delta x/2} \|_2$ will be approximately $\sqrt{2}$ times $\| \mathbf{v}_{\Delta x} \|_2$. Hence, for any smooth function, as $\Delta x \to 0$, $\| \mathbf{v}_{\Delta x} \|_2 \to \infty$. For this reason, we instead use the $\ell_{2,\Delta x}$ norm which retains all of the favorable properties of the ℓ_2 norm and which, for smooth functions approximates the $L_2(\mathbb{R})$ norm of the function.

Remark 2: We see in the above example that computationally, the consistency done rigorously is essentially the same as we did in Section 1.3. Using Definition 2.3.2, forces us to choose a norm. The choice of norm (which will ultimately determine the norm in which we converge) along with the consistency calculation shows what smoothness assumptions must be made. We will not always state carefully these smoothness conditions. *But it should be noted that the smoothness assumptions or proofs of the necessary smoothness are a part of the consistency calculations.*

Remark 3: In our previous example we expanded our functions about the index point (k,n), and it was reasonably obvious that this was the correct point about which to expand. However, that is not always the case. It does matter which point you expand about. The scheme is either consistent or it is not. If we choose the wrong point about which to expand, we may think that the scheme is not consistent (so we would generally discard it) or of lower order than it is (in which case, we might use smaller Δx's and/or Δt's than are necessary). Most often it is reasonably easy to determine about which point we should expand. Other times, for example the Crank-Nicolson scheme given in HW2.3.1(c) where we must expand about the point $(k, n+1/2)$, the decision about which point to expand must be made by carefully considering how we expect the difference scheme to approximate the partial differential equation. We next give a consistency argument which expands about a point other than (k,n). Another aspect of proving consistency that is illustrated in the example below is the difference in proving consistency for an implicit scheme from consistency for an explicit scheme. We see that proving consistency for implicit schemes can be very difficult.

Example 2.3.2 Discuss the consistency of the difference scheme

$$L_k^n u_k^n = \frac{u_k^{n+1} - u_k^n}{\Delta t} - \frac{\nu}{\Delta x^2} \delta^2 u_k^{n+1} = F_k^{n+1} \tag{2.3.14}$$

(where $F_k^{n+1} = F(k\Delta x, (n+1)\Delta t)$) with the partial differential equation $v_t = \nu v_{xx} + F$.
Solution: If we consider the difference scheme (2.3.14) which we refer to as a backward in time, centered in space or BTCS scheme (which is an **implicit scheme** because we cannot solve for u_k^{n+1} explicitly, hence, we can only solve for it implicitly), then it is logical to expand the difference equation about the point $(k, n+1)$. Performing this expansion (assuming that v is a solution to the partial differential equation) yields

$$
\begin{aligned}
L_k^n v_k^n - F_k^{n+1} &= \frac{v_k^{n+1} - v_k^n}{\Delta t} - \frac{\nu}{\Delta x^2}(v_{k+1}^{n+1} - 2v_k^{n+1} + v_{k-1}^{n+1}) - F_k^{n+1} \\
&= \frac{v_k^{n+1} - (v_k^{n+1} + (v_t)_k^{n+1}(-\Delta t) + (v_{tt})_k^{n+1}\frac{(-\Delta t)^2}{2!} + \cdots)}{\Delta t} \\
&\quad - \frac{\nu}{\Delta x^2}\left\{ v_k^{n+1} + (v_x)_k^{n+1}\Delta x + (v_{xx})_k^{n+1}\frac{\Delta x^2}{2!} + (v_{xxx})_k^{n+1}\frac{\Delta x^3}{3!} \right. \\
&\quad\quad + (v_{xxxx})_k^{n+1}\frac{\Delta x^4}{4!} + \cdots \\
&\quad\quad - 2v_k^{n+1} \\
&\quad\quad + v_k^{n+1} + (v_x)_k^{n+1}(-\Delta x) + (v_{xx})_k^{n+1}\frac{(-\Delta x)^2}{2!} + (v_{xxx})_k^{n+1}\frac{(-\Delta x)^3}{3!} \\
&\quad\quad \left. + (v_{xxxx})_k^{n+1}\frac{(-\Delta x)^4}{4!} + \cdots \right\} - F_k^{n+1} \\
&= (v_t)_k^{n+1} - \frac{\Delta t}{2}(v_{tt})_k^{n+1} + \cdots - \nu(v_{xx})_k^{n+1} \\
&\quad - 2\nu(v_{xxxx})_k^{n+1}\frac{\Delta x^2}{4!} + \cdots - F_k^{n+1} \\
&= -\frac{\Delta t}{2}(v_{tt})_k^{n+1} - 2\nu(v_{xxxx})_k^{n+1}\frac{\Delta x^2}{4!} + \cdots \tag{2.3.15} \\
&= \mathcal{O}(\Delta t) + \mathcal{O}(\Delta x^2).
\end{aligned}
$$

Thus we see that with the appropriate assumptions on the existence and boundedness of v_{tt} and v_{xxxx} near some point (x, t), the implicit scheme is pointwise consistent.

To see whether the scheme is consistent with respect to some norm, we must do some extra work. If we look at difference scheme (2.3.14) carefully, we see that it is not in the form and does not easily transform into the form of scheme (2.3.4). The above scheme was not solved for u^{n+1} before the above consistency argument was performed (which is not required to apply Definition 2.3.1). Instead of the form (2.3.4), difference scheme (2.3.14)(after we have multiplied through by Δt) is of the form

$$Q_1 \mathbf{u}^{n+1} = Q \mathbf{u}^n + \Delta t \mathbf{F}^{n+1} \tag{2.3.16}$$

where Q_1 is the "infinite matrix"

$$
Q_1 = \begin{pmatrix}
\ddots & \ddots & \ddots & \cdots & & & \\
\cdots & -r & 1+r & -r & 0 & \cdots & \\
\cdots & -r & 1+r & -r & 0 & \cdots & \\
& & \ddots & \ddots & \ddots & & \\
& \cdots & 0 & -r & 1+2r & -r & \cdots \\
& \cdots & 0 & -r & 1+r & -r & \cdots \\
& & & \cdots & \ddots & \ddots & \ddots
\end{pmatrix}, \tag{2.3.17}
$$

and $Q = I$, the identity. The consistency argument performed above gives a result that looks like

$$
Q_1 \mathbf{v}^{n+1} = Q\mathbf{v}^n + \Delta t \mathbf{F}^{n+1} + \Delta t \mathbf{r}^n \tag{2.3.18}
$$

where $\mathbf{r}^n$ is the error calculated in (2.3.15) which we refer to as the **residual error**. Hence, to satisfy either Definition 2.3.2 or 2.3.3, we must rewrite equation (2.3.18) as

$$
\mathbf{v}^{n+1} = Q_1^{-1} Q \mathbf{v}^n + \Delta t Q_1^{-1} \mathbf{F}^{n+1} + \Delta t Q_1^{-1} \mathbf{r}^n
$$

and consider

$$
\| \, \tau^n \, \| = \| \, Q_1^{-1} \mathbf{r}^n \, \| \, .
$$

Since

$$
\| \, Q_1^{-1} \mathbf{r}^n \, \| \leq \| \, Q_1^{-1} \, \| \, \| \, \mathbf{r}^n \, \|
$$

(where the norm acting on Q_1^{-1} is the operator norm induced by $\| \cdot \|$), we see that if $\| \, Q_1^{-1} \, \|$ is uniformly bounded as Δx and Δt approach zero, the consistency of the scheme with respect to the norm $\| \cdot \|$ is determined by $\mathbf{r}^n$.

If we consider the sup-norm and assume that certain derivatives of v are bounded (specifically v_{tt} and v_{xxxx}), then it is easy to see that difference scheme (2.3.14) is consistent. Specifically, the assumption of the bounded derivatives implies that $\| \, \mathbf{r}^n \, \|_\infty = \mathcal{O}(\Delta t) + \mathcal{O}(\Delta x^2)$. And since the operator Q_1 can be expressed as

$$
Q_1 \{\alpha_k\} = \{\beta_k\} \text{ where } \beta_k = -r\alpha_{k-1} + (1+2r)\alpha_k - r\alpha_{k+1},
$$

we can use the "backwards triangular inequality" and the fact that

$$
\| \, \{\alpha_k\} \, \|_\infty = \| \, \{\alpha_{k\pm1}\} \, \|_\infty
$$

(where by $\{\alpha_{k\pm1}\}$ we mean the sequence $\{\alpha_k\}$ where the indices have been translated by $+1$ or -1) to see that

$$
\begin{aligned}
\| \, Q_1\{\alpha_k\} \, \|_\infty &= \| \, \{-r\alpha_{k-1} + (1+2r)\alpha_k - r\alpha_{k+1}\} \, \|_\infty \\
&\geq (1+2r) \| \, \{\alpha_k\} \, \|_\infty - 2r \| \, \{\alpha_k\} \, \|_\infty \\
&= \| \, \{\alpha_k\} \, \|_\infty \, .
\end{aligned}
$$

Hence, Q_1 is bounded below, so Q_1^{-1} is bounded (i.e. $\| \, Q_1^{-1} \, \|_\infty < \infty$), ref. [12], page 86, ref. [7], page 101.

Thus, assuming the boundedness of the appropriate derivatives of the solution to the partial differential equation, difference scheme (2.3.14) is accurate of order $(2, 1)$ with respect to the sup-norm.

If we assume that v_{tt} and v_{xxxx} satisfy

$$
\sum_{k=-\infty}^{\infty} [(v_{tt})_k^n]^2 \, \Delta x < A < \infty
$$

and

$$\sum_{k=-\infty}^{\infty} [(v_{xxxx})_k^n]^2 \, \Delta x < B < \infty$$

for any Δx and Δt (which is the same assumption made in Example 2.3.1), then we see that difference scheme (2.3.14) is consistent with respect to the $\ell_{2,\Delta x}$ norm if Q_1^{-1} is bounded with respect to the operator norm induced by the $\ell_{2,\Delta x}$ norm.

To show (using basic principles) that Q_1^{-1} is bounded with respect to the $\ell_{2,\Delta x}$ norm is much more difficult than for the sup-norm. In Chapter 3 we shall obtain a result that will make this very easy, Proposition 3.1.9. For that reason, we delay proving that Q_1^{-1} is bounded until Chapter 3.

In either norm, we see that the implicit scheme has the same order of accuracy as the explicit scheme, $(2, 1)$.

Remark 1: In the Taylor series expansions used above, we used infinite expansions yet made assumptions on v_{tt} and v_{xxxx}. This is done for convenience. The arguments using the infinite Taylor series can always be replaced by arguments using a truncated Taylor series with remainder term. In either case, it is clear that v_{tt} and v_{xxxx} are the appropriate derivatives on which to make our assumptions.

Remark 2: We note that when we proved pointwise consistency, we assumed that certain derivatives were bounded in a neighborhood of (x, t) while when we proved consistency with respect to the sup-norm, we assumed that these derivatives were bounded on $\mathbb{R} \times [0, t_0]$ $(t_0 > t)$.

We emphasize that these are assumptions. When other norms are used, for example the $\ell_{2,\Delta x}$ norm, different but analogous assumptions will have to be made. These assumptions will generally be assumptions on the smoothness of the solution of the partial differential equation as were made in Example 2.3.2. The same problem will occur with regard to the dependence of the constant C on t. It will be important that this dependence is reasonably nice and predictable. This assumption will depend on the smoothness of the solution of the partial differential equation with respect to t. For most of the partial differential equations that we consider, it would be possible to prove apriori that the solutions satisfy the necessary conditions. Since these questions are better addressed in a course on partial differential equations and since we have already got enough problems, we will treat them as assumptions. Also we should realize that when the methods are used to solve more difficult problems that cannot be completely analyzed, the necessary smoothness *must* be treated as an assumption; and as a part of the solution procedure, the experimental part of solving such problems must take into account these assumptions.

Remark 3: We note that both Definitions 2.3.2 and 2.3.3 assume that the scheme is a two level scheme, where as, Definition 2.3.1 makes no such assumption. The two level assumption was used in Definitions 2.3.2 and 2.3.3 for convenience of notation. Multilevel analogs of these definitions could be given. We will not give these definitions, but we will still discuss and use (and assign problems) consistency for multilevel schemes.

Remark 4: The Taylor series calculations performed in the last several examples can be very tedious. Luckily, it is no longer necessary to do this work manually. The symbolic manipulators available, Maple, Mathematica, etc., are very capable of doing these calculations. For example, below (given in a pseudo code which should translate equally well to the languages of any of the available symbolic manipulators) we include the two steps necessary for doing the expansions performed in Example 2.3.2.

1. $IMP = ((v(x,t) - v(x,t-dt))/dt) - (nu/dx^2) * (u(x+dx,t) - 2 * u(x,t) + u(x-dx,t))$: Defines the function involved in the difference scheme where $x = k\Delta x$ and $t = (n+1)\Delta t$ (chosen that way because we are expanding about the point $(k\Delta x, (n+1)\Delta t)$.

2. $TIMP = Mtaylor(IMP, [dx, dt])$: Defines $TIMP$ to be the Taylor expansion of the function IMP about the point $(x,t) = (k\Delta x, (n+1)\Delta t)$ $(dx = 0$ and $dt = 0)$.

The expansion given for $TIMP$ should be enough to observe the order of r_k^n. If the expansion is too ugly (which does happen with the symbolic manipulators), the expression $TIMP$ may have to be massaged somewhat (certain terms can be collected, Taylor coefficients can be calculated, etc.). And, finally, any discussion of the relationship between the residual error r_k^n and the truncation error τ_k^n will still have to be done by people. (Sad to say, the symbolic manipulators will not prove for us that Q_1^{-1} is uniformly bounded.)

Remark 5: Some of the expansions necessary to do some of the consistency arguments in HW2.3.1 and HW2.3.5 involved two dimensional Taylor series expansions (fox example, parts (b) and (c) of HW2.3.1 and parts (a) and (c) of HW2.3.5). If $f : \mathbb{R}^2 \to \mathbb{R}$, then the Taylor expansion of $f(x+\Delta x, t+\Delta t)$ about the point (x,t) in Δx and Δt is given by

$$f(x+\Delta x, t+\Delta t) = f(x,t) + f_x(x,t)\Delta x + f_t(x,t)\Delta t$$
$$+ \frac{1}{2!}f_{xx}(x,t)\Delta x^2 + 2\frac{1}{2!}f_{xt}(x,t)\Delta x\Delta t + \frac{1}{2!}f_{tt}(x,t)\Delta t^2$$
$$+ \frac{1}{3!}f_{xxx}(x,t)\Delta x^2 + 3\frac{1}{3!}f_{xxt}(x,t)\Delta x^2\Delta t$$
$$+ 3\frac{1}{3!}f_{xtt}(x,t)\Delta x\Delta t^2 + \frac{1}{3!}f_{ttt}(x,t)\Delta t^2 + \cdots.$$

Remark 6: As we leave the topic of consistency and accuracy for initial–value schemes, we remark that it is difficult to decide how questions such as those given in HW2.3.1 should be answered. To give a complete analysis of consistency (including more than we have included concerning the assumptions on the smoothness of the solutions) is difficult. We suggest that when you discuss consistency and/or accuracy, you include the pointwise consistency result and discuss briefly the conditions necessary to obtain norm consistency (i.e. bounded derivatives for the sup-norm, $\ell_{2,\Delta x}$ bounds on the derivatives for the $\ell_{2,\Delta x}$-norm and uniform boundedness of Q_1^{-1} for implicit schemes. And, of course, if the scheme cannot be shown to be either pointwise or norm consistent, this explanation should be included.

HW 2.3.1 Determine the order of accuracy of the following difference equations to the given initial–value problems.
(a) Explicit scheme for heat equation with lower order term (FTCS).

$$u_k^{n+1} = u_k^n - \frac{a\Delta t}{2\Delta x}\delta_0 u_k^n + \frac{\nu\Delta t}{\Delta x^2}\delta^2 u_k^n$$

$$v_t + av_x = \nu v_{xx}$$

(b) Implicit scheme for heat equation with lower order term (BTCS).

$$u_k^{n+1} + \frac{a\Delta t}{2\Delta x}\delta_0 u_k^{n+1} - \frac{\nu\Delta t}{\Delta x^2}\delta^2 u_k^{n+1} = u_k^n$$

$$v_t + av_x = \nu v_{xx}$$

(c) Crank-Nicolson Scheme

$$u_k^{n+1} - \frac{\nu\Delta t}{2\Delta x^2}\delta^2 u_k^{n+1} = u_k^n + \frac{\nu\Delta t}{2\Delta x^2}\delta^2 u_k^n$$

$$v_t = \nu v_{xx}$$

Explain why it is logical to consider the consistency of this scheme at the point $(k\Delta x, (n+1/2)\Delta t)$ rather than at $(k\Delta x, n\Delta t)$ or $(k\Delta x, (n+1)\Delta t)$.
(d) Dufort-Frankel Scheme

$$u_k^{n+1} = \frac{2r}{1+2r}(u_{k+1}^n + u_{k-1}^n) + \frac{1-2r}{1+2r}u_k^{n-1}$$

where $r = \frac{\Delta t}{\Delta x^2}$.

$$v_t = v_{xx}$$

Is the any logical condition that can be placed on this scheme that will make it consistent?
(e) Forward-time, forward-space for a hyperbolic equation (FTFS).

$$u_k^{n+1} = u_k^n - \frac{a\Delta t}{\Delta x}(u_{k+1}^n - u_k^n)$$

$$v_t + av_x = 0$$

HW 2.3.2 (a) Show that the following difference scheme is a $\mathcal{O}(\Delta t) + \mathcal{O}(\Delta x^4)$ approximation of $v_t = \nu v_{xx}$ (where $r = \nu\Delta t/\Delta x^2$).

$$u_k^{n+1} = u_k^n + r\left(-\frac{1}{12}u_{k-2}^n + \frac{4}{3}u_{k-1}^n - \frac{5}{2}u_k^n + \frac{4}{3}u_{k+1}^n - \frac{1}{12}u_{k+2}^n\right)$$

Discuss the assumptions that must be made on the derivatives of the solution to the partial differential equation that are necessary to make the above statement true.

(b) Show that the following difference scheme is a $\mathcal{O}(\Delta t^2) + \mathcal{O}(\Delta x^4)$ approximation of $v_t + av_x = 0$ (where $R = a\Delta t/\Delta x$).

$$
u_k^{n+1} = u_k^n - \frac{R}{2}\delta_0 u_k^n + \frac{R}{12}\delta^2\delta_0 u_k^n + \frac{R^2}{2}\left(\frac{4}{3} + R^2\right)\delta^2 u_k^n
$$

$$
- \frac{R^2}{8}\left(\frac{1}{3} + R^2\right)\delta_0^2 u_k^n, \quad k = 1, 2, \cdots
$$

Discuss the assumptions necessary for the above statement to be true.

HW 2.3.3 Determine the order of accuracy of the following difference equations to the partial differential equation

$$
v_t + av_x = 0.
$$

(a) Leapfrog scheme $u_k^{n+1} = u_k^{n-1} - R\delta_0 u_k^n$

(b) $u_k^{n+1} = u_k^{n-1} - R\delta_0 u_k^n + \frac{R}{6}\delta^2\delta_0 u_k^n$

(c) $u_k^{n+1} = u_k^{n-1} - R\delta_0 u_k^n + \frac{R}{6}\delta^2\delta_0 u_k^n - \frac{R}{30}\delta^4\delta_0 u_k^n$ where $\delta^4 \doteq \delta^2\delta^2$.

(d) $u_k^{n+2} = u_k^{n-2} - \frac{2R}{3}\left(1 - \frac{1}{6}\delta^2\right)\delta_0\left(2u_k^{n+1} - u_k^n + 2u_k^{n-1}\right)$

2.3.2   Initial–Boundary–Value Problems

Just as convergence for initial–boundary–value problems had to be treated differently from initial–value problems, we must also be careful when considering the consistency for initial–boundary–value problems. Pointwise consistency will be the same as it was in Definition 2.3.1 except that we must now also analyze any boundary conditions that contain an approximation. This is explained below.

For norm consistency, as we did in Section 2.2.2, we consider a sequence of partitions of the interval $[0, 1]$ defined by a sequences of spatial increments $\{\Delta x_j\}$ and a sequence of the appropriate spaces, $\{X_j\}$, with norms $\{\| \cdot \|_j \}$. Then *Definitions 2.3.2 and 2.3.3 carry over to the initial–boundary–value problem* by replacing the norm in equations (2.3.5) and (2.3.6) by the sequence of norms $\| \cdot \|_j$. The important difference between the initial–value problem and initial–boundary–value problem is in writing the initial–boundary–value problem in some form that can be analyzed (say in a form like equation (2.3.4) or (2.3.16)) and set in some logical sequence of vector spaces. For example, it is easy to see that if we consider the explicit scheme (2.2.2) for an initial–boundary–value problem with Dirichlet boundary conditions, the scheme is pointwise consistent and accurate of order $(2,1)$ with respect to either the sup-norm or the $\ell_{2,\Delta x}$ norm (the finite dimensional sup-norm or the $\ell_{2,\Delta x}$ norm, of course).

Obviously for computational reasons, it is as important to approximate boundary conditions consistently as it is to approximate the partial differential equation consistently. We shall see later that when we apply consistency to get convergence, it will be important that the boundary conditions are approximated as an integrated part of our difference equation. We emphasize that it can be very important how we treat our boundary conditions. Throughout the rest of this text, we will be reminded repeatedly of the importance of the above statement.

To illustrate how we prove consistency for initial–boundary–value problems, we consider the following three examples where we again consider the explicit scheme (2.2.2), but now do so for a problem having a zero Dirichlet boundary condition at $x = 1$ and a zero Neumann boundary condition at $x = 0$. We begin by considering the second order approximation considered in Section 1.4, (1.4.12).

Example 2.3.3 Discuss the consistency of the difference scheme

$$u_k^{n+1} = (1 - 2r)u_k^n + r(u_{k+1}^n + u_{k-1}^n), \; k = 1, \cdots, M - 1 \qquad (2.3.19)$$

$$u_M^{n+1} = 0 \qquad (2.3.20)$$

$$u_0^{n+1} = (1 - 2r)u_0^n + 2ru_1^n \qquad (2.3.21)$$

to the initial–boundary–value problem

$$v_t = \nu v_{xx}, \quad x \in (0, 1), \quad t > 0 \qquad (2.3.22)$$

$$v(x, 0) = f(x), \; x \in [0, 1] \qquad (2.3.23)$$

$$v(1, t) = 0, \quad t > 0 \qquad (2.3.24)$$

$$v_x(0, t) = 0, \quad t > 0. \qquad (2.3.25)$$

Solution: We recall that we are approximating boundary condition (2.3.25) by

$$\frac{u_1^n - u_{-1}^n}{2\Delta x} = 0. \qquad (2.3.26)$$

We also recall that difference equation (2.3.19) is a $\mathcal{O}(\Delta t) + \mathcal{O}(\Delta x^2)$ approximation to the partial differential equation. Since (2.3.26) is a $\mathcal{O}(\Delta x^2)$ approximation to boundary condition (2.3.25), we see that difference scheme (2.3.19)–(2.3.21) is a $\mathcal{O}(\Delta t) + \mathcal{O}(\Delta x^2)$ pointwise approximation to initial–boundary–value problem (2.3.22)–(2.3.25).

To prove norm consistency, we write our difference scheme as

$$\mathbf{u}^{n+1} = Q\mathbf{u}^n, \qquad (2.3.27)$$

evaluate equation (2.3.27) with $u = v$ (where v is the solution to the initial–boundary–value problem) and write this result as

$$\mathbf{v}^{n+1} = Q\mathbf{v}^n + \Delta t \boldsymbol{\tau}^n. \qquad (2.3.28)$$

If we now use equations (2.3.19)–(2.3.21), we can write our difference scheme as a matrix equation of the form

$$
\begin{pmatrix} u_0^{n+1} \\ u_1^{n+1} \\ \cdots \\ u_{M-1}^{n+1} \end{pmatrix} = \begin{pmatrix} 1-2r & 2r & & & \cdots & \\ r & 1-2r & r & & \cdots & \\ 0 & r & 1-2r & r & \cdots & \\ & & \cdots & & & \\ & & \cdots & 0 & r & 1-2r \end{pmatrix} \begin{pmatrix} u_0^n \\ u_1^n \\ \cdots \\ u_{M-1}^n \end{pmatrix}.
$$

$$(2.3.29)$$

We note specifically that in equation (2.3.29), the Neumann boundary condition is represented by equation (2.3.21) and not by equation (2.3.26). This is necessary to be able to apply consistency Definitions 2.3.2 and 2.3.3.

Also, before we start our consistency analysis, we should decide on our sequence of spaces. From the form of equation (2.3.29), it is clear that the spaces we must use are the M-dimensional spaces consisting of the finite vectors $(u_0, \cdots, u_{M-1})^T$. We already know that τ_k^n is $\mathcal{O}(\Delta t) + \mathcal{O}(\Delta x^2)$ for $k = 1, \cdots, M - 1$. To examine the consistency of equation (2.3.21), we let v be the solution to the initial–boundary–value problem and note that

$$
\Delta t \tau_0^n = v_0^{n+1} - (1 - 2r)v_0^n - 2rv_1^n
$$

$$
= v_0^n + (v_t)_0^n \Delta t + (v_{tt})_0^n \frac{\Delta t^2}{2} + \cdots
$$

$$
- [(1 - 2r)v_0^n
$$

$$
+ 2r \left\{ v_0^n + (v_x)_0^n \Delta x + (v_{xx})_0^n \frac{\Delta x^2}{2} + (v_{xxx})_0^n \frac{\Delta x^3}{6} + \cdots \right\}]
$$

$$
= [(v_t)_0^n - \nu(v_{xx})_0^n] \Delta t - 2r\Delta x (v_x)_0^n + (v_{tt})_0^n \frac{\Delta t^2}{2} - \frac{\nu \Delta x \Delta t}{3}(v_{xxx})_0^n + \cdots.
$$

Using the facts that $(v_x)_0^n = 0$ and $(v_t - \nu v_{xx})_0^n = 0$, we see that

$$
\tau_0^n = \frac{\Delta t}{2}(v_{tt})_0^n - \Delta x \frac{\nu}{3}(v_{xxx})_0^n + \cdots. \qquad (2.3.30)
$$

For the points $k = 1, \cdots, M - 1$ the difference scheme is accurate to order $\mathcal{O}(\Delta t) + \mathcal{O}(\Delta x^2)$. But, since at $k = 0$ the scheme is only accurate to order $\mathcal{O}(\Delta t) + \mathcal{O}(\Delta x)$, difference scheme (2.3.19)–(2.3.21) is accurate to order $\mathcal{O}(\Delta t) + \mathcal{O}(\Delta x)$. Because of the way that the approximations for the partial differential equation and the boundary condition fit together, one order of accuracy with respect to Δx was lost.

Hence, we see that with the appropriate assumptions on certain derivatives of v, *difference scheme (2.3.19)–(2.3.21) will be consistent with respect to either the sup–norm or the $\ell_{2,\Delta x}$ norm and accurate to order $\mathcal{O}(\Delta t) + \mathcal{O}(\Delta x)$.*

As we stated earlier, we next consider the same initial–boundary–value problem, this time using the first order approximation to the Neumann boundary condition, (1.4.8), discussed in Section 1.4.

Example 2.3.4 Discuss the consistency of the difference scheme

$$
u_k^{n+1} = (1 - 2r)u_k^n + r(u_{k+1}^n + u_{k-1}^n), \quad k = 1, \cdots, M - 1 \qquad (2.3.31)
$$

$$
u_M^{n+1} = 0 \qquad (2.3.32)
$$

$$
u_0^n = u_1^n \qquad (2.3.33)
$$

to the initial–boundary–value problem

$$
v_t = \nu v_{xx}, \quad x \in (0,1), \quad t > 0 \qquad (2.3.34)
$$

$$
v(x,0) = f(x), \quad x \in [0.1] \qquad (2.3.35)
$$

$$
v(1,t) = 0, \quad t > 0 \qquad (2.3.36)
$$

$$
v_x(0,t) = 0, \quad t > 0. \qquad (2.3.37)
$$

Solution: This time we recall that the difference equation given at $k = 0$, equation (2.3.33), is obtained by using

$$\frac{u_1^n - u_0^n}{\Delta x} = 0 \tag{2.3.38}$$

to approximate the boundary condition (2.3.37).

Again we recall that difference equation (2.3.31) is a $\mathcal{O}(\Delta t) + \mathcal{O}(\Delta x^2)$ approximation of the partial differential equation and realize that equation (2.3.38) is a $\mathcal{O}(\Delta x)$ approximation to boundary condition (2.3.37). Thus we see that difference scheme (2.3.31)–(2.3.33) is a $\mathcal{O}(\Delta t) + \mathcal{O}(\Delta x)$ pointwise approximation to initial–boundary–value problem (2.3.34)–(2.3.37). And, considering the results of Example 2.3.3, we should realize that the order of approximation may be lower for norm consistency.

As stated in Example 2.3.3, to prove norm consistency we write our difference scheme as

$$\mathbf{u}^{n+1} = Q\mathbf{u}^n, \tag{2.3.39}$$

evaluate equation (2.3.39) with $u = v$ and write this result as

$$\mathbf{v}^{n+1} = Q\mathbf{v}^n + \Delta t \boldsymbol{\tau}^n. \tag{2.3.40}$$

We then check to see that $\| \boldsymbol{\tau}^n \| \to 0$ and how fast it goes to zero.

To write difference scheme (2.3.31)–(2.3.33) in the form of (2.3.39), we must first rewrite equation (2.3.33) in the proper form. We use equation (2.3.31) with $k = 1$ and equation (2.3.33) to write

$$
\begin{aligned}
u_1^{n+1} &= (1 - 2r)u_1^n + r(u_2^n + u_0^n) \\
&= (1 - r)u_1^n + ru_2^n.
\end{aligned} \tag{2.3.41}
$$

Then using equations (2.3.31)–(2.3.32) and (2.3.41) to represent the difference scheme, we see that we get a matrix equation of the form

$$
\begin{pmatrix} u_1^{n+1} \\ u_2^{n+1} \\ \cdots \\ u_{M-1}^{n+1} \end{pmatrix} =
\begin{pmatrix}
1 - r & r & 0 & \cdots & & \\
r & 1 - 2r & r & 0 & \cdots & \\
& \cdots & & & & \\
\cdots & 0 & r & 1 - 2r & r & \\
& \cdots & & 0 & r & 1 - 2r
\end{pmatrix}
\begin{pmatrix} u_1^n \\ u_2^n \\ \cdots \\ u_{M-1}^n \end{pmatrix}. \tag{2.3.42}
$$

If we now proceed as we did in Example 2.3.3, we see that

$$
\begin{aligned}
\Delta t \tau_1^n &= v_1^{n+1} - \{(1 - r)v_1^n + rv_2^n\} \\
&= v_1^n + (v_t)_1^n \Delta t + (v_{tt})_1^n \frac{\Delta t^2}{2} + \cdots \\
&\quad - \left[(1 - r)v_1^n + r\left\{ v_1^n + (v_x)_1^n \Delta x + (v_{xx})_1^n \frac{\Delta x^2}{2} + (v_{xxx})_1^n \frac{\Delta x^3}{6} + \cdots \right\} \right] \\
&= (v_t)_1^n \Delta t - r\Delta x (v_x)_1^n - r\frac{\Delta x^2}{2}(v_{xx})_1^n + \frac{\Delta t^2}{2}(v_{tt})_1^n \\
&\quad - r\frac{\Delta x^3}{6}(v_{xxx})_1^n + \cdots .
\end{aligned}
$$

Then since

$$0 = (v_x)_0^n = (v_x)_1^n + (v_{xx})_1^n(-\Delta x) + (v_{xxx})_1^n \frac{(-\Delta x)^2}{2} + \cdots ,$$

we have

$$(v_x)_1^n = \Delta x (v_{xx})_1^n - \frac{\Delta x^2}{2}(v_{xxx})_1^n + \cdots$$

and

$$
\begin{aligned}
\Delta t \tau_1^n &= (v_t)_1^n \Delta t - r\Delta x (v_x)_1^n - r\frac{\Delta x^2}{2}(v_{xx})_1^n + \frac{\Delta t^2}{2}(v_{tt})_1^n \\
&\quad - r\frac{\Delta x^3}{6}(v_{xxx})_1^n + \cdots . \\
&= (v_t)_1^n \Delta t - r\Delta x \{ \Delta x (v_{xx})_1^n - \frac{\Delta x^2}{2}(v_{xxx})_1^n + \cdots \} - r\frac{\Delta x^2}{2}(v_{xx})_1^n \\
&\quad + \frac{\Delta t^2}{2}(v_{tt})_1^n - r\frac{\Delta x^3}{6}(v_{xxx})_1^n + \cdots \\
&= \left[v_t - \nu \frac{3}{2} v_{xx} \right]_1^n \Delta t + \mathcal{O}(\Delta t \Delta x) + \mathcal{O}(\Delta t^2) \\
&= -\frac{\nu}{2}(v_{xx})_1^n \Delta t + \mathcal{O}(\Delta t \Delta x) + \mathcal{O}(\Delta t^2)
\end{aligned}
$$

(where we have used the facts that $r = \nu \Delta t / \Delta x^2$ and $[v_t - \nu v_{xx}]_1^n = 0$). We see that

$$
\tau_1^n = -\frac{\nu}{2}(v_{xx})_1^n + \mathcal{O}(\Delta x) + \mathcal{O}(\Delta t). \tag{2.3.43}
$$

Thus we see from equation (2.3.43) that *difference scheme (2.3.31)–(2.3.33) is not consistent according to Definition 2.3.2.*

Remark 1: We emphasize that though scheme (2.3.31)–(2.3.33) is not consistent by Definition 2.3.2, it is not necessarily a bad scheme. If we were to continue, we would see in Section 2.5 that because the scheme is not consistent, we cannot use the Lax Theorem to prove that the scheme converges. However, HW2.3.4 below shows experimentally that the scheme converges. The questions are: to what does it converge and does this alleged limit have anything to do with the initial–boundary–value problem (2.3.34)–(2.3.37)? We shall return to this question in Example 3.1.9 in Section 3.1.2.

Remark 2: We might wonder if we would have had better results if we had used $u_0^{n+1} = u_1^{n+1}$ as we did in (1.4.8) instead of $u_0^n = u_1^n$. The consistency calculations using $u_0^{n+1} = u_1^{n+1}$ will have the same problem as those done above. See HW2.3.6.

In our next example we consider a scheme very similar to difference scheme (2.3.31)–(2.3.33) on a different grid. This grid treats the region $[0, 1]$ like a vertex centered grid near the $x = 1$ boundary (the Dirichlet boundary condition) and a cell centered grid near the $x = 0$ boundary (the Neumann boundary condition) as shown in Figure 1.6.4. We shall see that this grid is very natural for treating an initial–boundary–value problem such as (2.3.34)–(2.3.37).

Example 2.3.5 Consider difference scheme

$$
u_k^{n+1} = r u_{k-1}^n + (1 - 2r)u_k^n + r u_{k+1}^n, \quad k = 1, \cdots, M - 1 \tag{2.3.44}
$$

$$
u_M^n = 0 \tag{2.3.45}
$$

$$
u_0^n = u_1^n \tag{2.3.46}
$$

on the grid $\{x_k : x_k = (k-1)\Delta x + \Delta x/2, \ k = 0, \cdots, M\}$ where $\Delta x = 2/(2M - 1)$. Discuss the consistency of difference scheme (2.3.44)–(2.3.46) with initial–boundary–value problem (2.3.34)–(2.3.37).

Solution: As usual, difference equation (2.3.44) will be pointwise consistent with partial differential equation (2.3.34), order $\mathcal{O}(\Delta t) + \mathcal{O}(\Delta x^2)$. Since the $x = 0$ boundary of the interval $[0, 1]$ occurs at $k = 1/2$, difference equation (2.3.46) will be pointwise consistent with the Neumann boundary condition (2.3.37), in this case $\mathcal{O}(\Delta x^2)$.

To be able to write difference scheme (2.3.44)–(2.3.46) in the form (2.3.39), we use equation (2.3.44) for $k = 2, \cdots, M - 1$ (as usual) and combine equation (2.3.44) for $k = 1$ with equation (2.3.46) to get

$$u_1^{n+1} = (1 - r)u_1^n + ru_2^n. \tag{2.3.47}$$

To then analyze the norm consistency of $\mathbf{u}^{n+1} = Q\mathbf{u}^n$, we consider the first component of the truncation vector, τ_1^n (of course, we know the truncation error at the other grid points). We see that

$$
\begin{aligned}
\Delta t \tau_1^n &= v_1^{n+1} - \{(1 - r)v_1^n + rv_2^n\} \\
&= v_1^n + (v_t)_1^n \Delta t + \mathcal{O}(\Delta t^2) \\
&\quad - \left\{(1 - r)v_1^n + r\left[v_1^n + (v_x)_1^n \Delta x + (v_{xx})_1^n \frac{\Delta x^2}{2} + \mathcal{O}(\Delta x^3)\right]\right\} \\
&= \left[v_t - \frac{\nu}{2}v_{xx}\right]_1^n \Delta t - r\Delta x (v_x)_1^n + \mathcal{O}(\Delta t^2) + \mathcal{O}(\Delta t \Delta x). \tag{2.3.48}
\end{aligned}
$$

We also have

$$
\begin{aligned}
0 &= (v_x)_{1/2}^n \\
&= (v_x)_1^n + (v_{xx})_1^n \left(\frac{-\Delta x}{2}\right) + \mathcal{O}(\Delta x^2).
\end{aligned}
$$

Hence,

$$(v_x)_1^n = \frac{\Delta x}{2}(v_{xx})_1^n + \mathcal{O}(\Delta x^2). \tag{2.3.49}$$

Combining equations (2.3.48) and (2.3.49), we have

$$
\begin{aligned}
\Delta t \tau_1^n &= \left[v_t - \nu v_{xx}\right]_1^n \Delta t + \mathcal{O}(\Delta t^2) + \mathcal{O}(\Delta t \Delta x). \\
&= \mathcal{O}(\Delta t^2) + \mathcal{O}(\Delta t \Delta x). \tag{2.3.50}
\end{aligned}
$$

Since $\Delta t \tau_k^n$, $k = 2, \cdots, M - 1$ is order $\mathcal{O}(\Delta t^2) + \mathcal{O}(\Delta t \Delta x^2)$ and $\Delta t \tau_1^n$ is order $\mathcal{O}(\Delta t^2) + \mathcal{O}(\Delta t \Delta x)$, *difference scheme (2.3.44)–(2.3.46) will be consistent order* $\mathcal{O}(\Delta t) + \mathcal{O}(\Delta x)$.

Remark 1: Thus we see that when we use a grid that is the logical grid for a Neumann boundary condition at $x = 0$, we get a consistent difference scheme. The grid defined in Example 2.3.5 is referred to as an **offset grid**.

Remark 2: The solution v to initial–boundary–value problem (2.3.34)–(2.3.37) will again satisfy $v_{xxx}(0, t) = 0$ for all t. Since the coefficient of the Δx term in the order argument for τ_1^n is $(v_{xxx})_1^n$ and

$$
\begin{aligned}
(v_{xxx})_1^n &= (v_{xxx})_{1/2}^n + (v_{xxxx})_{1/2}^n (\Delta x/2) + \cdots \\
&= (v_{xxxx})_{1/2}^n (\Delta x/2) + \cdots,
\end{aligned}
$$

difference scheme (2.3.44)–(2.3.46) is really accurate order $(2, 1)$.

HW 2.3.4 Use scheme (2.3.31)–(2.3.33) to solve problem (1.4.1)–(1.4.4) with $\nu = 1.0$. Use $M = 20$ ($\Delta t = 0.001$), and $M = 40$ ($\Delta t = 0.00025$). Find solutions at $t = 0.06$, $t = 0.1$ and $t = 0.9$. Compare and contrast the results with the exact answers and the first order results of HW 1.4.1.

Before we leave consistency, we include one last example that illustrates, as was the case with initial–value problems, proving consistency for implicit schemes for solving initial–boundary–value problems is more difficult than proving consistency for the analogous explicit schemes.

Example 2.3.6 Consider the implicit difference scheme (2.3.14) (with $F_k^{n+1} = 0$) along with boundary conditions $u_0^{n+1} = u_M^{n+1} = 0$. Show that this difference scheme is consistent with the initial–boundary–value problem

$$v_t = \nu v_{xx}, \ x \in (0, 1), \ t > 0$$
$$v(0, t) = v(1, t) = 0, \ t > 0.$$

Solution: As in the case of the implicit scheme for the initial–value problem, the difference scheme for this initial–boundary–value problem can be written as

$$Q_1 \mathbf{u}^{n+1} = Q \mathbf{u}^n. \tag{2.3.51}$$

However, in this case since equation (2.3.51) is a finite dimensional problem, it is a matrix equation where the k-th row of equation (2.3.51) is given by

$$-r u_{k-1}^{n+1} + (1 + 2r) u_k^{n+1} - r u_{k+1}^{n+1} = u_k^n.$$

Pointwise consistency is given by the same calculation that was done in Section 2.3.1. Again the truncation error, $\mathbf{r}^n$ (now an $(M_j - 1)$-vector), satisfies $\| \mathbf{r}^n \| = \mathcal{O}(\Delta t^2) + \mathcal{O}(\Delta x^2 \Delta t)$ (we have already multiplied through by the Δt) with either the sup-norm or the $\ell_{2, \Delta x}$ norm—depending on the assumptions on the derivatives of v.

And finally, to prove norm consistency or accuracy of order (p, q) we must again show that $\| Q_1^{-1} \|$ is uniformly bounded as Δt, $\Delta x \to 0$. Given that Q_1 is now a matrix, we might think that this problem should be easier than the infinite dimensional case. In reality, the proofs are generally as hard or harder. The proof for the case of the sup-norm can be done exactly as the case of the initial–value problem.

The proof that $\| Q_1^{-1} \|$ is bounded in the $\ell_{2, \Delta x}$ follows from the fact that

$$\| Q_1^{-1} \|_{2, \Delta x} = 1/|\lambda_{\min}|$$

where $\lambda_{\min}$ is the eigenvalue of Q_1 for which $\min |\lambda|$ is the smallest. Clearly, from (2.2.41) the eigenvalues of Q_1 are given by

$$\lambda = 1 + 2r - 2r \cos \frac{j\pi}{M}$$
$$= 1 + 4r \sin^2 \frac{j\pi}{2M}.$$

Thus, we see that

$$\| Q_1^{-1} \|_{2, \Delta x} = \frac{1}{\lambda_{\min}}$$
$$= \frac{1}{\min\{1 + 4r \sin^2 \frac{j\pi}{2M}\}}$$
$$\leq 1.$$

We should realize that the reason that we are able to handle this case is that the operator Q_1 is symmetric. Symmetry will often be a hypothesis for initial–boundary–value problems that will be necessary in order to obtain good results.

Remark: The statement made earlier that we usually do not want to use a difference boundary condition that does not accurately approximate our boundary condition is true. But it would be a lie to imply that this is not done. Often, for reasons of ease of coding, difference boundary conditions that are zeroth order approximations of the boundary conditions are used. Many of the results obtained using these "bad" approximations of the boundary conditions are good results. This is a time when the numerical analyst must be a good experimenter. Zeroth order approximations should be used very carefully.

HW 2.3.5 Determine the order of accuracy of the following difference equations to the given initial–boundary–value problems.
(a) Implicit scheme (BTCS) for an initial–boundary–value problem with a Neumann boundary condition and lower order terms.

$$u_k^{n+1} + \frac{a\Delta t}{2\Delta x}\delta_0 u_k^{n+1} - \frac{\nu\Delta t}{\Delta x^2}\delta^2 u_k^{n+1} = u_k^n, \ k = 0, \cdots, M-1$$
$$u_k^0 = f(k\Delta x), \ k = 1, \cdots, M$$
$$u_M^{n+1} = 0,$$
$$\frac{u_1^{n+1} - u_{-1}^{n+1}}{2\Delta x} = \alpha((n+1)\Delta t)$$

$$v_t + av_x = \nu v_{xx}, \ x \in (0,1), \ t > 0$$
$$v(x,0) = f(x), \ x \in [0,1]$$
$$v(1,t) = 0, \ t \geq 0$$
$$v_x(0,t) = \alpha(t), \ t \geq 0$$

(b) Implicit scheme (BTCS) for an initial–boundary–value problem with a Neumann boundary condition and lower order terms.

$$u_k^{n+1} + \frac{a\Delta t}{2\Delta x}\delta_0 u_k^{n+1} - \frac{\nu\Delta t}{\Delta x^2}\delta^2 u_k^{n+1} = u_k^n, \ k = 1, \cdots, M-1$$
$$u_k^0 = f(k\Delta x), \ k = 1, \cdots, M$$
$$u_M^{n+1} = 0,$$
$$\frac{u_1^{n+1} - u_0^{n+1}}{\Delta x} = \alpha((n+1)\Delta t)$$

$$v_t + av_x = \nu v_{xx}, \ x \in (0,1), \ t > 0$$
$$v(x,0) = f(x), \ x \in [0,1]$$
$$v(1,t) = 0, \ t \geq 0$$
$$v_x(0,t) = \alpha(t), \ t \geq 0$$

(c) Implicit scheme (BTCS) for an initial–boundary–value problem with a Neumann boundary condition and lower order terms.

$$u_k^{n+1} + \frac{a\Delta t}{2\Delta x}\delta_0 u_k^{n+1} - \frac{\nu\Delta t}{\Delta x^2}\delta^2 u_k^{n+1} = u_k^n, \ k = 1, \cdots, M-1$$
$$u_k^0 = f(k\Delta x), \ k = 1, \cdots, M$$
$$u_M^{n+1} = 0,$$
$$\frac{u_1^n - u_0^{n+1}}{\Delta x} = \alpha((n+1)\Delta t)$$

$$v_t + av_x = \nu v_{xx}, \ x \in (0,1), \ t > 0$$
$$v(x,0) = f(x), \ x \in [0,1]$$
$$v(1,t) = 0, \ t \geq 0$$
$$v_x(0,t) = \alpha(t), \ t \geq 0$$

HW 2.3.6 Show that if we use $u_0^{n+1} = u_1^{n+1} = (1 - 2r)u_1^n + r(u_2^n + u_0^n)$ to represent Neumann boundary condition (2.3.37), the resulting difference scheme will still not be norm consistent.

2.4 Stability

As we alluded to earlier, the theorem that we will prove and use will need consistency and stability for convergence. Since we have discussed both consistency and convergence, it seems clear that it's time to discuss stability. As we stated in the section on consistency, most of the schemes that are used are consistent. The major problem with proving convergence is to obtain stability. Though stability is much easier to establish than convergence, it is still often difficult to prove that a given scheme is stable.

2.4.1 Initial–Value Problems

One interpretation of stability of a difference scheme is that *for a stable difference scheme small errors in the initial conditions cause small errors in the solution.* As we shall see, the definition does allow the errors to grow, but limits them to grow no faster than exponential. Also, the definition of stability of a difference scheme is similar to the definition of well-posedness of a partial differential equation.

We define stability for a two level difference scheme of the form

$$\mathbf{u}^{n+1} = Q\mathbf{u}^n, \; n \geq 0, \tag{2.4.1}$$

which will generally be a difference scheme for solving a given initial–value problem on $\mathbb{R}$ which includes a homogeneous linear partial differential equation.

Definition 2.4.1 *The difference scheme (2.4.1) is said to be stable with respect to the norm $\|\cdot\|$ if there exist positive constants Δx_0 and Δt_0, and non-negative constants K and β so that*

$$\| \mathbf{u}^{n+1} \| \leq K e^{\beta t} \| \mathbf{u}^0 \|, \tag{2.4.2}$$

for $0 \leq t = (n+1)\Delta t, \; 0 < \Delta x \leq \Delta x_0$ and $0 < \Delta t \leq \Delta t_0$.

Remark 1: Notice that as with the definitions of convergence and consistency, the definition of stability is given in terms of a norm. As was also the case with convergence and consistency, this norm may differ depending on the situation. Also notice that the definition of stability does indeed allow the solution to grow. We should notice that *the solution can grow with time, not with the number of time steps.*

Remark 2: We also notice that stability is defined for a homogeneous difference scheme. As we shall see in Section 2.5, stability of the homogeneous equation, along with the correct consistency, is enough to prove convergence of the nonhomogeneous difference scheme. All of the contributions of the nonhomogeneous term will be contained in the truncation term $\boldsymbol{\tau}^n$. In fact, *when we discuss stability of a nonhomogeneous difference scheme, such as difference scheme (2.3.14), we consider the stability of the associated homogeneous scheme.*

Remark 3: And finally, we warn the reader that there are a variety of definitions of stability in the literature. Definition 2.4.1 happens to be one of the stronger definitions. One common definition is to require that condition (2.4.2) hold only for $(n+1)\Delta t \leq T$ for any T (where K and β depend on T). Another, more common, definition that is used is one that does not allow for exponential growth. Inequality (2.4.2) is replaced by

$$\| \mathbf{u}^{n+1} \| \leq K \| \mathbf{u}^0 \|, \tag{2.4.3}$$

(with or without the restriction $(n+1)\Delta t \leq T$.) Clearly, inequality (2.4.3) implies inequality (2.4.2). Also, inequality (2.4.2) along with the restriction $(n+1)\Delta t \leq T$ implies inequality (2.4.3). This latter definition of stability implies that the solutions to the difference equation must be bounded. Using the fact that the iterations must be bounded is a much nicer concept with which to work. However, very soon after a definition using inequality (2.4.3) is given, it must be expanded to include more general situations. We have merely included these more general situations in our first definition. But, when it is convenient (as it often is) to prove inequality (2.4.3) instead of (2.4.2), we will do so realizing that it is sufficient. *When we want to use stability based on inequality (2.4.3), we will refer to it as "Definition 2.4.1– (2.4.3)".*

One last characterization of stability that is often useful comes from considering the inequality (2.4.2). We state this in the following proposition.

Proposition 2.4.2 *The difference scheme (2.4.1) is stable with respect to the norm $\|\cdot\|$ if and only if there exists positive constants Δx_0 and Δt_0 and non-negative constants K and β so that*

$$\| Q^{n+1} \| \le K e^{\beta t} \tag{2.4.4}$$

for $0 \le t = (n+1)\Delta t$, $0 < \Delta x \le \Delta x_0$, and $0 < \Delta t \le \Delta t_0$.

Remark: We should note that the norm used in inequality (2.4.4) is an operator norm, not a vector norm. Operator norms were introduced in Section 2.2.3. As we shall see from the proof, the operator norm is the one induced from the norm defined on the space in which we are working.

Proof: Since

$$\mathbf{u}^{n+1} = Q\mathbf{u}^n = Q(Q\mathbf{u}^{n-1}) = Q^2\mathbf{u}^{n-1} = \cdots = Q^{n+1}\mathbf{u}^0,$$

expression (2.4.2) can be written as

$$\| \mathbf{u}^{n+1} \| = \| Q^{n+1}\mathbf{u}^0 \| \le K e^{\beta t} \| \mathbf{u}^0 \|,$$

or

$$\frac{\| Q^{n+1}\mathbf{u}^0 \|}{\| \mathbf{u}^0 \|} \le K e^{\beta t}.$$

($\Rightarrow$) By taking the supremum over both sides over all non-zero vectors $\mathbf{u}^0$, we get

$$\| Q^{n+1} \| \le K e^{\beta t}. \tag{2.4.5}$$

($\Leftarrow$) The fact that $\| Q^{n+1}\mathbf{u}^0 \| \le \| Q^{n+1} \| \| \mathbf{u}^0 \|$, and inequality (2.4.4) imply inequality (2.4.2) (stability).

As with convergence, it is difficult to prove stability directly. Of course, hopefully there are better methods for proving stability than for convergence (otherwise, taking this route for proving convergence would prove quite futile). To show how similar the convergence proof in Example 2.2.1 is to a stability proof, we first prove that difference scheme (2.2.2) is stable. We then include a stability proof for the difference scheme for a hyperbolic partial differential equation that we considered in HW2.3.1(e).

Example 2.4.1 Show that the difference scheme

$$u_k^{n+1} = (1 - 2r)u_k^n + r(u_{k+1}^n + u_{k-1}^n) \tag{2.4.6}$$

is stable with respect to the sup-norm.

Solution: The calculation that we do is essentially the same as we did in Example 2.2.1 (without the difference and order argument). We note that if $r \leq 1/2$,

$$\begin{aligned}
\mid u_k^{n+1} \mid &\leq (1 - 2r) \mid u_k^n \mid + r \mid u_{k+1}^n \mid + r \mid u_{k-1}^n \mid \\
&\leq \parallel \mathbf{u}^n \parallel_\infty .
\end{aligned}$$

If we then take the supremum over both sides (with respect to k), we get

$$\parallel \mathbf{u}^{n+1} \parallel_\infty \leq \parallel \mathbf{u}^n \parallel_\infty .$$

Hence inequality (2.4.3) is satisfied with $K = 1$ (or inequality (2.4.2) is satisfied with $K = 1$ and $\beta = 0$).

We note that for stability of the scheme (2.4.6) we have required that $r \leq 1/2$. In this case we say that the scheme is **conditionally stable** (where the condition is $r \leq 1/2$). In the case where no restrictions on the relationship between Δt and Δx are needed for stability, we say that the scheme is **stable** or **unconditionally stable**.

Example 2.4.2 Discuss the stability of the following difference scheme (FTFS).

$$u_k^{n+1} = u_k^n - a\frac{\Delta t}{\Delta x}(u_{k+1}^n - u_k^n). \tag{2.4.7}$$

Solution: We recall that we showed in HW 2.3.1(e) that difference scheme (2.4.7) was consistent with the hyperbolic partial differential equation

$$v_t + av_x = 0. \tag{2.4.8}$$

As is the case for most consistency arguments, depending on which assumptions we wish to make on the appropriate derivatives, the scheme is consistent with respect to either the sup-norm or the $\ell_{2,\Delta x}$ norm. In addition, we now require that $a < 0$.

We add also that we shall see in Chapter 5 that equation (2.4.8) and difference scheme (2.4.7) is an excellent model equation on which we shall base much of our work on hyperbolic equations. We shall use the equation and difference scheme often.

We rewrite difference scheme (2.4.7) (setting $R = a\Delta t/\Delta x$) as

$$u_k^{n+1} = (1 + R)u_k^n - Ru_{k+1}^n. \tag{2.4.9}$$

We then note that

$$\begin{aligned}
\sum_{k=-\infty}^{\infty} |u_k^{n+1}|^2 &= \sum_{k=-\infty}^{\infty} |(1 + R)u_k^n - Ru_{k+1}^n|^2 \\
\text{step 2} &\leq \sum_{k=-\infty}^{\infty} \big\{ |1 + R|^2|u_k^n|^2 + 2|1 + R||R||u_k^n||u_{k+1}^n| \\
&\qquad + |R|^2|u_{k+1}^n|^2 \big\} \\
\text{step 3} &\leq \sum_{k=-\infty}^{\infty} \big\{ |1 + R|^2|u_k^n|^2 + |1 + R||R|(|u_k^n|^2 + |u_{k+1}^n|^2) \\
&\qquad + |R|^2|u_{k+1}^n|^2 \big\} \\
\text{step 4} &= \sum_{k=-\infty}^{\infty} (|1 + R|^2 + 2|1 + R||R| + |R|^2)|u_k^n|^2
\end{aligned}$$

$$= (|1 + R| + |R|)^2 \sum_{k=-\infty}^{\infty} |u_k^n|^2.$$

We note that we used the fact that the geometric mean was less than or equal to the arithmetic mean to get from step 2 to step 3; and we renumbered the indices to get from step 3 to step 4. The above expression written in terms of ℓ_2 norms is

$$\| \mathbf{u}^{n+1} \|_2 \le K_1 \| \mathbf{u}^n \|_2, \qquad (2.4.10)$$

where $K_1 = |1 + R| + |R|$. We can apply this process n more times and get

$$\| \mathbf{u}^{n+1} \|_2 \le K_1^{n+1} \| \mathbf{u}^0 \|_2 . \qquad (2.4.11)$$

To prove stability we must compare this inequality to inequality (2.4.3) (or (2.4.2) with $\beta = 0$). Hence we see that for stability, we must find a constant K such that

$$(|1 + R| + |R|)^{n+1} \le K. \qquad (2.4.12)$$

This can be easily done by restricting R so that $|1 + R| + |R| \le 1$ and then choosing $K = 1$. Since $R = a\Delta t/\Delta x \le 0$, it is not hard to see that $|1 + R| - R \le 1$ implies that $|1 + R| \le 1 + R$. This implies that $-1 \le R \le 0$.

We note that the above analysis was done with respect to the ℓ_2 norm. However, as we have stated earlier, it makes no sense to apply these definitions with respect to the ℓ_2 norm. This is because we are using consistency, stability and convergence for sequences that are discretizations of functions that are solutions to partial differential equations and discrete functions that approximate these solutions. In these cases, as Δx approaches zero, the ℓ_2 norm of these functions goes to infinity. As usual, we instead work with the $\ell_{2,\Delta x}$ norm. The entire argument above can easily be changed to imply stability with respect to the $\ell_{2,\Delta x}$ norm by multiplying both sides of either inequality (2.4.10) or (2.4.11) by $\sqrt{\Delta x}$. We have then proved that difference scheme (2.4.7) is conditionally stable with respect to the $\ell_{2,\Delta x}$ norm. We note specifically that the condition for stability is given by $-1 \le R \le 0$ or $-1 \le a\Delta t/\Delta x \le 0$.

Remark: We might note that if $a > 0$, the entire argument through equation (2.4.11) still holds. But, if $a > 0$, there is no way to restrict K_1 to be less than or equal to one. In fact, in this case $K_1 > 1$.

HW 2.4.1 Show that for $| R | \le 1$, difference scheme

$$u_k^{n+1} = \frac{1}{2}(u_{k+1}^n + u_{k-1}^n) - \frac{R}{2}\delta_0 u_k^n$$

is stable with respect to the sup-norm.

2.4.2 Initial–Boundary–Value Problems

Before we leave the subject of stability and finally go on to the Lax Theorem, we must consider the case of initial–boundary–value problems. We shall return to this case later and do a more complete study. At this time, we shall include a large class of initial–boundary–value problems by considering them as we did for both convergence and consistency. That is, we assume that we have a sequence of partitions of our interval described by the sequence of increments, $\{\Delta x_j\}$ and a sequence of spaces $\{X_j\}$ with norms

$\| \cdot \|_j$. We then say that *the difference scheme for an initial–boundary–value problem is stable if it satisfies either inequality (2.4.2) or (2.4.3) with the norms replaced by $\| \cdot \|_j$.*

To illustrate the definition, we include the following example.

Example 2.4.3 Consider the initial–boundary–value problem

$$v_t = \nu v_{xx}, \ x \in (0,1), \ t > 0 \tag{2.4.13}$$
$$v(x,0) = f(x), \ x \in [0,1] \tag{2.4.14}$$
$$v(0,t) = v(1,t) = 0, \ t \geq 0 \tag{2.4.15}$$

along with the difference scheme ($\Delta x = 1/M$)

$$u_k^{n+1} = u_k^n + r\delta^2 u_k^n, \ k = 1, \cdots, M-1 \tag{2.4.16}$$
$$u_0^{n+1} = u_M^{n+1} = 0 \tag{2.4.17}$$
$$u_k^0 = f(k\Delta x), \ k = 0, \cdots, M. \tag{2.4.18}$$

Show that if $r \leq 1/2$, difference scheme (2.4.16)–(2.4.18) is stable.

Solution: The calculation is essentially the same as that done in Example 2.2.1 and is the same as done in Example 2.4.1. We consider any sequence of partitions of the interval $[0,1]$ defined by the sequence of increments $\{\Delta x_j\}$ and the associated spaces, $\{X_j\}$, and norms, $\{\| \cdot \|_j\}$. Specifically, we choose the space X_j to be the space of $(M_j - 1)$-vectors where $M_j \Delta x_j = 1$ and let $\| \cdot \|_j$ denote the sup-norm on X_j.

As we did before, we note that if $r \leq 1/2$,

$$\begin{aligned}
|u_k^{n+1}| &= |(1-2r)u_k^n + r(u_{k+1}^n + u_{k-1}^n)| \\
&\leq |(1-2r)||u_k^n| + r|u_{k+1}^n| + r|u_{k-1}^n| \\
&\leq (1-2r)\| \mathbf{u}^n \|_j + r\| \mathbf{u}^n \|_j + r\| \mathbf{u}^n \|_j \\
&= \| \mathbf{u}^n \|_j \ .
\end{aligned}$$

Then taking the maximum over k on both sides gives us

$$\| \mathbf{u}^{n+1} \|_j \leq \| \mathbf{u}^n \|_j \ . \tag{2.4.19}$$

If we apply inequality (2.4.19) repeatedly, we arrive at

$$\| \mathbf{u}^{n+1} \|_j \leq \| \mathbf{u}^0 \|_j,$$

so the difference scheme is stable using $K = 1$ and $\beta = 0$.

We should notice that it is equally easy to mimic Example 2.4.2 and prove that difference scheme (2.4.7) is stable with respect to the $\ell_{2,\Delta x}$ norm. We include that proof as a homework problem given below.

HW 2.4.2 Consider the initial–boundary–value problem

$$v_t + av_x = 0, \ x \in (0,1), \ t > 0 \tag{2.4.20}$$
$$v(x,0) = f(x), \ x \in [0,1] \tag{2.4.21}$$
$$v(1,t) = 0, \ t \geq 0, \tag{2.4.22}$$

where $a < 0$, along with the difference scheme ($\Delta x = 1/M$)

$$u_k^{n+1} = (1+R)u_k^n - Ru_{k+1}^n \ k = 0, \cdots, M-1 \tag{2.4.23}$$
$$u_M^{n+1} = 0 \tag{2.4.24}$$
$$u_k^0 = f(k\Delta x), \ k = 0, \cdots, M, \tag{2.4.25}$$

where $R = a\Delta t/\Delta x$. Show that if $|R| \leq 1$, difference scheme (2.4.23)–(2.4.25) is stable.

2.5 The Lax Theorem

2.5.1 Initial–Value Problems

Now that we have discussed convergence, consistency and stability, it is time to see how they are connected. As we have indicated previously, they are connected via the Lax Equivalence Theorem.

Theorem 2.5.1 Lax Equivalence Theorem *A consistent, two level difference scheme for a well-posed linear initial–value problem is convergent if and only if it is stable.*

Thus just as we have been promising, as long as we have a consistent scheme, convergence is synonymous with stability. Instead of trying to prove this theorem, we prove a slightly stronger version of half of the above theorem.

Theorem 2.5.2 Lax Theorem *If a two-level difference scheme*

$$\mathbf{u}^{n+1} = Q\mathbf{u}^n + \Delta t\mathbf{G}^n \qquad (2.5.1)$$

is accurate of order (p, q) in the norm $\|\cdot\|$ to a well-posed linear initial–value problem and is stable with respect to the norm $\|\cdot\|$, then it is convergent of order (p, q) with respect to the norm $\|\cdot\|$.

Remark: One of the hypotheses in both the Lax Equivalence Theorem and the Lax Theorem is that the initial–value problem be well-posed. An initial–value problem is well-posed if it depends continuously upon its initial conditions. Though the question of well-posedness is a very important question, we will not discuss it here. Most initial–value problems (and initial–boundary–value problems) that we wish to solve are well-posed. For a treatment of well-posedness, see either ref. [11] or ref. [5].

Proof: Let $v = v(x, t)$ denote the exact solution of the initial–value problem. Then since the difference scheme is accurate of order (p, q), we have

$$\mathbf{v}^{n+1} = Q\mathbf{v}^n + \Delta t\mathbf{G}^n + \Delta t\boldsymbol{\tau}^n \qquad (2.5.2)$$

with $\| \boldsymbol{\tau}^n \| = \mathcal{O}(\Delta x^p) + \mathcal{O}(\Delta t^q)$. Define $\mathbf{w}^j$ to be the difference $\mathbf{v}^j - \mathbf{u}^j$. Then $\mathbf{w}$ also satisfies

$$\mathbf{w}^{n+1} = Q\mathbf{w}^n + \Delta t\boldsymbol{\tau}^n. \qquad (2.5.3)$$

Applying equation (2.5.3) repeatedly gives

$$
\begin{aligned}
\mathbf{w}^{n+1} &= Q\mathbf{w}^n + \Delta t \boldsymbol{\tau}^n \\
&= Q(Q\mathbf{w}^{n-1} + \Delta t \boldsymbol{\tau}^{n-1}) + \Delta t \boldsymbol{\tau}^n \\
&= Q^2 \mathbf{w}^{n-1} + \Delta t Q \boldsymbol{\tau}^{n-1} + \Delta t \boldsymbol{\tau}^n \tag{2.5.4}
\end{aligned}
$$

$$
\cdots
$$

$$
= Q^{n+1} \mathbf{w}^0 + \Delta t \sum_{j=0}^{n} Q^j \boldsymbol{\tau}^{n-j}. \tag{2.5.5}
$$

Since $\mathbf{w}^0 = \boldsymbol{\theta}$, we have

$$
\mathbf{w}^{n+1} = \Delta t \sum_{j=0}^{n} Q^j \boldsymbol{\tau}^{n-j}. \tag{2.5.6}
$$

The fact that the difference scheme is stable implies that for any j,

$$
\| Q^j \| \leq K e^{\beta t}. \tag{2.5.7}
$$

Taking the norm of both sides of equation (2.5.6) and then using (2.5.7) yields

$$
\begin{aligned}
\| \mathbf{w}^{n+1} \| &\leq \Delta t \sum_{j=0}^{n} \| Q^j \| \| \boldsymbol{\tau}^{n-j} \| \\
&\leq \Delta t K \sum_{j=0}^{n} e^{\beta j \Delta t} \| \boldsymbol{\tau}^{n-j} \| \\
&\leq \Delta t K e^{\beta(n+1)\Delta t} \sum_{j=0}^{n} \| \boldsymbol{\tau}^{n-j} \| \\
&\leq (n+1)\Delta t K e^{\beta(n+1)\Delta t} C^*(t)(\Delta x^p + \Delta t^q), \tag{2.5.8}
\end{aligned}
$$

where $C^*(t) = \sup_{0 \leq s \leq t} C(s)$ where $C(s)$, $s = (n-j)\Delta t$, is the constant involved in the "big $\mathcal{O}$" expression for $\| \boldsymbol{\tau}^{n-j} \|$. As in the definition of convergence, Definition 2.2.2, t is chosen so that $(n+1)\Delta t \to t$ as $\Delta t \to 0$ (and of course, as $n \to \infty$). Thus, as Δx, $\Delta t \to 0$, expression (2.5.8) yields,

$$
(n+1)\Delta t K e^{\beta t} C^*(t)(\Delta x^p + \Delta t^q) \to t K e^{\beta t} C^*(t) \, 0 = 0. \tag{2.5.9}
$$

Of course, this is equivalent to $\| \mathbf{u}^{n+1} - \mathbf{v}^{n+1} \| \to 0$.

To see that the convergence is of order (p, q), we note that expression (2.5.8) can be rewritten as

$$
\begin{aligned}
\| \mathbf{w}^{n+1} \| &\leq K(t)(\Delta x^p + \Delta t^q) \\
&= \mathcal{O}(\Delta x^p) + \mathcal{O}(\Delta t^q).
\end{aligned}
$$

Remark 1: We note that the term $\mathbf{G}^n$ in both equations (2.5.1) and (2.5.2) are both the approximations to the source term. So, when we subtract to form an equation for $\mathbf{w}$, they subtract out. Any truncation error due to how well $\mathbf{G}^n$ approximates the source term is in the truncation term τ.

Remark 2: It should be noted in the above proof how important it is that the norms used in consistency and stability be the same and that the convergence is then given with respect to that norm. Also, we see that considering only pointwise consistency (which is done so often) is not sufficient. Since we have proved that if $r \leq 1/2$, the explicit scheme (2.4.6) is both accurate of order $(2,1)$ and stable with respect to the sup-norm, then by Theorem 2.5.2,

- *if $r \leq 1/2$, explicit scheme (2.4.6) is convergent of order $(2,1)$ with respect to the sup-norm.*

Also, it is not hard to see that since the explicit scheme for solving the hyperbolic partial differential equation $v_t + av_x = 0$, $a < 0$, (2.4.7), is both consistent of order $(1,1)$ (HW2.3.1(e)) and stable for $-1 \leq R$ (Example 2.4.2) with respect to the $\ell_{2,\Delta x}$ norm, then

- *if $-1 \leq R \leq 0$, difference scheme (2.4.7) is convergent of order $(1,1)$ with respect to the $\ell_{2,\Delta x}$ norm.*

Remark 3: In the above statements concerning convergence, we include restrictions on r and R. In those cases, the restrictions were due to the fact that we only had conditional stability. There will also have times when we only have conditional consistency. In these cases, we really have conditional convergence. These conditions do not effect the proof of Theorem 2.5.2. The most usual types of conditions we will have will be bounds on the r and/or R values as above, or a condition setting one of these values equal to a constant.

Remark 4: And finally, it would be nice to be able to prove the "other direction" of the Equivalence Theorem, but we will not. (For the proof of the "other direction," see ref. [9] or ref. [11].) However, whenever it is convenient, we will still use the "other direction." Specifically, we will use that fact that *if a consistent difference is not stable, then it does not converge.*

2.5.2 Initial–Boundary–Value Problems

As usual, after treating the initial–value problem, we return to treat the initial–boundary–value problem. We begin by assuming the same set-up that we have had in the past for the definitions of convergence, consistency and stability for initial–boundary–value problems. It is then easy to see that the proof of Theorem 2.5.2 still holds by merely adding a j subscript to each norm. Hence we have the following theorem.

Theorem 2.5.3 Lax Theorem *If a two level difference scheme is accurate of order (p, q) in the sequence of norms $\{\|\cdot\|_j\}$ to a well-posed linear initial–boundary–value problem and is stable with respect to the sequence of norms $\{\|\cdot\|_j\}$, then it is convergent of order (p, q) with respect to the sequence of norms $\{\|\cdot\|_j\}$.*

And finally, we note that using our previous proofs of consistency (Example 2.3.6) and stability (Example 2.4.3) of the explicit scheme with respect to the sup-norm, by Theorem 2.5.3, *if $r \leq 1/2$, difference scheme (2.4.16)–(2.4.18) is convergent with respect to the sup-norm.* In addition, using a slight variation of the consistency proved in HW2.3.1(e) and the stability proved in HW 2.4.2, we see that *if $-1 \leq R \leq 0$, explicit scheme (2.4.23)–(2.4.25) for approximating the solution to initial–boundary–value problem (2.4.20)–(2.4.22) is convergent with respect to the $\ell_{2,\Delta x}$ norm.* We should note that proving convergence via Theorem 2.5.3 is generally difficult. This is true both because of the difficulty we have with consistency (when we have Neumann boundary conditions) and the difficulty of proving stability.

Remark: A special note should be made of the numerical form of the initial–boundary–value problem considered in Theorem 2.5.3. Theorem 2.5.3 applies to initial–boundary–value problems that are written as

$$\mathbf{u}^{n+1} = Q\mathbf{u}^n + \Delta t \mathbf{G}^n. \tag{2.5.10}$$

As we saw in Example 2.3.4, it is not natural to write difference schemes for some initial–boundary–value problems in this form. The result of requiring our scheme to be written as (2.5.10) was that we cannot prove that difference scheme (2.3.31)–(2.3.33) converges on the usual grid (though we were able to prove in Example 2.3.5 that the same scheme converges on an offset grid). *The difficulties are caused by the fact that the discretizations of both the partial differential equation and the boundary condition must be written in the form of equation (2.5.10).*

In ref. [4], pages 514–522, the definition of stability and the Lax Theorem are presented where the discretized partial differential equation and boundary conditions are kept separate. The order of approximation in this case is not lowered by forcing the scheme to be written as (2.5.10). It appears that the problems encountered in Example 2.3.4 are circumvented by this theory.

It would be nice to be able to make precise comparisons between the results given in ref. [4] and in this chapter. However, the settings for the two theories make such a comparison difficult. Though the same words are used, it is not clear that they always have the same meanings. In Theorem 4 of ref. [4], page 527, they prove that for Dirichlet boundary conditions, their definition of stability is equivalent to a slight variation of our definition of stability. Though they imply that the result might be true for more general boundary conditions, it is not proved that the results of ref. [4] can be applied to difference scheme (2.3.31)–(2.3.33). We cite this reference so that these results can be used if the requirement of placing the difference scheme in the form (2.5.10) causes insurmountable problems. In Chapter 8, Part 2 we discuss convergence for initial–boundary–value schemes via the Gustafsson result, ref. [2]. The result due to Gustafsson shows that even though the order of accuracy of the boundary condition may be one lower than that of the difference scheme, we still get convergence of the order of the difference scheme. We will discuss this topic more fully in Chapter 8, Part 2.

2.6   Computational Interlude I

Before we continue on to methods for obtaining stability, it is time to return to some computing. It is not so much that it is the most logical time to include this segment on computing. It is included here so that the computational work will be spread out over a longer period of time. The results given in these Computational Interludes will generally be computational results or related to computational results. However, if it seems to be a logical time to introduce a more theoretical topic, we reserve that right to also include such topics in the Computational sections.

2.6.1   Review of Computational Results

We begin by discussing some of the results obtained in the exercises in Chapters 1 and 2. By this time, we have seen the basic explicit scheme often. Surely, if not by HW1.2.1, then by theoretical results in Chapter 2, we know that the explicit scheme (1.2.9) is only conditionally convergent. We shall see that the type of condition, $r \leq 1/2$, is common and generally acceptable in numerical partial differential equations. Hopefully, via HW1.3.1 we believe that the leapfrog scheme for the heat equation (1.3.9) is either not stable or has a very restrictive stability condition.

We saw in HW1.4.1 and HW1.4.2 that the explicit scheme along with either the first or second order approximation of the Neumann boundary condition does a good job of approximating the exact solution. However, we saw in Section 2.3.2, Example 2.3.4 that the first order Neumann boundary condition treatment along with the explicit difference scheme was not norm consistent with the initial–boundary–value problem (so we cannot use Theorem 2.5.3 to prove that the scheme is convergent). But in

HW1.4.2, HW1.5.10 and HW1.6.2, we saw that at least according to those experiments, the difference scheme does seem to "converge" to the analytic solution.

We will often have "differences" between what we can prove using the theory (at this time) and computational results. It is useful to separate "what we know to be true by numerical experiment" and "what we are able to prove." Both types of information are useful.

2.6.2 HW0.0.1

By now we have some results from HW1.5.1 where we considered a partial differential equation of the form

$$v_t + av_x = \nu v_{xx}. \tag{2.6.1}$$

We should notice that the first of our nonlinear problems, HW0.0.1, the viscous Burgers' equation, is not that different from the above equation. In fact, if we proceed as we did when introducing the scheme used in HW1.5.1, we see that a very logical difference approximation for the partial differential equation

$$v_t + vv_x = \nu v_{xx} \tag{2.6.2}$$

is

$$u_k^{n+1} = u_k^n - \frac{\Delta t}{2\Delta x} u_k^n \delta_0 u_k^n + \frac{\nu \Delta t}{\Delta x^2} \delta^2 u_k^n. \tag{2.6.3}$$

In fact the only difference between equation (2.6.3) and equation (1.5.4) is that since the coefficient of v_x is the function v in equation (2.6.2), the a term in equation (1.5.4) is replaced by u_k^n. It is very logical to approximate the term vv_x by

$$u_k^n \frac{u_{k+1}^n - u_{k-1}^n}{2\Delta x}.$$

Using scheme (2.6.3) as the difference approximation of equation (2.6.2) makes it clear that explicit schemes make it quite easy to handle nonlinearities. We shall see later just how easy it is to use (2.6.3) compared to the analogous implicit formulation.

Thus, now that we have difference scheme (2.6.3), you are able to return to the prelude and make your initial attempt at solving problem HW0.0.1.

HW 2.6.1 Discuss the consistency (nonlinear) of difference scheme (2.6.3) with partial differential equation (2.6.2).

2.6.3   Implicit Schemes

To this point, we have only computed with explicit schemes. The reason for this is that explicit schemes are so much easier to use and that we have not spent much time on implicit schemes. In Section 2.3 we introduced an implicit scheme, (2.3.14), for approximating the solution of the one dimensional heat equation, that is first order accurate in time and second order accurate in space. Also, in HW2.3.1(c) we introduced the Crank-Nicolson scheme for solving the same equation that was second order accurate in time and second order accurate in space. And finally, in HW2.3.1(b), we introduced an implicit scheme for approximating the solution to partial differential equation (2.6.1).

We shall illustrate what must be done to compute with these implicit schemes by working with the Crank-Nicolson scheme. We again consider the problem that we solved using the explicit scheme in Section 1.2:

$$v_t = \nu v_{xx}, \ x \in (0,1), \ t > 0 \tag{2.6.4}$$
$$v(x,0) = f(x), \ x \in [0,1] \tag{2.6.5}$$
$$v(0,t) = \alpha(t), \ v(1,t) = \beta(t), \ t \geq 0 \tag{2.6.6}$$

where $f(0) = \alpha(0)$, and $f(1) = \beta(0)$. If we apply the Crank-Nicolson scheme to this problem, we are left with the following finite difference scheme.

$$u_k^{n+1} - \frac{r}{2}\delta^2 u_k^{n+1} = u_k^n + \frac{r}{2}\delta^2 u_k^n,$$
$$k = 1, \cdots, M-1 \tag{2.6.7}$$
$$u_0^{n+1} = \alpha^{n+1}, \ u_M^{n+1} = \beta^{n+1}, \ n \geq 0 \tag{2.6.8}$$
$$u_k^0 = f(k\Delta x), \ k = 0, \cdots, M \tag{2.6.9}$$

where $r = \nu \Delta t / \Delta x^2$. The difference between using equations (2.6.7)–(2.6.9) to provide an approximation to problem (2.6.4)–(2.6.6) and the other methods that we have implemented is that *in this case we are not able to solve for u_k^{n+1} explicitly*. In this case we are left with a traditional system of equations of the form $Ax = b$ where x is the solution at the $(n+1)$st time level and b depends on u at the nth time level. And, of course, it is not just any A, the matrix we have has a very special structure.

If we first consider the first equation, $k = 1$, we see that we get

$$(1+r)u_1^{n+1} - \frac{r}{2}u_2^{n+1} = \frac{r}{2}u_0^n + (1-r)u_1^n + \frac{r}{2}u_2^n + \frac{r}{2}\alpha^{n+1}. \tag{2.6.10}$$

Notice that the last term on the right was on the left hand side in the difference equation (2.6.7). It has been transferred to the right side since it is a boundary condition and is known. Likewise, if we consider $k = M-1$, we get

$$-\frac{r}{2}u_{M-2}^{n+1} + (1+r)u_{M-1}^{n+1} = \frac{r}{2}u_{M-2}^n + (1-r)u_{M-1}^n + \frac{r}{2}u_M^n + \frac{r}{2}\beta^{n+1}. \tag{2.6.11}$$

And, of course, at all of the other grid points we have

$$-\frac{r}{2}u_{k-1}^{n+1} + (1+r)u_k^{n+1} - \frac{r}{2}u_{k+1}^{n+1} = \frac{r}{2}u_{k-1}^n + (1-r)u_k^n + \frac{r}{2}u_{k+1}^n,$$
$$k = 2, \cdots, M-2. \qquad (2.6.12)$$

Thus the solution scheme is to use the initial condition

$$u_k^0 = f(k\Delta x), \ k = 0, \cdots, M$$

to get started and then solve equations (2.6.10)–(2.6.12) at each time level. And as we stated earlier, (2.6.10)–(2.6.12) is just a linear system of equations of the form

$$Q_1 \mathbf{u}^{n+1} = \mathbf{r} \qquad (2.6.13)$$

where the coefficient matrix is the tri-diagonal matrix

$$Q_1 = \begin{pmatrix} 1+r & -\frac{r}{2} & 0 & 0 & \cdots & & & \\ -\frac{r}{2} & 1+r & -\frac{r}{2} & 0 & \cdots & & & \\ & & \cdots & & & & & \\ \cdots & 0 & -\frac{r}{2} & 1+r & -\frac{r}{2} & 0 & \cdots & \\ & & \cdots & & & & & \\ & & \cdots & 0 & -\frac{r}{2} & 1+r & -\frac{r}{2} \\ & & & \cdots & 0 & -\frac{r}{2} & 1+r \end{pmatrix},$$
$$(2.6.14)$$

the vector of unknowns is made up of our solution at the $M-1$ interior grid points at the $(n+1)$-st time level,

$$\mathbf{u}^{n+1} = \begin{pmatrix} u_1^{n+1} \\ u_2^{n+1} \\ \vdots \\ u_{M-1}^{n+1} \end{pmatrix}, \qquad (2.6.15)$$

and the right hand side vector $\mathbf{r}$ is given by

$$r_k = \frac{r}{2}u_{k-1}^n + (1-r)u_k^n + \frac{r}{2}u_{k+1}^n, \ k = 2, \cdots, M-2, \qquad (2.6.16)$$

and

$$r_1 = \frac{r}{2}u_0^n + (1-r)u_1^n + \frac{r}{2}u_2^n + \frac{r}{2}\alpha^{n+1} \qquad (2.6.17)$$
$$r_{M-1} = \frac{r}{2}u_{M-2}^n + (1-r)u_{M-1}^n + \frac{r}{2}u_M^n + \frac{r}{2}\beta^{n+1}. \qquad (2.6.18)$$

The right hand side of our system of equations could be written as a matrix times the vector of solutions at the nth time level (plus the two boundary conditions), but for solution purposes there is no good reason to do that.

The procedure for solving equation (2.6.13) is to write the augmented matrix

$$[Q_1|\mathbf{r}]$$

and use Gaussian elimination. If we write a general tri-diagonal matrix in the form $T[a, b, c]$ with the terms $a_2, \cdots, a_m$ on the subdiagonal, the terms $b_1, \cdots, b_m$ on the diagonal and the terms $c_1, \cdots, c_{m-1}$ on the superdiagonal, then the Gaussian elimination algorithm for solving

$$T[a, b, c]\mathbf{u} = \mathbf{r}$$

(sometimes referred to as the Thomas Algorithm) is as follows.

$$c_1' = c_1/b_1$$
$$r_1'' = r_1/b_1$$

For $j = 2, \cdots, m - 1$
$$b_j' = b_j - a_j c_{j-1}'$$
$$r_j' = r_j - a_j r_{j-1}''$$
$$c_j' = c_j/b_j'$$
$$r_j'' = r_j'/b_j'$$
Next j

$$b_m' = b_m - a_m c_{m-1}'$$
$$r_m' = r_m - a_m r_{m-1}''$$
$$r_m''' = r_m'/b_m'$$

For $j = m - 1, \cdots, 1$
$$r_j''' = r_j'' - c_j' r_{j+1}'''$$
Next j

Remark 1: The above algorithm should be written as a subroutine because we shall use it often.

Remark 2: Our problems will usually be nice enough that it would not be necessary to regenerate the a's, b's and c's at every call. But, since in the long run we want to use the methods developed here to solve more difficult problems, the subroutine should be designed to import values of a, b and c and write over them when convenient.

Remark 3: We should note that in the algorithm given above, each time we changed a term, we changed the expression for that term (we primed it). This is not necessary to do in your subroutine. We do not want to use all of that extra space. We, instead, write over the previous values.

Remark 4: Note that the solution will come out of the subroutine in place of the right hand side, **r**.

Now that we know how to solve tri-diagonal matrices, we are ready to approximately solve problem (2.6.4)–(2.6.6) using the Crank-Nicolson scheme. Returning to the pseudo code for solving time dependent partial differential equations given in Section 1.2.1, we see that the difference between the code needed to solve (2.6.4)–(2.6.6) by the Crank-Nicolson scheme and that using the explicit scheme is in the call to the solution scheme. The solution scheme will now consist of three calls.

Call Right Hand Side
Call Stencil
Call Trid

The Right Hand Side subroutine will obviously use equations (2.6.16)–(2.6.18) to build the right hand side of the equation that we must solve. Quite trivially, the Stencil subroutine will set $a_j = -r/2$, $j = 2, \cdots , m$, $b_j = 1 + r$, $j = 1 \cdots , m$ and $c_j = -r/2$, $j = 1, \cdots , m - 1$ where $m = M - 1$. And, of course, Trid will then solve the system of equations. We should notice that we really don't need an array for *unew* in this scheme. The solution at the new time step comes out of Trid in **r**. The solution can be output from **r** and then set equal to *uold*.

We are now ready to code implicit schemes. We should note that using any of the other implicit schemes (either with or without the lower order terms) is not much different from using the Crank-Nicolson scheme. Defining the right hand side is easier and defining the stencil is different. But, overall, it is the same code.

Remark: Earlier, when we considered explicit difference schemes, we considered higher order explicit difference schemes. It might be nice to have implicit schemes that have an order of accuracy higher than two. However, since higher order implicit schemes would require that we solve matrices more difficult than tridiagonals, we will not consider using them.

HW 2.6.2 Write codes to solve problem (2.6.4)–(2.6.6) using the implicit scheme (BTCS), (2.3.14), and the Crank-Nicolson scheme, (2.6.7). Use $f(x) = \sin 2\pi x$, $\alpha = \beta = 0$, $M = 10$ and $\nu = 1/6$. Find solutions at $t = 0.06$, $t = 0.1, t = 0.9$ and $t = 50$. For the first three values of t, use $\Delta t = 0.02$. To speed solution to the last value of t, use $\Delta t = 1.0$. Compare your solutions with the exact solutions and solutions obtained using the explicit scheme, HW1.2.1.

HW 2.6.3 Write a code to solve the problem involving the parabolic equation with lower order terms, (1.5.1)–(1.5.3), using the implicit scheme given in HW2.3.1(b). Use $f(x) = \sin 4\pi x$, $\nu = 1.0$, $a = 2$, $\Delta t = 0.001$ and $M = 20$. Find solutions at $t = 0.06$, $t = 0.1$ and $t = 0.9$. Compare the results with the solution found in HW1.5.1 and the exact solutions. Make an additional computation of u at $t = 0.9$ using $\Delta t = 0.1$. Compare this solution to the previous solution.

2.6.4   Neumann Boundary Conditions

Of course, once we are able to solve the problems in HW2.6.2 and 2.6.3 using an implicit scheme, we have to consider returning to problem (1.4.1)–(1.4.4) and see how well it works to treat Neumann boundary conditions implicitly. We approximate the partial differential equation by the implicit operator described in (2.3.14), and the initial condition and Dirichlet boundary condition as we did in(1.4.6)–(1.4.7). Hence, we have the equations

$$u_k^{n+1} - r\delta^2 u_k^{n+1} = u_k^n, \ k = 1, \cdots, M - 1 \tag{2.6.19}$$

$$u_k^0 = f(k\Delta x) = \cos\frac{\pi k\Delta x}{2}, \ k = 0, \cdots, M \tag{2.6.20}$$

$$u_M^{n+1} = 0, \ n \geq 0. \tag{2.6.21}$$

where $r = \nu\Delta t/\Delta x^2$. As we did in our explicit treatment of the Neumann boundary condition, we can treat the derivative in several ways. We should recall that we studied the order of approximation of the various treatments for the Neumann boundary condition in HW2.3.5(a)-(c). We will consider the following 0-th, 1-st and 2-nd approximation to the Neumann boundary condition.

$$\frac{u_1^n - u_0^{n+1}}{\Delta x} = 0 \tag{2.6.22}$$

$$\frac{u_1^{n+1} - u_0^{n+1}}{\Delta x} = 0 \tag{2.6.23}$$

$$\frac{u_1^{n+1} - u_{-1}^{n+1}}{2\Delta x} = 0 \tag{2.6.24}$$

0-th order
The 0-th order treatment is the easiest to implement. The implementation should be a slight variation of the code used to solve HW2.6.2. We solve equation (2.6.22) for

$$u_0^{n+1} = u_1^n \tag{2.6.25}$$

and use this condition as if it were a Dirichlet boundary condition. Using this 0-th order treatment of the boundary condition along with difference equations (2.6.19)–(2.6.21), leaves us with the following system of equations to solve.

$$
Q_1 \begin{pmatrix} u_1^{n+1} \\ u_2^{n+1} \\ \vdots \\ u_{M-1}^{n+1} \end{pmatrix} = \begin{pmatrix} u_1^n + r u_1^n \\ u_2^n \\ \vdots \\ u_{M-1}^n \end{pmatrix}
\tag{2.6.26}
$$

where Q_1 is the $M - 1 \times M - 1$ matrix

$$
Q_1 = \begin{pmatrix}
1+2r & -r & 0 & 0 & \cdots & & & \\
-r & 1+2r & -r & 0 & \cdots & & & \\
& & \cdots & & & & & \\
& \cdots & 0 & -r & 1+2r & -r & 0 & \cdots \\
& & & \cdots & & & & \\
& & & \cdots & 0 & -r & 1+2r & -r \\
& & & \cdots & & 0 & -r & 1+2r
\end{pmatrix}.
$$

We should note that the extra term in the first position of the right hand side, $r u_1^n$, is obtained as the difference equation at $k = 1$ "reaches" to the left and gets u_0^{n+1}. Our 0-th order treatment of the boundary condition, equation (2.18), replaces u_0^{n+1} by the known value u_1^n.

1-st order
The 1-st order treatment of the boundary condition provides us with an equation at the point $k = 0$,

$$
u_0^{n+1} - u_1^{n+1} = 0
\tag{2.6.27}
$$

that must be solved along with the equations that we get from the system (2.6.19)–(2.6.21). Treating derivative boundary conditions implicitly will generally give us an addition equation to solve. The system of equations that must now be solved will be of the form

$$
Q_1 \begin{pmatrix} u_0^{n+1} \\ u_1^{n+1} \\ u_2^{n+1} \\ \vdots \\ u_{M-1}^{n+1} \end{pmatrix} = \begin{pmatrix} 0 \\ u_1^n \\ u_2^n \\ \vdots \\ u_{M-1}^n \end{pmatrix},
\tag{2.6.28}
$$

where Q_1 is the $M \times M$ matrix

$$
Q_1 = \begin{pmatrix}
1 & -1 & 0 & \cdots & & & & \\
-r & 1+2r & -r & 0 & \cdots & & & \\
0 & -r & 1+2r & -r & 0 & \cdots & & \\
& & & \cdots & & & & \\
& \cdots & 0 & -r & 1+2r & -r & 0 & \cdots \\
& & & \cdots & & & & \\
& & & \cdots & 0 & -r & 1+2r & -r \\
& & & \cdots & & 0 & -r & 1+2r
\end{pmatrix}.
$$

$$(2.6.29)$$

Hence, whereas in the 0-th order case we had a system of $M - 1$ equations in $M - 1$ unknowns to solve, we now have a system of M equations in M unknowns.

2-nd order

The last case is the 2-nd order accurate approximation of the boundary condition, equation (2.6.24), which can be written as

$$u_{-1}^{n+1} - u_1^{n+1} = 0. \qquad (2.6.30)$$

We treat this boundary condition the same way we treated the 2-nd order boundary condition for explicit equations. Since by including equation (2.6.30) as our approximation to the Neumann boundary condition we have added another point, we must add another equation to our system. As in the case of the explicit scheme, we use our difference approximation of the partial differential equation at the point $k = 0$, i.e.

$$- r u_{-1}^{n+1} + (1+2r)u_0^{n+1} - r u_1^{n+1} = u_0^n. \qquad (2.6.31)$$

Using equation (2.6.30) to eliminate the u_{-1}^{n+1} term from equation (2.6.31), we are left with the following equation at the point $k = 0$.

$$(1+2r)u_0^{n+1} - 2r u_1^{n+1} = u_0^n \qquad (2.6.32)$$

Thus the matrix equation that must be solved to implement the 2-nd order approximation to the Neumann boundary condition is

$$
Q_1 \begin{pmatrix} u_0^{n+1} \\ u_1^{n+1} \\ u_2^{n+1} \\ \vdots \\ u_{M-1}^{n+1} \end{pmatrix} = \begin{pmatrix} u_0^n \\ u_1^n \\ u_2^n \\ \vdots \\ u_{M-1}^n \end{pmatrix}, \qquad (2.6.33)
$$

where Q_1 is the $M \times M$ matrix

$$
\begin{pmatrix}
1+2r & -2r & 0 & \cdots & & & & & \\
-r & 1+2r & -r & 0 & \cdots & & & & \\
0 & -r & 1+2r & -r & 0 & \cdots & & & \\
& & & \cdots & & & & & \\
& \cdots & 0 & -r & 1+2r & -r & 0 & \cdots & \\
& & & \cdots & & & & & \\
& & & \cdots & 0 & -r & 1+2r & -r & \\
& & & \cdots & 0 & & -r & 1+2r &
\end{pmatrix}.
$$

$$(2.6.34)$$

Again, as in the last case, we have a system of M equations in M unknowns to solve.

Remark: We note that neither of the matrices given in (2.6.29) and (2.6.34) are symmetric. Since it makes certain analyses more difficult and eliminates some solution techniques, this is very important to us. It should be noted that the problem involving boundary condition implementation (2.6.27) (1-st order) will yield a symmetric matrix if we use equation (2.6.27) to rewrite the system in the following manner. The equation associated with the first node ($k = 1$) is presently given by

$$- r u_0^{n+1} + (1 + 2r)u_1^{n+1} - ru_2^{n+1} = u_1^n. \tag{2.6.35}$$

If we use equation (2.6.27) to replace u_0^{n+1} by u_1^{n+1}, equation (2.6.35) becomes

$$(1 + r)u_1^{n+1} - ru_2^{n+1} = u_1^n. \tag{2.6.36}$$

Equation (2.6.36) can then be used to replace the first two equations given in the matrix equation (2.6.28) and the resulting matrix equation becomes

$$
Q_1 \begin{pmatrix} u_1^{n+1} \\ u_2^{n+1} \\ \vdots \\ u_{M-1}^{n+1} \end{pmatrix} = \begin{pmatrix} u_1^n \\ u_2^n \\ \vdots \\ u_{M-1}^n \end{pmatrix}, \tag{2.6.37}
$$

where Q_1 is the $M - 1 \times M - 1$ matrix

$$
Q_1 = \begin{pmatrix}
1+r & -r & 0 & \cdots & & & & \\
-r & 1+2r & -r & 0 & \cdots & & & \\
& & & \cdots & & & & \\
& \cdots & 0 & -r & 1+2r & -r & 0 & \cdots \\
& & & & \cdots & & & \\
& & & \cdots & 0 & -r & 1+2r & -r \\
& & & \cdots & 0 & & -r & 1+2r
\end{pmatrix}.
$$

Thus we see that this system of equations is completely equivalent to that given in equation (2.6.28) and, in this case, the coefficient matrix is symmetric.

There is no easy way to make the coefficient matrix associated with the second order implementation of the boundary condition symmetric. However, matrix (2.6.34) can by symmetrized. If we multiply matrix (2.6.34) by the diagonal matrix D with $1/2, 1, \cdots, 1$ on the diagonal, we see that the result is a symmetric matrix. Hence, we can multiply system (2.6.33) by the diagonal matrix D and get the system of equations (which will be equivalent to system (2.6.33))

$$DQ_1 \mathbf{u}^{n+1} = D\mathbf{r}$$

where the matrix DQ_1 is symmetric.

The implementations of the above three treatments of the Neumann boundary condition are not much different from the implementations for the Dirichlet boundary conditions. Obviously, the 0-th order case can be treated just as a Dirichlet problem with $\alpha^{n+1} = u_1^n$.

A little more must be done to implement the 1-st and 2-nd order treatments. The first difference is that we must expect, and lead the tri-diagonal subroutine to expect, one more equation than we had in the Dirichlet case. The major difference is that special care must be taken when filling the first row of the matrix and right hand side vector to take into account that the first equation in both of these cases (and the first element of the right hand side) does not look like the other equations. These differences must be taken into account in the Right Hand Side and Stencil subroutines.

HW 2.6.4 Solve the initial–boundary–value problem (1.4.1)–(1.4.4) using the implicit difference scheme (2.6.19)–(2.6.21) along with the 0-th, 1-st and 2-nd order approximation of the Neumann boundary condition. Use $M = 10$, $\Delta t = 0.004$ and $\nu = 1$. Find solutions at $t = 0.06$, $t = 0.1$ and $t = 0.9$. In addition, let $\Delta t = 1$. and find a solution at $t = 50$. Compare your results with those found in HW1.4.1.

2.6.5 Derivation of Implicit Schemes

In Chapters 1 and 2, when we introduced the explicit difference scheme, we derived the scheme in several different ways. Here, we include a short discussion illustrating that the implicit scheme can also be derived using the conservation law approach. We consider a vertex centered, uniform grid and the partial differential equation $v_t = \nu v_{xx} + F$. We return to the integral form of the conservation law, equation (1.6.9)

$$\int_{x_{k-1/2}}^{x_{k+1/2}} (v^{n+1} - v^n)dx = \int_{t_n}^{t_{n+1}} \int_{x_{k-1/2}}^{x_{k+1/2}} F(x,t)dxdt$$

$$+ \nu \int_{t_n}^{t_{n+1}} \left[(v_x)_{k+1/2} - (v_x)_{k-1/2} \right] dt, \qquad (2.6.38)$$

If we

1. integrate the term on the left as we did before by the midpoint rule,

2. integrate the nonhomogeneous term by the midpoint rule with respect to x and the trapezoidal rule with respect to t,

   and

3. approximate the differences of fluxes as we did in (1.6.13) and integrate the flux terms with respect to t by the trapezoidal rule,

we are left with the following equation.

$$\Delta x(v_k^{n+1} - v_k^n) = \frac{\Delta x \Delta t}{2}(F_k^n + F_k^{n+1}) + \frac{\nu \Delta t}{2\Delta x}\delta^2(v_k^n + v_k^{n+1})$$

$$+ \mathcal{O}(\Delta t^3 \Delta x) + \mathcal{O}(\Delta t \Delta x^3). \qquad (2.6.39)$$

Thus we see that when we drop the order terms, we are left with a nonhomogeneous version of the Crank-Nicolson scheme, (2.6.7),

$$u_k^{n+1} - \frac{r}{2}\delta^2 u_k^{n+1} = u_k^n + \frac{r}{2}\delta^2 u_k^n + \frac{1}{2}(F_k^n + F_k^{n+1}). \qquad (2.6.40)$$

If you note the order terms carefully, it is easy to see that the approximation will be $\mathcal{O}(\Delta t^2) + \mathcal{O}(\Delta x^2)$.

Since the order arguments deriving equation (2.6.39) were done carefully, it was easy to see that the scheme is order $\mathcal{O}(\Delta t^2) + \mathcal{O}(\Delta x^2)$. Because of our treatment of the nonhomogeneous term, we will perform a traditional consistency argument. We let $v = v(x,t)$ be a solution to the nonhomogeneous partial differential equation and note that

$$\tau_k^n = \frac{v_k^{n+1} - v_k^n}{\Delta t} - \frac{r}{2\Delta t}\delta^2(v_k^{n+1} + v_k^n) - \frac{1}{2}(F_k^n + F_k^{n+1})$$

$$= (v_t)_k^{n+1/2} + \mathcal{O}(\Delta t^2) - \nu(v_{xx})_k^{n+1/2} + \mathcal{O}(\Delta x^2) + \mathcal{O}(\Delta t^2)$$

$$- F_k^{n+1/2} - \frac{1}{4}(F_{tt})_k^{n+1/2}\Delta t^2 + \cdots .$$

Thus, if we assume for the moment that the implicit operator Q_1 is such that Q_1^{-1} is uniformly bounded (which we will see how to prove easily in Chapter 3), we see that the scheme accurate of order is $\mathcal{O}(\Delta t^2) + \mathcal{O}(\Delta x^2)$.

Since this result is the same that we claimed was easy earlier, why did we perform the calculation? We note in this example that part of the $\mathcal{O}(\Delta t^2)$ term depends on F. If instead of approximating the source term by $\frac{1}{2}(F_k^n + F_k^{n+1})$ (which came from trapezoidal integration and was very natural in the above derivation), we used $F_k^{n+1/2}$ (which would come from the midpoint rule of integration), the resulting scheme would be just as good (same order) or better (no order terms due to F). However, if we were not so careful and tried to approximate the source term by F_k^n (or F_k^{n+1}), then the above consistency calculation would show that the order of approximation of the scheme would be

$$\tau_k^n = \mathcal{O}(\Delta t^2) + \mathcal{O}(\Delta x^2) + \Delta t(F_t)_k^{n+1/2}.$$

In this case, the resulting difference scheme would be only order $\mathcal{O}(\Delta t) + \mathcal{O}(\Delta x^2)$ and *the loss of order would be due to the approximation used for the source term*.

To illustrate that we can apply the same approach to obtain implicit approximations to Neumann boundary conditions, we return to Section 1.6.1, the nonhomogeneous Neumann boundary condition (1.5.7) and the integral form of the conservation law (1.6.21)

$$\int_{x_0}^{x_{1/2}} v_t dx = \int_{x_0}^{x_{1/2}} F dx + \nu \left[(v_x)_{1/2} - (v_x)_0 \right]. \qquad (2.6.41)$$

If we proceed as we did in Section 1.6.1 except we now integrate with respect to time by the trapezoidal rule, we get

$$\frac{\Delta x}{2}(v_0^{n+1} - v_0^n) = \frac{\Delta x \Delta t}{4}(F_0^n + F_0^{n+1}) - \frac{\nu \Delta t}{2}(g^n + g^{n+1})$$

$$+\frac{\nu \Delta t}{2\Delta x}\delta_{x+}(v_0^n + v_0^{n+1}) + \mathcal{O}(\Delta t \Delta x^2) + \mathcal{O}(\Delta t^3 \Delta x) + \mathcal{O}(\Delta t^3). \quad (2.6.42)$$

Thus, we divide by Δx and obtain the following boundary condition at $k = 0$.

$$u_0^{n+1} - r\delta_{x+}u_0^{n+1} = u_0^n + r\delta_{x+}u_0^n$$

$$+ \frac{\Delta t}{2}(F_0^n + F_0^{n+1}) - r\Delta x(g^n + g^{n+1}). \qquad (2.6.43)$$

Again, as in Section 1.6.1, we are not able to divide equation (2.6.42) by $\Delta t \Delta x$ and make it look like the partial differential equation. As before, since equation (2.6.42) contains information about the diffusion process and the boundary condition, there is no reason that it should look like the partial differential equation.

HW 2.6.5 Discuss the consistency of difference scheme (2.6.7)–(2.6.9), $u_M^n = 0$ and (2.6.43). Show that when $r = \nu \Delta t / \Delta x^2$ is held constant, the scheme is order $\mathcal{O}(\Delta x^2)$.

HW 2.6.6 Use the conservation law approach to derive the implicit difference scheme (BTCS), (2.3.14).

HW 2.6.7 Use the conservation law approach to derive a $\mathcal{O}(\Delta t^2) + \mathcal{O}(\Delta x^2)$ implicit scheme for solving initial–boundary–value problem

$$v_t = \nu v_{xx} \ x \in (0,1), \ t > 0$$
$$v(1,t) = \beta(t) \ t \geq 0$$
$$v_x(0,t) = g(t) \ t \geq 0$$
$$v(x,0) = f(x) \ x \in [0,1].$$

Use the offset grid described in Figure 1.6.4.

3
Stability

In the previous chapter, we showed how important stability is for proving convergence of difference schemes. This chapter is devoted to proving stability of difference schemes. This is done largely by introducing tools that can be used to prove stability of difference schemes, such as the discrete Fourier transform, the Gerschgorin Circle Theorem and an assortment of basic propositions.

3.1 Analysis of Stability

3.1.1 Initial–Value Problems

When solving initial–value problems on the real line, a common analytical tool is to use the Fourier transform. For example, consider the problem

$$v_t = v_{xx}, \quad x \in \mathbb{R}, \ t > 0 \tag{3.1.1}$$
$$v(x,0) = f(x), \quad x \in \mathbb{R}. \tag{3.1.2}$$

If we define the Fourier transform of v to be

$$\hat{v}(\omega, t) = \frac{1}{\sqrt{2\pi}} \int_{-\infty}^{\infty} e^{-i\omega x} v(x,t)dx, \tag{3.1.3}$$

and take the transform of partial differential equation (3.1.1), we get

$$
\begin{aligned}
\hat{v}_t(\omega, t) &= \frac{1}{\sqrt{2\pi}} \int_{-\infty}^{\infty} e^{-i\omega x} v_t(x, t) dx \\
&= \frac{1}{\sqrt{2\pi}} \int_{-\infty}^{\infty} e^{-i\omega x} v_{xx}(x, t) dx \\
&= -\omega^2 \frac{1}{\sqrt{2\pi}} \int_{-\infty}^{\infty} e^{-i\omega x} v(x, t) dx \\
&= -\omega^2 \hat{v}(\omega, t).
\end{aligned}
$$

We should note that the second to the last step followed by integrating by parts twice (under the assumption that the solution to the partial differential equation was sufficiently nice at $\pm\infty$ so that the integrals exist and the terms evaluated at $\pm\infty$ are zero).

Hence, we see that the Fourier transform reduces the partial differential equation to an ordinary differential equation in transform space (the space of transformed functions). The technique is then to solve the ordinary differential equation in transform space and return to our solution space. The method by which we return to our solution space is to use the Fourier inversion formula, which is

$$
v(x, t) = \frac{1}{\sqrt{2\pi}} \int_{-\infty}^{\infty} e^{i\omega x} \hat{v}(\omega, t) d\omega. \tag{3.1.4}
$$

One property of the Fourier transform that we do not use directly in solving problems analytically by Fourier transforms, but is very important theoretically, is Parseval's Identity, that is,

$$
\| v \|_2 = \| \hat{v} \|_2,
$$

where $\|\cdot\|_2$ denotes the L_2 norm on $\mathbb{R}$ (Section 2.2.3). The natural setting to discuss Fourier transforms is in L_2 space. In this setting, the Fourier transform maps the space $L_2(\mathbb{R})$ onto $L_2(\mathbb{R})$ and the result of Parseval's Identity is that the norms of the function and its transform are equal in their respective spaces.

We can use essentially the same approach to analyze the stability of difference schemes for initial–value problems. We suppose that we are given a vector in ℓ_2, $\mathbf{u} = (\cdots, u_{-1}, u_0, u_1, \cdots)^T$, and define the **discrete Fourier transform** of $\mathbf{u}$ as follows.

Definition 3.1.1 *The discrete Fourier transform of* $\mathbf{u} \in \ell_2$ *is the function* $\hat{u} \in L_2[-\pi, \pi]$ *defined by*

$$
\hat{u}(\xi) = \frac{1}{\sqrt{2\pi}} \sum_{m=-\infty}^{\infty} e^{-im\xi} u_m \tag{3.1.5}
$$

for $\xi \in [-\pi, \pi]$.

We note that the ℓ_2 vectors that we will be using later will be $\ell_{2,\Delta x}$ vectors and will be the solutions to our difference scheme at time step n, $\mathbf{u}^n$.

Just as with the case of the continuous Fourier transform, there is both an inversion formula and a Parseval's Identity for the discrete Fourier transform, given below in Propositions 3.1.2 and 3.1.3.

Proposition 3.1.2 *If* $\mathbf{u} \in \ell_2$ *and* $\hat{u}$ *is the discrete Fourier transform of* $\mathbf{u}$, *then*

$$u_m = \frac{1}{\sqrt{2\pi}} \int_{-\pi}^{\pi} e^{im\xi} \hat{u}(\xi) d\xi. \tag{3.1.6}$$

Proposition 3.1.3 *If* $\mathbf{u} \in \ell_2$ *and* $\hat{u}$ *is the discrete Fourier transform of* $\mathbf{u}$, *then*

$$\| \hat{u} \|_2 = \| \mathbf{u} \|_2, \tag{3.1.7}$$

where the first norm is the L_2 norm on $[-\pi, \pi]$ and the second norm is the ℓ_2 norm.

Proof: Proposition 3.1.2 can be proved formally (a non-proof that is not concerned with any of the convergence questions) by the following calculation.

$$\frac{1}{\sqrt{2\pi}} \int_{-\pi}^{\pi} e^{ik\xi} \hat{u}(\xi) d\xi = \frac{1}{\sqrt{2\pi}} \int_{-\pi}^{\pi} e^{ik\xi} \frac{1}{\sqrt{2\pi}} \sum_{m=-\infty}^{\infty} e^{-im\xi} u_m d\xi$$

$$= \frac{1}{2\pi} \sum_{m=-\infty}^{\infty} u_m \int_{-\pi}^{\pi} e^{i(k-m)\xi} d\xi$$

$$= \frac{1}{2\pi} \sum_{\substack{m=-\infty \\ m \neq k}}^{\infty} u_m \left[\frac{e^{i(k-m)\xi}}{i(k-m)} \right]_{-\pi}^{\pi} + \frac{1}{2\pi} u_k \int_{-\pi}^{\pi} d\xi$$

$$= \frac{1}{2\pi} \sum_{\substack{m=-\infty \\ m \neq k}}^{\infty} u_m \frac{1}{i(k-m)} \left[e^{i(k-m)\pi} - e^{-i(k-m)\pi} \right] + u_k$$

$$= u_k.$$

Likewise, we can formally prove Parseval's Identity as follows.

$$\| \hat{u} \|_2^2 = \int_{-\pi}^{\pi} | \hat{u}(\xi) |^2 d\xi$$

$$= \int_{-\pi}^{\pi} \overline{\hat{u}(\xi)} \frac{1}{\sqrt{2\pi}} \sum_{m=-\infty}^{\infty} e^{-im\xi} u_m d\xi$$

$$= \frac{1}{\sqrt{2\pi}} \sum_{m=-\infty}^{\infty} u_m \int_{-\pi}^{\pi} e^{-im\xi} \overline{\hat{u}(\xi)} d\xi$$

$$= \sum_{m=-\infty}^{\infty} u_m \overline{\frac{1}{\sqrt{2\pi}} \int_{-\pi}^{\pi} e^{im\xi} \hat{u}(\xi) d\xi}$$

$$= \sum_{m=-\infty}^{\infty} u_m \overline{u_m}$$

$$= \parallel \mathbf{u} \parallel_2^2$$

The discrete Fourier transform and Parseval's Identity are two basic tools that we will use in our stability analyses. Recall that in our definition of stability, the inequality that was required in terms of the energy norm was of the form

$$\parallel \mathbf{u}^{n+1} \parallel_{2,\Delta x} \le Ke^{\beta(n+1)\Delta t} \parallel \mathbf{u}^0 \parallel_{2,\Delta x} . \tag{3.1.8}$$

But since

$$\parallel \mathbf{u} \parallel_{2,\Delta x} = \sqrt{\Delta x} \parallel \mathbf{u} \parallel_2$$
$$= \sqrt{\Delta x} \parallel \hat{u} \parallel_2,$$

if we can find a K and β to satisfy

$$\parallel \hat{u}^{n+1} \parallel_2 \le Ke^{\beta(n+1)\Delta t} \parallel \hat{u}^0 \parallel_2, \tag{3.1.9}$$

then the same K and β will also satisfy (3.1.8). *When inequality (3.1.9) holds, we say that the sequence $\{\hat{u}^n\}$ is stable in the transform space L_2.* The above results can be summarized in the following proposition.

Proposition 3.1.4 *The sequence $\{\mathbf{u}^n\}$ is stable in $\ell_{2,\Delta x}$ if and only if sequence $\{\hat{u}^n\}$ is stable in $L_2([-\pi, \pi])$.*

We notice that though the approach described above is very similar to the way that the Fourier transform is used to solve continuous initial–value problems, one aspect of using the discrete Fourier transform to analyze stability is much nicer than its continuous counterpart. You should notice that if we can find K and β for equation (3.1.9), there is no need to return to the original equation. In other words, *we do not have to work with the inverse transform* (in the case of the continuous problem, we usually do have to use the inversion formula and that is most often the most difficult part).

The last step in the process that we must consider is whether, as in the continuous case, taking the discrete Fourier transform will simplify the equation. The answer is often yes (otherwise, we would have been foolish to introduce the method) and will be illustrated by examples. We begin by analyzing the stability of the explicit scheme which we have used so often before.

Example 3.1.1 Analyze the stability of the difference scheme

$$u_k^{n+1} = ru_{k-1}^n + (1 - 2r)u_k^n + ru_{k+1}^n, \quad -\infty < k < \infty \qquad (3.1.10)$$

where $r = \nu\Delta t/\Delta x^2$.

Solution: We begin by taking the discrete Fourier transform of both sides of equation (3.1.10). We get

$$\hat{u}^{n+1}(\xi) = \frac{1}{\sqrt{2\pi}} \sum_{k=-\infty}^{\infty} e^{-ik\xi} u_k^{n+1} \qquad (3.1.11)$$

$$= \frac{1}{\sqrt{2\pi}} \sum_{k=-\infty}^{\infty} e^{-ik\xi} \{ru_{k-1}^n + (1 - 2r)u_k^n + ru_{k+1}^n\} \qquad (3.1.12)$$

$$= r\frac{1}{\sqrt{2\pi}} \sum_{k=-\infty}^{\infty} e^{-ik\xi} u_{k-1}^n + (1 - 2r)\frac{1}{\sqrt{2\pi}} \sum_{k=-\infty}^{\infty} e^{-ik\xi} u_k^n$$

$$\qquad + r\frac{1}{\sqrt{2\pi}} \sum_{k=-\infty}^{\infty} e^{-ik\xi} u_{k+1}^n \qquad (3.1.13)$$

$$= r\frac{1}{\sqrt{2\pi}} \sum_{k=-\infty}^{\infty} e^{-ik\xi} u_{k-1}^n + (1 - 2r)\hat{u}^n(\xi)$$

$$\qquad + r\frac{1}{\sqrt{2\pi}} \sum_{k=-\infty}^{\infty} e^{-ik\xi} u_{k+1}^n. \qquad (3.1.14)$$

By making the change of variables $m = k \pm 1$ we get,

$$\frac{1}{\sqrt{2\pi}} \sum_{k=-\infty}^{\infty} e^{-ik\xi} u_{k\pm 1}^n = \frac{1}{\sqrt{2\pi}} \sum_{m=-\infty}^{\infty} e^{-i(m\mp 1)\xi} u_m^n \qquad (3.1.15)$$

$$= e^{\pm i\xi} \frac{1}{\sqrt{2\pi}} \sum_{m=-\infty}^{\infty} e^{-im\xi} u_m^n \qquad (3.1.16)$$

$$= e^{\pm i\xi} \hat{u}(\xi). \qquad (3.1.17)$$

Thus using this expression in (3.1.14) leads us to

$$\hat{u}^{n+1}(\xi) = re^{-i\xi}\hat{u}^n(\xi) + (1 - 2r)\hat{u}^n(\xi) + re^{i\xi}\hat{u}^n(\xi)$$

$$= \{re^{-i\xi} + (1 - 2r) + re^{i\xi}\}\hat{u}^n(\xi)$$

$$= \{2r\cos\xi + (1 - 2r)\}\hat{u}^n(\xi)$$

$$= \left(1 - 4r\sin^2\frac{\xi}{2}\right)\hat{u}^n(\xi). \qquad (3.1.18)$$

The coefficient of $\hat{u}^n$ in equation (3.1.18),

$$\rho(\xi) = 1 - 4r\sin^2\frac{\xi}{2}, \qquad (3.1.19)$$

is called the **symbol** of difference scheme (3.1.10).

Thus we note that taking the discrete Fourier transform, just as was the case with the continuous analog, gets rid of the x derivatives and greatly simplifies the equation. If we apply the result of (3.1.18) $n + 1$ times, we get

$$\hat{u}^{n+1}(\xi) = \left(1 - 4r\sin^2\frac{\xi}{2}\right)^{n+1}\hat{u}^0(\xi). \qquad (3.1.20)$$

We note that if we restrict r so that

$$\left|1 - 4r\sin^2\frac{\xi}{2}\right| \leq 1, \qquad (3.1.21)$$

then we can choose $K = 1$ and $\beta = 0$ and satisfy inequality (3.1.9).

Thus from the earlier discussion, the difference scheme (3.1.10) is stable whenever (3.1.21) is satisfied. Since (3.1.21) is equivalent to

$$-1 \leq 1 - 4r \sin^2 \frac{\xi}{2} \leq 1, \tag{3.1.22}$$

it is easy to see that (3.1.21) is satisfied when

$$1 - 4r \sin^2 \frac{\xi}{2} \leq 1,$$

which is always true; and

$$-1 \leq 1 - 4r \sin^2 \frac{\xi}{2}$$

or

$$2 \geq 4r \sin^2 \frac{\xi}{2},$$

which is true when $r \leq 1/2$. Thus $r \leq 1/2$ is a sufficient condition for stability (and along with the consistency, for convergence).

We also note that if $r > 1/2$, then for at least some ξ (specifically, at least for $\xi = \pi$), we have

$$4r \sin^2 \frac{\xi}{2} > 2$$

or

$$\left| 1 - 4r \sin^2 \frac{\xi}{2} \right| > 1. \tag{3.1.23}$$

But if (3.1.23) is true, then

$$\left| 1 - 4r \sin^2 \frac{\xi}{2} \right|^{n+1} \tag{3.1.24}$$

is eventually greater than $Ke^{\beta(n+1)\Delta t}$ for any K and β. This is true since for any sequence of Δt (chosen so that $(n + 1)\Delta t \to t$) and choice of Δx (so that r remains constant), the expression (3.1.24) becomes unbounded while for sufficiently large values of n, $Ke^{\beta(n+1)\Delta t}$ will be bounded by $Ke^{\beta(t_0+1)}$ for some $t_0 > t$, t_0 near t. Thus, condition (3.1.21) (and hence, also $r \leq 1/2$) is also necessary for stability. Thus *the condition $r \leq 1/2$ is both necessary and sufficient for stability.* And finally, by the Lax Equivalence Theorem we see that *the condition $r \leq 1/2$ is both necessary and sufficient for convergence of difference scheme (3.1.10).*

Remark: We should note that there are assumptions on the convergence result proved here. The stability is with respect to the energy norm. The consistency, proved in Section 1.3, must necessarily also be with respect to the energy norm. This follows from the results of Example 2.3.2 if we add the assumptions concerning u_{tt} and u_{xxxx}. If these derivatives are sufficiently smooth so that the truncation error is in $\ell_{2,\Delta x}$, then the convergence proof is complete. We shall always treat the necessary smoothness as an assumption in this text.

Remark: In Example 3.1.1 we introduced the **symbol** of the difference operator as the analog to our difference operator in transform space. As we saw in Example 3.1.1 and will see in later calculations, ρ is a smooth function of ξ. In our work *we will assume that ρ is a continuous function of ξ*, and, often, we will use the fact that it has one or two derivatives.

Before we proceed to other stability analyses, let us set some notation and make note of some results we derived during the last example that we may be able to use often. To allow us to take the discrete Fourier transform without writing all of the summations, let us define the operator $\mathcal{F} : \ell_2 \rightarrow L_2([-\pi, \pi])$ as the discrete Fourier transform

$$\mathcal{F}(\mathbf{u}) = \frac{1}{\sqrt{2\pi}} \sum_{m=-\infty}^{\infty} e^{-im\xi} u_m. \tag{3.1.25}$$

The operator $\mathcal{F}$ has many properties but the two most important properties for our work are that $\mathcal{F}$ is linear and preserves the norm. If we then define the *shift* operators as

$$S_+ \mathbf{u} \quad = \{v_k\} \text{ where } v_k = u_{k+1}, \; k = 0, \pm 1, \cdots$$

$$\tag{3.1.26}$$

$$S_- \mathbf{u} \quad = \{v_k\} \text{ where } v_k = u_{k-1}, \; k = 0, \pm 1, \cdots,$$

we see that the result that we proved in (3.1.15)–(3.1.17) above can be stated as follows.

Proposition 3.1.5

$$\mathcal{F}(S_\pm \mathbf{u}) = e^{\pm i\xi} \mathcal{F}(\mathbf{u}) \tag{3.1.27}$$

We shall see that this result will make our stability analyses much easier.

Example 3.1.2 Consider the following partial differential equation:

$$v_t + a v_x = 0, \; a < 0. \tag{3.1.28}$$

Analyze the stability of the following difference equation approximation of the above partial differential equation:

$$u_k^{n+1} = (1 + R)u_k^n - R u_{k+1}^n, \; k = 0, \pm 1, \cdots \tag{3.1.29}$$

where $R = a\Delta t/\Delta x$.

Solution: We begin by taking the discrete Fourier transform of equation (3.1.29) to get

$$\hat{u}^{n+1} = (1 + R)\hat{u}^n - R e^{i\xi} \hat{u}^n$$
$$= [(1 + R) - R\cos\xi - iR\sin\xi]\hat{u}^n.$$

Then, in this case the symbol ρ is complex and given by

$$\rho(\xi) = 1 + R - R\cos\xi - iR\sin\xi.$$

As was the case in Example 3.1.1, $\hat{u}^{n+1} = \rho^{n+1}\hat{u}^0$. We must realize that since for this case ρ is complex, we must bound the **magnitude** of ρ by 1 in order to satisfy inequality (3.1.9) (with $K = 1$ and $\beta = 0$). Thus we calculate

$$| \rho |^2 = (1 + R)^2 - 2R(1 + R)\cos\xi + R^2. \tag{3.1.30}$$

Though it is possible to try to analyze the size of $| \rho |^2$ using trigonometry and inequalities as we did in the previous example, we choose to introduce another very basic method. We must determine the maximum and minimum value of $| \rho |^2$ for $\xi \in [-\pi, \pi]$. Hence, we proceed as if we were in our freshman calculus course, differentiate with respect to ξ, set it equal to zero and see that we have potential maximums at $\xi = 0$ and $\xi = \pm\pi$. If we then evaluate $|\rho(\xi)|$ at these values, we see that

$$|\rho(0)| = 1$$
$$|\rho(\pm\pi)| = |1+2R|.$$

To bound $|\rho(\pm\pi)|$ by one, we require that R satisfy $-1 \le 1 + 2R \le 1$. Then since $1 + 2R$ is always less than or equal to 1 ($R < 0$) and $-1 \le 1 + 2R$ is the same as $R \ge -1$, we see that difference scheme (3.1.29) is conditionally stable with condition $R \ge -1$. And of course, this stability with respect to the symbol can easily be translated back to stability in terms of the energy norm.

We next include a commonly used scheme for solving convection–diffusion equation. As we shall see, the difficulty increases quickly with the complexity of the scheme.

Example 3.1.3 Analyze the stability of the difference scheme

$$u_k^{n+1} = u_k^n - \frac{R}{2}\delta_0 u_k^n + r\delta^2 u_k^n. \tag{3.1.31}$$

Solution: Before we begin our analysis we note that difference scheme (3.1.31) is a scheme for approximating the solution to

$$v_t + av_x = \nu v_{xx}. \tag{3.1.32}$$

Difference scheme (3.1.31) is used often to solve partial differential equation (3.1.32). We shall see that the addition of the lower order term makes the analysis much more difficult. Also, *we see that if the convective term, av_x, dominates, this will not be a good scheme for approximating the solutions to partial differential equation (3.1.32).*

We begin by taking the discrete Fourier transform of equation (3.1.31) to get

$$\hat{u}^{n+1} = \hat{u}^n - \frac{R}{2}\left(e^{i\xi} - e^{-i\xi}\right)\hat{u}^n + r\left(e^{i\xi} - 2 + e^{-i\xi}\right)\hat{u}^n. \tag{3.1.33}$$

We can write equation (3.1.33) as $\hat{u}^{n+1} = \rho(\xi)\hat{u}^n$ where the symbol ρ is given by

$$\rho(\xi) = (1 - 2r) + 2r\cos\xi - iR\sin\xi. \tag{3.1.34}$$

Again, ρ is complex. We calculate

$$\begin{aligned}|\rho(\xi)|^2 &= (1 - 2r)^2 + 4r(1 - 2r)\cos\xi + 4r^2\cos^2\xi + R^2\sin^2\xi \\ &= (1 - 2r)^2 + R^2 + 4r(1 - 2r)\cos\xi + (4r^2 - R^2)\cos^2\xi. \end{aligned} \tag{3.1.35}$$

As we did in the last example, we differentiate $|\rho|^2$ with respect to ξ, set it equal to zero and see that we have potential maximums at $\xi = 0$, $\xi = \pm\pi$ and when

$$\cos\xi_0 = -\frac{2r(1 - 2r)}{4r^2 - R^2}. \tag{3.1.36}$$

Since $|\rho(0)| = 1$, we get no contribution to a stability limit from $\xi = 0$. Also, since $|\rho(\pm\pi)|^2 = (1 - 4r)^2$, as in Example 3.1.1, we must satisfy the condition $r \le 1/2$ for stability.

The last potential maximum is more difficult. For that reason, we treat it using several cases.

Case 1: $r^2 \geq \frac{R^2}{4}$

We note that in this case it could be said that the parabolic part of the equation is dominant. For this case, we can examine equation (3.1.35) and notice that (using the fact that we must also always satisfy $r \leq 1/2$)

$$|\rho|^2 \leq (1 - 2r)^2 + R^2 + 4r(1 - 2r) + 4r^2 - R^2 = 1.$$

Hence, for this case there will not be a maximum value greater than one at $\xi = \xi_0$ and the difference scheme is conditionally stable with condition $|R|/2 \leq r \leq 1/2$ which we refer to as stability condition S_1.

Case 2: $r^2 < \frac{R^2}{4}$

We see that in this case, the hyperbolic part of the partial differential equation dominates. To analyze this case, we subdivide our problem into two more cases.

Subcase 1: $-2r(1 - 2r)/(4r^2 - R^2) \geq 1$

For this case, the only solutions to equation (3.1.36) are $\xi_0 = 0$ (which we have already considered). Inequality $-2r(1 - 2r)/(4r^2 - R^2) \geq 1$, is equivalent to $r \geq R^2/2$. Thus for this case, difference scheme (3.1.31) is conditionally stable with condition

$$\frac{R^2}{2} \leq r \leq \frac{|R|}{2} \text{ and } r \leq \frac{1}{2}$$

which we refer to as stability condition S_2. We should note that when r is small (say ν is small), R must be made small (Δt must be made small) to satisfy this constraint.

Subcase 2: $-2r(1 - 2r)/(4r^2 - R^2) < 1$

This condition is equivalent to $r < R^2/2$. We must now investigate whether the solution to equation (3.1.36) will assume a maximum value greater than one. We substitute ξ_0 into equation (3.1.35) and see that

$$|\rho(\xi_0)|^2 = \frac{R^2}{4r^2 - R^2} \left\{ 4r - R^2 - 1 \right\}.$$

Setting

$$\frac{R^2}{4r^2 - R^2} \left\{ 4r - R^2 - 1 \right\} \leq 1,$$

we see that the restriction that $|\rho| \leq 1$ is equivalent to

$$0 \geq (R^2 - 2r)^2.$$

Thus, when $r < R^2/2$, the scheme will be unstable.

We see that for stability, either $|R|/2 \leq r \leq 1/2$ or $R^2/2 \leq r \leq 1/2$ will imply that $|R| \leq 1$. Since $|R| \leq 1$, $R^2/2 \leq |R|/2$. For a fixed R, from stability condition S_1 we have that the scheme is stable if $r \in [|R|/2, 1/2]$. From stability condition S_2 we have that the scheme is stable if $r \in [R^2/2, |R|/2]$. Hence, summarizing the results above, we see that *difference scheme (3.1.31) is stable if $R^2/2 \leq r \leq 1/2$*.

Remark 1: In the first case, when we require that $|R|/2 \leq r$, we are actually requiring that $\Delta x \leq 2\nu/|a|$. However, in the second case, requiring that $r < |R|/2$ is the same as requiring that $\Delta x > 2\nu/|a|$. Clearly, it is not possible to obtain stability satisfying this last inequality (because in the definition of stability, Definition 2.4.1, it was necessary that inequality (2.4.2) be satisfied for all Δx, $0 < \Delta x \leq \Delta x_0$. We obtain stability for difference scheme (3.1.31) here, because if we are originally in the second case, as $\Delta x \rightarrow 0$, we will always change into the first case and satisfy $|R|/2 \leq r$. However, this is not relevant when we are on a fixed grid and are computing.

Remark 2: We note that if ν is very small (as in the case in HW0.1) and a is not small, then the condition $r \geq R^2/2$ is very difficult to satisfy.

We next consider the stability analysis for a general implicit scheme.

Example 3.1.4 Analyze the stability of the difference scheme

$$-\alpha r u_{k-1}^{n+1} + (1 + 2\alpha r)u_k^{n+1} - \alpha r u_{k+1}^{n+1} = (1 - \alpha)r u_{k-1}^n + [1 - 2(1 - \alpha)r]u_k^n$$
$$+ (1 - \alpha)r u_{k+1}^n, \; k = 0, \pm 1, \cdots, \qquad (3.1.37)$$

where $\alpha \in [0, 1]$.

Solution: Before we begin analyzing the stability for this scheme, we should recognize that for $\alpha = 0$ the scheme is the standard FTCS explicit scheme, for $\alpha = 1$ the scheme is the BTCS implicit scheme and for $\alpha = \frac{1}{2}$ the scheme is the Crank-Nicolson scheme.

We begin by taking the discrete Fourier transform of equation (3.1.37) and get

$$-\alpha r e^{-i\xi}\hat{u}^{n+1} + (1 + 2\alpha r)\hat{u}^{n+1} - \alpha r e^{i\xi}\hat{u}^{n+1} = (1 - \alpha)r e^{-i\xi}\hat{u}^n$$
$$+ [1 - 2(1 - \alpha)r]\hat{u}^n + (1 - \alpha)r e^{i\xi}\hat{u}^n. \qquad (3.1.38)$$

Using the appropriate trigonometric identities allows us to write (3.1.38) as

$$\left(1 + 4\alpha r \sin^2 \frac{\xi}{2}\right)\hat{u}^{n+1} = \left[1 - 4(1 - \alpha)r \sin^2 \frac{\xi}{2}\right]\hat{u}^n,$$

or

$$\hat{u}^{n+1} = \rho(\xi)\hat{u}^n \qquad (3.1.39)$$

where

$$\rho(\xi) = \frac{1 - 4(1 - \alpha)r \sin^2 \frac{\xi}{2}}{1 + 4\alpha r \sin^2 \frac{\xi}{2}}. \qquad (3.1.40)$$

Of course, as with the previous examples, equation (3.1.39) can be applied $n + 1$ times to arrive at the equation

$$\hat{u}^{n+1} = [\rho(\xi)]^{n+1} \hat{u}^0. \qquad (3.1.41)$$

Thus as before we must obtain a bound for the coefficient of $\hat{u}^0$ in equation (3.1.41) and we shall do so by requiring that $|\rho(\xi)|$ be less than or equal to one.

If we now differentiate with respect to ξ and set the derivative equal to zero, we get

$$\rho' = \frac{-4r \sin \frac{\xi}{2} \cos \frac{\xi}{2}}{(1 + 4\alpha r \sin^2 \frac{\xi}{2})^2} = 0.$$

This equation implies that the critical points are at $\xi = 0, \pm\pi$. We note that $\rho(0) = 1$ and

$$\rho(\pm\pi) = \frac{1 - 4(1 - \alpha)r}{1 + 4\alpha r}.$$

It is easy to see that

$$\frac{1 - 4(1 - \alpha)r}{1 + 4\alpha r} \text{ is always } \leq 1.$$

Since

$$-1 \leq \frac{1 - 4(1 - \alpha)r}{1 + 4\alpha r} \qquad (3.1.42)$$

is the same as

$$4r(1 - 2\alpha) \leq 2,$$

it is easy to see that if $\alpha \geq 1/2$, inequality (3.1.42) is always satisfied and if $\alpha < 1/2$, inequality (3.1.42) is satisfied only if

$$r \leq \frac{1}{2(1 - 2\alpha)}. \qquad (3.1.43)$$

Hence we see that *if $\alpha \geq 1/2$, scheme (3.1.37) is unconditionally stable and if $\alpha < 1/2$, difference scheme (3.1.37) is conditionally stable, with the condition for stability given by (3.1.43).* As in the case of the explicit scheme in Example 3.1.1, when $\alpha < 1/2$, condition (3.1.43) is also necessary for stability.

We notice that the stability analyses that we have done have used $\beta = 0$. The reason for this is that we have not needed the extra growth in the norm of **u** that the exponential allows. To illustrate when it is necessary, we consider the following example.

Example 3.1.5 Consider the partial differential equation

$$v_t = v_{xx} + bv, \ t > 0, \ x \in \mathbb{R} \tag{3.1.44}$$

where $b > 0$. Analyze the convergence of the following difference scheme for the above partial differential equation.

$$u_k^{n+1} = ru_{k-1}^n + (1 - 2r + b\Delta t)u_k^n + ru_{k+1}^n, \ k = \pm 1, \cdots \tag{3.1.45}$$

Solution: Of course, our method for analyzing the convergence of scheme (3.1.45) is to consider the consistency and the stability of the scheme. It is fairly obvious that by using exactly the same expansions that we used for the explicit scheme (without the b) in equation (1.3.8), we see that scheme (3.1.45) is accurate of order $(2, 1)$ with the partial differential equation (3.1.44). Thus we are most interested in the stability of scheme (3.1.45).

As usual, we begin by taking the transform of difference scheme (3.1.45) and see that the symbol of the scheme is given by

$$\rho(\xi) = \left(1 - 4r \sin^2 \frac{\xi}{2}\right) + b\Delta t. \tag{3.1.46}$$

It should not surprise us that we might have to require that $r \leq 1/2$. It is easy to see that if we choose $r \leq 1/2$, we get

$$\left|1 - 4r \sin^2 \frac{\xi}{2} + b\Delta t\right| \leq 1 + b\Delta t \leq e^{b\Delta t}. \tag{3.1.47}$$

Hence we get

$$\| \hat{u}^{n+1} \|_2 \leq e^{b\Delta t} \| \hat{u}^n \|_2$$
$$\leq e^{b(n+1)\Delta t} \| \hat{u}^0 \|_2 \, .$$

Then

$$\| \mathbf{u}^{n+1} \|_{2,\Delta x} = \sqrt{\Delta x} \| \mathbf{u}^{n+1} \|_2$$
$$= \sqrt{\Delta x} \| \hat{u}^{n+1} \|_2$$
$$\leq \sqrt{\Delta x} e^{b(n+1)\Delta t} \| \hat{u}^0 \|_2$$
$$= \sqrt{\Delta x} e^{b(n+1)\Delta t} \| \mathbf{u}^0 \|_2$$
$$= e^{b(n+1)\Delta t} \| \mathbf{u}^0 \|_{2,\Delta x} \, .$$

Hence scheme (3.1.45) is stable (with $K = 1$ and $\beta = b$) with respect to the energy norm. Thus if we make the appropriate smoothness assumptions on the derivatives of the solution (so that our consistency is also with respect to the energy norm), then we have convergence of the scheme with respect to the energy norm.

The approach used in the stability proof in Example 3.1.5 is very useful. To make the results easier to use, we state the following two propositions.

Proposition 3.1.6 *The difference scheme*

$$\mathbf{u}^{n+1} = Q\mathbf{u}^n \tag{3.1.48}$$

is stable with respect to the $\ell_{2,\Delta x}$ norm if and only if there exist positive constants Δt_0 and Δx_0 and non–negative constants β and K so that

$$| \rho(\xi) |^{n+1} \leq K e^{\beta(n+1)\Delta t} \tag{3.1.49}$$

for $0 < \Delta t \leq \Delta t_0$, $0 < \Delta x \leq \Delta x_0$ and all $\xi \in [-\pi, \pi]$, and where ρ is the symbol of difference scheme (3.1.48).

Proof: The proof of Proposition 3.1.6 follows directly from Parseval's Identity, the definition of stability and the smoothness of ρ. In fact, Proposition 3.1.6 is really just a restatement of Proposition 3.1.4. To see this, we note that the ($\Leftarrow$) direction follows from multiplying inequality (3.1.49) by $\hat{u}^0(\xi)^2$, integrating from $-\pi$ to π and applying Proposition 3.1.4.

To prove the ($\Rightarrow$) direction, we assume false, i.e. for each K and β, there exists a $\xi_{K,\beta}$, such that

$$| \rho(\xi_{K,\beta}) |^{n+1} > K e^{\beta(n+1)\Delta t}. \tag{3.1.50}$$

Since ρ is assumed to be continuous, we can find an interval about $\xi_{K,\beta}$, $I_{K,\beta}$, so that inequality (3.1.50) is satisfied for all $\xi \in I_{K,\beta}$. If we then choose a function $\hat{u}^0$ which is zero outside of the interval $I_{K,\beta}$, multiplying the square of (3.1.50) by $\hat{u}^0(\xi)^2$ and integrating from $-\pi$ to π gives

$$\| \rho^{n+1}\hat{u}^0 \|^2 > \left(K e^{\beta(n+1)\Delta t} \right)^2 \| \hat{u}^0 \|^2 .$$

Since we can find a $\xi_{K,\beta}$ and $\hat{u}^0$ that does this for every K and β, this contradicts Proposition 3.1.4.

Proposition 3.1.7 *The difference scheme*

$$\mathbf{u}^{n+1} = Q\mathbf{u}^n \tag{3.1.51}$$

is stable with respect to the $\ell_{2,\Delta x}$ norm if and only if there exists positive constants Δt_0, Δx_0 and C so that

$$| \rho(\xi) | \leq 1 + C\Delta t \tag{3.1.52}$$

for $0 < \Delta t \leq \Delta t_0$, $0 < \Delta x \leq \Delta x_0$ and all $\xi \in [-\pi, \pi]$.

Proof: Note that when ρ satisfies inequality (3.1.52), it is said that ρ satisfies the **von Neumann condition**.

($\Leftarrow$) It is easy to see that since

$$| \rho(\xi) | \leq 1 + C\Delta t \leq e^{C\Delta t},$$

then

$$| \rho(\xi) |^{n+1} \leq e^{(n+1)C\Delta t}.$$

Hence, the difference scheme is stable by Proposition 3.1.6.

($\Rightarrow$) To prove the converse, we assume that the condition is not satisfied, i.e. we suppose that for every $C > 0$, there exists a $\xi_C \in [-\pi, \pi]$, such that

$$| \rho(\xi_C) | > 1 + C\Delta t.$$

Using a sequence $\{C_k\}$, such that $C_k \to \infty$ as $k \to \infty$, we get a sequence $\{\xi_k\}$ so that

$$| \rho(\xi_k) |^{n+1} \to \infty \text{ as } k \to \infty.$$

Thus $| \rho(\xi) |^{n+1}$ is surely not bounded. This contradicts Proposition 3.1.6.

Hence, we have shown that if the difference scheme does not satisfy the von Neumann condition, then it is not stable. Thus if it is stable, then it must satisfy the von Neumann condition, which is what we were to prove.

Remark 1: The reason that the term $e^{\beta(n+1)\Delta t}$ is needed in the definition of stability (instead of just requiring the constant K as in inequality (2.4.3)) and the C is needed in Proposition 3.1.7 is most often for proving stability of schemes involving a zero order derivative as in Example 3.1.5.

Remark 2: Here we include a much easier alternative stability analysis for difference scheme (3.1.31). We consider expression (3.1.34) and write $| \rho(\xi) |^2$ as

$$| \rho(\xi) |^2 = \left(1 - 4r \sin^2 \frac{\xi}{2}\right)^2 + R^2 \sin^2 \xi. \qquad (3.1.53)$$

If we require that r satisfy $r = r_0$ where r_0 is some constant such that $r_0 \le 1/2$, then (3.1.53) can be rewritten as

$$| \rho(\xi) |^2 = \left(1 - 4r_0 \sin^2 \frac{\xi}{2}\right)^2 + \frac{a^2 r_0}{\nu} \Delta t \sin^2 \xi. \qquad (3.1.54)$$

Since when $r_0 \le 1/2$, $(1 - 4r_0 \sin^2 \xi/2)^2 \le 1$ and $\sin^2 \xi \le 1$, equation (3.1.54) implies that

$$| \rho(\xi) |^2 \le 1 + C\Delta t \qquad (3.1.55)$$

where $C = a^2 r_0/\nu$. Hence, since the symbol of difference scheme (3.1.31) satisfies the von Neumann condition, the scheme is stable.

We must emphasize that the stability proved here is a very different stability than that proved in Example 3.1.3. Using the von Neumann condition with a nonzero C allows for exponential growth in the solution. The solution to this problem should not grow exponentially. Any growth that is seen is the scheme trying to go unstable. However, in this form, the scheme is special since the instability is less than or equal to an exponential. Even though no restriction on the relationship between r and R will allow for some bad growth, in the limit we will get convergence. We should also note that as Δt and Δx approach zero holding $r = r_0$, R will approach zero and eventually r and R will satisfy the condition $r^2 \geq R^2/4$ (in which case we get the bounded stability from the earlier calculation).

And, finally, we caution the reader. Using the stability described in this Remark to give confidence that the scheme is good can be very dangerous. If one then chooses a Δx and Δt and begins the computation, you may get bad results. The only promise is that as Δx and Δt approach zero, the computed solutions will approach the analytic solution. However, when it is convenient (when it is too difficult to prove stability via Definition 2.4.1–(2.4.3)), we will be willing to use the type of stability established in this Remark (and we will be careful with our calculations).

Stability for the class of schemes considered in Example 3.1.5 can be proved by the proposition given below. Hence, most often the β in the definition of stability and the C in Proposition 3.1.7 can be taken to be zero.

Proposition 3.1.8 *If the difference scheme*

$$\mathbf{u}^{n+1} = Q\mathbf{u}^n \tag{3.1.56}$$

is stable, then the difference scheme

$$\mathbf{u}^{n+1} = (Q + b\Delta t I)\mathbf{u}^n \tag{3.1.57}$$

is stable for any scalar b.

Proof: Obviously, the symbol for difference scheme (3.1.57) will be

$$\rho_1 = \rho + b\Delta t,$$

where ρ is the symbol for difference scheme (3.1.56). But if difference scheme (3.1.56) is stable, we know from Proposition 3.1.7 that ρ satisfies

$$|\rho| \leq 1 + C\Delta t$$

for some C. Hence, it is clear that ρ_1 satisfies

$$|\rho_1| \leq |\rho| + |b|\Delta t \leq 1 + (C + |b|)\Delta t.$$

So again using Proposition 3.1.7, we see that difference scheme (3.1.57) is also stable.

As we see above, the discrete Fourier transform is a very strong tool for proving stability in the $\ell_{2,\Delta x}$ norm. Another result we get from using the discrete Fourier transform involves consistency. Recall how difficult it was in Section 2.3.1 to prove consistency for implicit difference schemes. Below, in Proposition 3.1.9, we see that consistency of implicit difference equations with respect to the $\ell_{2,\Delta x}$ norm can be proved by calculating the order of $\mathbf{r}^n$ as we did in Example 2.3.2 and using the discrete Fourier transform to show that $\| Q_1^{-1} \|_{2,\Delta x}$ is uniformly bounded.

Proposition 3.1.9 *Consider an operator $Q_1 : \ell_{2,\Delta x} \to \ell_{2,\Delta x}$. Suppose that the symbol of Q_1, $\rho_1 = \rho_1(\xi)$, satisfies $\left| \frac{1}{\rho_1(\xi)} \right| \leq C$, $C > 0$, for all $\xi \in [-\pi, \pi]$, and all Δx and Δt. Then $\| Q_1^{-1} \|_{2,\Delta x} \leq C$.*

Proof: This proposition follows directly from Parseval's Identity, Proposition 3.1.3. We note that

$$
\begin{aligned}
\| Q_1 \|_{2,\Delta x} &= \sup_{\|u\|_{2,\Delta x}=1} \| Q_1 u \|_{2,\Delta x} \\
&= \sup_{\sqrt{\Delta x}\|u\|_2=1} \sqrt{\Delta x} \, \| Q_1 u \|_2 \\
&= \sup_{\|v\|_2=1} \| Q_1 v \|_2 \qquad \text{letting } u = v/\sqrt{\Delta x} \\
&= \sup_{\|\hat{u}\|_2=1} \| \rho_1(\xi)\hat{u} \|_2 \\
&\geq \frac{1}{C} \sup_{\|\hat{u}\|_2=1} \| \hat{u} \|_2 = \frac{1}{C}.
\end{aligned}
$$

Hence, since $\| Q_1 \|$ is bounded below by $1/C$, Q_1^{-1} exists and is bounded by C (ref. [7], page 101).

HW 3.1.1 Analyze the stability and convergence of the following difference schemes.
(a) FTBS for a hyperbolic equation.

$$u_k^{n+1} = Ru_{k-1}^n + (1 - R)u_k^n$$

where $R = a\Delta t/\Delta x$
(b) Higher order scheme presented in Section 1.5.3.

$$u_k^{n+1} = u_k^n + r\left[-\frac{1}{12}u_{k-2}^n + \frac{4}{3}u_{k-1}^n - \frac{5}{2}u_k^n + \frac{4}{3}u_{k+1}^n - \frac{1}{12}u_{k+2}^n \right]$$

where $r = \frac{\nu\Delta t}{\Delta x^2}$.

HW 3.1.2 Show that the following difference schemes for approximating the solution to

$$v_t + av_x = \nu v_{xx}$$

are unconditionally stable.
(a) $u_k^{n+1} + \frac{R}{2}\delta_0 u_k^{n+1} - r\delta^2 u_k^{n+1} = u_k^n$
(b) $u_k^{n+1} + \frac{R}{4}\delta_0 u_k^{n+1} - \frac{r}{2}\delta^2 u_k^{n+1} = u_k^n - \frac{R}{4}\delta_0 u_k^n + \frac{r}{2}\delta^2 u_k^n$

HW 3.1.3 Show that if $a < 0$ $(R < 0)$,

$$u_k^{n+1} + R\delta_+ u_k^{n+1} - r\delta^2 u_k^{n+1} = u_k^n$$

is stable.

3.1.2 Initial–Boundary–Value Problems

We now must discuss stability for initial–boundary–value problems. We shall discuss only problems that are bounded in each spatial variable. The first result that we obtain is a trivial result and probably the most used result. We recall that a difference scheme for an initial–boundary–value problem consists of a difference equation approximating the partial differential equation and difference equations approximating each boundary condition. If the difference scheme is unstable without considering the boundary conditions (i.e. considering the difference scheme as an initial–value scheme), then the scheme will also be unstable for the initial–boundary–value problem when the boundary condition equations are included. Hence, we obtain the following result.

Proposition 3.1.10 *Consider a difference scheme for an initial–boundary–value problem. The von Neumann condition for the difference scheme considered as a difference scheme for an initial–value problem is a necessary condition for stability.*

We illustrate the application of the above result with the following example.

Example 3.1.6 Consider the following initial–boundary–value problem:

$$v_t + av_x = 0, \ a < 0, \ x \in (0,1), \ t > 0 \tag{3.1.58}$$
$$v(1,t) = 0, \ t > 0 \tag{3.1.59}$$
$$v(x,0) = f(x), \ x \in [0,1]. \tag{3.1.60}$$

Find a necessary condition for stability (and, hence, convergence) of difference scheme

$$u_k^{n+1} = (1+R)u_k^n - Ru_{k+1}^n, \ k = 0\cdots, M-1 \tag{3.1.61}$$
$$u_M^{n+1} = 0 \tag{3.1.62}$$
$$u_k^0 = f(k\Delta x), \ k = 0,\cdots, M. \tag{3.1.63}$$

Solution: We note that by Proposition 3.1.10, if we consider difference equation (3.1.61) as a difference scheme for an initial–value problem, the stability of this scheme will be necessary for the stability of difference scheme (3.1.61)–(3.1.63). By Example 3.1.2, we know that difference scheme (3.1.61) is stable as an initial–value scheme if and only if $R = a\Delta t/\Delta x \geq -1$. Hence, $0 \geq R \geq -1$ is a necessary condition for the stability of difference scheme (3.1.61)–(3.1.63).

We must realize that the above result may not be very sharp. Clearly, the result does not take into account the boundary conditions. We must obtain stronger results. There are two main approaches to the topic; the matrix method and the GKSO method. We will discuss the matrix method at this time and delay the GKSO method until Chapter 8.

We began by illustrating the setting for initial–boundary–value problems. For example, consider the explicit scheme we used to solve HW1.2.1. We could write our solution scheme as the following matrix equation.

$$
\mathbf{u}^{n+1} =
\begin{pmatrix}
u_1^{n+1} \\
u_2^{n+1} \\
\vdots \\
u_{M-2}^{n+1} \\
u_{M-1}^{n+1}
\end{pmatrix}
= Q\mathbf{u}^n =
$$

$$
\begin{pmatrix}
1-2r & r & 0 & \cdots & & & & \\
r & 1-2r & r & 0 & \cdots & & & \\
 & & \cdots & & & & & \\
\cdots & 0 & r & 1-2r & r & 0 & \cdots & \\
 & & \cdots & & & & & \\
 & \cdots & & 0 & r & 1-2r & r & \\
 & & \cdots & & 0 & r & 1-2r &
\end{pmatrix}
\begin{pmatrix}
u_1^n \\
u_2^n \\
\vdots \\
u_{M-2}^n \\
u_{M-1}^n
\end{pmatrix}
$$

$$(3.1.64)$$

We solved the problem in HW1.2.1 by using $M = 10$. But, if we are going to discuss convergence, we need to consider the above equations for general M and Δt.

We recall that earlier when we discussed consistency, stability and convergence for difference schemes for initial–boundary–value problems, we used a sequence of finite dimensional norms. As was the case with initial–value problems, we will decide which norms we will use when we determine what techniques we can develop for proving stability (and hence, convergence). But for the discussion below, even though the subscript is not included on the norms, *we must realize that we are really working with a sequence of norms*.

As we did earlier, we consider a difference scheme for an initial–boundary–value problem of the form (3.1.64)

$$\mathbf{u}^{n+1} = Q\mathbf{u}^n. \tag{3.1.65}$$

We should note that this equation could be the matrix expression of an explicit scheme (just as was the case with (3.1.64)) or an implicit scheme in which case an equation of the form

$$Q_1\mathbf{u}^{n+1} = Q\mathbf{u}^n$$

has been transformed into an equation of the form

$$\mathbf{u}^{n+1} = Q_1^{-1}Q\mathbf{u}^n.$$

As was the case with initial–value problems, our approach is to use whatever tools are necessary to find a K and β that satisfy inequality (2.4.2). One approach is to use equation (3.1.65) in exactly the way we proved Proposition 2.4.2 to prove the following proposition.

Proposition 3.1.11 *Difference scheme (3.1.65) is stable if and only if there exist positive constants Δx_0 and Δt_0 and non-negative constants K and β so that*

$$\| Q^{n+1} \| \leq K e^{\beta(n+1)\Delta t}, \qquad (3.1.66)$$

for $0 < \Delta x \leq \Delta x_0$ and $0 < \Delta t \leq \Delta t_0$.

Thus our problem becomes one of computing the norm of Q^{n+1}. This is generally a difficult problem. If as in Section 2.2.3 we let $\sigma(Q)$ denote the spectral radius of the matrix Q ($\sigma(Q) = max\{|\lambda|: \lambda$ an eigenvalue of $Q\}$) and recall that $\sigma(Q) \leq \| Q \|_2$, we obtain the following result.

Proposition 3.1.12 $\sigma(Q) \leq 1 + C\Delta t$ *for some non-negative C is a necessary condition for stability of scheme (3.1.65) with respect to the $\ell_{2,\Delta x}$ norm.*

Thus we see that at least we can determine which schemes are bad (or taking a more positive view, which schemes are *possibly* good) by determining the eigenvalues of Q. In certain situations, we are able to do much better.

Proposition 3.1.13 *Suppose Q is symmetric. Then $\sigma(Q) \leq 1 + C\Delta t$ for some non-negative C is a necessary and sufficient condition for stability of difference scheme (3.1.65) with respect to the $\ell_{2,\Delta x}$ norm.*

Proof: The proof of Proposition 3.1.12 (and one direction of the proof of this proposition) follows immediately if we note that

$$\| Q^{n+1} \|_2 \geq \sigma(Q^{n+1}) = [\sigma(Q)]^{n+1}.$$

Then if $\sigma(Q) > 1 + C\Delta t$ for every C, Q will not be able to satisfy inequality (3.1.66).

The proof of Proposition 3.1.13 follows from the fact that for a symmetric matrix Q,

$$\| Q^{n+1} \|_2 = \sigma(Q^{n+1}) = [\sigma(Q)]^{n+1}.$$

Thus, inequality (3.1.66) can be satisfied with $K = 1$ and $\beta = C$.

Remark: The hypothesis in Proposition 3.1.13 that the matrix Q be symmetric can be relaxed to allow that Q be similar to a symmetric matrix. If Q is similar to $\tilde{Q}$ (there exists a matrix S such that $\tilde{Q} = S^{-1}QS$), then since $\det(Q - \lambda I) = \det(S^{-1}QS - \lambda I)$, Q and $\tilde{Q}$ have the same eigenvalues (same spectral radius). Then since $Q^n = \left(S\tilde{Q}S^{-1}\right)^n = S\tilde{Q}^n S^{-1}$,

$$\| Q^n \| \leq \| S \| \| \tilde{Q}^n \| \| S^{-1} \| = \| S \| \| S^{-1} \| [\sigma(Q)]^n.$$

Hence, we have proved the following result.

Proposition 3.1.14 *Suppose Q is similar to a symmetric matrix in such a way that $\| S \|$ and $\| S^{-1} \|$ are uniformly bounded where S is the similarity transformation. Then $\sigma(Q) \leq 1 + C\Delta t$ for some non-negative C is a necessary and sufficient condition for stability of difference scheme (3.1.65) with respect to the $\ell_{2,\Delta x}$ norm.*

Thus, we see above that we have an approach by which we can obtain either necessary or necessary and sufficient conditions for stability. We must realize, however, that it is not easy to determine the eigenvalues of a general matrix. We illustrate some of the results that we are able to obtain with the following examples.

Example 3.1.7 Analyze the stability of the difference scheme (3.1.64).

Solution: Using equation (2.2.41), we see that the eigenvalues for Q are

$$\lambda_j = 1 - 2r + 2r\cos\frac{j\pi}{M} = 1 - 4r\sin^2\frac{j\pi}{2M}.$$

Then the spectral radius of the matrix Q is equal to

$$\left| 1 - 4r\sin^2\frac{(M-1)\pi}{2M} \right|.$$

The requirement that the spectral radius be less than or equal to one implies that r must satisfy

$$r \leq \frac{1}{2\sin^2\frac{(M-1)\pi}{2M}}. \tag{3.1.67}$$

To satisfy condition (3.1.67) for all Δx, $0 < \Delta x \leq \Delta x_0$ for some Δx_0 (i.e. for all sufficiently large M), we must require that $r \leq 1/2$. Since Q is symmetric, by Proposition 3.1.13 the condition $r \leq 1/2$ is both a necessary and sufficient condition for stability of difference scheme (3.1.64).

Example 3.1.8 Analyze the stability of difference scheme (3.1.61)–(3.1.63).

Solution: If we write the difference scheme (3.1.61)–(3.1.63) in the form of equation (3.1.65), the $M \times M$ matrix Q is given by

$$Q = \begin{pmatrix} 1+R & -R & 0 & \cdots & \\ 0 & 1+R & -R & 0 & \cdots \\ & & \cdots & & \\ & \cdots & 0 & 1+R & -R \\ & & \cdots & 0 & 1+R \end{pmatrix}.$$

By using either formula (2.2.41) or the obvious observation that because Q is an upper triangular matrix all of its eigenvalues will be equal to $1 + R$, we find that the spectral radius of Q is given by $\sigma(Q) = |1 + R|$. By Proposition 3.1.12 the condition

$$-2 \leq R \leq 0 \tag{3.1.68}$$

is a necessary condition for the stability of difference scheme (3.1.61)–(3.1.63). Since Q is not symmetric, we cannot apply Proposition 3.1.13 to determine whether condition (3.1.68) is also sufficient But by Example 3.1.6 we have another necessary condition that is more restrictive than condition (3.1.68). Hence, condition (3.1.68) is clearly not sufficient.

Remark: It should be noted that in this example we get a poorer necessary condition from Proposition 3.1.12 than we get from Proposition 3.1.10.

We next include an example that analyzes two difference schemes dealing with Neumann boundary conditions.

Example 3.1.9 Discuss the stability and convergence of difference schemes (2.3.31–(2.3.33) and (2.3.44)–(2.3.46).

Solution: Before we begin, we recall that both difference schemes (2.3.31)–(2.3.33) and (2.3.44)–(2.3.46) (considered in Examples 2.3.4 and 2.3.5) use what looks like a first order approximation to the Neumann boundary condition at $x = 0$. However, the definition of the grid used in difference scheme (2.3.44)–(2.3.46) makes the approximation of the Neumann boundary condition a centered difference and, hence, second order. The results of Examples 2.3.4 and 2.3.5 were that difference scheme (2.3.31–(2.3.33) was not consistent with initial–boundary–value problem (2.3.34)–(2.3.37) (so the scheme can not be proved convergent by the Lax Theorem) and difference scheme (2.3.44)–(2.3.46) was consistent with initial–boundary–value problem (2.3.34–(2.3.37) (so it may be convergent).

We begin our discussion by writing difference scheme (2.3.44)–(2.3.46) in the form $\mathbf{u}^{n+1} = Q\mathbf{u}^n$. Using equations (2.3.44), (2.3.45) and (2.3.47) we see that difference scheme (2.3.44)–(2.3.46) can be written as

$$
\begin{pmatrix} u_1^{n+1} \\ u_2^{n+1} \\ \cdots \\ u_{M-1}^{n+1} \end{pmatrix} = \begin{pmatrix} 1-r & r & 0 & \cdots & \\ r & 1-2r & r & 0 & \cdots \\ & & \cdots & & \\ \cdots & 0 & r & 1-2r & r \\ & \cdots & 0 & r & 1-2r \end{pmatrix} \begin{pmatrix} u_1^n \\ u_2^n \\ \cdots \\ u_{M-1}^n \end{pmatrix}.
$$

$$(3.1.69)$$

We see that by comparing equation (3.1.69) with equation (2.3.42), the reason that we consider both of these schemes together in this example is that these difference schemes are identical. They have the same matrix Q.

Using the fact that $Q = I - rT_{N_1} D$ and equation (2.2.44), we see that the eigenvalues of the matrix Q are given by

$$
\begin{aligned}
\lambda_j &= 1 - 2r + 2\cos\frac{(2j-1)\pi}{2M-1} \\
&= 1 - 4r\sin^2\frac{(2j-1)\pi}{2(2M-1)}, \quad j = 1, \cdots, M-1.
\end{aligned}
$$

$$(3.1.70)$$

Then since $\sigma(Q) \le 1$ when $r \le 1/2$ and Q is symmetric, we see by Proposition 3.1.13 that the condition $r \le 1/2$ is a necessary and sufficient condition for the stability of difference scheme (2.3.44)–(2.3.46). (and since difference scheme (2.3.44)–(2.3.46) was consistent, $r \le 1/2$ is a necessary and sufficient condition for the convergence of difference scheme (2.3.44)–(2.3.46).

Remark 1: We note that difference scheme (2.3.44)–(2.3.46) is convergent (when $r \le 1/2$) while we were not able to prove that difference scheme (2.3.31)–(2.3.33) was convergent. In Section 2.3.2 where we considered the consistency of difference scheme (2.3.31)–(2.3.33), we noted that the experiment conducted in HW2.3.4 indicated that the scheme was convergent and wondered to what it converted. Since the numerical problem involved in difference scheme (2.3.31)–(2.3.33) is identical to that of difference scheme (2.3.44)–(2.3.46) (and this scheme converges), it is clear that the solution of difference scheme (2.3.31)–(2.3.33) does converge to the solution of initial–boundary–value problem (2.3.34)–(2.3.37), but on the grid given in Example 2.3.5. We must add, however, that difference scheme (2.3.31)–(2.3.33) *may* also converge to the solution of initial–boundary–value problem (2.3.34)–(2.3.37) on the grid considered in Example 2.3.4 (but we are unable to prove it).

Remark 2: In Section 2.2 we discussed that it made a difference in the consistency arguments about which point we did our expansions. Our choice then was to use either the time levels n, $n + 1$ or $n + 1/2$. The error in Example 2.3.4 is that we expanded our functions around the wrong spatial point. Obviously, instead of expanding our functions around the points in the most common grid, $x_k = k\Delta x$, $k = 0, \cdots, M$, in this case to see that the scheme was consistent, we would have to expand the functions around the points used in Example 2.3.5, $x_k = (k - 1)\Delta x + \Delta x/2$, $k = 0, \cdots, M$.

We conclude this section with the example that will use Proposition 3.1.14 to provide necessary and sufficient conditions for the stability and convergence of the problem involving the second order approximation of the Neumann boundary condition.

Example 3.1.10 Discuss the stability and convergence of difference scheme (2.3.19)–(2.3.21).

Solution: We see in Example 2.3.3 that difference scheme (2.3.19)–(2.3.21) can be written as $\mathbf{u}^{n+1} = Q\mathbf{u}^n$ where Q is given by

$$Q = \begin{pmatrix} 1 - 2r & 2r & & \cdots & \\ r & 1 - 2r & r & \cdots & \\ 0 & r & 1 - 2r & r & \cdots \\ & & \cdots & & \\ & & \cdots & 0 & r & 1 - 2r \end{pmatrix}.$$

Since Q can be written as $Q = I - rT_{N_2\,D}$ where $T_{N_2\,D}$ is given by (2.2.46), we see that the eigenvalues of Q are given by $\lambda_j = 1 - r\,(2 - 2\cos(2j + 1)\pi/2M) = 1 - 4r\sin^2(2j + 1)\pi/4M$, $j = 0, \cdots, M - 1$. We can then use Proposition 3.1.12 to get that $r \leq 1/2$ is a necessary condition for the stability of difference scheme (2.3.19)–(2.3.21).

However, noting that

$$S^{-1}QS = \begin{pmatrix} 1 - 2r & \sqrt{2}r & 0 & \cdots & \\ \sqrt{2}r & 1 - 2r & r & 0 & \cdots \\ & & \cdots & & \\ & & \cdots & 0 & r & 1 - 2r \end{pmatrix}$$

where S is the diagonal matrix with $\sqrt{2}, 1, \cdots, 1$ on the diagonal, we see that Q is similar to a symmetric matrix. Hence, by Proposition 3.1.14 we see that $r \leq 1/2$ *is a necessary and sufficient condition for the stability (and, hence convergence) of difference scheme (2.3.19)–(2.3.21).*

3.2   Finite Fourier Series and Stability

Often, instead of approaching stability for initial–boundary–value problems as we did in Section 3.1.2, the approach taken is to consider a discrete Fourier mode for the problem of the form

$$u_k^m = \xi^m e^{ijk\pi\Delta x} \tag{3.2.1}$$

where $0 \leq j \leq M$ and the superscript on the ξ term is a multiplicative exponent. If this general Fourier mode is inserted into the difference scheme (3.1.10) (the scheme associated with the matrix problem given by equation (3.1.64)), we get

$$
\begin{aligned}
u_k^{n+1} &= \xi^{n+1} e^{ijk\pi\Delta x} \\
&= r u_{k-1}^n + (1 - 2r) u_k^n + r u_{k+1}^n \\
&= r\xi^n e^{ij(k-1)\pi\Delta x} + (1 - 2r)\xi^n e^{ijk\pi\Delta x} + r\xi^n e^{ij(k+1)\pi\Delta x} \\
&= \xi^n e^{ijk\pi\Delta x} (r e^{-ij\pi\Delta x} + (1 - 2r) + r e^{ij\pi\Delta x}).
\end{aligned}
$$

Thus if we divide both sides of the above equation by $\xi^n e^{ijk\pi\Delta x}$, we get

$$
\begin{aligned}
\xi &= r e^{-ij\pi\Delta x} + (1 - 2r) + r e^{ij\pi\Delta x} \\
&= 1 - 2r(1 - \cos j\pi\Delta x) \\
&= 1 - 4r \sin^2 \frac{j\pi\Delta x}{2}.
\end{aligned}
$$

Then, the claim is that a necessary condition for stability is obtained by restricting r so that $|\xi| \leq 1$ (the ξ^n term will not grow without bound).

Remark 1: We should realize that the expression for ξ above is just a discrete version of the result obtained in Proposition 3.1.10, where we obtained a necessary condition for stability by considering the difference equation for the interior points of an initial–boundary–value problem as a scheme for an initial–value problem and applied the von Neumann criterion for stability (using the discrete Fourier transform). We shall refer to the process described above using the Fourier mode (3.2.1) as the **discrete von Neumann criterion for stability**.

Remark 2: We should also realize that the ξ found above is exactly the same as the eigenvalues of the matrix Q found in Example 3.1.7. Hence, obtaining a stability condition by requiring that $|\xi| \leq 1$ is the same as requiring that the eigenvalues of the matrix Q be less than or equal to one.

A common approach is then to make statements to the effect that discrete von Neumann stability analysis does not take into account the boundary conditions. By this, they mean that the scheme will be stable if the boundary conditions do not cause an instability. But, we must be very clear about the fact that the discrete von Neumann criterion gives us only a necessary condition.

The above method is clearly correct but is often used incorrectly. Also, the method often will give stronger results than just necessary conditions. The approach is really an obvious approach. When we applied the discrete Fourier transform to the initial–value problem, we argued that we did so because the analogous continuous Fourier transform was so useful for the analogous continuous initial–value problem. The common approach for solving continuous initial–boundary–value problems is to use a Fourier series. Thus, *it seems logical to try to use a finite Fourier series to help analyze the stability for initial–boundary–value problems.*

The basic approach using finite Fourier series is the same as its continuous counterpart. We consider functions f and g, periodic of period 2π. We allow f and g to be complex valued and define the following inner product which is analogous the integral inner product used with the continuous Fourier series.

$$(f, g) = \sum_{k=0}^{L} f(x_k)\overline{g(x_k)} \tag{3.2.2}$$

where

$$x_k = \frac{2\pi k}{L+1}, \quad k = 0, \cdots, L. \tag{3.2.3}$$

If we define

$$\phi_k = e^{ikx}, \tag{3.2.4}$$

then the sequence of ϕ's will satisfy the following.

Proposition 3.2.1

$$(\phi_j, \phi_k) = \begin{cases} L+1 & if \frac{j-k}{L+1} \ is \ an \ integer \\ 0 & otherwise \end{cases} \tag{3.2.5}$$

Proof: This can be seen to be the case by noting that

$$(\phi_j, \phi_k) = \sum_{m=0}^{L} e^{ijx_m} e^{-ikx_m}$$

$$= \sum_{m=0}^{L} e^{i(j-k)\frac{2\pi m}{L+1}}$$

$$= \sum_{m=0}^{L} \left(e^{i(j-k)\frac{2\pi}{L+1}} \right)^m$$

Thus we have a geometric sum with ratio

$$q = e^{i(j-k)\frac{2\pi}{L+1}}.$$

If $\frac{j-k}{L+1}$ is an integer, then $q = 1$ and the sum is $L+1$. Otherwise, the sum of the geometric series is

$$(\phi_j, \phi_k) = \frac{q^{L+1} - 1}{q - 1}$$

which is equal to zero since $q^{L+1} = e^{i(j-k)2\pi} = 1$.

Then, also just as is the case with the continuous Fourier series analog, we can represent a function in terms of the finite Fourier series as follows.

Theorem 3.2.2 *If the function f is defined on $\{x_0, \cdots, x_L\}$, then f can be written as*

$$f(x) = \sum_{j=-k_0}^{k_0+\theta} c_j e^{ijx} \tag{3.2.6}$$

where

$$c_j = \frac{(f, \phi_j)}{(\phi_j, \phi_j)} \tag{3.2.7}$$

and $\theta = 0$ and $k_0 = \frac{L}{2}$, if L is even, and $\theta = 1$ and $k_0 = \frac{L-1}{2}$ if L is odd.

Proof: Theorem 3.2.2 follows by using the same approach used so often in continuous Fourier series, that is, multiply both sides of equation (3.2.6) by e^{-imx}, $-k_0 \leq m \leq k_0 + \theta$ and then sum over x values from x_0 to x_L. We get

$$\sum_{p=0}^{L} f(x_p)e^{-imx_p} = \sum_{p=0}^{L} \sum_{j=-k_0}^{k_0+\theta} c_j e^{ijx_p} e^{-imx_p}$$

$$= \sum_{p=0}^{L} e^{-imx_p} \sum_{s=0}^{2k_0+\theta} c_{s-k_0} e^{i(s-k_0)x_p} \qquad (\text{letting } j = s - k_0)$$

$$= \sum_{s=0}^{L} c_{s-k_0}(\phi_{s-k_0}, \phi_m) \qquad (\text{since } 2k_0 + \theta = L)$$

$$= (L+1)c_m \qquad (\text{by } (3.2.5)).$$

So

$$c_m = \frac{1}{L+1} \sum_{p=0}^{L} f(x_p)e^{-imx_p} = \frac{(f, \phi_m)}{(\phi_m, \phi_m)}.$$

Also, we see that for $0 \leq m \leq L$

$$\sum_{j=-k_0}^{k_0+\theta} c_j e^{ijx_m} = \sum_{j=-k_0}^{k_0+\theta} \left(\frac{1}{L+1} \sum_{p=0}^{L} f(x_p)e^{-ijx_p} \right) e^{ijx_m}$$

$$= \frac{1}{L+1} \sum_{s=0}^{L} \sum_{p=0}^{L} f(x_p)e^{-i(s-k_0)x_p} e^{i(s-k_0)x_m} \quad (\text{let } j = s - k_0)$$

$$= \frac{1}{L+1} \sum_{p=0}^{L} f(x_p) \sum_{s=0}^{L} e^{isx_m} e^{-isx_p} \text{ (since } e^{ik_0 x_p} = e^{-ik_0 x_m} = 1)$$

$$= \frac{1}{L+1} \sum_{p=0}^{L} f(x_p) \sum_{s=0}^{L} e^{ix_s m} e^{-ix_s p}$$

$$= \frac{1}{L+1} \sum_{p=0}^{L} f(x_p)(\phi_m, \phi_p)$$

$$= \frac{1}{L+1} f(x_m)(\phi_m, \phi_m) = f(x_m).$$

Hopefully, by this time it is obvious that almost everything that is needed in finite Fourier series follows analogously to corresponding topics for continuous Fourier series. We should realize also, that though we did not develop them here, *it is also possible to get finite sine expansions for odd functions and finite cosine expansions for even functions.*

In the example below, we again return to the difference scheme (3.1.64) to illustrate how finite Fourier series can be used to solve difference equations for an initial–boundary–value problem.

Example 3.2.1 Use finite Fourier series to solve the following difference problem.

$$u_k^{n+1} = r u_{k-1}^n + (1 - 2r) u_k^n + r u_{k+1}^n, \cdot k = 1, \cdots, M - 1 \qquad (3.2.8)$$

$$u_0^{n+1} = 0 \qquad (3.2.9)$$

$$u_M^{n+1} = 0 \qquad (3.2.10)$$

$$u_k^0 = f(k \Delta x), \ k = 0, \cdots, M \qquad (3.2.11)$$

Solution: We should recall that the grid points for this problem are $k \Delta x = k/M, k = 0, \cdots, M$ so our problem will require some scaling to use the finite Fourier series presented above. Also, we should realize that, as in the continuous analog, we can either consider $[0, 1]$ as our interval for the finite Fourier series expansion and get a complete sine-cosine Fourier series representation of the solution, or use the fact that we have zero Dirichlet boundary conditions, extend the initial condition oddly to the interval $[-1, 1]$ and then proceed. Since the latter approach is the most common for the continuous problem (and we get a nicer result), this is the approach that we will use here. Thus, we begin by extending the initial condition oddly about $x = 0$ and then periodically to the entire real line. We should note that from $x = 0$ to $x = 2$ (which is the period of our functions), we will have $2M + 1$ (including both $x = -1.0$ and $x = 1.0$) points so the L used in the finite Fourier series above will be equal to $2M - 1$ (and hence, $k_0 = M - 1$ and $\theta = 1$). Hence, by Theorem 3.2.2 we have a solution of the form

$$u^n(x) = \sum_{j=-(M-1)}^{M} c_j^n e^{ij\pi x} \qquad (3.2.12)$$

where the superscript on the c term reminds us that c must depend on n (or t) and the π is the necessary scaling. If we then require that the function given in (3.2.12) satisfies equation (3.2.8) at each grid point, we get

$$u_k^{n+1} = \sum_{j=-(M-1)}^{M} c_j^{n+1} e^{ij\pi x_k} \tag{3.2.13}$$

$$= ru_{k-1}^n + (1 - 2r)u_k^n + ru_{k+1}^n \tag{3.2.14}$$

$$= r \sum_{j=-(M-1)}^{M} c_j^n e^{ij\pi x_{k-1}} + (1 - 2r) \sum_{j=-(M-1)}^{M} c_j^n e^{ij\pi x_k}$$

$$+ r \sum_{j=-(M-1)}^{M} c_j^n e^{ij\pi x_{k+1}} \tag{3.2.15}$$

$$= \sum_{j=-(M-1)}^{M} c_j^n (1 - 4r \sin^2 \frac{j\pi\Delta x}{2}) e^{ij\pi x_k}. \tag{3.2.16}$$

In the above calculation we consider lines (3.2.13) and (3.2.16) and note that they both are finite Fourier series expansions that are equal. The only way that this can occur is that the Fourier coefficients be equal. (This can also be seen by multiplying both sides by $e^{-ij\pi x_k}$ and summing over k from 0 to L.) Thus

$$c_j^{n+1} = c_j^n \left(1 - 4r \sin^2 \frac{j\pi\Delta x}{2} \right). \tag{3.2.17}$$

Equation (3.2.17) represents a functional equation for c_j^n as a function of n. We can easily find a solution to equation (3.2.17) by letting

$$\xi_j = \left(1 - 4r \sin^2 \frac{j\pi\Delta x}{2} \right) \tag{3.2.18}$$

and noting that

$$c_j^1 = c_j^0 \xi_j,$$
$$c_j^2 = c_j^1 \xi_j = c_j^0 \xi_j^2,$$
$$\cdots$$
$$c_j^{n+1} = c_j^0 \xi_j^{n+1}.$$

If we substitute this last expression into equation (3.2.12), we see that we have a solution of the form

$$u^{n+1}(x) = \sum_{j=-(M-1)}^{M} c_j^0 \left(1 - 4r \sin^2 \frac{j\pi\Delta x}{2} \right)^{n+1} e^{ij\pi x}. \tag{3.2.19}$$

The final step of the solution process is to determine c_j^0, $j = -(M - 1), \cdots, M$, so that the expansion (3.2.19) satisfies $u^0(k\Delta x) = u_k^0$, $k = -M, \cdots, M$, where u_k^0 is the given initial function. If we let $n = -1$ in equation (3.2.19) and evaluate the equation at the points $x_k = k\Delta x$, $k = -(M - 1), \cdots, M$, we get

$$u^0(k\Delta x) = u_k^0 = \sum_{j=-(M-1)}^{M} c_j^0 e^{ij\pi k\Delta x}. \tag{3.2.20}$$

Thus, we see that the constants c_k^0, $k = -(M - 1), \cdots, M$ are just the finite Fourier coefficients of the given function u_k^0, $k = 0, \cdots, M$. If we multiply both sides of equation (3.2.20) by $e^{-im\pi k\Delta x}$ and sum from $k = -(M - 1)$ to M, we get

$$c_m^0 = \frac{1}{2M} \sum_{k=-(M-1)}^{M} u_k^0 e^{-im\pi k\Delta x}.$$

Remark: We should realize that if we do not take advantage of the fact that we have a zero Dirichlet boundary condition in this problem (and, hence, do not extend the function oddly about $x = 0$), we can still use the finite Fourier series to obtain a solution. See HW3.2.1. In that case, the solution is expanded in terms of basis functions of the form $e^{2\pi ijx}$. The reason that we are able to logically expand the solution in terms of the basis functions $e^{i\pi jx}$ in Example 3.2.1 is that we are using the fact that the functions satisfy the zero Dirichlet boundary condition. Obviously, the solution found in HW3.2.1 does not contain the $e^{i\pi x}$ mode. But, based on Example 3.2.1, the solution could contain the $e^{i\pi x}$ mode. (In fact, the initial condition in Example 3.2.1 could be fixed so that the solution contained only the $e^{i\pi x}$ mode.) Theorem 3.2.2 shows that though an expansion in terms of basis functions $e^{2\pi ijx}$ does not explicitly contain the $e^{i\pi x}$ mode, the basis can resolve functions containing that mode.

HW 3.2.1 Solve problem (3.2.8)–(3.2.11) by expanding u^0 periodically in $\mathbb{R}$ (*without* extending u^0 oddly about $x = 0$). Compare your solution to that obtained in Example 3.2.1.

We see in equation (3.2.19) that *the discrete von Neumann criterion is just that condition that restricts the finite Fourier coefficients for any mode from growing without bound.*

At this time the relationship between the discrete von Neumann criterion and the finite Fourier series solution seems clear. However, it is not clear why both of the expressions involved in the discrete von Neumann criterion and the coefficient of the finite Fourier series turn out to be the eigenvalues of Q. We next show the relationship between the finite Fourier series solution to the problem and the eigenvalues and eigenvectors of the matrix Q. We show that it is quite natural to have the eigenvalues and eigenvectors involved in the finite Fourier series solution to the problem.

To better understand what is happening (or make it more mysterious), we take the form of solution (3.2.19) two steps further. First, since u^{n+1} must be real, we see that

$$c^0_{-j} = \overline{c^0_j}.$$

In addition, since we extended our functions oddly about $x = 0$, we see that by setting $u^0_k = -u^0_{-k}$ and using the fact that $u^0_0 = u^0_M = 0$, we get

$$c^0_0 = c^0_M = 0 \text{ and } c^0_j = -c^0_{-j},\ j = 1, \cdots, M - 1.$$

Thus just as in the continuous case, we see that c^0_j must be a pure imaginary number and we can combine terms in equation (3.2.19) and write u^{n+1} as

$$u^{n+1}(x) = -2 \sum_{j=1}^{M-1} \tilde{c}^0_j (1 - 4r \sin^2 \frac{j\pi \Delta x}{2})^{n+1} \sin j\pi x \qquad (3.2.21)$$

where $\tilde{c}_j^0 = -ic_j^0$. If we set

$$
\mathbf{u}_j = \begin{pmatrix} \sin j\pi\Delta x \\ \sin j\pi 2\Delta x \\ \cdots \\ \sin j\pi(M-1)\Delta x \end{pmatrix}, \tag{3.2.22}
$$

we see that we can write a vector form of the solution (3.2.21) evaluated at x_m, $m = 1, \cdots, M-1$ as

$$
\mathbf{u}^{n+1} = -2 \sum_{j=1}^{M-1} \tilde{c}_j^0 \left(1 - 4r\sin^2\frac{j\pi\Delta x}{2}\right)^{n+1} \mathbf{u}_j. \tag{3.2.23}
$$

We now have an alternative form for the solution to the problem. If we rewrite difference scheme (3.2.8)–(3.2.11) in the form $\mathbf{u}^{n+1} = Q\mathbf{u}^n$ (where Q will be given as in equation (3.1.64)), we see that equation (3.2.23), which is completely equivalent to solution (3.2.19), is a solution given in terms of the eigenvalues and eigenvectors of the matrix Q. (The vectors $\mathbf{u}_j$, $j = 1 \cdots, M-1$ can be seen to be the eigenvectors of Q by using formula (2.2.42).)

The relationship above, between the finite Fourier series solution and the eigenvalues and eigenvectors of the matrix Q, can be explained easily by considering a more direct solution approach to problem (3.2.8)–(3.2.11). As we would do with the analogous analytic problem, we will use separation of variables to solve problem (3.2.8)–(3.2.11).

Example 3.2.2 Using discrete separation of variables, solve initial–boundary–value problem (3.2.8)–(3.2.11).

Solution: We assume that u_k^n can be written as $u_k^n = X_k T^n$, insert $u_k^n = X_k T^n$ into equation (3.2.8) and separate the variables to get

$$
\frac{T^{n+1}}{T^n} = \frac{rX_{k-1} + (1-2r)X_k + rX_{k+1}}{X_k}. \tag{3.2.24}
$$

Then using the usual separation of variables reasoning that since the left hand side of equation (3.2.24) is a function of n alone and the right hand side of equation (3.2.24) is a function of k alone, then both sides must be equal to a constant, say λ. Thus the equations generated by (3.2.24) are

$$
rX_{k-1} + (1-2r)X_k + rX_{k+1} = \lambda X_k, \quad k = 1, \cdots, M-1 \tag{3.2.25}
$$

and

$$
T^{n+1} = \lambda T^n. \tag{3.2.26}
$$

Inserting $u_k^n = X_k T^n$ into the boundary conditions of the problem give boundary conditions

$$
X_0 = X_M = 0. \tag{3.2.27}
$$

The separation of variables approach is to solve equations (3.2.25), (3.2.27) for X_k and λ (which is equivalent to finding the eigenvalues and eigenvectors of the matrix Q: $\lambda_j = 1 - 4r\sin^2\frac{j\pi\Delta x}{2}$ and

$$
\mathbf{u}_j = [\sin j\pi\Delta x, \cdots, \sin jk\pi\Delta x, \cdots, \sin j\pi(M-1)\Delta x]^T, \quad j = 1, \cdots, M-1),
$$

use equation (3.2.26) to determine T^n (which will be

$$T^n = \lambda_j^n T^0 = (1 - 4r \sin^2 \frac{j\pi\Delta x}{2})^n T^0),$$

set

$$u_k^n = \sum_{j=1}^{M-1} d_j T^n X_k = \sum_{j=1}^{M-1} d_j T^0 \lambda_j^n \sin \frac{kj\pi}{M},$$

and then use the initial condition to determine the $d_j T^0$'s. (See ref. [1], page 379.) Using this approach on this problem, we will get solution (3.2.23). Moreover, we now see that it is not just a coincidence that our finite Fourier solution contains both the eigenvalues and eigenvectors of Q. And, more generally, if we try to **analytically solve** a general equation of the form $\mathbf{u}^{n+1} = Q\mathbf{u}^n$ by separation of variables, the scheme will depend on the eigenvalues and eigenvectors of the matrix Q.

So far in this section we have discovered the following.

(i) The discrete von Neumann criterion for stability is to require that

$$| \lambda_j |=| 1 - 4r \sin^2 \frac{j\pi\Delta x}{2} | \le 1, \ j = 1, \cdots, M-1$$

for all Δx that satisfies $0 < \Delta x \le \Delta x_0$ for some Δx_0.

(ii) Item (i) is the same as requiring that the coefficients of our finite Fourier solution be bounded by one.

(iii) Likewise, Item (i) is the same as requiring that the eigenvalues of the matrix Q be bounded by one.

Clearly, based on what we did in Example 3.1.7, since the matrix Q is symmetric we can use Item (iii) above and Proposition 3.1.13 to see that the discrete von Neumann criterion actually gives a necessary and sufficient condition for stability. However, this can also be readily seen using the fact that expression (3.2.19) is an exact solution to the initial–boundary–value problem (3.2.8)–(3.2.11) and Item (ii) above. (See HW3.2.3.)

Hence, we see that the discrete von Neumann analysis gives us a necessary and sufficient condition for stability of a difference scheme for an initial–boundary–value problem. As we claimed earlier, it is often stated that the discrete von Neumann analysis does not include instabilities due to boundary conditions. Clearly, as we saw above, a discrete von Neumann analysis can include the boundary conditions. *We next try to explain the relationship between a discrete von Neumann analysis and instabilities due to boundary conditions: why the discrete von Neumann criterion sometimes gives us necessary and sufficient conditions for stability and often gives us only necessary conditions for stability.* We consider several more examples.

Example 3.2.3 Perform a discrete von Neumann stability analysis of the following difference scheme.

$$u_k^{n+1} = ru_{k-1}^n + (1 - 2r)u_k^n + ru_{k+1}^n, \ k = 1, \cdots, M-1 \qquad (3.2.28)$$

$$u_M^n = 0 \qquad (3.2.29)$$

$$u_0^n = u_1^n \qquad (3.2.30)$$

Solution: Substituting the discrete Fourier mode

$$u_k^n = \xi^n e^{ijk\pi\Delta x}$$

into equation (3.2.28), we get

$$\xi = 1 - 4r \sin^2 \frac{j\pi\Delta x}{2}. \tag{3.2.31}$$

Then the condition that $|\xi| \leq 1$ requires that r satisfy $r \leq 1/2$.

Remark 1: We begin by noting that the above stability condition is the same as that found in Example 3.1.9. In Example 3.1.9 we found that $r \leq 1/2$ is necessary and sufficient for the stability of difference scheme (2.3.31)–(2.3.33) (which is the same as difference scheme (3.2.28)–(3.2.30)). Hence, the discrete von Neumann analysis for this problem gave a condition that is both necessary and sufficient for the stability of the scheme. In our previous example we found that there were reasons that the discrete von Neumann analysis gave us a necessary and sufficient condition for stability (the fact that ξ was an expression for the eigenvalues of Q or by direct proof as in HW3.2.3). Are there any good reasons that the discrete von Neumann analysis performed in Example 3.2.3 gives a condition that is both necessary and sufficient for stability? Or is it a coincidence that the stability condition found in Example 3.2.3 is the same as that found in Example 3.1.9? *The answer is that it is just a coincidence and a discrete von Neumann stability analysis of difference scheme (3.2.28)–(3.2.30) should yield only a necessary condition for stability.*

Remark 2: Unlike the difference scheme analyzed in the beginning of this section, the discrete von Neumann analysis performed in Example 3.2.3 does not find the eigenvalues of the matrix Q. In Example 3.1.9 the eigenvalues of Q were found to be

$$\lambda_j = 1 - 4r \sin^2 \frac{(2j-1)\pi}{2(2M-1)} = 1 - 4r \sin^2 \frac{(2j-1)\pi\Delta x}{4}, \ j = 1, \cdots, M-1$$

whereas in Example 3.2.3,

$$\xi = 1 - 4r \sin^2 \frac{j\pi\Delta x}{2}.$$

Remark 3: The finite Fourier mode (3.2.1) satisfies both boundary conditions in problem (3.2.8)–(3.2.11). It does not satisfy boundary condition (3.2.30) in problem (3.2.28)–(3.2.30). Hence, (3.2.1) is not a Fourier mode of problem (3.2.28)–(3.2.30).

Example 3.2.4 Solve problem (3.2.28)–(3.2.30).

Solution: We begin by looking for a separated solution of the form

$$u_k^n = X_k T^n. \tag{3.2.32}$$

Substituting (3.2.32) into equation (3.2.28) yields

$$\frac{T^{n+1}}{T^n} = \frac{rX_{k-1} + (1 - 2r)X_k + rX_{k+1}}{X_k}. \tag{3.2.33}$$

Again using the usual separation of variables argument that the left hand side of the equation is a function of only n and the right hand side of the equation is a function of only k, we are left the the following pair of equations.

$$T^{n+1} = \lambda T^n \tag{3.2.34}$$

$$rX_{k-1} + (1 - 2r)X_k + rX_{k+1} = \lambda X_k \tag{3.2.35}$$

Inserting expression (3.2.32) into the boundary conditions (3.2.29) and (3.2.30) gives

$$X_0 - X_1 = X_M = 0. \tag{3.2.36}$$

As usual, the solution process is to solve equations (3.2.35) and (3.2.36) for X_k and λ, and then solve equation (3.2.34). If we replace X_0 in equation (3.2.35), $k = 1$, by X_1, we see that solving (3.2.35), (3.2.36) is equivalent to solving $Q\mathbf{X} = \lambda \mathbf{X}$ where Q is given by

$$Q = \begin{pmatrix} 1 - r & r & 0 & \cdots & \\ r & 1 - 2r & r & 0 & \cdots \\ & & \cdots & & \\ \cdots & 0 & r & 1 - 2r & r \\ & \cdots & 0 & r & 1 - 2r \end{pmatrix}. \tag{3.2.37}$$

Since $Q = I - rT_{N_1 D}$ (from (2.2.43)), the solution to equations (3.2.35), (3.2.36) is

$$\lambda_j = 1 - 4r \sin^2 \frac{(2j - 1)\pi}{2(2M - 1)}, \; j = 1, \cdots, M - 1 \tag{3.2.38}$$

$$X_k = \cos \frac{(2j - 1)(2k - 1)\pi}{2(2M - 1)}, \; j = 1, \cdots, M - 1, \; k = 1, \cdots, M - 1 \tag{3.2.39}$$

$$X_0 = X_1. \tag{3.2.40}$$

The solution to equation (3.2.34) is

$$T^n = \lambda_j^n T^0, \; n \geq 0. \tag{3.2.41}$$

And, finally, the solution to problem (3.2.28)–(3.2.30) is given by

$$u_k^n = \sum_{j=1}^{M-1} d_j T^0 \lambda_j^n \cos \frac{(2j - 1)(2k - 1)\pi}{2(2M - 1)}, \; k = 1, \cdots, M - 1 \tag{3.2.42}$$

where the $T^0 d_j$ terms are determined by the initial condition.

Remark: We saw in Example 3.2.3 that the discrete von Neumann analysis did not find the eigenvalue of the matrix Q. In the last example, we see that generalized finite Fourier series solution of problem (3.2.28)–(3.2.30) is not given in terms of basis functions of the form of (3.2.1). *The reason that the discrete von Neumann stability analysis done in Example 3.2.3 did not prove that $r \leq 1/2$ was also sufficient for stability was that the analysis was performed with the wrong basis functions, $e^{ijk\pi\Delta x}$, instead of $e^{i(2j-1)(2k-1)\pi/2(2M-1)}$.* We note that if we had done a "generalized" discrete von Neumann stability analysis using appropriate basis functions instead of the traditional discrete von Neumann stability analysis that was done in Example 3.2.3, we would get necessary and sufficient conditions for stability (see HW3.2.5).

To make it perfectly clear that the discrete von Neumann analysis will not always (for good reasons or bad) give us a condition that is both necessary and sufficient for stability, we include an example where the discrete von Neumann analysis gives a necessary condition for stability that is not sufficient. We will see another example in HW5.6.10 where the boundary condition (in that case it will be a numerical boundary condition) will completely determine whether the scheme is stable or unstable.

Example 3.2.5 Perform a discrete von Neumann stability analysis for the difference scheme

$$u_k^{n+1} = ru_{k-1}^n + (1 - 2r)u_k^n + ru_{k+1}^n, \ k = 0, \cdots, M - 1 \qquad (3.2.43)$$

$$\frac{u_1^n - u_{-1}^n}{2\Delta x} = 10u_0^n \qquad (3.2.44)$$

$$u_M^n = 0. \qquad (3.2.45)$$

Solution: If we insert the discrete Fourier mode (3.2.1) into equation (3.2.43) and solve for ξ, we will get $\xi = 1 - 4r\sin^2 j\pi\Delta x/2$ and find that to bound ξ by one, we will have to require that $r \leq 1/2$. Hence, the condition given by discrete von Neumann stability criterion is $r \leq 1/2$. In Example 3.3.1 we will find that $r \leq 1/(2 + 10\Delta x)$ will be a necessary condition for stability for difference scheme (3.2.43)–(3.2.45). Hence, clearly the condition $r \leq 1/2$ cannot be sufficient.

Hopefully by now we have enough examples to describe what is happening. We next try to explain what the relationship is between stability and the discrete von Neumann stability criterion and why it is this way. Consider the following problem.

$$v_t = v_{xx}, \ t > 0, \ x \in (0, 1) \qquad (3.2.46)$$

$$-10v(0, t) + v_x(0, t) = v(1, t) = 0, \ t \geq 0 \qquad (3.2.47)$$

$$v(x, 0) = f(x), \ x \in [0, 1] \qquad (3.2.48)$$

To solve problem (3.2.46)–(3.2.48) analytically, it is necessary to use a generalized Fourier expansion instead of a Fourier series. If we tried to force a solution by Fourier series, the resulting "solution" would not satisfy the boundary condition $-10v(0, t) + v_x(0, t) = 0$. The reason for this is that Fourier series can only represent periodic functions. If the periodic extension of a function is not continuous, the Fourier series will converge to the average of the left and right hand limits of the function at the points of discontinuity. This is the phenomena that happens when a problem that has the *wrong boundary conditions* is solved using Fourier series.

The exact same thing happens when you solve a difference equation by a finite Fourier series. If we write a finite difference scheme for problem

(3.2.46)–(3.2.48),

$$u_k^{n+1} = u_k^n + r\delta^2 u_k^n, \ k = 0, \cdots, M - 1 \quad (3.2.49)$$

$$-10u_0^n + \frac{u_1^n - u_{-1}^n}{2\Delta x} = 0 \quad (3.2.50)$$

$$u_M^n = 0 \quad (3.2.51)$$

$$u_k^n = f(k\Delta x), \quad (3.2.52)$$

and tried to "force" a finite Fourier series solution of problem (3.2.49)–(3.2.52), the resulting solution would not generally satisfy the boundary condition (3.2.50). Hence, *any stability analysis that is done based on using the Fourier mode (3.2.1) will not provide any stability information related to boundary condition (3.2.50).*

However, if instead of the usual sines and cosines we used a generalized finite Fourier series (the form of which could be found using discrete separation of variables on problem (3.2.49)–(3.2.52)), the resulting solution would satisfy problem (3.2.49)–(3.2.52). Then, if a discrete von Neumann stability analysis were performed with these generalized basis functions, the stability results obtained would be both necessary and sufficient. This is what we saw happen in Example 3.2.4 and HW3.2.5 for difference scheme (3.2.28)–(3.2.30).

Thus the obvious result is that *if the discrete initial–boundary–value problem can be solved using a finite Fourier series, then the discrete von Neumann criterion is both a necessary and sufficient condition for stability.*

The obvious approach for other problems, say (3.2.49)–(3.2.52), is to perform a generalized discrete von Neumann analysis on them based on a generalized Fourier basis rather than the exponential (as is done in HW3.2.5). That approach is nice in principle but will not generally work. The reason is that in order to find the generalized Fourier basis functions to use in the stability analysis, we would have to solve the eigenvalue problem involved in the separation of variables solution scheme. But, if we could find these eigenvalues, we would prove stability using either Proposition 3.1.12 or 3.1.13. The eigenvalue problems for these more difficult schemes are usually such that we can write neither the eigenvalues nor the eigenvectors in closed form (so we could not determine which generalized Fourier basis functions to use).

The next question is for what class of initial–boundary–value problems will the discrete von Neumann criterion provide sufficient conditions for stability. A logical guess would be for those problems with Dirichlet and Neumann boundary conditions. This appears to be false by the results given in Example 3.2.3. Suppose we have the heat equation and a zero Neumann boundary condition at $x = 0$. The approach used to solve this problem analytically is to extend the problem evenly about $x = 0$ and solve the problem on a bigger domain (for analytic problems, usually on all of $\mathbb{R}$). The Neumann boundary condition is then satisfied because of the even symmetry about $x = 0$ given at the beginning of the problem and preserved by the partial differential equation.

For difference operators we must be more careful. If again we have the heat equation and a zero Neumann boundary condition at $x = 0$, and we use the first order approximation of the derivative in this boundary condition, we saw that the discrete von Neumann analysis does not show that the von Neumann criterion is sufficient for stability of the difference operator plus boundary conditions (and the problem is not solvable by standard finite Fourier series). This is (as we shall see in HW3.2.2(c)) because the even extension of the first order Neumann boundary condition is not consistent with the usual difference operator used to approximate the heat equation. More specifically, if we had a zero Dirichlet condition at $x = 1$ and proceeded to extend our initial condition evenly about $x = 0$, oddly about $x = 1$ and then periodically to the whole real line (and used the usual explicit difference approximation to the heat equation), the finite Fourier series solution would satisfy the difference equation, the Dirichlet boundary condition and the *second order approximation of the Neumann boundary condition*. To solve the system of difference equations using the first order Neumann boundary condition, we must be very careful how we handle the boundary condition. In Example 3.2.4 we solved this problem using discrete separation of variables, and in Example 3.1.9 we used Proposition 3.1.13 to prove that the condition $r \leq 1/2$ is both necessary and sufficient for stability. However, we can only get the necessary condition by using the standard discrete von Neumann stability analysis.

The theorem that describes the situation is as follows.

Theorem 3.2.3 *A scheme for solving an initial–value problem with a periodic initial condition is stable with respect to the finite dimensional $\ell_{2,\Delta x}$ norm on one period if and only if it satisfies the discrete von Neumann stability criterion.*

Remark 1: We must be careful about how we interpret the statement of the above theorem. Earlier we tried to make it very clear that we did not get a sufficient condition for stability of difference scheme (3.2.28)–(3.2.30) by the discrete von Neumann analysis. However, it is not difficult to see that if we consider difference scheme (3.2.28)–(3.2.30) on the grid $\{x_k : x_k = (k-1)\Delta x + \Delta x/2, \ k = 0, \cdots, M\}$ and extend our initial condition evenly about $x = 0$ (and, of course, oddly about $x = 1$ and periodically to $\mathbb{R}$), then solving equation (3.2.8), $-\infty < k < \infty$, is equivalent to solving (3.2.8)–(3.2.11). The reason for this apparent contradiction is that *the statement of Theorem 3.2.3 is grid dependent*. Theorem 3.2.3 assumes that the scheme is defined on a grid of the form $\{x_k : x_k = k\Delta x, \ -\infty < k < \infty\}$. Initial-value problems with periodic initial conditions defined on other grids will still be stable if and only if they satisfy a discrete von Neumann criterion; however, the discrete von Neumann criterion will have to be based on an exponential that is slightly different from that given in (3.2.1).

Remark 2: The above theorem seems to be misplaced because it applies to initial–value problems. The approach used to predict stability using the discrete von Neumann criterion (using finite Fourier series) is a method which is based on extending our initial–boundary–value problem to one which is periodic on the real line. We emphasize again that for systems of difference equations, the extensions must be done carefully so that the solution to the extended problem satisfies the original boundary conditions.

Remark 3: It should be made clear that the norm in which the scheme (in Theorem 3.2.3) is stable cannot be the $\ell_{2,\Delta x}$ norm on the whole real line. The periodic functions are not in $\ell_{2,\Delta x}$. The norm used is the finite dimensional energy norm on one period.

Remark 4: One of the benefits of the finite Fourier series description of solutions of difference equations and stability is that it shows how the processes can be considered in terms of the actions on particular modes of the expansion. Much of the intuition involving Fourier modes in continuous problems can be carried over to involve discrete Fourier modes for solutions of difference equations.

In summary,

- the discrete von Neumann stability analysis will most often provide only a necessary condition for stability,

- when the problem is solvable by a finite Fourier series, the von Neumann criterion is both necessary and sufficient for stability,

  and

- using the discrete von Neumann stability analysis to obtain necessary conditions for stability is the same as using the von Neumann condition for the initial–value scheme and Proposition 3.1.10.

HW 3.2.2 (a) Consider the difference scheme

$$u_k^{n+1} = r u_{k-1}^n + (1 - 2r) u_k^n + r u_{k+1}^n,$$

$$k = 1, \cdots, M - 1 \tag{3.2.53}$$

$$u_M^{n+1} = 0 \tag{3.2.54}$$

$$u_1^{n+1} - u_{-1}^{n+1} = 0 \tag{3.2.55}$$

$$u_k^0, \qquad k = 0, \cdots, M \quad \text{given.} \tag{3.2.56}$$

Solve the above scheme by reflecting the equation evenly at $x = 0$ and using a finite Fourier series as we did in Example 3.2.1.

(b) Show that the solution found in part (a) on $[-1, 1]$ satisfies the system of difference equations (3.2.53)–(3.2.55) (especially the Neumann boundary condition).

(c) Show that the solution found in part (a) will not satisfy the first order Neumann boundary condition $u_1^{n+1} - u_0^{n+1} = 0$.

(d) Show that the von Neumann condition will be sufficient for stability for this problem.

HW 3.2.3 Use the norm induced by inner product (3.2.2) ($\| u \| = \sqrt{(u,u)}$) and the solution to the initial–boundary–value problem (3.2.8)–(3.2.11) given by equation (3.2.19) to prove that $| \lambda_j | \leq 1$, $j = 1, \cdots, M - 1$ implies that $\| u^{n+1} \| \leq \| u^0 \|$ (or $| \lambda_j | \leq 1$, $j = 1, \cdots, M - 1$ implies that difference scheme (3.2.8)–(3.2.11) is stable).

HW 3.2.4 Prove that if the discrete initial–boundary–value problem can be solved using finite Fourier series, then the discrete von Neumann criterion is necessary and sufficient for stability.

HW 3.2.5 Perform a "generalized" discrete von Neumann stability analysis on problem (3.2.28)–(3.2.30), using basis functions of the form

$$u_k^n = \xi^n e^{\frac{i(2j-1)(2k-1)\pi}{2(2M-1)}}.$$

HW 3.2.6 Consider the function $f(x) = x(1 - x)$, $x \in [0, 1]$.

(a) Expand f in terms of a finite Fourier series based on a grid $x_k = k\Delta x$, $k = 0, \cdots, 10$ with $\Delta x = 0.1$.

(b) Expand f in terms of a sine series (extend f oddly about $x = 0$) based on the grid given in part (a).

(c) Expand f in terms of a cosine series (extend f evenly about $x = 0$) based on the grid given in part (a).

3.3   Gerschgorin Circle Theorem

At this time, for initial–boundary–value problems, the matrix stability approach is all we have. And we must realize that about the only way to get information on the norm of the matrix is by finding the eigenvalues. Obviously, it is often difficult to find the eigenvalues of a matrix. One tool for obtaining bounds on the size of the eigenvalues is the Gerschgorin Circle Theorem.

Let

$$Q = [a_{ij}]_{L \times L}$$

denote a general $L \times L$ matrix. Define

$$\rho_s = \sum_{\substack{j=1 \\ j \neq s}}^{L} | a_{sj} | \tag{3.3.1}$$

to be the sum of the absolute values of the elements in the sth row except for the diagonal element. We then have the following theorem.

Theorem 3.3.1 *For each eigenvalue of Q, λ, there exists an s such that*

$$| \lambda - a_{ss} | \le \rho_s.$$

Proof: Let λ be an eigenvalue of Q and let $\mathbf{u}$ denote a corresponding eigenvector. Let

$$\mathbf{u} = \begin{pmatrix} u_1 \\ \vdots \\ u_L \end{pmatrix}$$

and choose s so that $| u_s | \ge | u_j |$ for $j = 1, \cdots, L$. Since λ is an eigenvalue and $\mathbf{u}$ is an eigenvector,

$$Q\mathbf{u} = \lambda\mathbf{u}. \tag{3.3.2}$$

If we divide the s-th row of equation (3.3.2) by u_s, we get

$$\lambda = a_{s1}\frac{u_1}{u_s} + a_{s2}\frac{u_2}{u_s} + \cdots + a_{ss} + \cdots + a_{sL}\frac{u_L}{u_s}. \tag{3.3.3}$$

Subtracting a_{ss} from both sides of equation (3.3.3), taking the magnitude of both sides and using the triangular inequality then gives

$$| \lambda - a_{ss} | \le \sum_{\substack{j=1 \\ j \ne s}}^{L} | a_{sj} | = \rho_s,$$

which is what we were to prove.

We next demonstrate how to use the Gerschgorin Circle Theorem via the following examples. We should make a special note that the scheme presented in Example 3.3.1 is a general version of the scheme discussed in Section 3.2, (3.2.49)–(3.2.52) (where we decided that the scheme was one for which a stability analysis was very difficult).

Example 3.3.1 Consider the initial–boundary–value problem

$$\begin{align}
v_t &= \nu v_{xx}, \ x \in (0,1), \ t > 0 \tag{3.3.4} \\
v_x(0,t) &= h_1[v(0,t) - v_1], \ t \ge 0 \tag{3.3.5} \\
v_x(1,t) &= -h_2[v(1,t) - v_2], t \ge 0 \tag{3.3.6} \\
v(x,0) &= f(x), \ x \in [0,1], \tag{3.3.7}
\end{align}$$

where h_j, v_j are constant and $h_j \geq 0$ for $j = 1, 2$, and the following difference scheme for obtaining an approximation to the solution of problem (3.3.4)–(3.3.7),

$$\frac{u_1^n - u_{-1}^n}{2\Delta x} = h_1[u_0^n - v_1] \qquad (3.3.8)$$

$$u_k^{n+1} = ru_{k-1}^n + (1 - 2r)u_k^n + ru_{k+1}^n, \quad k = 0, \cdots, M \qquad (3.3.9)$$

$$\frac{u_{M+1}^n - u_{M-1}^n}{2\Delta x} = -h_2[u_M^n - v_2]. \qquad (3.3.10)$$

Analyze the stability of the difference scheme (3.3.8)–(3.3.10).

Solution: If we proceed as we did in Section 1.4 with second order accurate Neumann boundary conditions, we solve equations (3.3.8) and (3.3.10) for u_{-1}^n and u_{M+1}^n, respectively, and use them to eliminate u_{-1}^n and u_{M+1}^n from equation (3.3.9) with $k = 0$ and $k = M$. The revised difference scheme can then be written as

$$u_0^{n+1} = (1 - 2r - 2r\Delta x h_1)u_0^n + 2ru_1^n + 2r\Delta x v_1 h_1 \qquad (3.3.11)$$

$$u_k^{n+1} = ru_{k-1}^n + (1 - 2r)u_k^n + ru_{k+1}^n, \quad k = 1, \cdots, M - 1 \qquad (3.3.12)$$

$$u_M^{n+1} = 2ru_{M-1}^n + (1 - 2r - 2r\Delta x h_2)u_M^n + 2r\Delta x v_2 h_2. \qquad (3.3.13)$$

Hence the scheme can be written in matrix form as

$$\mathbf{u}^{n+1} = \begin{pmatrix} u_0^{n+1} \\ u_1^{n+1} \\ \cdots \\ u_{M-1}^{n+1} \\ u_M^{n+1} \end{pmatrix} = Q\mathbf{u}^n + \mathbf{G} \qquad (3.3.14)$$

$$= \begin{pmatrix} 1 - 2r - 2r\Delta x h_1 & 2r & 0 & \cdots & \\ r & 1 - 2r & r & 0 & \cdots \\ & & \cdots & & \\ \cdots & 0 & r & 1 - 2r & r \\ & \cdots & 0 & 2r & 1 - 2r - 2r\Delta x h_2 \end{pmatrix} \begin{pmatrix} u_0^n \\ u_1^n \\ \cdots \\ u_{M-1}^n \\ u_M^n \end{pmatrix}$$

$$+ \begin{pmatrix} 2r\Delta x v_1 h_1 \\ 0 \\ \cdots \\ 0 \\ 2r\Delta x v_2 h_2 \end{pmatrix}. \qquad (3.3.15)$$

We shall analyze the stability by applying Gerschgorin's Theorem.

As it has been the case so often before, since Q is nonsymmetric, any results we do obtain will only be necessary conditions for stability. By Gerschgorin's Theorem, if λ is an eigenvalue of Q, then there exists an s such that $| a_{ss} - \lambda | \leq \rho_s$. Hence,

$$| \lambda - (1 - 2r - 2r\Delta x h_1) | \leq 2r \qquad (3.3.16)$$

$$| \lambda - (1 - 2r) | \leq 2r \qquad (3.3.17)$$

or

$$| \lambda - (1 - 2r - 2r\Delta x h_2) | \leq 2r. \qquad (3.3.18)$$

Using the "backwards triangular inequality," inequalities (3.3.16)–(3.3.18) give the following inequalities.

$$| \lambda | - | (1 - 2r - 2r\Delta x h_1) | \leq 2r \qquad (3.3.19)$$

$$| \lambda | - | (1 - 2r) | \leq 2r \qquad (3.3.20)$$

or

$$| \lambda | - | (1 - 2r - 2r\Delta x h_2) | \leq 2r \qquad (3.3.21)$$

We note that inequalities (3.3.19)–(3.3.21) imply that

$$|\lambda| \leq 2r + |(1 - 2r - 2r\Delta xh_1)|$$
$$|\lambda| \leq 2r + |(1 - 2r)|$$
or
$$|\lambda| \leq 2r + |(1 - 2r - 2r\Delta xh_2)|.$$

Thus if λ is an eigenvalue of Q, we know that λ satisfies at least one of the above inequalities. To insure that $|\lambda| \leq 1$ for all eigenvalues of Q, we restrict r so that each of the upper limits are less than or equal to 1. Restricting the right hand sides of each of the above inequalities to be less than or equal to one leaves us with

$$|1 - 2r - 2r\Delta xh_1| \leq 1 - 2r \tag{3.3.22}$$

$$|1 - 2r| \leq 1 - 2r \tag{3.3.23}$$

$$|1 - 2r - 2r\Delta xh_2| \leq 1 - 2r. \tag{3.3.24}$$

We first note that to satisfy inequality (3.3.23), we must restrict r so that $r \leq 1/2$. With this restriction, the right hand sides of inequalities (3.3.22) and (3.3.24) are now greater than or equal to zero. These two conditions can then be written as

$$-(1 - 2r) \leq 1 - 2r - 2r\Delta xh_1 \leq 1 - 2r$$
and
$$-(1 - 2r) \leq 1 - 2r - 2r\Delta xh_2 \leq 1 - 2r.$$

The right hand side of the above two inequalities is obviously satisfied. The left hand side inequalities imply that

$$r \leq \frac{1}{2 + \Delta xh_1}$$
and
$$r \leq \frac{1}{2 + \Delta xh_2}.$$

Thus the necessary condition for stability is

$$r \leq \min\{\frac{1}{2 + h_1\Delta x}, \frac{1}{2 + h_2\Delta x}\}. \tag{3.3.25}$$

Of course, we realize that if we used the von Neumann condition and Proposition 3.1.12, we would get $r \leq 1/2$ as a necessary condition for stability. Thus, though both methods (the eigenvalue method and the von Neumann condition) give us necessary conditions for stability, it is clear that *including the effect of the boundary conditions gives us a more restrictive necessary condition.* We note that when we consider condition (3.3.25) for all sufficiently small Δx, $0 < \Delta x \leq \Delta x_0$ (as is necessary to satisfy Definition 2.4.1), the condition reduces to

$$r \leq \min\{\frac{1}{2 + h_1\Delta x_0}, \frac{1}{2 + h_2\Delta x_0}\}.$$

Example 3.3.2 Use Gerschgorin's Theorem to investigate the stability of the Crank-Nicolson scheme (2.6.7)–(2.6.9).

Solution: We begin by writing the scheme in the form

$$Q_1\mathbf{u}^{n+1} = Q\mathbf{u}^n \tag{3.3.26}$$

where

$$Q_1 = \begin{pmatrix} 1+r & -\frac{r}{2} & 0 & & \cdots \\ -\frac{r}{2} & 1+r & -\frac{r}{2} & 0 & \cdots \\ & & \cdots & & \\ & 0 & -\frac{r}{2} & 1+r & -\frac{r}{2} \\ & & 0 & -\frac{r}{2} & 1+r \end{pmatrix}, \tag{3.3.27}$$

and

$$Q = \begin{pmatrix} 1-r & \frac{r}{2} & 0 & & \cdots \\ \frac{r}{2} & 1-r & \frac{r}{2} & 0 & \cdots \\ & & \cdots & & \\ & 0 & \frac{r}{2} & 1-r & \frac{r}{2} \\ & & 0 & \frac{r}{2} & 1-r \end{pmatrix}. \tag{3.3.28}$$

We then notice that if we let

$$B = \begin{pmatrix} 1+r & -\frac{r}{2} & 0 & & \cdots \\ -\frac{r}{2} & 1+r & -\frac{r}{2} & 0 & \cdots \\ & & \cdots & & \\ & 0 & -\frac{r}{2} & 1+r & -\frac{r}{2} \\ & & 0 & -\frac{r}{2} & 1+r \end{pmatrix}, \tag{3.3.29}$$

the system in (3.3.26) can be written as

$$B\mathbf{u}^{n+1} = (2I - B)\mathbf{u}^n,$$

which when multiplied by B^{-1} yields

$$\mathbf{u}^{n+1} = \tilde{Q}\mathbf{u}^n = (2B^{-1} - I)\mathbf{u}^n. \tag{3.3.30}$$

If λ is an eigenvalue of B, then $\mu = \frac{2}{\lambda} - 1$ is an eigenvalue of $\tilde{Q}$. In order than the eigenvalues of $\tilde{Q}$ be less than or equal to 1 in magnitude, we must have

$$\left| \frac{2}{\lambda} - 1 \right| \leq 1.$$

If λ is negative, this is impossible. If λ is positive, this is the same as $\lambda \geq 1$.

By Gerschgorin's Theorem, if λ is an eigenvalue of B, then

$$| \lambda - (1+r) | \leq \frac{r}{2} \tag{3.3.31}$$

or

$$| \lambda - (1+r) | \leq r. \tag{3.3.32}$$

Obviously, if λ satisfies (3.3.31), it satisfies (3.3.32.) Thus, we restrict r so that (3.3.32) is satisfied. But inequality (3.3.32) is equivalent to

$$1 \leq \lambda \leq 1 + 2r.$$

Thus we see that λ is always greater than or equal to 1. Also, we see that λ can never be negative. This then implies that the eigenvalues of $\tilde{Q}$ are always less than or equal to 1 in magnitude.

We should also note that since B is symmetric and $BB^{-1} = I$,

$$I = (BB^{-1})^T = (B^{-1})^T B^T = (B^{-1})^T B,$$

and $(B^{-1})^T = B^{-1}$. Then the symmetry of B and the fact that $\tilde{Q} = 2B^{-1} - I$ implies that $\tilde{Q}$ is symmetric. The symmetry of $\tilde{Q}$ then implies that we have both a necessary and sufficient condition for stability and, hence, that the Crank-Nicolson scheme is unconditionally stable.

Remark: We note in Example 3.3.1 that we were very careful to use the magnitude of the eigenvalue λ at all times and not assume that λ is real. Because the matrix Q is not symmetric, it is not clear whether or not the eigenvalues are real. However, in Example 3.3.2, since the matrix B is symmetric, we know that the eigenvalues of B are real and we proceed using that fact. As usual, the non-symmetry of the matrix makes everything more difficult.

Hopefully by now we see that there are various ways to prove or find conditions that will imply stability. All of these methods are tools that are helpful in obtaining the necessary bounds needed for stability. We next proceed to other aspects of numerical partial differential equations (multi-dimensional problems, hyperbolic problems, elliptic problems, etc.). However, in most of these discussions, we will return to discuss the stability of our particular schemes.

3.4 Computational Interlude II

3.4.1 Review of Computational Results

Since we last talked about implementations and computer codes, we have seen some surprising results. For example, when we compare explicit and implicit solutions to the same problem (and since these schemes are the same order, we feel that the errors should be approximately the same), we find that they do not seem comparable. See for example HW2.6.2 and HW2.6.3. Likewise, when we use the 0-th, 1-st and 2-nd order approximation to the zero Neumann condition, we find that sometimes the results get worse as the order of approximation gets higher (HW2.6.4). We must remember that the order arguments (and the order of convergence result of the Lax Theorem, Theorem 2.5.2) do involve a multiplicative constant. Because two methods are both 2-nd order in Δx does not necessarily mean that the constants will be the same. In addition, often we obtain results that are better than we deserve. Because these results are more than we should expect, they make other results look bad or wrong.

3.4.1.1 Neumann Boundary Conditions

Specifically, consider the 0-th order approximation to the zero Neumann condition treated in HW2.6.4. Though we would think that no one should ever want to compute with a 0-th order scheme, it is done often and this example shows that by luck excellent results can be obtained using such an approximation. Since the problem is an easy problem that we can solve exactly, the reason for these good results in this case can easily be studied.

Using a Taylor series expansion, we see that

$$\frac{u_1^n - u_0^{n+1}}{\Delta x} = (u_x)_0^{n+1} - (u_t)_0^{n+1}\frac{\Delta t}{\Delta x} + (u_{xx})_0^{n+1}\frac{\Delta x}{2}$$
$$- (u_{xt})_0^{n+1}\Delta t + (u_{tt})_0^{n+1}\frac{\Delta t^2}{2\Delta x} + \cdots. \qquad (3.4.1)$$

Obviously, the "bad" term is the $\frac{\Delta t}{\Delta x}$ term. But if we know what we are doing, the bad affects of this term can be minimized. Consider the obvious approach of taking $\Delta t = \Delta x^2$. If this is done, then the scheme is still first order in both Δx and Δt and conditionally consistent. Actually, in HW2.6.4 we used $\Delta x = 0.1$ and $\Delta t = 0.004$, so it should not surprise us that a $\Delta t/\Delta x$ should not hurt the calculation.

Furthermore, since the solution to the problem is

$$v(x, t) = e^{-\frac{\pi^2 t}{4}} \cos\frac{\pi x}{2}, \qquad (3.4.2)$$

we see that

$$(v_t)_0^{n+1} = -\frac{\pi^2}{4} e^{-\frac{\pi^2 t}{4}} \qquad (3.4.3)$$

where $t = (n + 1)\Delta t$. Then because

$$(v_t)_0^{n+1}\frac{\Delta t}{\Delta x} = \begin{cases} -0.085 & \text{when } t = 0.06 \\ -0.077 & \text{when } t = 0.1 \\ -0.014 & \text{when } t = 0.8 \end{cases}, \qquad (3.4.4)$$

we see that the results using this 0-th order approximation should be good. Furthermore, if we take the first two error terms, we notice that

$$-(v_t)_0^{n+1}\frac{\Delta t}{\Delta x} + (v_{xx})_0^{n+1}\frac{\Delta x}{2} = \begin{cases} 0.059 & \text{when } t = 0.06 \\ 0.053 & \text{when } t = 0.1 \\ 0.0094 & \text{when } t = 0.8. \end{cases}$$

Hence, we see that the error due to the 0-th order (with which we were already pleased), is made even better when we consider two terms.

The effects of the other terms in the order argument can also be easily computed. The above computation was done to show how a supposedly inaccurate scheme can work to give good results.

3.4.1.2 Explicit Verses Implicit Schemes

Another aspect of the computations done with the implicit schemes in Computational Interlude I that we must discuss is the choice of Δt. We found in Chapter 1 (and again in Chapter 3) that with explicit methods, if we choose Δt (or really $r = \nu\Delta t/\Delta x^2$) too large, then the scheme becomes unstable. In the Computational Interlude I and again in Chapter 3, we found that for implicit methods we can generally choose Δt (or r) to be as large as we wish, and the scheme will remain stable. This property of implicit methods allows them to be both well used and abused. There are times when the restriction placed on the time step for explicit schemes that is necessary to obtain stability (for a given spatial grid) is unrealistically severe. A computation is no good if it never gets done (in either your life time or during the life time of your computer and/or computer account). In this case, it is often advantageous to use an implicit method.

But, there is also the situation where the calculation is taking too long (whatever the measure of too long is) using an explicit scheme, so a larger time step and an implicit method are used that give inaccurate results. *When choosing the larger time steps that implicit methods allow, we must always be aware that the error increases with increasing time step. We must always balance the allowable error verses the speed of computation.*

It is not difficult (but it differs depending on what type of machine is being used) to compare the computer time required of one explicit time step verses one implicit time step. The ratios are usually between 1–12 and 1–30. For the latter case, it means that the implicit time step must be 30 times the explicit time step before computer time is saved using an implicit scheme. The question is then whether the accuracy obtained using this larger time step is sufficient. We shall see that with two and three dimensional problems, it is not difficult to justify using implicit schemes rather than explicit schemes (and still maintain sufficient accuracy) because the stability restriction on the time step for the explicit schemes becomes more severe as the dimension increases.

Another situation where implicit schemes are used with great success is to compute steady solutions. It is sometimes advantageous to solve the time dependent problem badly, trying to get to the steady limit as fast as possible, rather than solving the steady problem directly. In this case, either very large time steps or varying time steps (some small and some large) have been used successfully.

3.4.2   HW0.0.1

And finally, we return to HW0.0.1. When we tried to solve this problem using the explicit scheme (2.6.3), we found that for smaller values of ν, the solution "blew up." If we did our "numerical experiment" carefully, we found that even when we chose $r \leq 1/2$, the computation blew up. We must realize that we cannot perform a stability analysis of the scheme (2.6.3) to see why it was unstable. In light of the results given in Example 3.1.3 (which is an excellent model equation for the viscous Burgers' equation), this should not surprise us. From Example 3.1.3 we know that $R^2/2 < r \leq 1/2$ is a necessary condition for convergence. In our application of difference scheme (2.6.3), we choose the $\nu, \Delta t$ and Δx so that r is fixed. But the a

term we use is really varying. Thus we must satisfy conditions for any a that satisfies $\mid a \mid\le 1$ (since $\mid u_k^n \mid\le 1$). We see that it is very difficult to satisfy $R^2/2 \le r$, when ν is small.

HW 3.4.1 Use the matrix–eigenvalue approach (as used in Example 3.1.7) to investigate the stability of the initial–boundary–value scheme

$$u_k^{n+1} = u_k^n - \frac{R}{2}\delta_0 u_k^n + r\delta^2 u_k^n \tag{3.4.5}$$

$$u_0^{n+1} = 0 \tag{3.4.6}$$

$$u_M^{n+1} = 0. \tag{3.4.7}$$

Since we have seen that implicit methods are generally unconditionally stable (where the explicit schemes have stability conditions), it should be logical to try an analog of the scheme used in HW2.3.1(b) and HW2.6.3 (i.e. an implicit scheme).

Thus we see that if we write a logical implicit scheme for solving the viscous Burgers' equation (a BTCS scheme), we get

$$\frac{u_k^{n+1} - u_k^n}{\Delta t} + u_k^{n+1}\frac{u_{k+1}^{n+1} - u_{k-1}^{n+1}}{2\Delta x} = \frac{\nu}{\Delta x^2}\delta^2 u_k^{n+1}. \tag{3.4.8}$$

Obviously, equation (3.4.8) is not as nice as other equations we have been solving in that it is nonlinear. There is no "nice" way to solve for u_k^{n+1}. There are, however, several ways to approximate the solution of equation (3.4.8).

Lag Nonlinear Term The easiest and maybe the most common method is to "lag" part of the nonlinear term. That is, approximate equation (3.4.8) by

$$\frac{u_k^{n+1} - u_k^n}{\Delta t} + u_k^n\frac{u_{k+1}^{n+1} - u_{k-1}^{n+1}}{2\Delta x} = \frac{\nu}{\Delta x^2}\delta^2 u_k^{n+1}. \tag{3.4.9}$$

By replacing the one term u_k^{n+1} by u_k^n, we have made the equation linear and it can now easily be solved using the Thomas Algorithm, page 87 (Trid).

Linearize About Previous Time Step The second most common method is to linearize the equation by linearizing about the previous time step. Linearization about the previous time step is accomplished by setting

$$u_k^{n+1} = u_k^n + \Delta u_k \tag{3.4.10}$$

and replacing u_k^{n+1} in equation (3.4.8) by equation (3.4.10). The result is

$$\frac{\Delta u_k}{\Delta t} + (u_k^n + \Delta u_k)\frac{(u_{k+1}^n - u_{k-1}^n) + (\Delta u_{k+1} - \Delta u_{k-1})}{2\Delta x}$$
$$= \frac{\nu}{\Delta x^2}\delta^2 u_k^n + \frac{\nu}{\Delta x^2}\delta^2 \Delta u_k. \tag{3.4.11}$$

If we then approximate equation (3.4.8) by eliminating any terms in equation (3.4.11) of the form $(\Delta u)^2$ or higher, we are left with the following equation.

$$\frac{\Delta u_k}{\Delta t} + \frac{1}{2\Delta x}(\delta_0 u_k^n)\Delta u_k + \frac{1}{2\Delta x}u_k^n\delta_0\Delta u_k - \frac{\nu}{\Delta x^2}\delta^2\Delta u_k$$
$$= -\frac{1}{2\Delta x}u_k^n\delta_0 u_k^n + \frac{\nu}{\Delta x^2}\delta^2 u_k^n \tag{3.4.12}$$

Of course when we solve equation (3.4.12) for Δu_k, we then add it to u_k^n to get our approximation for u_k^{n+1}. We might note that equation (3.4.12) can be written in the form

$$(-\frac{R}{2}u_k^n - r)\Delta u_{k-1} + (1 + 2r + \frac{R}{2}\delta_0 u_k^n)\Delta u_k + (\frac{R}{2}u_k^n - r)\Delta u_{k+1}$$
$$= -\frac{R}{2}u_k^n\delta_0 u_k^n + r\delta^2 u_k^n \tag{3.4.13}$$

where as usual $r = \nu\Delta t/\Delta x^2$ and $R = \Delta t/\Delta x$.

Newton's Method And finally, the most obvious method but also the most difficult method is to solve equation (3.4.8) using Newton's method. To apply Newton's method to equation (3.4.8), we began be making it clear what we mean by an L dimensional Newton's method. We suppose we have a function $\mathbf{f} : \mathbb{R}^L \to \mathbb{R}^L$ (where $\mathbf{f}$ is a vector of functions) and denote the derivative of $\mathbf{f}$ at the point $\mathbf{x}$ (where of course, $\mathbf{x}$ is an L-vector) by the $L \times L$ matrix

$$\mathbf{f}'(\mathbf{x}) = \left[\frac{\partial f_i}{\partial x_j}\right]_{L \times L}. \tag{3.4.14}$$

If $\mathbf{x}_0$ is some initial guess to a solution of the equation

$$\mathbf{f}(\mathbf{x}) = \boldsymbol{\theta}, \tag{3.4.15}$$

then the Newton's iteration for solving equation (3.4.15) is given by

$$\text{solve } \mathbf{f}'(\mathbf{x}_k)\Delta\mathbf{x} = -\mathbf{f}(\mathbf{x}_k) \text{ for } \Delta\mathbf{x} \tag{3.4.16}$$

and

$$\text{set } \mathbf{x}_{k+1} = \mathbf{x}_k + \Delta\mathbf{x}, \tag{3.4.17}$$

for $k = 0, \cdots$. Of course there are convergence results for Newton's method, most notably of these being that *for a sufficiently good initial guess, Newton's method will convergence quadratically.*

To apply Newton's method to solving equation (3.4.8) on the interval $[0, 1]$ with zero Dirichlet boundary conditions, we begin by writing equation (3.4.8) in the form

$$u_k^{n+1} + \frac{R}{2}u_k^{n+1}\left(u_{k+1}^{n+1} - u_{k-1}^{n+1}\right) - r\delta^2 u_k^{n+1} - u_k^n = 0. \qquad (3.4.18)$$

We then define $\mathbf{f} : \mathbb{R}^{M-1} \to \mathbb{R}^{M-1}$ by setting

$$\mathbf{u} = \begin{pmatrix} u_1 \\ \cdots \\ u_{M-1} \end{pmatrix} = \begin{pmatrix} u_1^{n+1} \\ \cdots \\ u_{M-1}^{n+1} \end{pmatrix}$$

and then defining the k-th term of the function $\mathbf{f}$, f_k as

$$f_k(\mathbf{u}) = u_k + \frac{R}{2}u_k(u_{k+1} - u_{k-1}) - r(u_{k-1} - 2u_k + u_{k+1}) - u_k^n. \qquad (3.4.19)$$

We see that $\mathbf{f}'(\mathbf{u}) =$

$$\begin{pmatrix} 1 + \frac{R}{2}\delta_0 u_1 + 2r & \frac{R}{2}u_1 - r & 0 & \cdots & \\ & \cdots & \cdots & & \\ \cdots \ 0 & -\frac{R}{2}u_k - r & 1 + \frac{R}{2}\delta_0 u_k + 2r & \frac{R}{2}u_k - r & 0 \ \cdots \\ & \cdots & \cdots & & \\ & \cdots & 0 & -\frac{R}{2}u_{M-1} - r & 1 + \frac{R}{2}\delta_0 u_{M-1} + 2r \end{pmatrix}.$$

$$(3.4.20)$$

To solve a difference scheme such as (3.4.18), we return to the basic scheme given in Section 1.2.1 and replace the "Solution Scheme" subroutine. The Solution Scheme subroutine necessary to apply Newton's method will be as follows.

1. Call Right Hand Side

Call Stencil

Call Trid

$unew = unew + RHS$

If RHS is sufficiently small, go to 2

Otherwise, go to 1

2. Continue

We note that the Right Hand Side subroutine calculates $-\mathbf{f}(\mathbf{u}_k)$ from equation (3.4.19) and puts it in RHS (called $\mathbf{r}$ in Section 2.6.3). Take special note that we use *uold* as our initial guess. During the first step, all of the values used in formula (3.4.19) are from *uold*. In later steps, the first three terms of formula (3.4.19) will come from the last Newton iterate, *unew*. The last term in formula (3.4.19) will always be *uold*.

Next the Stencil subroutine uses the expressions in (3.4.20) to fill the matrix (the diagonals of which were called a, b and c in Section 2.6.3). Recall that Trid returns the solution in the RHS array. That is why the size of RHS is tested to determine whether to iterate longer ($\Delta \mathbf{x}$) and why RHS is added onto *unew* (as in equation (3.4.17)).

Remark 1: In the above solution algorithm, when we test to see whether RHS is sufficiently small we have several decisions to make. We must first decide how small is sufficient. Of course, this is up to the user and depends on the situation. For some applications, 10^{-4} or 10^{-5} is sufficient, whereas in other applications a tolerance of 10^{-10} may be necessary. Since this iteration is expensive, it is important not to use a smaller tolerance than is necessary. We must also decide how to measure $RHS = \Delta \mathbf{x}$. The two most obvious ways to measure the size of RHS is to use either the sup-norm or the ℓ_2 norm. There are times when one of these is better than the other, but usually, either one or the other or both are sufficient.

Remark 2: The above algorithm is a difficult way to take a time step. We should have confidence that if it is done this way and we are still having trouble with our scheme, it is probably not because we are not solving equation (3.4.8) sufficiently well.

Remark 3: We note that the obvious choice for the initial guess of u_k^{n+1} is u_k^n. If this is done, the the method of linearizing about the n-th time step is equivalent to doing one step of Newton's method.

The above three methods are probably the three most common methods for handling general nonlinearities. We should now be able to return to HW0.0.1 and try solving the problem using an implicit scheme. It is suggested that all three methods be tried. Since the method of linearizing about the previous time step is contained in the Newton's iteration method (if the correct choice of initial guess is used), these two methods can surely be combined.

We might mention that other implicit schemes could also be used. For example, it would not be difficult to write a nonlinear Crank-Nicolson analog to equation (3.4.8) in which case the linearized equations that are eventually solved would look more like the Crank-Nicolson scheme.

And finally, we claimed that we should or might try an implicit scheme because implicit schemes are generally unconditionally stable. It makes sense to do a stability analysis on the analogous linear scheme before too much work is done. Again, since the coefficient a in the nonlinear equation can be either positive or negative, we must do our stability analysis carefully so as to include both positive and negative a's. We consider the difference scheme

$$(-r - R/2)u_{k-1}^{n+1} + (1 + 2r)u_k^{n+1} + (-r + R/2)u_{k+1}^{n+1} = u_k^n \quad (3.4.21)$$
$$u_0^{n+1} = 0 \quad (3.4.22)$$
$$u_M^{n+1} = 0. \quad (3.4.23)$$

Of course, this scheme can be written as

$$Q_1 u^{n+1} = \begin{pmatrix} 1 + 2r & -r + R/2 & 0 & \cdots \\ -r - R/2 & 1 + 2r & -r + R/2 & 0 & \cdots \\ & & \cdots & & \\ & \cdots & 0 & -r - R/2 & 1 + 2r \end{pmatrix} u^{n+1}$$
$$= u^n. \quad (3.4.24)$$

We should also notice that we can write the scheme as $u^{n+1} = Qu^n$, where $Q = Q_1^{-1}$. And finally, we should note that since Q_1 is not symmetric, Q will also be not symmetric. Thus, *any stability results we obtain will only be necessary conditions for stability.*

The way we compute the eigenvalues of Q is to realize that the eigenvalues of Q are reciprocals of the eigenvalues of Q_1. Using (2.2.41), the eigenvalues of Q_1 can be written as

$$\lambda_j = 1 + 2r + 2(R/2 - r)\sqrt{\frac{-r - R/2}{-r + R/2}} \cos \frac{\pi j}{M}, \quad j = 1, \cdots, M - 1. \quad (3.4.25)$$

If $r > | R | /2$,

$$\lambda_j = 1 + 2r - 2\sqrt{r^2 - R^2/4} \cos \frac{\pi j}{M}, \quad j = 1, \cdots, M - 1. \quad (3.4.26)$$

As with the explicit scheme (HW3.4.1), it is clear that the maximum and minimum eigenvalues will occur at λ_1 and λ_{M-1}. Since

$$\sqrt{r^2 - R^2/4} \cos \pi/M < r,$$

$\lambda_1 \geq 1$. And, since $\cos \pi(M - 1)/M < 0$, $\lambda_{M-1} \geq 1$ also.

If $R/2 > r$, then

$$\lambda_j = 1 + 2r + 2i\sqrt{R^2/4 - r^2} \cos \frac{\pi j}{M}.$$

Then it is clear that

$$| \lambda_j |^2 = (1 + 2r)^2 + 4\sqrt{R^2/4 - r^2} \cos^2 \frac{\pi j}{M} > 1, \quad (3.4.27)$$

for all $j = 1, \cdots, M - 1$.

And if $r <| R | /2$ and $R < 0$, we see that

$$\lambda_j = 1 + 2r - 2i\sqrt{R^2/4 - r^2}\cos\frac{\pi j}{M}.$$

Since in this case $| \lambda_j |^2$ is the same as in equation (3.4.27), the magnitude is again greater than one.

And finally, if $r = R/2$, it is clear that the eigenvalues are given by $\lambda_j = 1 + 2r$, for $j = 1, \cdots, M - 1$. Hence, again all of the eigenvalues are greater than one.

Since the magnitude of all of the eigenvalues of Q_1 are greater than or equal to one, the magnitude of all of the eigenvalues of Q are less than or equal to one. Thus there is no condition on r and R that will necessarily cause an instability.

HW 3.4.2 Discuss the stability of the following Crank-Nicolson scheme for the one dimensional parabolic equation with lower order terms and zero Dirichlet boundary conditions.

$$u_k^{n+1} - u_k^n + \frac{R}{4}\delta_0(u_k^{n+1} + u_k^n) = \frac{r}{2}\delta^2(u_k^{n+1} + u_k^n) \qquad (3.4.28)$$
$$u_0^{n+1} = 0$$
$$u_M^{n+1} = 0$$

Hint: The scheme can be written as $Q_1 \mathbf{u}^{n+1} = Q\mathbf{u}^n$. We want the eigenvalues of $Q_1^{-1}Q$ which are the same as the generalized eigenvalues of (Q_1, Q), i.e. the eigenvalues and eigenvectors that satisfy $Q\mathbf{u} = \lambda Q_1\mathbf{u}$. But this can be written as $\lambda Q_1\mathbf{u} = \lambda\{B + I\}\mathbf{u} = Q\mathbf{u} = \{-B + I\}\mathbf{u}$ where B is the tridiagonal matrix with $-\frac{r}{2} - \frac{R}{2}$, r and $-\frac{r}{2} + \frac{R}{2}$ on the subdiagonal, diagonal and superdiagonal, respectively. Using the fact that we know the eigenvalues of B, we can then derive an expression for the eigenvalues of $Q_1^{-1}Q$.

4

Parabolic Equations

4.1 Introduction

Parabolic partial differential equations are a large, important class of equations. Obviously, most of the one dimensional work that we have done introducing difference methods, convergence, stability etc. has been directed toward parabolic equations. Of course, most of this was done with the model heat equation, but the heat equation is an excellent model for the class of parabolic equations.

This chapter will address topics related to parabolic equations that were not previously covered. The major topic will be difference schemes for two and three dimensional parabolic partial differential equations. As a part of this discussion, we will also discuss convergence, stability, etc. for two and three dimensional difference schemes.

Before we proceed, we would like to emphasize that we will not introduce any new tools for proving convergence. The approach we will use is the same as in the case of one dimensional difference schemes. Our basic tool will be the Lax Theorem, Theorem 2.5.2. Again, the approach we will use for proving stability of difference schemes for initial–value problems will be the discrete Fourier transform. As in the case with one dimensional schemes, using this approach we obtain necessary and sufficient conditions for convergence.

Likewise, as in the case with difference schemes for one dimensional initial–boundary–value problems, the stability of the difference scheme without consideration of the boundary conditions (the von Neumann criteria) will be necessary for convergence. Again, at this time our only approach for obtaining necessary and sufficient conditions for convergence for initial–boundary–value schemes is to be able to compute the eigenvalues of the matrix operator. And, this approach gives us necessary and sufficient conditions for convergence only when the matrix operator is symmetric or similar to a symmetric matrix.

4.2   Two Dimensional Parabolic Equations

We begin our discussion of multi-dimensional problems by considering the model parabolic problem

$$v_t = \nu(v_{xx} + v_{yy}) + F(x,y,t), \ (x,y) \in R, \ t > 0 \qquad (4.2.1)$$
$$v(x,y,t) = g(x,y,t) \text{ on } \partial R, \ t > 0 \qquad\qquad\qquad (4.2.2)$$
$$v(x,y,0) = f(x,y) \ (x,y) \in \overline{R}. \qquad\qquad\qquad\qquad (4.2.3)$$

As is the case with the one dimensional problems, our approach will be to cover the region $\overline{R}$ with a grid and approximate our problem by a problem defined on that grid. Of course, in this case, the grid will be a two dimensional grid. Also, if we consider a non-rectangular region, R, working with the grid (especially near and on the boundaries) will be difficult. We will discuss complex geometries in Chapter 11. In this chapter, we will concentrate our discussion on solving problems on rectangles, the half plane or whole plane.

Consider $R = [0,1] \times [0,1]$. To cover $[0,1] \times [0,1]$ with a grid, we must choose a Δx and a Δy (or an M_x and M_y). Doing so, we obtain a grid of the form shown in Figure 4.2.1. Denote the grid covering R by G_R. We should mention that often it is convenient (in that the computations and analysis are more palatable) to let $\Delta x = \Delta y$. At times we shall do this, but we will try to restrain ourselves from doing it too often.

If we are to approximate our problem and our solution on a grid such as that shown in Figure 4.2.1, we must have some relatively consistent notation by which we denote points and functions defined on such grids. We shall use the convention of using an ordered pair of indices (j,k) to denote the point $(j\Delta x, k\Delta y)$ in R, where $j = 0, \cdots, M_x$ and $k = 0, \cdots, M_y$. A function $v = v(x,y,t)$ approximated at the (j,k) grid point and the nth time level will be denoted by u_{jk}^n.

If we then consider the point $(j,k) \in G_R$ shown in Figure 4.2.2, we see that we can approximate partial differential equation (4.2.1) in a manner completely analogous to the one dimensional approach.

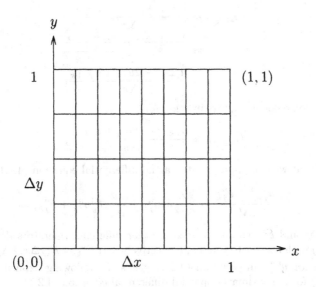

FIGURE 4.2.1. Two dimensional grid on the region $[0, 1] \times [0, 1]$.

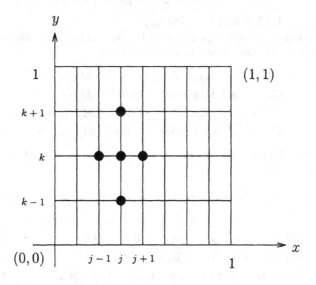

FIGURE 4.2.2. Stencil for approximating $v_{xx} + v_{yy}$ on a two dimensional grid.

We note that

$$(v_{xx})_{jk} \approx \frac{(v_x)_{j+\frac{1}{2}k} - (v_x)_{j-\frac{1}{2}k}}{\Delta x}$$

$$\approx \frac{\frac{u_{j+1\,k} - u_{jk}}{\Delta x} - \frac{u_{jk} - u_{j-1\,k}}{\Delta x}}{\Delta x}$$

$$= \frac{u_{j+1\,k} - 2u_{jk} + u_{j-1\,k}}{\Delta x^2}.$$

Using the analogous approximation for $(v_{yy})_{jk}$,

$$(v_{yy})_{jk} \approx \frac{u_{j\,k+1} - 2u_{jk} + u_{j\,k-1}}{\Delta y^2},$$

we see that we can approximate partial differential equation (4.2.1) by

$$\frac{u_{jk}^{n+1} - u_{jk}^n}{\Delta t} = \frac{\nu}{\Delta x^2}\delta_x^2 u_{jk}^n + \frac{\nu}{\Delta y^2}\delta_y^2 u_{jk}^n + F_{jk}^n, \qquad (4.2.4)$$

where δ_x^2 and δ_y^2 denote the second order difference operators defined in (1.3.13) with respect to j and k, respectively and $F_{jk}^n = F(j\Delta x, k\Delta y, n\Delta t)$. Solving for u_{jk}^{n+1} in equation (4.2.4) yields the following explicit scheme (FTCS) for approximating partial differential equation (4.2.1)

$$u_{jk}^{n+1} = u_{jk}^n + (r_x\delta_x^2 + r_y\delta_y^2)u_{jk}^n + \Delta t F_{jk}^n, \qquad (4.2.5)$$

where $r_x = \nu\Delta t/\Delta x^2$ and $r_y = \nu\Delta t/\Delta y^2$.

The boundary condition (4.2.2) with $R = [0,1] \times [0,1]$ are handled much as was done in the one dimensional case. We set

$$u_{0k}^n = g(0, k\Delta y, n\Delta t), \quad k = 0, \cdots, M_y, \ n \geq 0 \qquad (4.2.6)$$
$$u_{M_x\,k}^n = g(1, k\Delta y, n\Delta t), \quad k = 0, \cdots, M_y, \ n \geq 0 \qquad (4.2.7)$$
$$u_{j0}^n = g(j\Delta x, 0, n\Delta t), \quad j = 0, \cdots, M_x, \ n \geq 0 \qquad (4.2.8)$$
$$u_{j\,M_y}^n = g(j\Delta x, 1, n\Delta t), \quad j = 0, \cdots, M_x, \ n \geq 0. \qquad (4.2.9)$$

With the above discussion, we are now prepared to solve the following homework problems.

HW 4.2.1 Write a code to use (4.2.5) and (4.2.6)–(4.2.9) to solve

$$v_t = \nu(v_{xx} + v_{yy}), \quad (x,y) \in (0,1) \times (0,1), \ t > 0 \qquad (4.2.10)$$
$$v(0,y,t) = v(1,y,t) = 0, \quad y \in [0,1], \ t \geq 0 \qquad (4.2.11)$$
$$v(x,0,t) = v(x,1,t) = 0, \quad x \in [0,1], \ t \geq 0 \qquad (4.2.12)$$
$$v(x,y,0) = \sin \pi x \sin 2\pi y, \quad (x,y) \in [0,1] \times [0,1]. \qquad (4.2.13)$$

Use $M_x = M_y = 20$, $\nu = 1.0$, $\Delta t = 0.0005$ and calculate solutions at $t = 0.06$, $t = 0.1$ and $t = 0.9$. Calculate the solutions a second time using $\Delta t = 0.001$. Compare your solutions with each other and the exact solution.

HW 4.2.2 Redo problem HW4.2.1 using $M_x = 10$ and $M_y = 20$.

HW 4.2.3 Use difference scheme (4.2.5)–(4.2.9) to approximate the solution of initial–boundary–value problem (4.2.1)–(4.2.3) where $R = (0, 1) \times (0, 1)$, $F(x, t) = \sin t \sin 2\pi x \sin 4\pi y$, $g = 0$, $f = 0$ and $\nu = 0.5$. Use $M_x = M_y = 20$ and $\Delta t = 0.0005$. Give the results at times $t = 1.5$, $t = 4.5$ and $t = 6.25$.

HW 4.2.4 Redo HW4.2.3 using $M_x = M_y = 100$ and $\Delta t = 0.00002$.

HW 4.2.5 Use difference scheme (4.2.5)–(4.2.9) to approximate the solution to problem (4.2.1)–(4.2.3) where $R = (0, 1) \times (0, 1)$, $F = 0$, $g(0, y, t) = \sin t \sin \pi y$, $g(x, 0, t) = \sin t \sin \pi x$, $g(1, y, t) = g(x, 1, t) = 0$, $f = 0$ and $\nu = 1.0$. Use $M_x = M_y = 20$ and $\Delta t = 0.001$. Give results at times $t = 1.5$, $t = 4.5$ and $t = 6.25$.

HW 4.2.6 Redo HW4.2.5 using $M_x = M_y = 100$ and $\Delta t = 0.00002$.

We might mention that the code for the above problem should be very similar to that used in one dimensional problems. Again, it is worthwhile to take some time to write good initialization and output subroutines because you will get to use them often. Though graphical output is still very useful for observing results defined on a two dimensional domain, three dimensional graphics are not nearly as easy to use as the analogous two dimensional plots used in earlier chapters. For this reason, special care should be taken when you design your output routine. One useful form of output is to provide two dimensional slices of the solution (holding either x or y constant while the other varies).

4.2.1 Neumann Boundary Conditions

Neumann boundary conditions for two dimensional problems are also handled in the same manner as for one dimensional problems. For example, if we consider

$$v_t = \nu \nabla^2 v$$
$$= \nu(v_{xx} + v_{yy}), \ (x, y) \in (0, 1) \times (0, 1), \ t > 0 \quad (4.2.14)$$
$$v(0, y, t) = v(x, 0, t)$$
$$= v(x, 1, t) = 0, \ x \in [0, 1], \ y \in [0, 1], \ t \geq 0 \quad (4.2.15)$$
$$v_x(1, y, t) = -\pi e^{-5\nu\pi^2 t} \sin 2\pi y, \ y \in [0, 1], \ t \geq 0 \quad (4.2.16)$$
$$v(x, y, 0) = \sin \pi x \sin 2\pi y, \ (x, y) \in [0, 1] \times [0, 1], \quad (4.2.17)$$

we see that on $x = 1$, we have the Neumann boundary condition

$$v_x(1, y, t) = -\pi e^{-5\nu\pi^2 t} \sin 2\pi y. \quad (4.2.18)$$

If we use a first order difference scheme to approximate the derivative in equation (4.2.18), we get

$$\frac{u_{M_x\,k}^{n+1} - u_{M_x-1\,k}^{n+1}}{\Delta x} = -\pi e^{-5\nu\pi^2(n+1)\Delta t} \sin 2\pi k \Delta y, \quad k = 0, \cdots, M_y,$$

$$(4.2.19)$$

or

$$u_{M_x\,k}^{n+1} = u_{M_x-1\,k}^{n+1} - \pi \Delta x e^{-5\nu\pi^2(n+1)\Delta t} \sin 2\pi k \Delta y, \quad k = 0, \cdots, M_y.$$

$$(4.2.20)$$

We can then use difference scheme (4.2.5) along with boundary conditions (4.2.6), (4.2.8), (4.2.9) and (4.2.20) to solve initial–boundary–value problem (4.2.14)–(4.2.17).

If we instead use the second order approximation of the Neumann boundary condition, we get

$$u_{M_x+1\,k}^{n} = u_{M_x-1\,k}^{n} - 2\pi \Delta x e^{-5\nu\pi^2 n\Delta t} \sin 2\pi k \Delta y, \quad k = 0, \cdots, M_y,$$

$$(4.2.21)$$

as an approximation to Neumann boundary condition (4.2.18). As we did in the one dimensional case, we also require that equation (4.2.5) be satisfied at $j = M_x$,

$$u_{M_x\,k}^{n+1} = u_{M_x\,k}^{n} + \nu \Delta t \left(\frac{1}{\Delta x^2} \delta_x^2 + \frac{1}{\Delta y^2} \delta_y^2 \right) u_{M_x\,k}^{n}, \quad k = 0, \cdots, M_y.$$

$$(4.2.22)$$

If we then use equation (4.2.21) to eliminate the $u_{M_x+1\,k}^{n}$ term from equation (4.2.22), we are left with

$$u_{M_x\,k}^{n+1} = u_{M_x\,k}^{n} + 2r_x(u_{M_x-1\,k}^{n} - u_{M_x\,k}^{n}) + r_y \delta_y^2 u_{M_x\,k}^{n}$$

$$- 2\pi\nu \frac{\Delta t}{\Delta x} e^{-5\nu\pi^2 n\Delta t} \sin 2\pi k \Delta y, \quad k = 1, \cdots, M_y - 1. \quad (4.2.23)$$

Then, difference scheme (4.2.5) along with boundary conditions (4.2.6), (4.2.8), (4.2.9) and (4.2.23) gives us a second order accurate scheme for solving problem (4.2.14)–(4.2.17).

Remark: We should mention that we are using the terms first order accurate and second order accurate loosely above. As we saw for the one dimensional case, the first order accurate approximation of the Neumann boundary condition along with the difference scheme leaves us with a 0-th order accurate scheme for the problem via Definition 2.3.3 (unless we consider the scheme with respect to an offset grid). A similar reduction of order of accuracy occurred with the second order accurate approximation of the Neumann boundary condition. This will also be the case for two dimensional problems. Here, we again refer to them as the first and second order approximations in that they approximate the Neumann boundary condition to that order.

HW 4.2.7 Solve problem (4.2.14)–(4.2.17) using both the first and second order treatments of the Neumann boundary condition. Use $M_x = M_y = 10$, $\nu = 1.0$, $\Delta t = 0.001$ and calculate solutions at $t = 0.06$, $t = 0.1$ and $t = 0.9$.

HW 4.2.8 Resolve problem (4.2.14)–(4.2.17) using $M_x = M_y = 20$ (and, $\Delta t = 0.0005$) and $M_x = M_y = 40$ (and, $\Delta t = 0.0001$). Compare your solutions with each other, the solution found in HW4.2.7 and the exact solution.

4.2.2 Derivation of Difference Equations

As we did for one dimensional problems in Chapter 1, difference schemes for approximating two and three dimensional partial differential equations can also be derived based on finite volumes and conservation laws. Since the two and three dimensional derivations look somewhat different from their one dimensional counterpart, we shall derive a two dimensional difference equation. Specifically, that we do not get too boring, instead of rederiving difference equation (4.2.5) again, we shall derive a difference equation different from (4.2.5) that approximates partial differential equation (4.2.1).

We could, as we did in Chapter 1, use the derivation of the partial differential equation to exhibit the desired form of the conservation law. Instead, we assume that R is an arbitrary region in the plane (within the domain of the physical problem), denote the boundary of R by ∂R and give the integral form of the conservation law as

$$\int_R \frac{\partial v}{\partial t} dA = \int_R F(\mathbf{x}, t) dA + \int_{\partial R} \nu \frac{\partial v}{\partial \mathbf{n}} dS, \qquad (4.2.24)$$

$$= \int_R F(\mathbf{x}, t) dA + \int_{\partial R} \nu \mathbf{n} \cdot \nabla v dS \qquad (4.2.25)$$

where $\mathbf{n}$ is the outward normal. We should note that equation (4.2.24) is the same as equation (1.6.2) except that the flux term is now an integral. The reason for this is that in the one dimensional case the boundary was just two points, so that the flux integral analogous to the one in equation (4.2.24) was just the evaluation at the two points (with a sign due to the direction of the normal derivative). We should realize that we can also obtain equations (4.2.24) and (4.2.25) from partial differential equation (4.2.1) by integrating both sides of equation (4.2.1) over R and applying Green's Theorem to the Laplacian term.

We apply equation (4.2.25) to the cell centered at the (j, k) grid point as is shown in Figure 4.2.3. We then obtain the following version of the integral form of the conservation law.

$$\int_{y_{k-1/2}}^{y_{k+1/2}} \int_{x_{j-1/2}}^{x_{j+1/2}} v_t dx dy = \int_{y_{k-1/2}}^{y_{k+1/2}} \int_{x_{j-1/2}}^{x_{j+1/2}} F(x, y, t) dx dy$$

$$+\nu \int_{y_{k-1/2}}^{y_{k+1/2}} v_x(x_{j+1/2}, y, t)dy - \nu \int_{y_{k-1/2}}^{y_{k+1/2}} v_x(x_{j-1/2}, y, t)dy$$

$$+\nu \int_{x_{j-1/2}}^{x_{j+1/2}} v_y(x, y_{k+1/2}, t)dx - \nu \int_{x_{j-1/2}}^{x_{j+1/2}} v_y(x, y_{k-1/2}, t)dx \tag{4.2.26}$$

If we then integrate equation (4.2.26) from t_n to t_{n+1}, we get

$$\int_{y_{k-1/2}}^{y_{k+1/2}} \int_{x_{j-1/2}}^{x_{j+1/2}} (v^{n+1} - v^n)dxdy = \int_{t_n}^{t_{n+1}} \int_{y_{k-1/2}}^{y_{k+1/2}} \int_{x_{j-1/2}}^{x_{j+1/2}} F(x, y, t)dxdydt$$

$$+\nu \int_{t_n}^{t_{n+1}} \int_{y_{k-1/2}}^{y_{k+1/2}} v_x(x_{j+1/2}, y, t)dydt - \nu \int_{t_n}^{t_{n+1}} \int_{y_{k-1/2}}^{y_{k+1/2}} v_x(x_{j-1/2}, y, t)dydt$$

$$+\nu \int_{t_n}^{t_{n+1}} \int_{x_{j-1/2}}^{x_{j+1/2}} v_y(x, y_{k+1/2}, t)dxdt - \nu \int_{t_n}^{t_{n+1}} \int_{x_{j-1/2}}^{x_{j+1/2}} v_y(x, y_{k-1/2}, t)dxdt.$$

$$\tag{4.2.27}$$

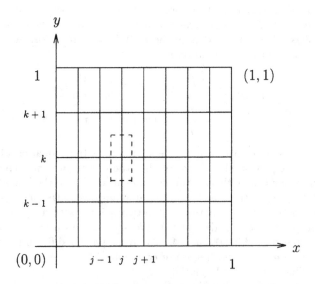

FIGURE 4.2.3. Cell centered at grid point (j, k).

We now proceed to derive a difference equation that approximates partial differential equation (4.2.14) by approximating the integral form of the conservation law (4.2.27). We begin by approximating the first two integrals by using the midpoint rule with respect to x and y and the trapezoidal rule with respect to t to get

$$\Delta x \Delta y (v_{jk}^{n+1} - v_{jk}^n) + \mathcal{O}(\Delta x^3 \Delta y \Delta t) + \mathcal{O}(\Delta x \Delta y^3 \Delta t) = \frac{\Delta t \Delta x \Delta y}{2}(F_{jk}^n + F_{jk}^{n+1})$$

$$+ \mathcal{O}(\Delta x^3 \Delta y \Delta t^2) + \mathcal{O}(\Delta x \Delta y^3 \Delta t^2) + \mathcal{O}(\Delta x \Delta y \Delta t^3)$$

$$+ \nu \int_{t_n}^{t_{n+1}} \int_{y_{k-1/2}}^{y_{k+1/2}} v_x(x_{j+1/2}, y, t) dy dt - \nu \int_{t_n}^{t_{n+1}} \int_{y_{k-1/2}}^{y_{k+1/2}} v_x(x_{j-1/2}, y, t) dy dt$$

$$+ \nu \int_{t_n}^{t_{n+1}} \int_{x_{j-1/2}}^{x_{j+1/2}} v_y(x, y_{k+1/2}, t) dx dt - \nu \int_{t_n}^{t_{n+1}} \int_{x_{j-1/2}}^{x_{j+1/2}} v_y(x, y_{k-1/2}, t) dx dt.$$

$$(4.2.28)$$

If we were to now approximate the derivatives in each of the last four integrals by centered differences, it would appear that we are left with a first order (in Δx and Δy) scheme. However, if we approximate $v_x(x_{j+1/2}, y, t) - v_x(x_{j-1/2}, y, t)$ instead, due to the symmetry, it is clear that we are left with a second order scheme (as was the case with our one dimensional derivation). Hence, approximating the derivatives in the last four integrals using centered differences, approximating the integrals with respect to x and y by the midpoint rule and the integrals with respect to t by the trapezoidal rule leaves us with

$$\Delta x \Delta y (v_{jk}^{n+1} - v_{jk}^n) + \mathcal{O}(\Delta x^3 \Delta y \Delta t) + \mathcal{O}(\Delta x \Delta y^3 \Delta t) = \frac{\Delta t \Delta x \Delta y}{2}(F_{jk}^n + F_{jk}^{n+1})$$

$$+ \mathcal{O}(\Delta x^3 \Delta y \Delta t^2) + \mathcal{O}(\Delta x \Delta y^3 \Delta t^2) + \mathcal{O}(\Delta x \Delta y \Delta t^3)$$

$$+ \nu \frac{\Delta t}{2} \Delta y \frac{1}{\Delta x} \delta_x^2 (v_{jk}^n + v_{jk}^{n+1}) + \mathcal{O}(\Delta t \Delta x^3 \Delta y)$$

$$+ \nu \frac{\Delta t}{2} \Delta x \frac{1}{\Delta y} \delta_y^2 (v_{jk}^n + v_{jk}^{n+1}) + \mathcal{O}(\Delta t \Delta x \Delta y^3).$$

$$(4.2.29)$$

Thus, if we divide by $\Delta x \Delta y$ and approximate equation (4.2.29) by ignoring the order terms, we get the following two dimensional Crank-Nicolson scheme for approximating partial differential equation (4.2.1)

$$\left(1 - \frac{r_x}{2}\delta_x^2 - \frac{r_y}{2}\delta_y^2\right)u_{jk}^{n+1} = u_{jk}^n + \frac{\Delta t}{2}(F_{jk}^n + F_{jk}^{n+1})$$

$$+ \frac{r_x}{2}\delta_x^2 u_{jk}^n + \frac{r_y}{2}\delta_y^2 u_{jk}^n.$$

$$(4.2.30)$$

We see above that the conservation law derivation of difference schemes for two dimensional schemes (and, hopefully, also for three dimensional schemes) follows in much the same manner as those for one dimensional schemes. The major difference is that for one dimensional schemes, the flux terms were derivatives evaluated at a point, whereas for two and three dimensional schemes the flux terms will be line and surface integrals, respectively. As in the case with one dimensional schemes, *if we treat the nonhomogeneous term poorly (say, use the rectangular rule as the integration scheme instead of the trapezoidal rule), the order of the scheme will be lowered.*

4.3   Convergence, Consistency, Stability

Before we proceed with more calculations based on two dimensional difference schemes, we should consider the convergence, consistency or stability of the schemes. We begin our discussion with the fact that our previous definitions of convergence (Definition 2.2.2), consistency (Definitions 2.3.2 and 2.3.3) and stability (Definition 2.4.1) will still hold for two and three dimensional partial differential equations and difference schemes. Likewise, the Lax Theorem (Theorem 2.5.2) and Lax Equivalence Theorem (Theorem 2.5.1) still hold. The only difference from the results for one dimensional problems is that the spatial norms will now be two or three dimensional (as we discussed in Section 2.2.3).

Our first concern is consistency. Most of the consistency arguments are the same as in the one dimensional case except for the fact that the Taylor series expansions are now two or three dimensional. For the case of implicit schemes of the form $Q_1 \mathbf{u}^{n+1} = Q\mathbf{u}^n$, to obtain norm consistency (which is what is necessary to use the Lax Theorem) we must still prove that Q_1^{-1} is uniformly bounded. Since Q_1 will be more complicated for two and three dimensional problems than it was for one dimensional problems, proving Q_1^{-1} uniformly bounded is more difficult for two and three dimensional problems than it was for one dimensional problems. As in the one dimensional case, when we are using the $\ell_{2,\Delta x}$ norm, we can apply the two or three dimensional version of Proposition 3.1.9 and easily obtain the uniform boundedness.

Based on the above discussion and using methods that we already have, it is not hard to see that difference scheme (4.2.5) is accurate order $\mathcal{O}(\Delta t) + \mathcal{O}(\Delta x^2) + \mathcal{O}(\Delta y^2)$ and difference scheme (4.2.30) is accurate order $\mathcal{O}(\Delta t^2) + \mathcal{O}(\Delta x^2) + \mathcal{O}(\Delta y^2)$ (where we realize to obtain the norm accuracy for difference scheme (4.2.30) it is still necessary to prove that the appropriate Q_1^{-1} is uniformly bounded). We should remember that when a consistency calculation is necessary, we can let one of the algebraic manipulators do the work for us (Maple, Mathematica, etc.) (much more work is saved using one of these packages to do two and three dimensional consistency analyses than was saved using an algebraic manipulator to do one dimensional consistency analyses).

Hopefully, it is clear that as in the case of one dimensional difference schemes, for two and three dimensional difference schemes the most difficult step to obtain convergence via the Lax Theorem is proving stability. For this reason, we now proceed to study stability for two and three dimensional schemes.

As a first step, let us consider difference scheme (4.2.5) as a scheme for an initial–value problem. We must realize that the stability analysis provided by considering (4.2.5) for an initial–value problem will at least provide us with necessary conditions for convergence for the initial–boundary–value problems discussed earlier.

4.3.1   Stability of Initial–Value Schemes

As in the case for stability of one dimensional difference schemes for initial–value problems, we shall use the discrete Fourier transform as a tool for proving stability. When we consider difference schemes having two spatial dimensions, we must consider the two dimensional discrete Fourier transform. For a sequence $\mathbf{u} = \{u_{j\,k}\}, -\infty < j, k < \infty$, we define

$$\mathcal{F}(\mathbf{u})(\xi,\eta) = \hat{u}(\xi,\eta) = \frac{1}{2\pi} \sum_{j,k=-\infty}^{\infty} e^{-ij\xi - ik\eta} u_{j\,k}. \qquad (4.3.1)$$

It is not hard to see that taking a two dimensional discrete Fourier transform is completely analogous to taking two one dimensional discrete Fourier transforms, one with respect to each of the two variables. With this approach, we then also know that we also have an inverse transform. Since we noted in Chapter 3 that the inverse transform was never really used, we will not give the inverse transform here. However, it is very important that

- Parseval's Identity holds for the two dimensional discrete Fourier transform, i.e. $\parallel \mathbf{u} \parallel_2 = \parallel \hat{\mathbf{u}} \parallel_2$ where the norms are the two dimensional ℓ_2 and $L_2\left([-\pi,\pi] \times [-\pi,\pi]\right)$ norms, respectively.

- $\mathcal{F}(S_{x\pm}\mathbf{u}) = e^{\pm i\xi}\mathcal{F}(\mathbf{u})$ and $\mathcal{F}(S_{y\pm}\mathbf{u}) = e^{\pm i\eta}\mathcal{F}(\mathbf{u})$, where $S_{x\pm}$ and $S_{y\pm}$ are the obvious two dimensional extensions of the shift operators.

Example 4.3.1 Analyze the stability of difference scheme (4.2.5).

Solution: To analyze the stability of difference scheme (4.2.5) as a scheme for solving an initial–value problem, we proceed as we did in the one dimensional case. We take the discrete Fourier transform of equation (4.2.5) (without the nonhomogeneous term F) to get

$$\begin{aligned}
\hat{u}^{n+1}(\xi,\eta) &= \hat{u}^n(\xi,\eta) + r_x(e^{-i\xi}\hat{u}^n(\xi,\eta) - 2\hat{u}^n(\xi,\eta) + e^{i\xi}\hat{u}^n(\xi,\eta)) \\
&\quad + r_y(e^{-i\eta}\hat{u}^n(\xi,\eta) - 2u^n(\xi,\eta) + e^{i\eta}\hat{u}^n(\xi,\eta)) \qquad (4.3.2) \\
&= \big[1 + r_x(e^{-i\xi} - 2 + e^{i\xi}) \\
&\quad + r_y(e^{-i\eta} - 2 + e^{i\eta})\big]\hat{u}(\xi,\eta). \qquad (4.3.3)
\end{aligned}$$

As before we have an expression

$$\hat{u}^{n+1} = \rho(\xi, \eta)\hat{u}^n \tag{4.3.4}$$

where ρ is again referred to as the symbol of the difference scheme and we have not written the (ξ, η) arguments in the $\hat{u}$ functions. (When it is clear what is meant, we will not include the arguments in the ρ expression.) As in the one dimensional case, the ρ term is the multiplier between steps n and $n + 1$. Applying (4.3.4) $n + 1$ times yields

$$\hat{u}^{n+1} = \rho^{n+1}\hat{u}^0. \tag{4.3.5}$$

Using exactly the same rationale used in the one dimensional case (the number of variables in ρ does not alter the argument at all), we see that if

$$\mid \rho(\xi, \eta) \mid \leq 1 + C\Delta t, \tag{4.3.6}$$

then

$$\mid \rho^{n+1} \mid \leq e^{C(n+1)\Delta t}$$

and

$$\parallel \hat{u}^{n+1} \parallel_2 \leq e^{C(n+1)\Delta t} \parallel \hat{u}^0 \parallel_2 . \tag{4.3.7}$$

Then equation (4.3.7), along with Parseval's Identity, yields

$$\parallel \mathbf{u}^{n+1} \parallel_{2,\Delta x,\Delta y} \leq e^{C(n+1)\Delta t} \parallel \mathbf{u}^n \parallel_{2,\Delta x,\Delta y},$$

where the two dimensional energy $\ell_{2,\Delta x}$ norm, $\parallel \mathbf{u} \parallel_{2,\Delta x,\Delta y}$, is as defined in Section 2.2.3.

Thus we see that all we have to do to obtain stability is to show that the symbol ρ satisfies (4.3.6). Since we do not have the "bu" term in our partial differential equation, we should suspect that C can be chosen to be zero. We find where the function $\mid \rho \mid$ has its maximum values by finding the maximum and minimum values of ρ. It is clear that ρ can be written as

$$\begin{aligned} \rho &= 1 + 2r_x(\cos\xi - 1) + 2r_y(\cos\eta - 1) \\ &= 1 - 4r_x \sin^2\frac{\xi}{2} - 4r_y \sin^2\frac{\eta}{2}. \end{aligned} \tag{4.3.8}$$

Differentiating equation (4.3.8) with respect to ξ and η, setting these derivatives equal to zero and solving for ξ and η, we find the the critical points for this function occur at all combinations of $\xi = -\pi, 0, \pi$ and $\eta = -\pi, 0, \pi$.

It is easy to see that the maximum of

$$\rho = 1 \quad \text{occurs at } (\xi, \eta) = (0, 0)$$

and the minimum of

$$\rho = 1 - 4r_x - 4r_y \quad \text{occurs at } (\xi, \eta) = (\pi, \pi).$$

Then requiring that ρ satisfy $\rho \geq -1$ yields the stability condition

$$r_x + r_y \leq \frac{1}{2}. \tag{4.3.9}$$

Hence, we see that *difference scheme (4.2.5) is conditional stable.*

We should note that when $\Delta x = \Delta y$, then the condition for stability becomes $r \leq 1/4$. Thus the two dimensional stability condition for the explicit scheme is more severe than the one dimensional analog.

Example 4.3.2 Discuss the stability of the difference scheme

$$u_{j\,k}^{n+1} = \left(1 - \frac{R_x}{2}\delta_{x0} - \frac{R_y}{2}\delta_{y0} + r_x\delta_x^2 + r_y\delta_y^2\right)u_{j\,k}^n. \tag{4.3.10}$$

Solution: Before we begin we should realize that difference scheme (4.3.10) is a scheme for approximating the solution to a partial differential equation of the form

$$v_t + av_x + bv_y = \nu_1 v_{xx} + \nu_2 v_{yy}.$$

We should also note that we analyzed the one dimensional analog to difference scheme (4.3.10) in Example 3.1.3. It is very difficult (maybe impossible) to perform the calculations done in Example 3.1.3 for this two dimensional scheme. Instead, we return to the approach described in Remark 2, page 109. We begin by taking the discrete Fourier transform of equation (4.3.10) and see that the symbol of difference scheme (4.3.10) is given by

$$\rho(\xi, \eta) = 1 - 4r_x \sin^2\frac{\xi}{2} - 4r_y \sin^2\frac{\eta}{2} - i(R_x \sin\xi + R_y \sin\eta). \tag{4.3.11}$$

The magnitude of the symbol can then be written as

$$|\rho|^2 = \left(1 - 4r_x \sin^2\frac{\xi}{2} - 4r_y \sin^2\frac{\eta}{2}\right)^2 + (R_x \sin\xi + R_y \sin\eta)^2. \tag{4.3.12}$$

If we then choose r_x and r_y fixed constants satisfying $r_x + r_y \leq 1/2$, we use the results of Example 4.3.1 and the approach used in Remark 2, page 109, to get

$$|\rho|^2 \leq 1 + \left(\frac{a^2 r_x}{\nu_1} + \frac{2\,|ab|\,\sqrt{r_x r_y}}{\sqrt{\nu_1 \nu_2}} + \frac{b^2 r_y}{\nu_2}\right)\Delta t.$$

Hence, difference scheme (4.3.10) is conditionally stable. We emphasize, however, that just as was the case with the stability proved in Remark 2, page 109, we must be careful with the stability proved above. Again we are proving the symbol is bounded by an exponential when we should be able to show that it is bounded by a constant. We know that as Δt, Δx and Δy approach zero, everything will be fine. If we choose any particular Δt, Δx and Δy satisfying $r_x + r_y \leq 1/2$ and begin computing, we may get bad results (exponential growth with time). However, we do know that if we keep r_x and r_y fixed and make Δt smaller, we will eventually get good results.

Example 4.3.3 Discuss the stability of the two dimensional Crank-Nicolson scheme, (4.2.30).
Solution: We begin by taking the two dimensional discrete Fourier transform of equation (4.2.30) (without the nonhomogeneous term) to get

$$\left(1 + 2r_x \sin^2\frac{\xi}{2} + 2r_y \sin^2\frac{\eta}{2}\right)\hat{u}^{n+1} = \left(1 - 2r_x \sin^2\frac{\xi}{2} - 2r_y \sin^2\frac{\eta}{2}\right)\hat{u}^n. \tag{4.3.13}$$

Thus we see that the symbol for difference scheme (4.2.30) is given by

$$\rho(\xi, \eta) = \frac{1 - 2r_x \sin^2\frac{\xi}{2} - 2r_y \sin^2\frac{\eta}{2}}{1 + 2r_x \sin^2\frac{\xi}{2} + 2r_y \sin^2\frac{\eta}{2}}. \tag{4.3.14}$$

Since

$$\left|\frac{1-r}{1+r}\right| \leq 1$$

for any $r \geq 0$, clearly $|\rho(\xi, \eta)| \leq 1$ and difference scheme (4.2.30) is unconditionally stable.

Remark: We should note in the above example, since

$$\frac{1}{1 + 2r_x \sin^2 \frac{\xi}{2} + 2r_y \sin^2 \frac{\eta}{2}} \leq 1,$$

by the two dimensional version of Proposition 3.1.9, Q_1^{-1} is uniformly bounded, difference scheme (4.2.30) is norm accurate of order $\mathcal{O}(\Delta t^2) + \mathcal{O}(\Delta x^2) + \mathcal{O}(\Delta y^2)$ and by the stability result found in Example 4.3.3, the Crank-Nicolson scheme, (4.2.30), is convergent.

HW 4.3.1 Discuss the stability of the following BTCS difference scheme.

$$u_{jk}^{n+1} - r_x \delta_x^2 u_{jk}^{n+1} - r_y \delta_y^2 u_{jk}^{n+1} = u_{jk}^n \qquad (4.3.15)$$

HW 4.3.2 Discuss the stability of the following difference scheme for approximating the solution to partial differential equation.

$$v_t + av_x + bv_y = \nu_1 v_{xx} + \nu_2 v_{yy}$$

$$u_{jk}^{n+1} + \frac{R_x}{2} \delta_{x0} u_{jk}^{n+1} + \frac{R_y}{2} \delta_{0y} u_{jk}^{n+1} - r_x \delta_x^2 u_{jk}^{n+1} - r_y \delta_y^2 u_{jk}^{n+1} = u_{jk}^n$$

HW 4.3.3 Discuss the consistency, stability and convergence of the following difference scheme.

$$u_{jk}^{n+1} = u_{jk}^n + r_x \left(-\frac{1}{6} u_{j-2k}^n + \frac{5}{3} u_{j-1k}^n - 3u_{jk}^n + \frac{5}{3} u_{j+1k}^n - \frac{1}{6} u_{j+2k}^n \right)$$

$$+ r_y \left(-\frac{1}{6} u_{jk-2}^n + \frac{5}{3} u_{jk-1}^n - 3u_{jk}^n + \frac{5}{3} u_{jk+1}^n - \frac{1}{6} u_{jk+2}^n \right) \qquad (4.3.16)$$

4.3.2   Stability of Initial–Boundary–Value Schemes

In reality, though we can do a reasonably good job analyzing the consistency, stability and convergence of schemes for initial–value problems, we must realize that we do not put these schemes on the computer. Thus we must consider the convergence of schemes for initial–boundary–value problems. Suppose we begin by considering difference scheme (4.2.5)–(4.2.9). If we recall our work on initial–boundary–value problems in one dimension, we realize that our results depend on finding the norm of the iteration matrix via its eigenvalues. To write the scheme (4.2.5)–(4.2.9) in matrix form, we must first consider how to write our vector of solution values. There are many logical ways (and many of them are useful for some purpose) to order the values. To begin with, we choose one of the most obvious orderings and set

$$\mathbf{u}^n = [u_{11}^n \cdots u_{M_x-11}^n u_{12}^n \cdots u_{M_x-12}^n \cdots u_{M_x-1 M_y-1}^n]^T. \qquad (4.3.17)$$

Of course, we must realize that the reason that the indices started at 1 (instead of 0) and ended at $M_x - 1$ or $M_y - 1$ (instead of M_x or M_y) is because we have Dirichlet boundary conditions, so the values with indices of 0, M_x and M_y are not needed. If we consider the ordering given in (4.3.17) and carefully examine (4.2.5)–(4.2.9), we see that the scheme can be written in matrix form as

$$\mathbf{u}^{n+1} = Q\mathbf{u}^n + \Delta t \mathbf{G}^n \qquad (4.3.18)$$

where

$$Q = \begin{pmatrix} B & r_y I & \Theta & \cdots & \\ r_y I & B & r_y I & \Theta & \cdots \\ & \cdots & \cdots & \cdots & \\ & \cdots & \Theta & r_y I & B \end{pmatrix}_{(M_y-1)\times(M_y-1)} \qquad (4.3.19)$$

B is the $(M_x - 1) \times (M_x - 1)$ matrix

$$\begin{pmatrix} 1 - 2r_x - 2r_y & r_x & 0 & \cdots & \\ r_x & 1 - 2r_x - 2r_y & r_x & 0 & \cdots \\ & \cdots & \cdots & \cdots & \\ \cdots & 0 & r_x & 1 - 2r_x - 2r_y & r_x \\ \cdots & & 0 & r_x & 1 - 2r_x - 2r_y \end{pmatrix}$$

$$(4.3.20)$$

I is the $(M_x - 1) \times (M_x - 1)$ identity, Θ is the $(M_x - 1) \times (M_x - 1)$ zero matrix and $\mathbf{G}^n = [G_{jk}^n]$ where G_{jk}^n is given by

$$\begin{cases} F_{jk}^n & \text{if } j \neq 1, \; k \neq 1, \; j \neq M_x - 1, \; k \neq M_y - 1 \\ F_{jk}^n + r_x u_{0k}^n + r_y u_{j0}^n & \text{if } j = k = 1 \\ F_{jk}^n + r_y u_{j0}^n & \text{if } k = 1, \text{ and } j \neq 1 \text{ or } M_x - 1 \\ F_{jk}^n + r_x u_{M_x k}^n + r_y u_{j0}^n & \text{if } k = 1 \text{ and } j = M_x - 1 \\ F_{jk}^n + r_x u_{0k}^n & \text{if } j = 1, \text{ and } k \neq 1 \text{ or } M_y - 1 \\ F_{jk}^n + r_x u_{M_x k}^n & \text{if } j = M_x - 1, \text{ and } k \neq 1 \text{ or } M_y - 1 \\ F_{jk}^n + r_x u_{0k}^n + r_y u_{jM_y}^n & \text{if } j = 1 \text{ and } k = M_y - 1 \\ F_{jk}^n + r_x u_{M_x k}^n + r_y u_{jM_y}^n & \text{if } j = M_x - 1 \text{ and } k = M_y - 1 \end{cases} \qquad (4.3.21)$$

We suggest that the reader write out the matrix Q and nonhomogeneous term to get a better feeling of what it looks like. To try to give a more intuitive look at the matrix, in Figure 4.3.1 we include a version of the matrix where we have chosen $M_x = 5$ and $M_y = 6$.

The two approachs that we have for obtaining stability results for initial–boundary–value problems such as that described by the matrix equation (4.3.18) are

- to treat the scheme as an initial–value scheme and obtain the resulting necessary conditions; or

$$
\left(\begin{array}{cccc|cccc|cccc|cccc|cccc}
d & r_x & 0 & 0 & r_y & 0 & 0 & 0 & 0 & 0 & 0 & 0 & 0 & 0 & 0 & 0 & 0 & 0 & 0 & 0 \\
r_x & d & r_x & 0 & 0 & r_y & 0 & 0 & 0 & 0 & 0 & 0 & 0 & 0 & 0 & 0 & 0 & 0 & 0 & 0 \\
0 & r_x & d & r_x & 0 & 0 & r_y & 0 & 0 & 0 & 0 & 0 & 0 & 0 & 0 & 0 & 0 & 0 & 0 & 0 \\
0 & 0 & r_x & d & \mathbf{0} & 0 & 0 & r_y & 0 & 0 & 0 & 0 & 0 & 0 & 0 & 0 & 0 & 0 & 0 & 0 \\
\hline
r_y & 0 & 0 & \mathbf{0} & d & r_x & 0 & 0 & r_y & 0 & 0 & 0 & 0 & 0 & 0 & 0 & 0 & 0 & 0 & 0 \\
0 & r_y & 0 & 0 & r_x & d & r_x & 0 & 0 & r_y & 0 & 0 & 0 & 0 & 0 & 0 & 0 & 0 & 0 & 0 \\
0 & 0 & r_y & 0 & 0 & r_x & d & r_x & 0 & 0 & r_y & 0 & 0 & 0 & 0 & 0 & 0 & 0 & 0 & 0 \\
0 & 0 & 0 & r_y & 0 & 0 & r_x & d & \mathbf{0} & 0 & 0 & r_y & 0 & 0 & 0 & 0 & 0 & 0 & 0 & 0 \\
\hline
0 & 0 & 0 & 0 & r_y & 0 & 0 & \mathbf{0} & d & r_x & 0 & 0 & r_y & 0 & 0 & 0 & 0 & 0 & 0 & 0 \\
0 & 0 & 0 & 0 & 0 & r_y & 0 & 0 & r_x & d & r_x & 0 & 0 & r_y & 0 & 0 & 0 & 0 & 0 & 0 \\
0 & 0 & 0 & 0 & 0 & 0 & r_y & 0 & 0 & r_x & d & r_x & 0 & 0 & r_y & 0 & 0 & 0 & 0 & 0 \\
0 & 0 & 0 & 0 & 0 & 0 & 0 & r_y & 0 & 0 & r_x & d & \mathbf{0} & 0 & 0 & r_y & 0 & 0 & 0 & 0 \\
\hline
0 & 0 & 0 & 0 & 0 & 0 & 0 & 0 & r_y & 0 & 0 & \mathbf{0} & d & r_x & 0 & 0 & r_y & 0 & 0 & 0 \\
0 & 0 & 0 & 0 & 0 & 0 & 0 & 0 & 0 & r_y & 0 & 0 & r_x & d & r_x & 0 & 0 & r_y & 0 & 0 \\
0 & 0 & 0 & 0 & 0 & 0 & 0 & 0 & 0 & 0 & r_y & 0 & 0 & r_x & d & r_x & 0 & 0 & r_y & 0 \\
0 & 0 & 0 & 0 & 0 & 0 & 0 & 0 & 0 & 0 & 0 & r_y & 0 & 0 & r_x & d & \mathbf{0} & 0 & 0 & r_y \\
\hline
0 & 0 & 0 & 0 & 0 & 0 & 0 & 0 & 0 & 0 & 0 & 0 & r_y & 0 & 0 & \mathbf{0} & d & r_x & 0 & 0 \\
0 & 0 & 0 & 0 & 0 & 0 & 0 & 0 & 0 & 0 & 0 & 0 & 0 & r_y & 0 & 0 & r_x & d & r_x & 0 \\
0 & 0 & 0 & 0 & 0 & 0 & 0 & 0 & 0 & 0 & 0 & 0 & 0 & 0 & r_y & 0 & 0 & r_x & d & r_x \\
0 & 0 & 0 & 0 & 0 & 0 & 0 & 0 & 0 & 0 & 0 & 0 & 0 & 0 & 0 & r_y & 0 & 0 & r_x & d
\end{array}\right)
$$

FIGURE 4.3.1. Matrix Q, (4.3.19), where $M_x = 5$ and $M_y = 6$, $d = 1 - 2r_x - 2r_y$ and the bold face zeros denote the special zeros that are necessary due to the boundaries. The partitioning analogous to that described in equation (4.3.19) is included.

- to calculate the spectral radius of Q and get necessary and sufficient conditions or necessary conditions based on whether or not Q is symmetric (or similar to a symmetric matrix).

It is generally difficult to calculate the eigenvalues of a matrix such as the one given in (4.3.18). We will obtain some general methods later that will give us a method for calculating eigenvalues of large difficult matrices. *Often, we will have to be content with obtaining only a necessary condition* (by treating the scheme as an initial–value problem and applying the two dimensional analog of Proposition 3.1.10).

The situation is relatively nice for the scheme described by (4.3.18). If we consider the approximation to the partial differential equation in difference scheme (4.3.18) as an initial–value scheme, we see from the work done in Section 4.3.1 for difference scheme (4.2.5) (which is the approximation to the partial differential equation used in scheme (4.3.18)) and Proposition 3.1.10, $r_x + r_y \leq 1/2$ *is a necessary condition for convergence of difference scheme* (4.3.18). Since we have Dirichlet boundary conditions, the problem is solvable by finite Fourier series. Then the discrete von Neumann stability criteria will apply and will calculate the eigenvalues of Q to give both necessary and sufficient conditions for stability (since Q is obviously symmetric). To apply the discrete von Neumann criteria to a two dimensional problem, we let

$$u^n_{jk} = \xi^n e^{ijp\pi\Delta x + ikq\pi\Delta y}. \tag{4.3.22}$$

If we replace u^n_{jk} and u^{n+1}_{jk} in (4.2.5) (excluding the F^n_{jk} term) by expressions of the form given by equation (4.3.22), we get

$$\xi^{n+1} = \xi^n \left[1 - 4r_x \sin^2 \frac{p\pi\Delta x}{2} - 4r_y \sin^2 \frac{q\pi\Delta y}{2} \right]. \tag{4.3.23}$$

Dividing by ξ^n yields

$$\xi = 1 - 4r_x \sin^2 \frac{p\pi\Delta x}{2} - 4r_y \sin^2 \frac{q\pi\Delta y}{2}. \tag{4.3.24}$$

Of course, as we did earlier, $\mid \xi \mid \leq 1$ whenever $r_x + r_y \leq 1/2$.

Since Q is a symmetric matrix, the condition is necessary and sufficient for stability. Then, since the scheme is consistent $(\mathcal{O}(\Delta t) + \mathcal{O}(\Delta x^2) + \mathcal{O}(\Delta y^2))$, the scheme is also convergent.

We close this section with an example illustrating the application of the two dimensional discrete von Neumann criteria to obtain a necessary and sufficient condition for stability of an implicit scheme. We note that essentially this same analysis also applies to the Crank-Nicolson scheme.

Example 4.3.4 Discuss the stability of the implicit difference scheme

$$u^{n+1}_{jk} - r_x \delta^2_x u^{n+1}_{jk} - r_y \delta^2_y u^{n+1}_{jk} = u^n_{jk} + \Delta t F^{n+1}_{jk} \tag{4.3.25}$$

along with boundary conditions (4.2.6)–(4.2.9).

Solution: We begin by recalling that in HW4.3.1 we found that difference scheme (4.3.25) considered as an initial–value scheme was unconditionally stable. Hence, "no condition is necessary for stability" or we have no reason to expect that difference scheme (4.3.25), (4.2.6)–(4.2.9) will be unstable. Since $Q = Q_1^{-1}$ is clearly symmetric (Q_1 is symmetric and $Q_1 Q_1^{-1} = I$ implies $I = (Q_1^{-1})^T Q_1^T = (Q_1^{-1})^T Q_1$. So $Q_1^{-1} = (Q_1^{-1})^T$.) and we have Dirichlet boundary conditions, we realize that this problem can be solved by finite Fourier series so that the von Neumann analysis will give us both necessary and sufficient conditions for stability.

If we proceed as we did in the explicit case and carefully write this scheme as a matrix equation, we get

$$Q_1 \mathbf{u}^{n+1} = \mathbf{u}^n + \Delta t \mathbf{G}^{n+1} \qquad (4.3.26)$$

where both $\mathbf{u}^{n+1}$ and $\mathbf{u}^n$ are given in the same order as the vector in (4.3.17),

$$Q_1 = \begin{pmatrix} B & -r_y I & \Theta & \cdots \\ -r_y I & B & -r_y I & \Theta & \cdots \\ \cdots & \cdots & \cdots & \cdots & \cdots \\ & \cdots & \Theta & -r_y I & B \end{pmatrix}_{(M_y-1)\times(M_y-1)} \qquad (4.3.27)$$

B is the $(M_x - 1) \times (M_x - 1)$ matrix

$$B = \begin{pmatrix} 1+2r_x+2r_y & -r_x & 0 & \cdots \\ -r_x & 1+2r_x+r_y & -r_x & 0 & \cdots \\ \cdots & \cdots & \cdots & \cdots \\ & \cdots & 0 & -r_x & 1+2r_x+2r_y \end{pmatrix} \qquad (4.3.28)$$

and $\mathbf{G}^n$, as in the case of the explicit scheme, contains the nonhomogeneous terms and the boundary conditions.

If we compare this scheme with the explicit scheme, it is not difficult to see that substituting expression (4.3.22) into equation (4.3.25) (excluding the nonhomogeneous term) will give

$$\xi = \frac{1}{1 + 4r_x \sin^2 \frac{p\pi\Delta x}{2} + 4r_y \sin^2 \frac{q\pi\Delta y}{2}}.$$

It is easy to see that $0 \leq \xi \leq 1$. Thus, as in the one dimensional case, the implicit scheme is unconditionally stable. We should mention that *the unconditional stability here is specifically for Dirichlet boundary conditions*. It is also the case that the scheme is unconditionally stable for second order Neumann boundary conditions. For other boundary conditions, the above analysis only shows that we do not know that it is unstable for any value of r_x and r_y.

HW 4.3.4 Discuss the stability (and, hence, convergence) for the Crank-Nicolson scheme (specifically, (4.2.30),(4.2.6)–(4.2.9)) as a solution scheme for the initial–boundary–value problem (4.2.1)–(4.2.3) with $\overline{R} = [0,1] \times [0,1]$.

4.4   Alternating Direction Implicit Schemes

Now that we know that the implicit difference scheme (4.3.25) is unconditionally stable, we consider how we might solve the matrix equation

(4.3.26). To use the above implicit scheme (and it is not hard to see that the two dimensional Crank-Nicolson scheme would have the same form), we must solve a broadly banded matrix equation. We do not generally want to try to solve an equation such as (4.3.26) by Gaussian elimination because, in general, Gaussian elimination is too expensive to use for solving the matrix equations associated with difference schemes in two or more dimensions.

Another approach is to use any of the iterative schemes to solve equation (4.3.26) (Gauss Seidel, Jacobi, SOR, multigrid, etc.). Iterative solvers can be and are sometimes used. Generally, iterative solvers are expensive to use to solve equation (4.3.26) at each time step (and, more so, the three dimensional analog to equation (4.3.26)). We will introduce and test using some of the iterative solvers for solving time dependent partial differential equations in Chapter 10.

A way around solving an equation such as equation (4.3.26) is to use an alternating direction implicit (ADI) scheme. As we shall see, using an ADI scheme can be interpreted as either approximating the solution to equation (4.3.26) or as using equation (4.3.26) to help us develop a new (better) scheme. In either case, ADI schemes will give us all of the advantages of implicit schemes, and computationally will only require that we solve tridiagonal matrices. Thus, ADI schemes provide us with a class of computationally efficient, implicit schemes for two and three dimensional problems. We introduce the general topic of ADI schemes by developing several of the most common schemes.

4.4.1   Peaceman-Rachford Scheme

4.4.2   Initial–Value Problems

Though we will eventually be interested in approximating solution to initial–boundary–value problems (say by considering difference equation (4.3.25) along with boundary conditions (4.2.6)–(4.2.9)), we begin by considering an approximation to an initial–value problem consisting of the homogeneous version of difference equation (4.3.25) along with an initial condition. We want to devise a scheme that will either solve or approximately solve equation (4.3.25) (again with $F = 0$) and be computationally efficient.

One way to try to make the difference equation nicer is to evaluate one of the spatial derivatives implicitly and the other explicitly. Computationally, this scheme would be nice because we would be left with a tridiagonal matrix to solve. If, for example, we evaluate the derivative with respect to x implicitly, the derivative with respect to y explicitly and use a time step of $\Delta t/2$ (why we do this will become clear very soon), we get the scheme

$$\frac{u_{jk}^{n+\frac{1}{2}} - u_{jk}^n}{\Delta t/2} = \frac{\nu}{\Delta x^2}\delta_x^2 u_{jk}^{n+\frac{1}{2}} + \frac{\nu}{\Delta y^2}\delta_y^2 u_{jk}^n. \qquad (4.4.1)$$

We rewrite difference scheme (4.4.1) as

$$\left(1 - \frac{r_x}{2}\delta_x^2\right)u_{jk}^{n+\frac{1}{2}} = \left(1 + \frac{r_y}{2}\delta_y^2\right)u_{jk}^n. \tag{4.4.2}$$

If we take the discrete Fourier transform of equation (4.4.2), we get

$$\left(1 + 2r_x \sin^2\frac{\xi}{2}\right)\hat{u}^{n+\frac{1}{2}} = \left(1 - 2r_y \sin^2\frac{\eta}{2}\right)\hat{u}^n. \tag{4.4.3}$$

It is easy to see that the symbol of the difference operator,

$$\rho(\xi, \eta) = \frac{1 - 2r_y \sin^2\frac{\eta}{2}}{1 + 2r_x \sin^2\frac{\xi}{2}}, \tag{4.4.4}$$

is less than or equal to one. Using the fact that the minimum value of ρ occurs at $(\xi, \eta) = (0, \pm\pi)$, we see that scheme (4.4.2) is conditionally stable with a stability limit of

$$r_y \leq 1.$$

(If we had used a full Δt time step, this condition would be $r_y \leq 1/2$.) Having this stability condition, we cannot justify having to solve a tridiagonal matrix.

By examining ρ given in equation (4.4.4) carefully, we see that if we take another half step (now we see why we used $\Delta t/2$ in equation (4.4.1)) of the form

$$\frac{u_{jk}^{n+1} - u_{jk}^{n+\frac{1}{2}}}{\Delta t/2} = \frac{\nu}{\Delta x^2}\delta_x^2 u_{jk}^{n+\frac{1}{2}} + \frac{\nu}{\Delta y^2}\delta_y^2 u_{jk}^{n+1}, \tag{4.4.5}$$

(this time explicit in x and implicit in y) or

$$\left(1 - \frac{r_y}{2}\delta_y^2\right)u_{jk}^{n+1} = \left(1 + \frac{r_x}{2}\delta_x^2\right)u_{jk}^{n+\frac{1}{2}}, \tag{4.4.6}$$

and take the discrete Fourier transform of the result, we get

$$\left(1 + 2r_y \sin^2\frac{\eta}{2}\right)\hat{u}^{n+1} = \left(1 - 2r_x \sin^2\frac{\xi}{2}\right)\hat{u}^{n+\frac{1}{2}}$$
$$= \left(1 - 2r_x \sin^2\frac{\xi}{2}\right)\frac{1 - 2r_y \sin^2\frac{\eta}{2}}{1 + 2r_x \sin^2\frac{\xi}{2}}\hat{u}^n. \tag{4.4.7}$$

Thus, the symbol of the two step difference scheme (4.4.2),(4.4.6) is

$$\rho(\xi, \eta) = \frac{(1 - 2r_x \sin^2\frac{\xi}{2})(1 - 2r_y \sin^2\frac{\eta}{2})}{(1 + 2r_x \sin^2\frac{\xi}{2})(1 + 2r_y \sin^2\frac{\eta}{2})}. \tag{4.4.8}$$

It is not hard to see that ρ satsifies $|\rho| \leq 1$. Hence, *the two step scheme (4.4.2),(4.4.6), called the* **Peaceman-Rachford scheme**, *is unconditionally stable.*

4.4.2.1   Consistency

We next consider the accuracy (consistency) of scheme (4.4.2),(4.4.6). Considering the method by which we motived the scheme, we should guess that the scheme would be $\mathcal{O}(\Delta t) + \mathcal{O}(\Delta x^2) + \mathcal{O}(\Delta y^2)$ (which would be wrong). If we operate on both sides of equation (4.4.2) by $(1 + \frac{r_x}{2}\delta_x^2)$, we get

$$\left(1 + \frac{r_x}{2}\delta_x^2\right)\left(1 - \frac{r_x}{2}\delta_x^2\right)u_{jk}^{n+\frac{1}{2}} = \left(1 + \frac{r_x}{2}\delta_x^2\right)\left(1 + \frac{r_y}{2}\delta_y^2\right)u_{jk}^n. \qquad (4.4.9)$$

Since the product of the two operators on the left hand side commute, equation (4.4.9) can be written as

$$\left(1 - \frac{r_x}{2}\delta_x^2\right)\left(1 + \frac{r_x}{2}\delta_x^2\right)u_{jk}^{n+\frac{1}{2}} = \left(1 + \frac{r_x}{2}\delta_x^2\right)\left(1 + \frac{r_y}{2}\delta_y^2\right)u_{jk}^n. \qquad (4.4.10)$$

Equation (4.4.6) can be used to eliminate the $u_{jk}^{n+\frac{1}{2}}$ term and lets us write the Peaceman-Rachford scheme as

$$\left(1 - \frac{r_x}{2}\delta_x^2\right)\left(1 - \frac{r_y}{2}\delta_y^2\right)u_{jk}^{n+1} = \left(1 + \frac{r_x}{2}\delta_x^2\right)\left(1 + \frac{r_y}{2}\delta_y^2\right)u_{jk}^n. \qquad (4.4.11)$$

Expanding the terms in equation (4.4.11) (being careful to not commute some operator that might not want to commute), we see that form (4.4.11) of the Peaceman-Rachford scheme is equivalent to

$$\frac{u_{jk}^{n+1} - u_{jk}^n}{\Delta t} = \frac{\nu}{2}\frac{\delta_x^2}{\Delta x^2}(u_{jk}^n + u_{jk}^{n+1}) + \frac{\nu}{2}\frac{\delta_y^2}{\Delta y^2}(u_{jk}^n + u_{jk}^{n+1})$$
$$- \frac{\nu^2\Delta t}{4}\frac{\delta_x^2}{\Delta x^2}\frac{\delta_y^2}{\Delta y^2}(u_{jk}^{n+1} - u_{jk}^n). \qquad (4.4.12)$$

Notice that (4.4.12) makes it clear that the two step difference scheme (4.4.2),(4.4.6) is very much like the two dimensional Crank-Nicolson scheme (except for the last term). If we expand the last term in a Taylor series expansion, we see that

$$\frac{\delta_x^2}{\Delta x^2}\frac{\delta_y^2}{\Delta y^2}(u_{jk}^{n+1} - u_{jk}^n) = \Delta t\left(\frac{\partial^5 u}{\partial t \partial^2 x \partial^2 y}\right)_{jk}^n + \mathcal{O}(\Delta t \Delta x^2)$$
$$+ \mathcal{O}(\Delta t \Delta y^2) + \mathcal{O}(\Delta t^2). \qquad (4.4.13)$$

Hence, since the term expanded above is multiplied by Δt in equation (4.4.12), *the two step difference scheme (4.4.2),(4.4.6) is (like Crank-Nicolson) second order accurate in $\Delta t, \Delta x$ and Δy.* Of course, since we have a consistent, stable scheme, we have a convergent scheme.

Recall that we were looking for an unconditionally stable, computationally efficient scheme. We would not have introduced the Peaceman-Rachford scheme if it did not satisfy these criteria. However, we delay showing how and why the Peaceman-Rachford scheme is computationally efficient until Section 4.4.3.4.

We might notice that once we have written the Peaceman-Rachford scheme in the form (4.4.11), there is at least one other obvious way to split this equation so as to make the computations efficient (other than as (4.4.2) and (4.4.6)). If we split equation (4.4.11) into

$$\left(1 - \frac{r_x}{2}\delta_x^2\right)u_{jk}^* = \left(1 + \frac{r_x}{2}\delta_x^2\right)\left(1 + \frac{r_y}{2}\delta_y^2\right)u_{jk}^n \qquad (4.4.14)$$

$$\left(1 - \frac{r_y}{2}\delta_y^2\right)u_{jk}^{n+1} = u_{jk}^*, \qquad (4.4.15)$$

the scheme is referred to as the **D'Yakonov scheme**.

4.4.2.2   Approximate Factorization

Before discussing other aspects of the Peaceman-Rachford scheme, we pause to show an alternate formulation of the scheme. We begin with the two dimensional Crank-Nicolson scheme,

$$\left(1 - \frac{r_x}{2}\delta_x^2 - \frac{r_y}{2}\delta_y^2\right)u_{jk}^{n+1} = \left(1 + \frac{r_x}{2}\delta_x^2 + \frac{r_y}{2}\delta_y^2\right)u_{jk}^n, \qquad (4.4.16)$$

Examining the left hand side of this equation, we see that we might want to factor the left hand side of equation (4.4.16) into

$$\left(1 - \frac{r_x}{2}\delta_x^2\right)\left(1 - \frac{r_y}{2}\delta_y^2\right)u_{jk}^{n+1}.$$

In doing this factorization, we have added in a term of the form

$$\frac{r_x r_y}{4}\delta_x^2\delta_y^2 u_{jk}^{n+1}. \qquad (4.4.17)$$

We might note that if the above factorization were sufficiently accurate, we would not care about the fact that we have not factored the right hand side. The right hand side all depends on u_{jk}^n and can be easily built for the solution process. In adding a term of the form of (4.4.17), we have added a term that is like

$$\frac{\Delta t^2}{4}\left(\frac{\partial^4 u}{\partial^2 x \partial^2 y}\right)_{jk}^{n+1} + \text{higher order terms.}$$

Since equation (4.4.16) must be divided by Δt to get it in the form of the partial differential equation to perform a consistency analysis, adding a term of the form (4.4.17) will lower the order of the Crank-Nicolson scheme from second to first order in Δt.

If we also add a term to the right hand side of the form,

$$\frac{r_x r_y}{4}\delta_x^2\delta_y^2 u_{jk}^n,$$

the net result will be the same as adding a term of the form

$$\frac{r_x r_y}{4} \delta_x^2 \delta_y^2 (u_{jk}^{n+1} - u_{jk}^n).$$

Since we saw in equation (4.4.13) that the above term will be order $\mathcal{O}(\Delta t^3)$ (remembering that we must still divide by one Δt), if we add these two terms to equation (4.4.16), the resulting equation will have the same order of accuracy as Crank-Nicolson. Hence, we add the two terms to equation (4.4.16), factor both sides of the equation and write the resulting equation as equation (4.4.11). Schemes that are developed using this approximate factorization are called **approximate factorization schemes**. The general class of approximate factorization schemes is a large and important class of schemes and we shall return to the concept later.

4.4.3   Initial–Boundary–Value Problems

As usual, it is more difficult to consider initial–boundary–value problems than it is to consider initial–value problems. Of course, the stability results for the Peaceman-Rachford initial–value problem scheme from Section 4.4.2 provide necessary conditions for stability (and, hence, convergence) of the initial–boundary–value problem Peaceman-Rachford scheme. Also, as before, if our matrix is symmetric and we can find the eigenvalues of the matrix (which we can—at least for Dirichlet and Neumann boundary conditions), we obtain necessary and sufficient conditions for stability and convergence. We begin with a discussion of how to apply boundary conditions for ADI schemes.

4.4.3.1   Dirichlet Boundary Conditions

One of the special difficulties encountered with the implementation of factored schemes such as (4.4.2), (4.4.6) or (4.4.14)–(4.4.15) is that of boundary conditions. The difficulties are not necessarily in the boundary conditions for u_{jk}^n or u_{jk}^{n+1} (though we must show as much care as we usually do for these boundary conditions), but in the boundary conditions for the functions $u_{jk}^{n+\frac{1}{2}}$ or u_{jk}^*. It is not at all difficult (and it is all too common) to naively choose boundary conditions for these functions that will lower the order of the scheme.

For example, if we were to apply the D'Yakonov scheme, (4.4.14)–(4.4.15), to an initial–boundary–value problem (4.2.1)–(4.2.3) (on $R = [0,1] \times [0,1]$) and use boundary conditions

$$u_{0k}^* = g(0, k\Delta y, (n+1)\Delta t)$$
$$u_{M_x k}^* = g(1, k\Delta y, (n+1)\Delta t)$$
$$u_{j0}^* = g(j\Delta x, 0, (n+1)\Delta t)$$
$$u_{j M_y}^* = g(j\Delta x, 1, (n+1)\Delta t)$$

it is easy to see that from equation (4.4.15)

$$u_{jk}^{n+1} - u_{jk}^* = \frac{\nu \Delta t}{2} \frac{1}{\Delta y^2} \delta_y^2 u_{jk}^{n+1} = \frac{\nu \Delta t}{2} [(u_{yy})_{jk}^{n+1} + \mathcal{O}(\Delta y^2)].$$

Hence, *though difference scheme (4.4.14)–(4.4.15) is second order in time and space, using* $u_{0k}^* = g(0, k\Delta y, (n+1)\Delta t)$ *would be only first order in time.* To keep the scheme second order, we must use the definition of u_{jk}^* given by equation (4.4.15) and at $j = 0$ use the boundary condition

$$u_{0k}^* = (1 - \frac{r_y}{2}\delta_y^2)u_{0k}^{n+1}$$

$$= -\frac{r_y}{2}u_{0k+1}^{n+1} + (1 + r_y)u_{0k}^{n+1} - \frac{r_y}{2}u_{0k-1}^{n+1}$$

$$= -\frac{r_y}{2}g(0, (k+1)\Delta y, (n+1)\Delta t) + (1 + r_y)g(0, k\Delta y, (n+1)\Delta t)$$

$$- \frac{r_y}{2}g(0, (k-1)\Delta y, (n+1)\Delta t).$$

And, of course, we must use an analogous boundary condition at $j = M_x$.

For the case of the Peaceman-Rachford scheme, it is logical (based on the derivation given in Section 4.4.1) to use

$$u_{0k}^{n+\frac{1}{2}} = g(0, k\Delta y, (n + \frac{1}{2})\Delta t) \tag{4.4.18}$$

as the boundary condition at $j = 0$. If we add the left side of equation (4.4.6) to the right side of equation (4.4.2) and solve for $u_{jk}^{n+\frac{1}{2}}$, we get

$$u_{jk}^{n+\frac{1}{2}} = \frac{1}{2}\left(1 - \frac{r_y}{2}\delta_y^2\right)u_{jk}^{n+1} + \frac{1}{2}\left(1 + \frac{r_y}{2}\delta_y^2\right)u_{jk}^n. \tag{4.4.19}$$

Thus, we see that if we are using the Peaceman-Rachford scheme with Dirichlet boundary conditions, we should use the following boundary conditions for the half-step, $u^{n+\frac{1}{2}}$, in equation (4.4.2)

$$u_{0k}^{n+\frac{1}{2}} = \frac{1}{2}\left(1 - \frac{r_y}{2}\delta_y^2\right)g(0, k\Delta y, (n+1)\Delta t)$$

$$+ \frac{1}{2}\left(1 + \frac{r_y}{2}\delta_y^2\right)g(0, k\Delta y, n\Delta t) \tag{4.4.20}$$

$$u_{M_x k}^{n+\frac{1}{2}} = \frac{1}{2}\left(1 - \frac{r_y}{2}\delta_y^2\right)g(1, k\Delta y, (n+1)\Delta t)$$

$$+ \frac{1}{2}\left(1 + \frac{r_y}{2}\delta_y^2\right)g(1, k\Delta y, n\Delta t). \tag{4.4.21}$$

We note that when the boundary conditions are independent of time, boundary conditions (4.4.20)–(4.4.21) reduce to

$$u_{0k}^{n+\frac{1}{2}} = g(0, k\Delta y) \tag{4.4.22}$$

$$u_{M_x k}^{n+\frac{1}{2}} = g(1, k\Delta y). \tag{4.4.23}$$

Also, it is not difficult to show that boundary condition (4.4.18) approximates boundary condition (4.4.20) to order Δt^2.(See HW4.4.1.) Hence, using boundary conditions of the form (4.4.18) instead of (4.4.20)–(4.4.21) would not lower the order of the difference scheme and *it is possible to use either (4.4.18) or (4.4.20)–(4.4.21) as the Dirichlet boundary condition for the Peaceman-Rachford scheme.*

4.4.3.2   Neumann Boundary Conditions

To apply an approximation to a Neumann boundary condition along with a factored scheme, we proceed in a manner similar to that used for Dirichlet boundary conditions. If we have a Neumann boundary condition on a boundary where we need u_{jk}^{n+1} (such as with equations (4.4.6) or (4.4.15)), we proceed as we have done before. Consider partial differential equation (4.2.1) on $R = (0,1) \times (0,1)$ along with Dirichlet boundary conditions $v(x,y,t) = g(x,y,t)$ for $x = 0$, $x = 1$ and $y = 1$, and a Neumann boundary condition

$$v_y(x,0,t) = g^N(x,t). \tag{4.4.24}$$

If we use the first order approximation of Neumann boundary condition (4.4.24), $u_{j0}^{n+1} = u_{j1}^{n+1} - \Delta y g^N(j\Delta x, (n+1)\Delta t)$, and difference equation (4.4.6) or (4.4.15) at $k = 1$, the difference scheme can be rewritten so as not to include the u_{j0}^{n+1} term.

If we use the second order approximation, we apply the difference equation (4.4.6) or (4.4.15) at the boundary, $k = 0$, and use the second order approximation of boundary condition (4.4.24) to eliminate the -1 index. For example, for the Peaceman-Rachford scheme, use

$$u_{j-1}^{n+1} = u_{j1}^{n+1} - 2\Delta y g^N(j\Delta x, (n+1)\Delta t)$$

and

$$\left(1 - \frac{r_y}{2}\delta_y^2\right)u_{j0}^{n+1} = \left(1 + \frac{r_x}{2}\delta_x^2\right)u_{j0}^{n+\frac{1}{2}}$$

to get

$$u_{j0}^{n+1} - r_y(u_{j1}^{n+1} - u_{j0}^{n+1}) = \left(1 + \frac{r_x}{2}\delta_x^2\right)u_{j0}^{n+\frac{1}{2}} - \nu\frac{\Delta t}{\Delta y}g^N(j\Delta x, (n+1)\Delta t). \tag{4.4.25}$$

Then for each j, we solve equation (4.4.25) along with equation (4.4.6) for $k = 1, \cdots, M_y - 1$ as a part of our solution procedure (where solving equation (4.4.2) with Dirichlet boundary conditions is the other part of our solution procedure).

If we have a Neumann boundary condition at $j = 0$ and/or $j = M_x$ and not at either $k = 0$ or $k = M_y$, we can interchange the orders of solutions, solve

$$\left(1 - \frac{r_y}{2}\delta_y^2\right)u_{jk}^{n+\frac{1}{2}} = \left(1 + \frac{r_x}{2}\delta_x^2\right)u_{jk}^n$$

and

$$\left(1 - \frac{r_x}{2}\delta_x^2\right)u_{jk}^{n+1} = \left(1 + \frac{r_y}{2}\delta_y^2\right)u_{jk}^{n+\frac{1}{2}}$$

and handle this case in the same manner as that above. There was never a special reason for considering the x-direction (equation (4.4.2)) first in difference scheme (4.4.2), (4.4.6).

We must be more careful when we have a Neumann boundary condition on enough of the boundary that will force us to treat either equation (4.4.2) or equation (4.4.14) with a Neumann boundary condition (or for some reason we do not want to interchange the orders as above). Suppose, for example, that we have a boundary condition $v_x(0, y, t) = g^N(y, t)$ and must apply this condition as a part of solving equation (4.4.14) of the D'Yakonov scheme. If we are content with using the first order approximation of the boundary condition, we get

$$u_{1k}^{n+1} - u_{0k}^{n+1} = \Delta x g^N(k\Delta y, (n+1)\Delta t). \tag{4.4.26}$$

If we operate on equation (4.4.26) by $(1 - \frac{r_y}{2}\delta_y^2)$, we see that boundary condition (4.4.26) transforms into a boundary condition equation for u^* of the form

$$u_{1k}^* - u_{0k}^* = g_k^{*(n+1)} = \Delta x\left(1 - \frac{r_y}{2}\delta_y^2\right)g^N(k\Delta y, (n+1)\Delta t) \tag{4.4.27}$$

or

$$u_{0k}^* = u_{1k}^* - g_k^{*(n+1)}.$$

This boundary condition can be used just as the first order approximation of u_{j0}^{n+1} was used earlier.

If a second order approximation of the Neumann boundary condition is desired, the approach is the same. Again consider boundary condition (4.4.24) in a problem that we wish to solve using a D'Yakonov scheme. We approximate the boundary condition at the $(n+1)st$ time step by

$$u_{-1k}^{n+1} = u_{1k}^{n+1} - 2\Delta x g^N(k\Delta y, (n+1)\Delta t)$$

and operate on it by $(1 - \frac{r_y}{2}\delta_y^2)$ to transform it into a boundary condition for u^*,

$$u_{-1k}^* = u_{1k}^* - 2\Delta x\left(1 - \frac{r_y}{2}\delta_y^2\right)g^N(k\Delta y, (n+1)\Delta t) \tag{4.4.28}$$

$$= u_{1k}^* - g_k^{*(n+1)}.$$

Then, as is usually the case with the second order accurate Neumann boundary conditions, we apply equation (4.4.14) at $j = 0$, use the transformed boundary condition (4.4.28) to eliminate the -1 subscript and get

$$(1 + r_x)u_{0k}^* - r_x u_{1k}^* = \left(1 + \frac{r_x}{2}\delta_x^2\right)\left(1 + \frac{r_y}{2}\delta_y^2\right)u_{0k}^n - \frac{r_x}{2}g_k^{*(n+1)}.$$

Hopefully, we have described carefully how boundary conditions should be determined for the Peaceman-Rachford scheme, the D'Yakonov scheme or most other factored schemes. We return to this topic in the next section when we discuss how these boundary conditions should be implemented as a part of the scheme.

Before we proceed, we must admit that the most common implementation of Neumann boundary conditions on such schemes is that of using the 0-th order approximation obtained by using something like

$$u_{0\,k}^{n+1} = u_{1\,k}^n - \Delta x g^N(k\Delta y, (n+1)\Delta t)$$

to approximate a boundary condition of the form $v_x(0, y, t) = g^N(y, t)$. Often, such an approximation is sufficiently good. Use of such an approximation should be done with care (and good numerical experimental techniques).

HW 4.4.1 Show that boundary condition (4.4.18) approximates boundary condition (4.4.20) to order Δt^2 (and, hence, would not lower the order of the scheme).

HW 4.4.2 Derive second order accurate Neumann boundary conditions for both u^{n+1} and u^* in D'Yakonov's scheme.

4.4.3.3   Symmetry

To see if or when the matrix might be symmetric so that the scheme will also be unconditionally stable for initial–boundary–value problems (remember we promised that the stability calculation using finite Fourier series would be essentially the same as that done by the discrete Fourier transform) with Dirichlet boundary conditions, we must eliminate $u_{j\,k}^{n+\frac{1}{2}}$ from equations (4.4.2) and (4.4.6). The easiest way that this can be done is to first write both equations (4.4.2) and (4.4.6) as matrix equations of the form

$$Q_1 \mathbf{u}^{n+\frac{1}{2}} = Q \mathbf{u}^n \tag{4.4.29}$$

$$Q_1' \mathbf{u}^{n+1} = Q' \mathbf{u}^{n+\frac{1}{2}}. \tag{4.4.30}$$

If we examine difference equation (4.4.2) carefully, we see that Q_1 is a tridiagonal matrix with $1 + r_x$ on the diagonal and $-r_x/2$ on the super and sub-diagonals, except for the "special zeros." These "special zeros" will occur in the $(M_x - 1)th$, $2(M_x - 1)th$, etc. rows on the super-diagonal and in the $M_x th$, $(2(M_x - 1) + 1)th$, etc. rows on the sub-diagonal. Hence Q_1 will be symmetric.

The matrix Q will be a matrix with $1 + r_y$ on the diagonal, $r_y/2$ on the $(M_x - 1)th$ super-diagonal (starting in the $M_x th$ column) and an $r_y/2$

on the $(M_x - 1)th$ sub-diagonal (starting in the $M_x th$ row). Thus Q is also clearly symmetric. Since Q'_1 and Q' are almost the same as Q and Q_1 (except for a sign change that is done everywhere), Q'_1 and Q' are also both symmetric.

If we eliminate $\mathbf{u}^{n+\frac{1}{2}}$ from equations (4.4.29)–(4.4.30), we see that

$$\mathbf{u}^{n+1} = (Q'_1)^{-1} Q' Q_1^{-1} Q \mathbf{u}^n$$

where each matrix in the product is symmetric. Since these products commute, the product is symmetric and the Peaceman-Rachford scheme will be unconditionally stable for suitably nice boundary conditions.

4.4.3.4   Implementation

Dirichlet Boundary Conditions

We are now interested in implementing the Peaceman-Rachford scheme. If we proceed as we indicated in our introduction to the scheme, it is easy to see that if we consider the unknowns, u_{jk}, being ordered in the (j, k) direction (meaning the j index runs first, followed by the k index, as in (4.3.17)), equation (4.4.2) along with Dirichlet boundary conditions can easily be solved by our tridiagonal matrix solver, TRID. We emphasize that (with Dirichlet boundary conditions) this will be an $(M_x - 1)(M_y - 1) \times (M_x - 1)(M_y - 1)$ tridiagonal matrix. In addition, we remind the reader of the "special zeros" that are necessary on the sub and super diagonals due to the x boundaries.

If we then recall our earlier discussion of the matrix form of equation (4.4.6), we see that the matrix can be written as

$$Q'_1 \mathbf{u}^{n+1} = Q' \mathbf{u}^{n+\frac{1}{2}}, \qquad (4.4.31)$$

where

$$Q'_1 = \begin{pmatrix} 1+r_y & 0 & \cdots & 0 & -\frac{r_y}{2} & 0 & \cdots & & & \\ & & & & \cdots & & & & & \\ -\frac{r_y}{2} & 0 & \cdots & 0 & 1+r_y & 0 & \cdots & 0 & -\frac{r_y}{2} & 0 & \cdots \\ & & & & \cdots & & & & & \\ & & \cdots & 0 & -\frac{r_y}{2} & 0 & \cdots & 0 & 1+r_y \end{pmatrix}.$$

$$(4.4.32)$$

Thus we see that the matrix Q'_1 is clearly not tridiagonal.

If we reorder our unknowns in the (k, j) order (index k runs first, j second), then the matrix representation of equation (4.4.6) along with Dirichlet boundary conditions will be a tridiagonal matrix. Thus we see that between solving equations (4.4.2) and (4.4.6) (and again before we return to solve equation (4.4.2) at the new time step), a transpose of our data occurs.

This transpose need not literally be done. Instead, after equation (4.4.2) has been solved, the right hand side of equation (4.4.6) (and the matrix, if the matrix coefficients depended on j and k) is constructed in the transposed order. Then after equation (4.4.6) is solved, the output (which will come out of Trid in the right hand side array in the (k, j) order) can be placed in *unew* in a (j, k) order.

The implementation of the Peaceman-Rachford and D'Yakonov schemes is very similar to our other implementations. As was the case in Section 1.2.1, the main calls in our code will be to an initialization subroutine (which is now two dimensional and should already have been written for the two dimensional explicit computations), a solution scheme and an output scheme (which also now must take into account that we have two dimensional results and should also already have been written). As in Section 2.6.3, the solution scheme must consist of a right hand side calculation, a stencil calculation and a call to Trid. Since for these two dimensional problems we are solving two tridiagonal matrices, we generally have two of each of the above in our solution scheme. For example, if we consider the Peaceman-Rachford scheme where we do not worry about the boundary conditions (or just consider zero Dirichlet conditions), the solution scheme might look like

Call Right Hand Side:(4.4.2)

Call Stencil:(4.4.2)

Call Trid

Call Right Hand Side:(4.4.6)

Call Stencil:(4.4.6)

Call Trid

where the subroutines are as follows.

1. Subroutine Right Hand Side:(4.4.2) would look approximately like the following.

   $kk = 0$

   For $k = 1, \cdots, M_y - 1$

   For $j = 1, \cdots, M_x - 1$

   $\quad kk = kk + 1$

   $\quad RHS(kk) = (1 + \frac{r_y}{2}\delta_y^2)u^n_{j\,k}$: formula on the right hand side of
   equation (4.4.2).

   Next j

   Next k

2. Subroutine Stencil:(4.4.2) would fill the diagonals of our tridiagonal matrix with $-r_x/2$ on the subdiagonal (along with zeros in the $M_x th$ row, $(2M_x - 1)th$ row, etc.), $1 + r_x$ on the main diagonal and $-r_x/2$ on the super-diagonal (with zeros in the $(M_x - 1)th$ row, $2(M_x - 1)th$ row, etc.).

3. Trid solves equation (4.4.2) with the solution, $u^{n+\frac{1}{2}}$, given in the (j, k) order in the RHS array.

4. Subroutine Right Hand Side:(4.4.6) looks approximately like the following.

   $kk = 0$

   For $j = 1, \cdots, M_x - 1$

   For $k = 1, \cdots, M_y - 1$

       $kk = kk + 1$

       $RHS(kk) = (1 + \frac{r_x}{2}\delta_x^2)u_{jk}^{n+\frac{1}{2}}$: formula on the right hand side of equation (4.4.6).

   Next k

   Next j

5. Subroutine Stencil:(4.4.6) again fills the three diagonals of the tridiagonal matrix with $-r_y/2$ (and appropriate zeros), $1 + r_y$ and $-r_y/2$ (and again the appropriate zeros).

6. Trid will solve equation (4.4.6) with the solution, u^{n+1}, given in the (k, j) order in the RHS array.

7. The RHS array can be read into a u^{n+1} array as follows.

   $kk = 0$

   For $j = 1, \cdots, M_x - 1$

   For $k = 1, \cdots, M_y - 1$

       $kk = kk + 1$

       $u_{jk}^{n+1} = RHS(kk)$

   Next k

   Next j

And, finally, if we have nonzero Dirichlet boundary conditions, we must include the nonhomogeneous terms in the appropriate places in the RHS arrays.

One last very important aspect about implementation is related to the method used to make our resulting matrices tridiagonal matrices. In the discussions above, we solve equations (4.4.2) and (4.4.6) as two large ($M_x -$

$1)(M_y - 1) \times (M_x - 1)(M_y - 1)$ tridiagonals. The form of equation (4.4.2) is such that though the right hand side must difference in the y direction, the matrix equation only involves solving due to differences in the x direction. Thus, the $(M_x - 1)(M_y - 1) \times (M_x - 1)(M_y - 1)$ matrix equation (4.4.2) can be decomposed into $M_y - 1$ matrix equations of the form of equation (4.4.2).

One of the ways to see this very clearly is to write out a large version of equation (4.4.2) and note that the zeros in the matrix are in the right places so that the system becomes

$$
\begin{pmatrix} B & \Theta & \cdots \\ \Theta & B & \Theta & \cdots \\ \cdots & \cdots \\ \cdots & \Theta & B \end{pmatrix}
\begin{pmatrix} \mathbf{u}_1^{n+\frac{1}{2}} \\ \vdots \\ \mathbf{u}_{M_y-1}^{n+\frac{1}{2}} \end{pmatrix}
=
\begin{pmatrix} \mathbf{r}_1 \\ \vdots \\ \mathbf{r}_{M_y-1} \end{pmatrix},
\qquad (4.4.33)
$$

where

$$
B = \begin{pmatrix}
1 + r_x & -\frac{r_x}{2} & 0 & \cdots \\
-\frac{r_x}{2} & 1 + r_x & -\frac{r_x}{2} & 0 & \cdots \\
& & \cdots \\
\cdots & & 0 & -\frac{r_x}{2} & 1 + r_x
\end{pmatrix},
\qquad (4.4.34)
$$

$$
\mathbf{u}_k^{n+\frac{1}{2}} = \begin{pmatrix} u_{1\,k}^{n+\frac{1}{2}} \\ \vdots \\ u_{M_x-1\,k}^{n+\frac{1}{2}} \end{pmatrix}
\qquad (4.4.35)
$$

and

$$
\mathbf{r}_k = \begin{pmatrix} r_{1\,k} \\ \vdots \\ r_{M_x-1\,k} \end{pmatrix}.
\qquad (4.4.36)
$$

To try to make it easier, in Figure 4.4.1 we have include the coefficient matrix for a case with $M_x = 5$ and $M_y = 6$. It is easy to see that solving equation (4.4.2) or (4.4.33) can be reduced to solving

$$
B\mathbf{u}_k^{n+\frac{1}{2}} = \mathbf{r}_k, \quad k = 1, \cdots, M_y - 1.
\qquad (4.4.37)
$$

Each of these $M_y - 1$ systems of equations corresponds to solving equation (4.4.2) on a fixed $y = y_k = k\Delta y$ line. Solving equation (4.4.2) now consists of solving $M_y - 1$ small $(M_x - 1 \times M_x - 1)$ systems.

Solving equation (4.4.6) can be approached in the same manner. Hence, solving equation (4.4.6) can be done by solving $M_x - 1\ M_y - 1 \times M_y - 1$ tridiagonal systems of equations.

To be more specific, suppose we are again solving a problem of the form (4.2.5)–(4.2.9) with $R = (0, 1) \times (0, 1)$. Then the system that must be solved (equation (4.4.2)) on the kth coordinate line ($y = k\Delta y$) is of the form

$$B\mathbf{u}_k^{n+\frac{1}{2}} = \mathbf{r}_k \tag{4.4.38}$$

where B and $\mathbf{u}_k^{n+\frac{1}{2}}$ are as in (4.4.34) and (4.4.35), respectively,

$$\mathbf{r}_k = \begin{pmatrix} (1 + \frac{r_y}{2}\delta_y^2)u_{1\,k}^n + \frac{r_x}{2}u_{0\,k}^{n+\frac{1}{2}} \\ (1 + \frac{r_y}{2}\delta_y^2)u_{2\,k}^n \\ \cdots \\ (1 + \frac{r_y}{2}\delta_y^2)u_{M_x-2\,k}^n \\ (1 + \frac{r_y}{2}\delta_y^2)u_{M_x-1\,k}^n + \frac{r_x}{2}u_{M_x\,k}^{n+\frac{1}{2}} \end{pmatrix}, \tag{4.4.39}$$

and $u_{j\,k}^{n+\frac{1}{2}}$, $j = 0$ or $j = M_x$ are given by (4.4.20) and (4.4.21). This equation then holds for each k, $k = 1, \cdots, M_y - 1$.

Likewise, the equations that must be solved in the y direction on the jth coordinate line ($x_j = j\Delta x$), equation (4.4.6), has the form

$$Q_2\mathbf{u}_j^{n+1} = \mathbf{r}_j, \tag{4.4.40}$$

where

$$Q_2 = \begin{pmatrix} 1 + r_y & -\frac{r_y}{2} & 0 & \cdots & \\ -\frac{r_y}{2} & 1 + r_y & -\frac{r_y}{2} & 0 & \cdots \\ & \cdots & & & \\ \cdots & 0 & -\frac{r_y}{2} & 1 + r_y & -\frac{r_y}{2} \\ & \cdots & 0 & -\frac{r_y}{2} & 1 + r_y \end{pmatrix}, \tag{4.4.41}$$

$$\mathbf{u}_j^{n+1} = \begin{pmatrix} u_{j\,1}^{n+1} \\ \cdots \\ u_{j\,M_y-1}^{n+1} \end{pmatrix}, \tag{4.4.42}$$

$$\mathbf{r}_j = \begin{pmatrix} (1 + \frac{r_x}{2}\delta_x^2)u_{j\,1}^{n+\frac{1}{2}} + \frac{r_y}{2}g_{j\,0}^{n+1} \\ (1 + \frac{r_x}{2}\delta_x^2)u_{j\,2}^{n+\frac{1}{2}} \\ \cdots \\ (1 + \frac{r_x}{2}\delta_x^2)u_{j\,M_y-2}^{n+\frac{1}{2}} \\ (1 + \frac{r_x}{2}\delta_x^2)u_{j\,M_y-1}^{1+\frac{1}{2}} + \frac{r_y}{2}g_{j\,M_y}^{n+1} \end{pmatrix} \tag{4.4.43}$$

and $g_{j\,k}^{n+1} = g(j\Delta x, k\Delta y, (n + 1)\Delta t)$ for $k = 0$ and $k = M_y$.

The implementation of solving equations (4.4.2) and (4.4.6) by solving equations (4.4.38), $k = 1, \cdots, M_y-1$ and (4.4.40), $j = 1, \cdots, M_x-1$ is only slightly different from that given earlier for the large matrix. The solution scheme will now take the form

$$
\left(\begin{array}{cccc|cccc|cccc|cccc|cccc}
d & r & 0 & 0 & 0 & 0 & 0 & 0 & 0 & 0 & 0 & 0 & 0 & 0 & 0 & 0 & 0 & 0 & 0 & 0 \\
r & d & r & 0 & 0 & 0 & 0 & 0 & 0 & 0 & 0 & 0 & 0 & 0 & 0 & 0 & 0 & 0 & 0 & 0 \\
0 & r & d & r & 0 & 0 & 0 & 0 & 0 & 0 & 0 & 0 & 0 & 0 & 0 & 0 & 0 & 0 & 0 & 0 \\
0 & 0 & r & d & \mathbf{0} & 0 & 0 & 0 & 0 & 0 & 0 & 0 & 0 & 0 & 0 & 0 & 0 & 0 & 0 & 0 \\
\hline
0 & 0 & 0 & \mathbf{0} & d & r & 0 & 0 & 0 & 0 & 0 & 0 & 0 & 0 & 0 & 0 & 0 & 0 & 0 & 0 \\
0 & 0 & 0 & 0 & r & d & r & 0 & 0 & 0 & 0 & 0 & 0 & 0 & 0 & 0 & 0 & 0 & 0 & 0 \\
0 & 0 & 0 & 0 & 0 & r & d & r & 0 & 0 & 0 & 0 & 0 & 0 & 0 & 0 & 0 & 0 & 0 & 0 \\
0 & 0 & 0 & 0 & 0 & 0 & r & d & \mathbf{0} & 0 & 0 & 0 & 0 & 0 & 0 & 0 & 0 & 0 & 0 & 0 \\
\hline
0 & 0 & 0 & 0 & 0 & 0 & 0 & \mathbf{0} & d & r & 0 & 0 & 0 & 0 & 0 & 0 & 0 & 0 & 0 & 0 \\
0 & 0 & 0 & 0 & 0 & 0 & 0 & 0 & r & d & r & 0 & 0 & 0 & 0 & 0 & 0 & 0 & 0 & 0 \\
0 & 0 & 0 & 0 & 0 & 0 & 0 & 0 & 0 & r & d & r & 0 & 0 & 0 & 0 & 0 & 0 & 0 & 0 \\
0 & 0 & 0 & 0 & 0 & 0 & 0 & 0 & 0 & 0 & r & d & \mathbf{0} & 0 & 0 & 0 & 0 & 0 & 0 & 0 \\
\hline
0 & 0 & 0 & 0 & 0 & 0 & 0 & 0 & 0 & 0 & 0 & \mathbf{0} & d & r & 0 & 0 & 0 & 0 & 0 & 0 \\
0 & 0 & 0 & 0 & 0 & 0 & 0 & 0 & 0 & 0 & 0 & 0 & r & d & r & 0 & 0 & 0 & 0 & 0 \\
0 & 0 & 0 & 0 & 0 & 0 & 0 & 0 & 0 & 0 & 0 & 0 & 0 & r & d & r & 0 & 0 & 0 & 0 \\
0 & 0 & 0 & 0 & 0 & 0 & 0 & 0 & 0 & 0 & 0 & 0 & 0 & 0 & r & d & \mathbf{0} & 0 & 0 & 0 \\
\hline
0 & 0 & 0 & 0 & 0 & 0 & 0 & 0 & 0 & 0 & 0 & 0 & 0 & 0 & 0 & \mathbf{0} & d & r & 0 & 0 \\
0 & 0 & 0 & 0 & 0 & 0 & 0 & 0 & 0 & 0 & 0 & 0 & 0 & 0 & 0 & 0 & r & d & r & 0 \\
0 & 0 & 0 & 0 & 0 & 0 & 0 & 0 & 0 & 0 & 0 & 0 & 0 & 0 & 0 & 0 & 0 & r & d & r \\
0 & 0 & 0 & 0 & 0 & 0 & 0 & 0 & 0 & 0 & 0 & 0 & 0 & 0 & 0 & 0 & 0 & 0 & r & d
\end{array}\right)
$$

FIGURE 4.4.1. Coefficient matrix for equation (4.4.33) for the case when $M_x = 5$, $M_y = 6$, $d = 1 + r_x$, $r = -r_x/2$ and the **bold face** zeros denote the special zeros that are necessary due to the boundaries. The lines included show the partitioning given by the form of the matrix given in equation (4.4.33).

For $k = 1, \cdots , M_y - 1$

> Call Right Hand Side:(4.4.39)
>
> Call Stencil:(4.4.34)
>
> Call Trid

Next k

For $j = 1, \cdots , M_x - 1$

> Call Right Hand Side:(4.4.43)
>
> Call Stencil:(4.4.41)
>
> Call Trid

Next j

For the above implementation, subroutines Right Hand Side:(4.4.39) and Stencil:(4.4.34) contain loops (from $j = 1, \cdots , M_x - 1$) that fill the right hand side vector and the coefficient matrix for the first series of matrix equations; and subroutines Right Hand Side:(4.4.43) and Stencil:(4.4.41) contain loops (from $k = 1, \cdots , M_y - 1$) that fill the right hand side vector and the coefficient matrix for the second series of matrix equations. As always with ADI schemes, we must be careful that the solutions of the equations (4.4.38), $k = 1, \cdots , M_y - 1$, assume that the data is in a (j, k) ordering, while the solutions of the equations (4.4.40), $j = 1, \cdots , M_x - 1$ assumes that the data is in a (k, j) ordering.

And lastly, the act of solving equations (4.4.2) and (4.4.6) by solving many small tridiagonals makes our job somewhat easier because when we fill the stencil, we do not have to decide where the "special zeros" must be placed. The real importance of the above observation, however, is that when solving equation (4.4.2), the solving of the $M_y - 1$ small tridiagonal matrices can literally be done at the same time. The small systems of equations are completely independent. For example, we notice that equation (4.4.38) depends only on the y index k. When this equation is being solved, it does not have to know that other coordinate lines exist (except in the right hand side, but that all depends on u at the nth time level). Hence, *solving equations (4.4.2) and (4.4.6) by solving "many tridiagonals at the same time" is one of the many ways to use the capabilities of vector and parallel computers or clusters of workstations for approximating the solutions of partial differential equations.*

Neumann Boundary Conditions

Most of the material presented for Dirichlet boundary conditions is also relevant to the implementation of Neumann boundary conditions. In this section we will give a short discussion of some special situations encountered when we implement difference schemes for problems involving Neumann boundary conditions. Suppose we consider, as we did in Section

4.4.3.2, partial differential equation (4.2.1) on $R = (0,1) \times (0,1)$ along with Dirichlet boundary conditions on $x = 0$, $x = 1$ and $y = 1$, and Neumann boundary condition (4.4.24) on $y = 0$. For this discussion, we consider the Peaceman-Rachford scheme, (4.4.2), (4.4.6), and assume that we are solving the problem using the approach of solving equation (4.4.2) on each of the k lines and solving equation (4.4.6) on each of the j lines (as was described above).

Equation (4.4.2)

The left hand side of equation (4.4.2) (the matrix) is very easy when the Neumann condition is at $y = 0$. The left hand side of the equation does not know that any other lines exist. It only has differences in the x direction. Hence, the left hand side of equation (4.4.2) is the same as it was for Dirichlet boundary conditions. Since the right hand side of equation (4.4.2) contains y differences, *the Neumann boundary condition will affect the right hand side.* If we use a first order accurate approximation of the Neumann boundary condition, we solve only on lines $k = 1, \cdots, M_y - 1$. If the $k = 0$ line was updated at the end of the last time step (as it always should be), there is no problem computing the right hand side of equation (4.4.2). When we are solving on the $k = 1$ row, the right hand side equation reaches to the $k = 0$ row, but since it reaches at the n-th time level, the scheme does not know or care that we have a Neumann boundary condition at $y = 0$.

If we use the second order approximation of the Neumann boundary condition, then we must solve on the $k = 0$ line. Again, the left hand side doesn't care. When we compute the right hand side on the $k = 0$ line, the difference formula reaches to $k = -1$. Thus *the formula for the second order approximation of the Neumann boundary condition must be used to compute the right hand side of equation (4.4.2) when $k = 0$.* The second order approximation of boundary condition (4.4.24) will give

$$u_{j-1}^n = u_{j1}^n - 2\Delta y g^N (j\Delta x, n\Delta t)$$

and the right hand side of equation (4.4.2), $k = 0$ can be written as

$$(1 - r_y)u_{j0}^n + r_y u_{j1}^n - \Delta y r_y g^N (j\Delta x, n\Delta t).$$

Another approach for implementing the calculating using a second order approximation of the Neumann boundary condition is to include room in the u arrays for the "ghost line." Then if u_{j-1}^n is updated at the end of the calculation at the n-th time level, the right hand side calculation can reach to $k = -1$ without a problem.

Equation (4.4.6)

Though the Neumann boundary condition at $y = 0$ probably affects the solution of equation (4.4.6) more and is theoretically more difficult to implement, the solution of equation (4.4.6) should be easy for us. *For any* *j line, the Neumann boundary condition at $y = 0$ will not affect the right hand side of equation (4.4.6), but will affect the left hand side.* However, the way that the left hand side will be affected will be exactly the same way a one dimensional implicit scheme for a problem with a Neumann boundary condition at $x = 0$ was affected. Hence, we return to Section 2.6.4 to see the following.

- If we use a first order approximation, for each $j = 1, \cdots, M_x - 1$, we still solve the equation from $k = 1, \cdots, M_y - 1$ and use the approximation

$$u_{j\,0}^{n+1} = u_{j\,1}^{n+1} - \Delta y g^N(j\Delta x, (n+1)\Delta t)$$

  to change the first diagonal element of our tridiagonal matrix (to $1 + r_y/2$) and subtract $\Delta y \frac{r_y}{2} g^N(j\Delta x, (n+1)\Delta t)$ from the right hand side.

- If we use a second order approximation, for each $j = 1, \cdots, M_x - 1$, we solve the equation from $k = 0, \cdots, M_y - 1$ where we obtain the equation at $k = 0$ by using difference equation (4.4.6) at $k = 0$ and eliminate the $u_{j\,-1}^{n+1}$ term by using the second order approximation of the boundary condition,

$$u_{j\,-1}^{n+1} = u_{j\,1}^{n+1} - 2\Delta y g^N(j\Delta x, (n+1)\Delta t),$$

  i.e. we use equation (4.4.25).

The implementation of these methods for solving problems involving Neumann boundary conditions for other ADI schemes is very similar. If one prefers to solve equations (4.4.2) and (4.4.6) as two large tridiagonals, the only difference is that the equations discussed above must be placed in the correct places in the right hand side and in the matrix stencil. The difficulty of finding the correct places to alter the matrix and right hand side is one of the reasons that makes solving on separate lines attractive.

In summary, we see that *the Peaceman-Rachford scheme is a second order scheme (in all variables) that is unconditionally stable (for Dirichlet and second order accurate Neumann boundary conditions).* The solution procedure involves solving tridiagonal matrices. Though these are more expensive to solve than using the explicit scheme, if the larger Δt's made available by using an unconditionally stable scheme are appropriate for required accuracy, then the Peaceman-Rachford scheme is more efficient to use than the explicit scheme.

HW 4.4.3 Solve the initial–boundary–value problem given in HW4.2.1 using the Peaceman-Rachford scheme and the same M_x, M_y and ν. Use $\Delta t = 0.01$. Compare your solutions with those of HW4.2.1.

HW 4.4.4 Redo HW4.4.3 using $M_x = 10$ and $M_y = 20$.

HW 4.4.5 Redo HW4.2.7 and HW4.2.8 replacing the explicit difference scheme by the Peaceman-Rachford scheme and using the same Neumann boundary condition treatments, the same M_x and M_y, and $\Delta t = 0.01$. Compare your solutions with the solutions found in HW4.2.7 and 4.2.8.

HW 4.4.6 (a) Consider the initial–value problem

$$v_t = \nu(v_{xx} + v_{yy}), \ x, \ y \in \mathbb{R} \qquad (4.4.44)$$

$$v(x, y, 0) = f(x, y), \ x, \ y \in \mathbb{R} \qquad (4.4.45)$$

along with the **locally one dimensional scheme** (lod scheme) for solving (4.4.44)–(4.4.45)

$$(1 - r_x \delta_x^2)u_{jk}^{n+\frac{1}{2}} = u_{jk}^n \qquad (4.4.46)$$

$$(1 - r_y \delta_y^2)u_{jk}^{n+1} = u_{jk}^{n+\frac{1}{2}}. \qquad (4.4.47)$$

Discuss the consistency and stability of the locally one dimensional difference scheme (4.4.46)–(4.4.47).

(b) Solve the problem given in HW4.2.1 using the locally one dimensional scheme given in part (a), the same M_x, M_y and ν. Use $\Delta t = 0.01$. Compare you solutions with those found in HW4.2.1 and HW4.4.3.

HW 4.4.7 Develop a locally one dimensional Crank-Nicolson scheme for solving initial–value problem (4.4.44)–(4.4.45) analogous to that given in (4.4.46)–(4.4.47). Analyze the consistency and stability of the scheme and use the scheme to solve the problem described in part (b) of HW4.4.6.

4.4.4 Douglas-Rachford Scheme

Even though we introduced the Peaceman-Rachford scheme by starting with the implicit scheme, we saw that it was the approximate factorization of the Crank-Nicolson scheme. We proceed now to develop the Douglas-Rachford scheme by actually approximately factoring the implicit scheme (BTCS). If we write the implicit scheme, (4.3.25) (without the nonhomogeneous term), as

$$(1 - r_x \delta_x^2 - r_y \delta_y^2)u_{jk}^{n+1} = u_{jk}^n, \qquad (4.4.48)$$

we see that the left hand side naturally factors into

$$(1 - r_x \delta_x^2)(1 - r_y \delta_y^2)u_{jk}^{n+1}. \qquad (4.4.49)$$

Thus to factor equation (4.4.48), we must add a term of the form

$$r_x r_y \delta_x^2 \delta_y^2 u_{jk}^{n+1}. \qquad (4.4.50)$$

As we did with the approximate factorization of the Crank-Nicolson scheme to get the Peaceman-Rachford scheme, we counter adding the term (4.4.50) by adding the term

$$r_x r_y \delta_x^2 \delta_y^2 u_{j\,k}^n \tag{4.4.51}$$

to the right hand side. As in Section 4.4.2.2, since the term that is being added ((4.4.50) minus (4.4.51)) is a higher order term, *the resulting difference equation has the same order accuracy as the implicit scheme (first order in Δt and second order in Δx and Δy).* Thus, if we add (4.4.50) to the left hand side, (4.4.51) to the right hand side and factor, the **Douglas-Rachford scheme** can be written as

$$(1 - r_x \delta_x^2)(1 - r_y \delta_y^2) u_{j\,k}^{n+1} = (1 + r_x r_y \delta_x^2 \delta_y^2) u_{j\,k}^n \tag{4.4.52}$$

The form of the Douglas-Rachford scheme used for computations is

$$(1 - r_x \delta_x^2) u_{j\,k}^* = (1 + r_y \delta_y^2) u_{j\,k}^n \tag{4.4.53}$$

$$(1 - r_y \delta_y^2) u_{j\,k}^{n+1} = u_{j\,k}^* - r_y \delta_y^2 u_{j\,k}^n, \tag{4.4.54}$$

which is equivalent to equation (4.4.52).

As with the Peaceman-Rachford and D'Yakonov schemes, we must select the boundary conditions for u^* carefully. Solving for u^* from equation (4.4.54) yields

$$u_{j\,k}^* = (1 - r_y \delta_y^2) u_{j\,k}^{n+1} + r_y \delta_y^2 u_{j\,k}^n. \tag{4.4.55}$$

Then the boundary conditions for u^* at $j = 0$ and $j = M_x$ can be given in terms of the given Dirichlet boundary conditions of u by

$$u_{0\,k}^* = (1 - r_y \delta_y^2) u_{0\,k}^{n+1} + r_y \delta_y^2 u_{0\,k}^n \tag{4.4.56}$$

and

$$u_{M_x\,k}^* = (1 - r_y \delta_y^2) u_{M_x\,k}^{n+1} + r_y \delta_y^2 u_{M_x\,k}^n. \tag{4.4.57}$$

As was the case with the Peaceman-Rachford scheme, when the boundary conditions do not depend on time, the boundary conditions for u^* reduce to those for u.

Treatment of Neumann boundary conditions for the Douglas-Rachford scheme follow the derivation given for Neumann boundary conditions given in Section 4.4.3.2. For example, if the Neumann boundary condition

$$v_x(0, y, t) = g(0, y, t)$$

is approximated by the first order approximation

$$u_{0\,k}^{n+1} = u_{1\,k}^{n+1} - \Delta x g(0, k\Delta y, (n+1)\Delta t), \tag{4.4.58}$$

we operate on equation (4.4.58) by $(1 - r_y \delta_y^2)$ to get

$$(1 - r_y \delta_y^2) u_{0\,k}^{n+1} = (1 - r_y \delta_y^2) u_{1\,k}^{n+1} - \Delta x (1 - r_y \delta_y^2) g(0, k\Delta x, (n+1)\Delta t). \tag{4.4.59}$$

If we then add $r_y \delta_y^2 u_{0\,k}^n$ to both sides, and add and subtract $r_y \delta_y^2 u_{1\,k}^n$ from the right hand side, we get

$$u_{0\,k}^* = u_{1\,k}^* + r_y \delta_y^2 (u_{0\,k}^n - u_{1\,k}^n) - \Delta x (1 - r_y \delta_y^2) g(0, k\Delta y, (n+1)\Delta t). \tag{4.4.60}$$

And, of course, a Neumann boundary condition at $j = M_x$ can be handled in a similar fashion.

And, finally, the implementation of the Douglas-Rachford scheme is essentially the same as that of the Peaceman-Rachford scheme or almost any other ADI scheme. It is not difficult to change a Peaceman-Rachford code (using either the large tridiagonal or the many small tridiagonal version) into a Douglas-Rachford code.

4.4.4.1 Stability of Douglas-Rachford Scheme

If we proceed to analyze the stability of the Douglas-Rachford scheme as an initial–value problem scheme (remembering that the results we get will be the same as the discrete von Neumann stability criteria for the initial–boundary–value problem), we can use the discrete Fourier transform as we did for the Peaceman-Rachford scheme. If we transform equations (4.4.53)–(4.4.54) and eliminate u^* from the equations, we are left with

$$\hat{u}^{n+1}(\xi, \eta) = \rho(\xi, \eta) \hat{u}^n(\xi, \eta) \tag{4.4.61}$$

$$= \frac{1 + 16 r_x r_y \sin^2 \frac{\xi}{2} \sin^2 \frac{\eta}{2}}{(1 + 4 r_x \sin^2 \frac{\xi}{2})(1 + 4 r_y \sin^2 \frac{\eta}{2})} \hat{u}^n(\xi, \eta). \tag{4.4.62}$$

It is easy to see that the symbol of the scheme, ρ, is greater than or equal to zero. By expanding the denominator, it is equally easy to see that $\rho \le 1$.

HW 4.4.8 Solve the problem given in HW4.2.1 using the Douglas-Rachford scheme. Use the same M_x, M_y and ν and use $\Delta t = 0.01$.

4.4.5 Nonhomogeneous ADI Schemes

As with any other difference scheme, there are times when we want to use an ADI scheme to solve a problem with a nonhomogeneous partial differential equation. There is often confusion where, in the algorithm, the nonhomogeneous terms should go. Most often, the approach seems to be to guess where the nonhomogeneous terms should go and then, maybe, perform a consistency analysis on the resulting scheme. Of course, if the consistency analysis is done properly, this approach is adequate. The order analysis is a very important step. *It is very easy to work hard to get a higher order scheme and then lower the order by improperly including the nonhomogeneous term.*

In this section, we shall describe a methodology for including the nonhomogeneous terms that is based on conservation law formulations of the scheme. We begin by considering partial differential equation (4.2.1) and the integral form of the conservation equation used in Section 4.2.2, equation (4.2.27),

$$
\int_{y_{k-1/2}}^{y_{k+1/2}} \int_{x_{j-1/2}}^{x_{j+1/2}} (v^{n+1} - v^n) dx dy = \int_{t_n}^{t_{n+1}} \int_{y_{k-1/2}}^{y_{k+1/2}} \int_{x_{j-1/2}}^{x_{j+1/2}} F(x,y,t) dx dy dt
$$

$$
+ \nu \int_{t_n}^{t_{n+1}} \int_{y_{k-1/2}}^{y_{k+1/2}} v_x(x_{j+1/2}, y, t) dy dt - \nu \int_{t_n}^{t_{n+1}} \int_{y_{k-1/2}}^{y_{k+1/2}} v_x(x_{j-1/2}, y, t) dy dt
$$

$$
+ \nu \int_{t_n}^{t_{n+1}} \int_{x_{j-1/2}}^{x_{j+1/2}} v_y(x, y_{k+1/2}, t) dx dt - \nu \int_{t_n}^{t_{n+1}} \int_{x_{j-1/2}}^{x_{j+1/2}} v_y(x, y_{k-1/2}, t) dx dt.
$$

$$(4.4.63)$$

If we then integrate the first term by the midpoint rule, approximate the flux terms as we did in Section 4.2.2, integrate the spatial integral of the x-directional flux term by the midpoint rule and integrate the spatial integral of the y-directional flux term by the trapezoidal rule, we are left with

$$
\Delta x \Delta y (v_{jk}^{n+1} - v_{jk}^n) + \mathcal{O}(\Delta x^3 \Delta y \Delta t) + \mathcal{O}(\Delta x \Delta y^3 \Delta t) =
$$

$$
+ \int_{t_n}^{t_{n+1}} \int_{y_{k-1/2}}^{y_{k+1/2}} \int_{x_{j-1/2}}^{x_{j+1/2}} F(x,y,t) dx dy dt
$$

$$
+ \nu \Delta t \Delta y \frac{1}{\Delta x} \delta_x^2 v_{jk}^{n+1/2} + \mathcal{O}(\Delta t \Delta x \Delta y^3) + \mathcal{O}(\Delta t^3 \Delta x \Delta y)
$$

$$
+ \nu \frac{\Delta t}{2} \Delta x \frac{1}{\Delta y} \delta_y^2 (v_{jk}^n + v_{jk}^{n+1}) + \mathcal{O}(\Delta t \Delta x^3 \Delta y) + \mathcal{O}(\Delta t^3 \Delta x \Delta y). \quad (4.4.64)
$$

We note that *if we did not have the nonhomogeneous term*, we could add and subtract a $v_{jk}^{n+1/2} \Delta x \Delta y$ term to get

$$
\Delta x \Delta y (v_{jk}^{n+1} - v_{jk}^{n+1/2}) + \Delta x \Delta y (v_{jk}^{n+1/2} - v_{jk}^n) + \mathcal{O}(\Delta x^3 \Delta y \Delta t)
$$

$$
+ \mathcal{O}(\Delta x \Delta y^3 \Delta t) = \nu \Delta t \Delta y \frac{1}{\Delta x} \delta_x^2 v_{jk}^{n+1/2} + \mathcal{O}(\Delta t \Delta x \Delta y^3)
$$

$$
+ \mathcal{O}(\Delta t^3 \Delta x \Delta y) + \nu \frac{\Delta t}{2} \Delta x \frac{1}{\Delta y} \delta_y^2 v_{jk}^n
$$

$$
+ \nu \frac{\Delta t}{2} \Delta x \frac{1}{\Delta y} \delta_y^2 v_{jk}^{n+1} + \mathcal{O}(\Delta t \Delta x^3 \Delta y) \quad + \mathcal{O}(\Delta t^3 \Delta x \Delta y). \quad (4.4.65)
$$

If we then drop the order terms, equate the first term on the left hand side of the equation with one half of the first term on the right hand side of the equation plus the third term; and equate the second term on the left hand side of the equation with one half of the first term on the right hand side of the equation plus the second term, we are left with

$$\Delta x \Delta y (u_{jk}^{n+1} - u_{jk}^{n+1/2}) = \frac{1}{2} \nu \Delta t \Delta y \frac{1}{\Delta x} \delta_x^2 u_{jk}^{n+1/2} + \nu \frac{\Delta t}{2} \Delta x \frac{1}{\Delta y} \delta_y^2 u_{jk}^{n+1}$$

(4.4.66)

$$\Delta x \Delta y (u_{jk}^{n+1/2} - u_{jk}^{n}) = \frac{1}{2} \nu \Delta t \Delta y \frac{1}{\Delta x} \delta_x^2 u_{jk}^{n+1/2} + \nu \frac{\Delta t}{2} \Delta x \frac{1}{\Delta y} \delta_y^2 u_{jk}^{n}$$

(4.4.67)

Note that if we can find a u^n, $u^{n+1/2}$ and a u^{n+1} that satisfies equations (4.4.66)–(4.4.67), then the same u's will satisfy equation (4.4.65) up to the order terms (which can be seen by adding equations (4.4.66) and (4.4.67) together. We also see that equations (4.4.66)–(4.4.67) are equivalent to the Peaceman-Rachford equations, (4.4.2), (4.4.6). Of course, *the above argument is not a straightforward approach for deriving the Peaceman-Rachford scheme*. But, since we now have a conservation law approach to deriving the Peaceman-Rachford scheme, *it is easy to return to equation (4.4.64) to decide how to treat the nonhomogeneous term*.

We see that we must approximate the nonhomogeneous integral at least to the order $(\mathcal{O}(\Delta t^3 \Delta x \Delta y) + \mathcal{O}(\Delta t \Delta x^3 \Delta y) + \mathcal{O}(\Delta t \Delta x \Delta y^3))$ or else the resulting order of the approximating scheme will be lowered. Hence, if we approximate the nonhomogeneous integral by the midpoint rule with respect to x and y, and either the midpoint rule or the trapezoidal rule with respect to t, we will get the proper order of approximation. Hence, dropping the orders we are left with

$$\Delta x \Delta y (u_{jk}^{n+1} - u_{jk}^{n}) = \Delta t \Delta x \Delta y F_{jk}^{n+1/2} + \nu \Delta t \Delta y \frac{1}{\Delta x} \delta_x^2 u_{jk}^{n+1/2}$$
$$+ \nu \frac{\Delta t}{2} \Delta x \frac{1}{\Delta y} \delta_y^2 (u_{jk}^{n} + u_{jk}^{n-1}) \qquad (4.4.68)$$

or

$$\Delta x \Delta y (u_{jk}^{n+1} - u_{jk}^{n}) = \frac{\Delta t}{2} \Delta x \Delta y (F_{jk}^{n} + F_{jk}^{n+1}) + \nu \Delta t \Delta y \frac{1}{\Delta x} \delta_x^2 u_{jk}^{n+1/2}$$
$$+ \nu \frac{\Delta t}{2} \Delta x \frac{1}{\Delta y} \delta_y^2 (u_{jk}^{n} + u_{jk}^{n+1}). \qquad (4.4.69)$$

We then decompose either of these approximate conservation laws as we did above for the homogeneous case. We see that for equation (4.4.68), we can include the nonhomogeneous term with either of the equations or we can include half of it with each equation (or, include any portions that eventually adds to one with each equation—though that might be silly).

Any of these ways, if the resulting equations are solved and added together, equation (4.4.68) will be satisfied and the conservation law will be satisfied to the appropriate order.

Likewise, the two nonhomogeneous terms given in equation (4.4.69) can be included with either of the equations in any fashion so that when they are added back together we again get equation (4.4.69). The most logical forms resulting from applying (4.4.68) and (4.4.69) to obtain nonhomogeneous Peaceman-Rachford schemes are the following.

$$\left(1 - \frac{r_x}{2}\delta_x^2\right)u_{jk}^{n+\frac{1}{2}} = \left(1 + \frac{r_y}{2}\delta_y^2\right)u_{jk}^n + \frac{\Delta t}{2}F_{jk}^{n+1/2} \tag{4.4.70}$$

$$\left(1 - \frac{r_y}{2}\delta_y^2\right)u_{jk}^{n+1} = \left(1 + \frac{r_x}{2}\delta_x^2\right)u_{jk}^{n+\frac{1}{2}} + \frac{\Delta t}{2}F_{jk}^{n+1/2} \tag{4.4.71}$$

and

$$\left(1 - \frac{r_x}{2}\delta_x^2\right)u_{jk}^{n+\frac{1}{2}} = \left(1 + \frac{r_y}{2}\delta_y^2\right)u_{jk}^n + \frac{\Delta t}{2}F_{jk}^n \tag{4.4.72}$$

$$\left(1 - \frac{r_y}{2}\delta_y^2\right)u_{jk}^{n+1} = \left(1 + \frac{r_x}{2}\delta_x^2\right)u_{jk}^{n+\frac{1}{2}} + \frac{\Delta t}{2}F_{jk}^{n+1} \tag{4.4.73}$$

Before we leave this section, we briefly show that the same approach can be used with the D'Yakonov scheme. It is hoped that with these two examples in hand, the reader should be able to include the nonhomogeneous term in just about any scheme. We return to the integral form of the conservation equation (4.4.63). If we proceed as we did before, except now we integrate both flux terms with respect to t by the trapezoidal rule, we get

$$\Delta x \Delta y(v_{jk}^{n+1} - v_{jk}^n) + \mathcal{O}(\Delta x^3 \Delta y \Delta t) + \mathcal{O}(\Delta x \Delta y^3 \Delta t) =$$

$$+ \int_{t_n}^{t_{n+1}} \int_{y_{k-1/2}}^{y_{k+1/2}} \int_{x_{j-1/2}}^{x_{j+1/2}} F(x,y,t)dxdydt$$

$$+ \nu\frac{\Delta t}{2}\Delta y\frac{1}{\Delta x}\delta_x^2(v_{jk}^n + v_{jk}^{n+1}) + \mathcal{O}(\Delta t \Delta x \Delta y^3) + \mathcal{O}(\Delta t^3 \Delta x \Delta y)$$

$$+ \nu\frac{\Delta t}{2}\Delta x\frac{1}{\Delta y}\delta_y^2(v_{jk}^n + v_{jk}^{n+1}) + \mathcal{O}(\Delta t \Delta x^3 \Delta y) + \mathcal{O}(\Delta t^3 \Delta x \Delta y). \tag{4.4.74}$$

If we divide by $\Delta x \Delta y$, add $\frac{r_x r_y}{4}\delta_x^2\delta_y^2(v_{jk}^{n+1} - v_{jk}^n)$ to the left hand side of equation (4.4.74) (it is a higher order term) and set $v_{jk}^* = (1 - \frac{r_y}{2}\delta_y^2)v_{jk}^{n+1}$, it is not hard to see that equation (4.4.74) is equivalent to the pair of equations

$$v_{jk}^{n+1} - \frac{r_x}{2}\delta_x^2 v_{jk}^{n+1} - \frac{r_y}{2}\delta_y^2 v_{jk}^{n+1} + \frac{r_x r_y}{4}\delta_x^2\delta_y^2 v_{jk}^{n+1} = v_{jk}^n$$

$$+ \frac{r_x}{2}\delta_x^2 v_{jk}^n + \frac{r_y}{2}\delta_y^2 v_{jk}^n + \frac{r_x r_y}{4}\delta_x^2\delta_y^2 v_{jk}^n + \mathcal{O}(\Delta t \Delta x^2) + \mathcal{O}(\Delta t \Delta y^2)$$

$$+\mathcal{O}(\Delta t^3) + \frac{1}{\Delta x \Delta y} \int_{t_n}^{t_{n+1}} \int_{y_{k-1/2}}^{y_{k+1/2}} \int_{x_{j-1/2}}^{x_{j+1/2}} F(x,y,t)dxdydt \quad (4.4.75)$$

$$(1 - \frac{r_y}{2}\delta_y^2)v_{jk}^{n+1} = v_{jk}^*. \quad (4.4.76)$$

Note that using the v^* notation introduced in equation (4.4.76), the four terms on the left hand side of equation (4.4.75) are equal to

$$\left(1 - \frac{r_x}{2}\delta_x^2\right)v_{jk}^*.$$

Then a careful examination of equation (4.4.75) will show that if we drop the nonhomogeneous term and the order terms, system (4.4.75)–(4.4.76) is equivalent to the D'Yakonov scheme, (4.4.14)–(4.4.15). Of course, *we do not present the above derivation with the thought that it is a logical derivation of the D'Yakonov scheme. It is a very artificial derivation. But, the above derivation does make it clear how accurately we must approximate the nonhomogeneous term and where that term should be in the nonhomogeneous scheme.* Thus, we integrate the nonhomogeneous term by the midpoint rule with respect to x and y and with the trapezoidal rule with respect to t (the midpoint rule could have been used for t also) and see that the nonhomogeneous D'Yakonov scheme can be written as

$$\left(1 - \frac{r_x}{2}\delta_x^2\right)u_{jk}^* = \left(1 + \frac{r_x}{2}\delta_x^2\right)\left(1 + \frac{r_y}{2}\delta_y^2\right)u_{jk}^n + \frac{\Delta t}{2}(F_{jk}^n + F_{jk}^{n+1}) \quad (4.4.77)$$

$$\left(1 - \frac{r_y}{2}\delta_y^2\right)u_{jk}^{n+1} = u_{jk}^*. \quad (4.4.78)$$

And, as we have seen before, the above nonhomogeneous term could also be replaced by a midpoint rule evaluation with respect to t. Note that we would never want to include any part of the nonhomogeneous term in equation (4.4.78).

We leave this section with the following summary. *It is not always clear where the nonhomogeneous term should go in an ADI scheme. The accuracy can be lowered or destroyed by using the wrong nonhomogeneous term or placing the nonhomogeneous term in the wrong place. A conservation law derivation of the scheme (no matter how artificial it may be) is a strong tool for describing the nonhomogeneous term.*

HW 4.4.9 Show that the following variation of the nonhomogeneous D'Yakonov scheme

$$\left(1 - \frac{r_x}{2}\delta_x^2\right)u_{jk}^* = \left(1 + \frac{r_x}{2}\delta_x^2\right)\left(1 + \frac{r_y}{2}\delta_y^2\right)u_{jk}^n + \frac{\Delta t}{2}F_{jk}^n \quad (4.4.79)$$

$$\left(1 - \frac{r_y}{2}\delta_y^2\right)u_{jk}^{n+1} = u_{jk}^* + \frac{\Delta t}{2}F_{jk}^{n+1} \quad (4.4.80)$$

is only first order with respect to Δt.

HW 4.4.10 Use the nonhomogeneous version of the Peaceman-Rachford scheme (4.4.72)–(4.4.73) to solve

$$v_t = \nu(v_{xx} + v_{yy}) + F(x,t), \ (x,y) \in (0,1) \times (0,1), \ t > 0 \quad (4.4.81)$$
$$v(0,y,t) = v(1,y,t) = 0, \ y \in [0,1], \ t \geq 0 \quad (4.4.82)$$
$$v(x,0,t) = v(x,1,t) = 0, \ x \in [0,1], \ t \geq 0 \quad (4.4.83)$$
$$v(x,y,0) = 0, \ (x,y) \in [0,1] \times [0,1], \quad (4.4.84)$$

where $F(x,t) = \sin 5\pi t \sin 2\pi x \sin \pi y$. Use $M_x = M_y = 20$, $\Delta t = 0.01$ and $\nu = 1.0$. Plot your results at times $t = 0.06$, $t = 0.1$, $t = 0.2$, $t = 0.4$ and $t = 0.8$.

4.4.6   Three Dimensional Schemes

We next include a brief discussion of three dimensional schemes for parabolic equations. We make the discussion brief because, by this time, we assume the reader can do most of the necessary calculations. As in one and two dimensions, we consider the partial differential equation

$$v_t = \nu \nabla^2 v + F(x,y,z,t) \quad (4.4.85)$$

and the obvious FTCS explicit scheme for approximating the solution of equation (4.4.85)

$$u^{n+1}_{j\,k\,\ell} = u^n_{j\,k\,\ell} + (r_x \delta^2_x + r_y \delta^2_y + r_z \delta^2_z) u^n_{j\,k\,\ell} + F^n_{j\,k\,\ell}. \quad (4.4.86)$$

It should not be surprising that difference scheme (4.4.86) is a $\mathcal{O}(\Delta t) + \mathcal{O}(\Delta x^2) + \mathcal{O}(\Delta y^2) + \mathcal{O}(\Delta z^2)$ order approximation of partial differential equation (4.4.85). It should also not be surprising to learn that difference scheme (4.4.86) is conditionally stable, with stability condition $r_x + r_y + r_z \leq 1/2$. Thus, we see that the stability condition for the three dimensional explicit scheme is again more restrictive than the two dimensional scheme. For example, if $\Delta x = \Delta y = \Delta z$, then ν, Δt and Δx must satisfy $r \leq 1/6$.

It is equally easy to obtain unconditionally stable schemes in three dimensions as it was in one or two dimensions. Thus both the three dimensional BTCS scheme

$$u^{n+1}_{j\,k\,\ell} - (r_x \delta^2_x + r_y \delta^2_y + r_z \delta^2_z) u^{n+1}_{j\,k\,\ell} = u^n_{j\,k\,\ell} + F^n_{j\,k\,\ell} \quad (4.4.87)$$

and the three dimensional Crank-Nicolson scheme

$$u^{n+1}_{j\,k\,\ell} - \frac{1}{2}(r_x \delta^2_x + r_y \delta^2_y + r_z \delta^2_z) u^{n+1}_{j\,k\,\ell} = u^n_{j\,k\,\ell} + \frac{1}{2}(r_x \delta^2_x + r_y \delta^2_y + r_z \delta^2_z) u^n_{j\,k\,\ell}$$
$$+ \frac{1}{2}(F^n_{j\,k\,\ell} + F^{n+1}_{j\,k}) \quad (4.4.88)$$

are both unconditionally stable schemes which are $\mathcal{O}(\Delta t) + \mathcal{O}(\Delta x^2) + \mathcal{O}(\Delta y^2) + \mathcal{O}(\Delta z^2)$ and $\mathcal{O}(\Delta t^2) + \mathcal{O}(\Delta x^2) + \mathcal{O}(\Delta y^2) + \mathcal{O}(\Delta z^2)$, respectively. But, it should be very clear that these equations are even much more difficult to solve than the two dimensional analogs.

Thus, as we did in two dimensions, we resort to ADI schemes to find implicit schemes that are reasonable for computational purposes. Of course, a logical approach would be to begin with a three dimensional Peaceman-Rachford scheme (along with the rationalization of using $\Delta t/3$ as time steps). But, as we see in HW4.4.11, *the three dimensional Peaceman-Rachford scheme* **is not** *unconditionally stable*. To make matters worse, we also see in HW4.4.11 that the three dimensional Peaceman-Rachford scheme is only first order accurate in time.

To obtain a usable three dimensional ADI scheme, we approximately factor the three dimensional Crank-Nicolson scheme, (4.4.88). The natural factorization of the left hand side of equation (4.4.88) is

$$\left(1 - \frac{r_x}{2}\delta_x^2\right)\left(1 - \frac{r_y}{2}\delta_y^2\right)\left(1 - \frac{r_z}{2}\delta_z^2\right)u_{jk\ell}^{n+1}. \tag{4.4.89}$$

If we add

$$\left[\frac{r_x r_y}{4}\delta_x^2\delta_y^2 + \frac{r_x r_z}{4}\delta_x^2\delta_z^2 + \frac{r_y r_z}{4}\delta_y^2\delta_z^2\right](u_{jk\ell}^{n+1} - u_{jk\ell}^n) - \frac{r_x r_y r_z}{8}\delta_x^2\delta_y^2\delta_z^2(u_{jk\ell}^{n+1} + u_{jk\ell}^n)$$

to the left hand side of equation (4.4.88) (as in the two dimensional approximately factored scheme, these will all be higher order terms) and distribute the appropriate half of the above term to the right hand side, we are left with

$$\left(1 - \frac{r_x}{2}\delta_x^2\right)\left(1 - \frac{r_y}{2}\delta_y^2\right)\left(1 - \frac{r_z}{2}\delta_z^2\right)u_{jk\ell}^{n+1} =$$
$$\left(1 + \frac{r_x}{2}\delta_x^2\right)\left(1 + \frac{r_y}{2}\delta_y^2\right)\left(1 + \frac{r_z}{2}\delta_z^2\right)u_{jk\ell}^n + \frac{1}{2}(F_{jk}^n + F_{jk}^{n+1}). \tag{4.4.90}$$

This equation could be solved in a variety of ways. We next rewrite equation (4.4.90) in what is called the **delta formulation**. We note that equation (4.4.90) is equivalent to

$$\left(1 - \frac{r_x}{2}\delta_x^2\right)\left(1 - \frac{r_y}{2}\delta_y^2\right)\left(1 - \frac{r_z}{2}\delta_z^2\right)(u_{jk\ell}^{n+1} - u_{jk\ell}^n) = (r_x\delta_x^2 + r_y\delta_y^2 + r_z\delta_z^2)u_{jk\ell}^n$$

$$+ \frac{r_x r_y r_z}{8}\delta_x^2\delta_y^2\delta_z^2 u_{jk\ell}^n + \frac{1}{2}(F_{jk}^n + F_{jk}^{n+1}). \tag{4.4.91}$$

(The easiest way to see this equivalence is to expand both equations (4.4.90) and (4.4.91) and compare.) We can then drop the $\delta_x^2\delta_y^2\delta_z^2$ term (because it is higher order) and get the following form of the **Douglas-Gunn scheme**

$$\left(1 - \frac{r_x}{2}\delta_x^2\right)\Delta u^* = (r_x\delta_x^2 + r_y\delta_y^2 + r_z\delta_z^2)u_{jk\ell}^n + \frac{1}{2}(F_{jk}^n + F_{jk}^{n+1}) \quad (4.4.92)$$

$$\left(1 - \frac{r_y}{2}\delta_y^2\right)\Delta u^{**} = \Delta u^* \quad (4.4.93)$$

$$\left(1 - \frac{r_z}{2}\delta_z^2\right)\Delta u = \Delta u^{**} \quad (4.4.94)$$

$$\Delta u = u_{jk\ell}^{n+1} - u_{jk\ell}^n. \quad (4.4.95)$$

Reviewing the derivation of difference scheme (4.4.92)–(4.4.95), it is clear that the Douglas-Gunn scheme is accurate of order $\mathcal{O}(\Delta t^2) + \mathcal{O}(\Delta x^2) + \mathcal{O}(\Delta y^2) + \mathcal{O}(\Delta z^2)$. To see that the scheme is unconditionally stable, we take the discrete Fourier transforms of the nonhomogeneous version of equations (4.4.92)–(4.4.95) to get

$$\left(1 + 2r_x \sin^2 \frac{\xi}{2}\right)\widehat{\Delta u^*} = \left(-4r_x \sin^2 \frac{\xi}{2} - 4r_y \sin^2 \frac{\eta}{2} - 4r_z \sin^2 \frac{\zeta}{2}\right)\hat{u}^n \quad (4.4.96)$$

$$\left(1 + 2r_y \sin^2 \frac{\eta}{2}\right)\widehat{\Delta u^{**}} = \widehat{\Delta u^*} \quad (4.4.97)$$

$$\left(1 + 2r_z \sin^2 \frac{\zeta}{2}\right)\widehat{\Delta u} = \widehat{\Delta u^{**}} \quad (4.4.98)$$

$$\widehat{\Delta u} = \hat{u}^{n+1} - \hat{u}^n. \quad (4.4.99)$$

Solving equations (4.4.96)–(4.4.99) for $\hat{u}^{n+1}$ as a function of $\hat{u}^n$, leaves us with

$$\hat{u}^{n+1} = \left[1 - 2r_x \sin^2 \frac{\xi}{2} - 2r_y \sin^2 \frac{\eta}{2} - 2r_z \sin^2 \frac{\zeta}{2} + 4r_x r_y \sin^2 \frac{\xi}{2} \sin^2 \frac{\eta}{2}\right.$$

$$+ 4r_x r_z \sin^2 \frac{\xi}{2} \sin^2 \frac{\zeta}{2} + 4r_y r_z \sin^2 \frac{\eta}{2} \sin \frac{\zeta}{2}$$

$$\left. + 8r_x r_y r_z \sin^2 \frac{\xi}{2} \sin^2 \frac{\eta}{2} \sin^2 \frac{\zeta}{2}\right] / \left[\left(1 + 2r_x \sin^2 \frac{\xi}{2}\right)\right.$$

$$\left. \left(1 + 2r_y \sin^2 \frac{\eta}{2}\right)\left(1 + 2r_z \sin^2 \frac{\zeta}{2}\right)\right]\hat{u}^n \quad (4.4.100)$$

$$= \rho(\xi, \eta, \zeta)\hat{u}^n. \quad (4.4.101)$$

The above expression is in the general form

$$\frac{1 - a - b - c + d + e + f + g}{1 + a + b + c + d + e + f + g},$$

where $a, \cdots, g$ are all positive and it is easy to see that

$$-1 \leq \frac{1 - a - b - c + d + e + f + g}{1 + a + b + c + d + e + f + g} \leq 1.$$

Likewise, $|\rho(\xi, \eta, \zeta)| \leq 1$. Hence, the Douglas-Gunn scheme is uncondi-tionally stable. Since it is both consistent and unconditionally stable, *the Douglas-Gunn scheme is convergent.*

HW 4.4.11 Show that the three dimensional analog of the Peaceman-Rachford scheme,

$$\left(1 - \frac{r_x}{3}\delta_x^2\right)u_{jk}^{n+\frac{1}{3}} = \left(1 + \frac{r_y}{3}\delta_y^2 + \frac{r_z}{3}\delta_z^2\right)u_{jk}^n \qquad (4.4.102)$$

$$\left(1 - \frac{r_y}{3}\delta_y^2\right)u_{jk}^{n+\frac{2}{3}} = \left(1 + \frac{r_x}{3}\delta_x^2 + \frac{r_z}{3}\delta_z^2\right)u_{jk}^{n+\frac{1}{3}} \qquad (4.4.103)$$

$$\left(1 - \frac{r_z}{3}\delta_z^2\right)u_{jk}^{n+1} = \left(1 + \frac{r_x}{3}\delta_x^2 + \frac{r_y}{3}\delta_y^2\right)u_{jk}^{n+2/3}, \qquad (4.4.104)$$

is conditionally stable and order $\mathcal{O}(\Delta t) + \mathcal{O}(\Delta x^2) + \mathcal{O}(\Delta y^2) + \mathcal{O}(\Delta z^2)$.

4.5   Polar Coordinates

Looking at the problems considered in this book, one might believe that the only coordinate system that is used in the area of numerical partial differential equations is the Cartesian coordinate system. Generally this approach is not too bad because most methods that we have considered and that we will consider transfer to other coordinate systems trivially. There can be some special problems that arise in certain coordinate systems, and specifically, when we consider equations in polar coordinates we find some special problems. For this reason, we develop a numerical scheme for a two dimensional parabolic equation in polar coordinates. Hopefully, the work done in this section will allow the reader to transfer most of our techniques from Cartesian coordinates to polar coordinates.

We begin by considering the partial differential equation

$$v_t = \frac{1}{r}(rv_r)_r + \frac{1}{r^2}v_{\theta\theta} + F(r, \theta, t)\ 0 \leq r < 1,\ 0 \leq \theta < 2\pi,\ t > 0 \tag{4.5.1}$$

along with boundary and initial conditions

$$v(1, \theta, t) = g(\theta, t)\ 0 \leq \theta < 2\pi,\ t > 0 \tag{4.5.2}$$

$$v(r, \theta, 0) = f(r, \theta)\ 0 \leq r \leq 1,\ 0 \leq \theta < 2\pi. \tag{4.5.3}$$

We should realize that we have a continuity condition at $\theta = 2\pi$ of the form

$$v(r, 0, t) = \lim_{\theta \to 2\pi} v(r, \theta, t).$$

We will generally refer to this condition as

$$v(r, 0, t) = v(r, 2\pi, t). \tag{4.5.4}$$

Though *this condition is not a boundary condition*, in many numerical methods (as well as in many analytic methods) this continuity condition will be used very much like a boundary condition.

In Figure 4.5.1 we place a polar coordinate grid on the domain of our problem. We will number our grid points as (r_j, θ_k), $j = 0, \cdots, M_r$, $k = 0, \cdots, M_\theta$, where $r_0 = 0$, $r_{M_r} = 1$, $\theta_0 = 0$ and $\theta_{M_\theta} = 2\pi$. At times it is convenient to represent the grid as in Figure 4.5.2. As usual, we denote function values defined at the (r_j, θ_k) grid point at $t = n\Delta t$ by u_{jk}^n. However, since *there is only one point associated with $j = 0$*, we use both notations to refer to a function defined at the origin, u_{0k}^n and u_0^n, understanding that for all k, $u_{0k}^n = u_0^n$. And, finally, the continuity condition described above looks like a continuity condition on the grid given in Figure 4.5.1 at $\theta = 0$ (and $\theta = 2\pi$) and is given as $u_{jM_\theta}^n = u_{j0}^n$. On the grid given in Figure 4.5.2, this condition looks very much like a periodic boundary condition.

We now proceed to approximate our problem (4.5.1)–(4.5.3) (along with (4.5.4)) by a difference scheme. As usual, we approximate v_t by

$$(v_t)_{jk}^n \approx \frac{u_{jk}^{n+1} - u_{jk}^n}{\Delta t}.$$

Though we have not done any approximations with variable coefficients (in equation (4.5.1) the $1/r$ and $1/r^2$ terms are just variable coefficients) it is not difficult to see that the $v_{\theta\theta}$ term can be approximated by

$$\left(\frac{1}{r^2}v_{\theta\theta}\right)_{jk}^n \approx \frac{1}{r_j^2}\frac{1}{\Delta\theta^2}\delta_\theta^2 u_{jk}^n, \ j = 1, \cdots, M_r - 1, \ k = 1, \cdots, M_\theta - 1. \tag{4.5.5}$$

Note that we did not claim that equation (4.5.5) held for either $j = 0$ or $k = 0$. We treat the $j = 0$ point later (it is a very important point). To apply equation (4.5.5) at $k = 0$ we must use the continuity condition (4.5.4) and reach to $k = M_\theta - 1$, i.e.

$$\left(\frac{1}{r^2}v_{\theta\theta}\right)_{j0}^n \approx \frac{1}{r_j^2}\frac{1}{\Delta\theta^2}(u_{j1}^n - 2u_{j0}^n + u_{jM_\theta-1}^n). \tag{4.5.6}$$

To approximate the $(rv_r)_r/r$ term for $j = 1, \cdots, M_r - 1$, we return to the approach used in Chapter 1 when we first introduced an approximation of the second partial derivative. We see that

$$\left(\frac{1}{r}(rv_r)_r\right)_{jk}^n \approx \frac{1}{r_j}\frac{(rv_r)_{j+1/2k}^n - (rv_r)_{j-1/2k}^n}{\Delta r}$$

$$\approx \frac{1}{r_j}\frac{r_{j+1/2}\frac{u_{j+1k}^n - u_{jk}^n}{\Delta r} - r_{j-1/2}\frac{u_{jk}^n - u_{j-1k}^n}{\Delta r}}{\Delta r}$$

$$\approx \frac{1}{r_j}\frac{1}{\Delta r^2}\left[r_{j+1/2}(u_{j+1k}^n - u_{jk}^n) - r_{j-1/2}(u_{jk}^n - u_{j-1k}^n)\right].$$

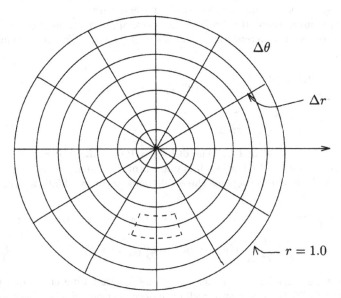

FIGURE 4.5.1. Two dimensional polar coordinate grid on the region $0 \leq r \leq 1$, $0 \leq \theta \leq 2\pi$.

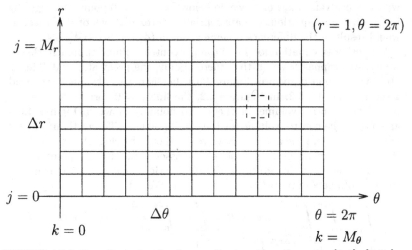

FIGURE 4.5.2. Two dimensional polar coordinate grid on the region $[0, 1] \times [0, 2\pi]$.

Of course it would also be logical to rewrite $(rv_r)_r/r$ as $v_r/r + v_{rr}$ and approximate this form. However, since we do know the value of $r_{j+1/2}$ and $r_{j-1/2}$, the approach used above will be more accurate.

We are almost done. If we collect the above approximations together, we get the following difference approximation of partial differential equation (4.5.1).

$$\frac{u_{jk}^{n+1} - u_{jk}^n}{\Delta t} = \frac{1}{r_j} \frac{1}{\Delta r^2} \left[r_{j+1/2}(u_{j+1\,k}^n - u_{jk}^n) - r_{j-1/2}(u_{jk}^n - u_{j-1\,k}^n) \right]$$

$$+ \frac{1}{r_j^2} \frac{1}{\Delta\theta^2} \delta_\theta^2 u_{jk}^n + F_{jk}^n \tag{4.5.7}$$

We note that equation (4.5.7) holds for $j = 1, \cdots, M_r - 1$, $k = 1, \cdots, M_\theta - 1$ (where for the $j = 1$ equation, it is important to remember that $u_{0k}^n = u_0^n$ for all k). At $k = 0$, we replace the θ difference term by (4.5.6) and get

$$\frac{u_{j0}^{n+1} - u_{j0}^n}{\Delta t} = \frac{1}{r_j} \frac{1}{\Delta r^2} \left[r_{j+1/2}(u_{j+1\,0}^n - u_{j0}^n) - r_{j-1/2}(u_{j0}^n - u_{j-1\,0}^n) \right]$$

$$+ \frac{1}{r_j^2} \frac{1}{\Delta\theta^2} (u_{j1}^n - 2u_{j0}^n + u_{j\,M_\theta-1}^n) + F_{j0}^n. \tag{4.5.8}$$

Before we consider $j = 0$, let us emphasize that difference equations (4.5.7) and (4.5.8) could just as well have been derived using a conservation law approach. In both Figures 4.5.1 and 4.5.2 we have included a general control volume. If we had proceeded as we have in the past and started with a conservation law or as we do below for the $j = 0$ point and begin by integrating our equation, the appropriate approximations of the integrals would result in the difference scheme given in (4.5.7). See HW4.5.4.

We derive an equation for $j = 0$ using a control volume approach. Before we do, we emphasize again that there is only one point at $j = 0$. This is obvious in the grid given in Figure 4.5.1, but is not so obvious in the grid given in Figure 4.5.2. In Figure 4.5.2, the line $j = 0$ corresponds to one point. The control volume for the $j = 0$ point is shown in Figures 4.5.3 and 4.5.4. The control volume for the $j = 0$ point is different than that for any other point in the domain. However, it should be clear that the volume shown is the only logical control volume associated with the point $j = 0$. We now proceed to integrate partial differential equation (4.5.1) over the control volume and from $t = t_n$ to $t = t_{n+1}$. We get

$$\int_{t_n}^{t_{n+1}} \int_0^{2\pi} \int_0^{\Delta r/2} v_t r\,dr\,d\theta\,dt = \int_{t_n}^{t_{n+1}} \int_0^{2\pi} \int_0^{\Delta r/2} F(r,\theta,t) r\,dr\,d\theta\,dt$$

$$+ \int_{t_n}^{t_{n+1}} \int_0^{2\pi} \int_0^{\Delta r/2} \left[\frac{1}{r}(rv_r)_r + \frac{1}{r^2} v_{\theta\theta} \right] r\,dr\,d\theta\,dt. \tag{4.5.9}$$

If we integrate the first term with respect to t and apply the Divergence Theorem to the third term (of course, the polar coordinate form of the

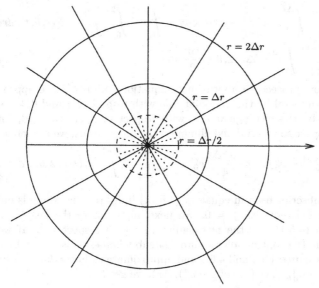

FIGURE 4.5.3. Control volume of the point $r = 0$ when the grid is given in polar coordinates (analogous to 4.5.1).

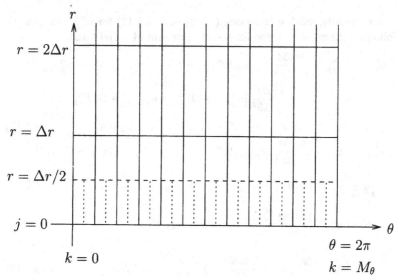

FIGURE 4.5.4. Control volume of the point $r = 0$ when the grid is given as a rectangle (analogous to 4.5.2).

Divergence Theorem, ref. [8], page 351), we get

$$\int_0^{2\pi} \int_0^{\Delta r/2} (v^{n+1} - v^n) r \, dr \, d\theta = \int_{t_n}^{t_{n+1}} \int_0^{2\pi} \int_0^{\Delta r/2} F(r, \theta, t) r \, dr \, d\theta \, dt$$

$$+ \int_{t_n}^{t_{n+1}} \int_0^{2\pi} v_r(\Delta r/2, \theta, t) \frac{\Delta r}{2} \, d\theta \, dt. \tag{4.5.10}$$

We now proceed to approximate equation (4.5.10). If we approximate the first integral by the midpoint rule with respect to r and θ, the second integral by the rectangular rule with respect to t and the midpoint rule with respect to r and θ and eliminate the order terms, we are left with

$$\pi \frac{\Delta r^2}{4} (u_0^{n+1} - u_0^n) \approx \pi \frac{\Delta r^2 \Delta t}{4} F_0^n + \frac{\Delta r}{2} \int_{t_n}^{t_{n+1}} \int_0^{2\pi} v_r(\Delta r/2, \theta, t) \, d\theta \, dt. \tag{4.5.11}$$

The 0 subscript used in equation (4.5.11) indicates that there is only one point associated with $j = 0$. We next approximate the last integral in equation (4.5.11) by the rectangular rule with respect to t. If we then subdivide the control volume into M_θ sub-volumes as is shown by dotted lines in Figures 4.5.3 and 4.5.4 and approximate v_r in the k-th sub-volume by $(u_{1k}^n - u_{0k}^n)/\Delta r$ (or $(u_{1k}^n - u_0^n)/\Delta r$), we get

$$\pi \frac{\Delta r^2}{4} (u_0^{n+1} - u_0^n) = \pi \frac{\Delta r^2 \Delta t}{4} F_0^n + \frac{\Delta r}{2} \Delta t \sum_{k=0}^{M_\theta - 1} \frac{u_{1k}^n - u_0^n}{\Delta r} \Delta \theta. \tag{4.5.12}$$

And, finally, solving equations (4.5.7) and (4.5.12) for u_{jk}^{n+1}, we get the following difference scheme for solving problem (4.5.1)–(4.5.3).

$$u_{jk}^{n+1} = u_{jk}^n + \frac{1}{r_j} \frac{\Delta t}{\Delta r^2} \left[r_{j+1/2}(u_{j+1k}^n - u_{jk}^n) - r_{j-1/2}(u_{jk}^n - u_{j-1k}^n) \right]$$

$$+ \frac{1}{r_j^2} \frac{\Delta t}{\Delta \theta^2} \left[u_{jk+1}^n - 2u_{jk}^n + u_{jk-1}^n \right] + \Delta t F_{jk}^n,$$

$$j = 1, \cdots, M_r - 1, \; k = 1, \cdots, M_\theta - 1 \tag{4.5.13}$$

$$u_{j0}^{n+1} = u_{j0}^n + \frac{1}{r_j} \frac{\Delta t}{\Delta r^2} \left[r_{j+1/2}(u_{j+10}^n - u_{j0}^n) - r_{j-1/2}(u_{j0}^n - u_{j-10}^n) \right]$$

$$+ \frac{1}{r_j^2} \frac{\Delta t}{\Delta \theta^2} \left[u_{j1}^n - 2u_{j0}^n + u_{jM_\theta - 1}^n \right] + \Delta t F_{j0}^n, \; j = 1, \cdots, M_r - 1 \tag{4.5.14}$$

$$u_0^{n+1} = u_0^n + \frac{2\Delta \theta \Delta t}{\pi \Delta r^2} \sum_{k=0}^{M_\theta - 1} (u_{1k}^n - u_0^n) + \Delta t F_0^n$$

$$= \left(1 - \frac{4\Delta t}{\Delta r^2} \right) u_0^n + \frac{2\Delta \theta \Delta t}{\pi \Delta r} \sum_{k=0}^{M_\theta - 1} u_{1k}^n + \Delta t F_0^n \tag{4.5.15}$$

$$u_{M_r k}^{n+1} = g_k^{n+1}, \; k = 0, \cdots, M_\theta - 1 \tag{4.5.16}$$

$$u_{jk}^0 = f_{jk}, \; j = 0, \cdots, M_r - 1, \; k = 0, \cdots, M_\theta - 1 \tag{4.5.17}$$

Implementation of difference scheme (4.5.13)–(4.5.17) is not very different from implementing other two dimensional explicit schemes. As usual, in the solution phase of the code, you must loop through the equations (4.5.13)–(4.5.15). However, because it is an explicit scheme (the right hand side depends only on the previous time step), we do not have to worry about in which order these computations are performed. Equation (4.5.13) (looping from $j = 1$ to $j = M_r - 1$ and from $k = 1$ to $k = M_\theta - 1$) is used to compute u^{n+1} at "most" of the points. Equation (4.5.14) (looping from $j = 1$ to $j = M_r - 1$) is used to compute u^{n+1} at the points along the $\theta = 0$ axis and equation (4.5.15) is used to compute u_0^{n+1}. Your previously written initialization routine should be almost what you need as an initialization routine here. You should be more careful with your output routine. Any routine that outputs data will still work fine. More care must be taken to get graphical output in polar coordinates.

HW 4.5.1 Solve the following initial–boundary–value problem in polar coordinates.

$$v_t = \frac{1}{r}(rv_r)_r + \frac{1}{r^2}v_{\theta\theta} \ \ 0 \le r < 1,\ 0 \le \theta < 2\pi,\ t > 0 \qquad (4.5.18)$$
$$v(1,\theta,t) = \sin 4\theta \sin t \ 0 \le \theta < 2\pi,\ t \ge 0 \qquad (4.5.19)$$
$$v(r,\theta,0) = 0,\ 0 \le r \le 1,\ 0 \le \theta < 2\pi \qquad (4.5.20)$$

Use $M_r = 20$, $M_\theta = 32$, $\Delta t = 0.001$ and plot solutions at $t = 0.1$, $t = 0.5$, $t = 1.5$, $t = 3$ and $t = 6.2$.

HW 4.5.2 Solve the following initial–boundary–value problem in polar coordinates.

$$v_t = \frac{1}{r}(rv_r)_r + \frac{1}{r^2}v_{\theta\theta} \ \ 0 \le r < 1,\ 0 \le \theta < 2\pi,\ t > 0 \qquad (4.5.21)$$
$$v(1,\theta,t) = 0 \ 0 \le \theta < 2\pi,\ t \ge 0 \qquad (4.5.22)$$
$$v(r,\theta,0) = (1 - r^2)\sin 2\theta \ 0 \le r \le 1 \ 0 \le \theta < 2\pi \qquad (4.5.23)$$

Use $M_r = 20$, $M_\theta = 32$, $\Delta t = 0.001$ and plot solutions at $t = 0.1$, $t = 0.5$ and $t = 1.0$.

HW 4.5.3 Solve the following initial–boundary–value problem in polar coordinates.

$$v_t = \frac{1}{r}(rv_r)_r + \frac{1}{r^2}v_{\theta\theta} + (1 - r^2)\sin t$$
$$0 \le r < 1,\ 0 \le \theta < 2\pi,\ t > 0 \qquad (4.5.24)$$
$$v(1,\theta,t) = 0 \ \ 0 \le \theta < 2\pi,\ t \ge 0 \qquad (4.5.25)$$
$$v(r,\theta,0) = 0 \ \ 0 \le r \le 1 \ 0 \le \theta < 2\pi \qquad (4.5.26)$$

Use $M_r = 20$, $M_\theta = 32$, $\Delta t = 0.001$ and plot solutions at $t = 0.1$, $t = 0.5$, $t = 1.5$, $t = 3$ and $t = 6.2$.

HW 4.5.4 Use the control volume pictured in Figure 4.5.1 to derive difference equation (4.5.13). Use an analogous control volume at point $(j\Delta r, 0)$ to derive difference equation (4.5.14).

Remark 1: If we wanted to use an implicit scheme to solve our problem instead of the explicit scheme derived above, the approach would be the same. Following the above derivations with an $n + 1$ replacing the n, the implicit analogs to equations (4.5.13)–(4.5.14) would be

$$u_{jk}^{n+1} - \frac{1}{r_j}\frac{\Delta t}{\Delta r^2}\left[r_{j+1/2}(u_{j+1\,k}^{n+1} - u_{jk}^{n+1}) - r_{j-1/2}(u_{jk}^{n+1} - u_{j-1\,k}^{n+1})\right]$$

$$- \frac{1}{r_j^2}\frac{\Delta t}{\Delta\theta^2}\left[u_{j\,k+1}^{n+1} - 2u_{jk}^{n+1} + u_{j\,k-1}^{n+1}\right] = u_{jk}^n + \Delta t F_{jk}^{n+1},$$

$$j = 1, \cdots, M_r - 1, \; k = 1, \cdots, M_\theta - 1 \qquad (4.5.27)$$

$$u_{j0}^{n+1} - \frac{1}{r_j}\frac{\Delta t}{\Delta r^2}\left[r_{j+1/2}(u_{j+1\,0}^{n+1} - u_{j0}^{n+1}) - r_{j-1/2}(u_{j0}^{n+1} - u_{j-1\,0}^{n+1})\right]$$

$$- \frac{1}{r_j^2}\frac{\Delta t}{\Delta\theta^2}\left[u_{j1}^{n+1} - 2u_{j0}^{n+1} + u_{j\,M_\theta-1}^{n+1}\right] = u_{j0}^n + \Delta t F_{j0}^{n+1},$$

$$j = 1, \cdots, M_r - 1 \qquad (4.5.28)$$

To derive the implicit version of equation (4.5.15), we treat the forcing term and the flux term in the derivation implicitly instead of explicitly (still approximating the integral with respect to t by the rectangular rule, but now using the right hand end point $t_{n+1} = (n + 1)\Delta t$ instead of the left hand end point). We then find the following equation.

$$\left(1 + \frac{4\Delta t}{\Delta r^2}\right)u_0^{n+1} - \frac{2\Delta t\Delta\theta}{\pi\Delta r^2}\sum_{k=0}^{M_\theta-1} u_{1\,k}^{n+1} = u_0^n + \Delta t F_0^{n+1} \qquad (4.5.29)$$

Thus to solve problem (4.5.1)–(4.5.3) using an implicit scheme, we must solve equations (4.5.27)–(4.5.29) along with boundary condition (4.5.16) and initial condition (4.5.17). As with implicit schemes for most two dimensional problems, we do not expect to get a very nice matrix. However, in this case the matrix is worse than usual. If we order the variables as

$$u_0, \; u_{1\,0}, \; \cdots, \; u_{1\,M_\theta-1}, \; u_{2\,0}, \; \cdots, \; u_{M_r-1\,M_\theta-1},$$

we see that the equation that we must solve is of the form

$$Q_1\mathbf{u}^{n+1} = \mathbf{u}^n + \mathbf{G}^{n+1} \qquad (4.5.30)$$

where

$$
Q_1 = \begin{pmatrix}
\alpha & \mathbf{r}^T & \boldsymbol{\theta}^T & \cdots & & & & \\
\mathbf{c} & T_1 & -\gamma_1 I & \ominus & \cdots & & & \\
\boldsymbol{\theta} & -\alpha_2 I & T_2 & -\gamma_2 I & \ominus & \cdots & & \\
\boldsymbol{\theta} & \ominus & -\alpha_3 I & T_3 & -\gamma_3 I & \ominus & & \cdots \\
\vdots & \vdots & \ddots & \ddots & \ddots & \ddots & & \\
\boldsymbol{\theta} & \ominus & \cdots & \ominus & -\alpha_{M_r-2}I & T_{M_r-2} & -\gamma_{M_r-2}I \\
\boldsymbol{\theta} & \ominus & \cdots & \cdots & \ominus & -\alpha_{M_r-1}I & T_{M_r-1}
\end{pmatrix}
$$

$$(4.5.31)$$

$\alpha = 1 + 4\Delta t/\Delta r^2$, $\mathbf{r}^T = [a \cdots a]$ ($\mathbf{r}$ an M_θ vector), $a = -2\Delta t\Delta\theta/\pi\Delta r^2$,
$\mathbf{c} = [-\alpha_1 \cdots -\alpha_1]^T$ ($\mathbf{c}$ an M_θ vector),

$$
\begin{aligned}
\alpha_j &= r_{j-1/2}\Delta t/r_j\Delta r^2, \\
\gamma_j &= r_{j+1/2}\Delta t/r_j\Delta r^2, \\
\beta_j &= 1 + ((r_{j-1/2} + r_{j+1/2})\Delta t/r_j\Delta r^2) + (2\Delta t/r_j^2\Delta\theta^2), \\
\epsilon_j &= \Delta t/r_j^2\Delta\theta^2
\end{aligned}
$$

and

$$
T_j = \begin{pmatrix}
\beta_j & -\epsilon_j & 0 & \cdots & & 0 & -\epsilon_j \\
-\epsilon_j & \beta_j & -\epsilon_j & 0\cdots & & & \\
0 & -\epsilon_j & \beta_j & -\epsilon_j & 0\cdots & & \\
\ddots & \ddots & \ddots & \ddots & \ddots & & \\
0 & \cdots & 0 & -\epsilon_j & \beta_j & -\epsilon_j \\
-\epsilon_j & 0 & \cdots & 0 & -\epsilon_j & \beta_j
\end{pmatrix} . \qquad (4.5.32)
$$

To make it easier to visualize what matrix Q_1 is like, in Figure 4.5.5 we include a version of matrix Q_1 with $M_\theta = 5$ and $M_r = 4$.

Thus we see that the basic matrix is broadly banded like most matrices for two dimensional schemes and has nonzero terms in the diagonal blocks due to the periodicity in θ. In addition, the first row and first column are all messed up due to the $r = 0$ grid point. This matrix equation is generally a bad matrix equation to have to solve. But there are tricks or methods that make solving equation (4.5.30) reasonable. The Sherman-Morrison Algorithm can be used to handle both the bad first row and column and the periodicity. We do not introduce this solution method at this time. We instead show how to solve this problem in Section 6.10.3 after the Sherman-Morrison Algorithm has been introduced and used.in a nicer (easier) setting in Section 5.6.1 and again in Section 6.3.2.3. If the reader has a specific interest in this problem, the description of the Sherman-Morrison Algorithm in Section 5.6.1 and the application of the algorithm

$$
\left(\begin{array}{c|ccccc|ccccc|ccccc}
\alpha & a & a & a & a & a & 0 & 0 & 0 & 0 & 0 & 0 & 0 & 0 & 0 & 0 \\
\hline
-\alpha_1 & \beta_1 & -\epsilon_1 & 0 & 0 & -\epsilon_1 & -\gamma_1 & 0 & 0 & 0 & 0 & 0 & 0 & 0 & 0 & 0 \\
-\alpha_1 & -\epsilon_1 & \beta_1 & -\epsilon_1 & 0 & 0 & 0 & -\gamma_1 & 0 & 0 & 0 & 0 & 0 & 0 & 0 & 0 \\
-\alpha_1 & 0 & -\epsilon_1 & \beta_1 & -\epsilon_1 & 0 & 0 & 0 & -\gamma_1 & 0 & 0 & 0 & 0 & 0 & 0 & 0 \\
-\alpha_1 & 0 & 0 & -\epsilon_1 & \beta_1 & -\epsilon_1 & 0 & 0 & 0 & -\gamma_1 & 0 & 0 & 0 & 0 & 0 & 0 \\
-\alpha_1 & -\epsilon_1 & 0 & 0 & -\epsilon_1 & \beta_1 & \mathbf{0} & 0 & 0 & 0 & -\gamma_1 & 0 & 0 & 0 & 0 & 0 \\
\hline
0 & -\alpha_2 & 0 & 0 & 0 & \mathbf{0} & \beta_2 & -\epsilon_2 & 0 & 0 & -\epsilon_2 & -\gamma_2 & 0 & 0 & 0 & 0 \\
0 & 0 & -\alpha_2 & 0 & 0 & 0 & -\epsilon_2 & \beta_2 & -\epsilon_2 & 0 & 0 & 0 & -\gamma_2 & 0 & 0 & 0 \\
0 & 0 & 0 & -\alpha_2 & 0 & 0 & 0 & -\epsilon_2 & \beta_2 & -\epsilon_2 & 0 & 0 & 0 & -\gamma_2 & 0 & 0 \\
0 & 0 & 0 & 0 & -\alpha_2 & 0 & 0 & 0 & -\epsilon_2 & \beta_2 & -\epsilon_2 & 0 & 0 & 0 & -\gamma_2 & 0 \\
0 & 0 & 0 & 0 & 0 & -\alpha_2 & -\epsilon_2 & 0 & 0 & -\epsilon_2 & \beta_2 & \mathbf{0} & 0 & 0 & 0 & -\gamma_2 \\
\hline
0 & 0 & 0 & 0 & 0 & 0 & -\alpha_3 & 0 & 0 & 0 & \mathbf{0} & \beta_3 & -\epsilon_3 & 0 & 0 & -\epsilon_3 \\
0 & 0 & 0 & 0 & 0 & 0 & 0 & -\alpha_3 & 0 & 0 & 0 & -\epsilon_3 & \beta_3 & -\epsilon_3 & 0 & 0 \\
0 & 0 & 0 & 0 & 0 & 0 & 0 & 0 & -\alpha_3 & 0 & 0 & 0 & -\epsilon_3 & \beta_3 & -\epsilon_3 & 0 \\
0 & 0 & 0 & 0 & 0 & 0 & 0 & 0 & 0 & -\alpha_3 & 0 & 0 & 0 & -\epsilon_3 & \beta_3 & -\epsilon_3 \\
0 & 0 & 0 & 0 & 0 & 0 & 0 & 0 & 0 & 0 & -\alpha_3 & -\epsilon_3 & 0 & 0 & -\epsilon_3 & \beta_3
\end{array}\right)
$$

FIGURE 4.5.5. Matrix Q_1, (4.5.31), where $M_\theta = 5$ and $M_r = 4$. The horizontal and vertical lines partition the matrix analogous to the representation given in equation (4.5.31).

to polar coordinate problems in Section 6.10.3 are independent of the other material and can be read now.

It should be emphasized that when we write the implicit scheme as we do in (4.5.30), the $\mathbf{G}$ vector on the right hand side includes a variety of terms from equations (4.5.27)–(4.5.29). Obviously, $\mathbf{G}^n$ includes the $\Delta t \mathbf{F}^n$'s terms and does not include the $\mathbf{u}^n$ (since it appears explicitly in equation (4.5.30)). The $\mathbf{G}^n$ term will also include any boundary conditions due to the difference equations reaching to $j = M_r$. There will be no other boundary conditions involved in the problem (other than the periodic boundary conditions which we handle as a part of the equations).

Remark 2: If we have a problem to solve involving polar coordinates that does not involve the origin, the solution scheme is similar but easier. The grid will be as is given in Figure 4.5.1 except that there will be an inner circle of points that are excluded. The explicit scheme will involve using difference equations (4.5.13)–(4.5.14) plus boundary conditions and initial conditions (and not equation (4.5.15)). The implicit scheme will give us a matrix equation that looks like the equation considered in Remark 1, (4.5.30), except that the first row and first column will not be there. This problem will still have the nonzero terms in the corners of the diagonal blocks due to the periodicity with respect to θ.

Remark 3: We also mention that if we consider a time independent problem (as we will do in Chapter 10), the approach will be the same. As was the case for the implicit scheme, the matrix will have one row and one column that contains many nonzero terms. In Chapter 10 where we generally use iterative schemes for solving the matrix equations associated with elliptic problems, this bad row will not cause any special difficulties. Iterative methods could also be used to solve problem (4.5.27)–(4.5.29) along with its initial and boundary conditions.

HW 4.5.5 Derive a Crank-Nicolson scheme for solving problem (4.5.1)–(4.5.3).

5
Hyperbolic Equations

5.1 Introduction

Hyperbolic partial differential equations are used to model a large and extremely important collection of phenomena. This includes aerodynamic flows, flows of fluids and contaminants through a porous media, atmospheric flows, etc. The solutions to hyperbolic equations tend to be more complex and interesting than those to parabolic and elliptic equations. The interesting characteristics of the solutions to hyperbolic equations follows from the fact that the phenomena modeled by hyperbolic equations are generally difficult and interesting. Of course, most of the applications mentioned above involve nonlinear systems of equations. However, most techniques for solving nonlinear systems are generalizations of or are otherwise related to some linear technique. For this reason, we begin by considering linear systems of hyperbolic equations.

A general linear, constant coefficient one dimensional system of hyperbolic partial differential equations defined on $\mathbb{R}$ can be written in the form

$$\mathbf{v}_t = A\mathbf{v}_x, \ x \in \mathbb{R}, \ t > 0, \tag{5.1.1}$$

where $\mathbf{v}$ is a vector of unknowns $\mathbf{v} = [v_1 \ \cdots \ v_K]^T$ and A is a diagonalizable $K \times K$ matrix. When A is assumed to be diagonalizable, system (5.1.1) is said to be **strongly hyperbolic**. We will refer to this as being hyperbolic and note that A will be diagonalizable when A is symmetric (system (5.1.1)is **symmetric hyperbolic**), if A has K distinct eigenvalues (system (5.1.1) is **strictly hyperbolic**) or if A has K linearly independent eigenvectors. Any of these hypotheses will be sufficient for the results in this text.

Because the matrix A is diagonalizable, there exists a matrix S (where S^{-1} is the matrix of eigenvectors of A) that will diagonalize A, i.e. $D = SAS^{-1}$ where D is a diagonal matrix having the eigenvalues of A, μ_j, $j = 1, \cdots, K$, on the diagonal. Then since

$$
\begin{aligned}
S\mathbf{v}_t &= SA\mathbf{v}_x \\
&= SAS^{-1}S\mathbf{v}_x \\
&= DS\mathbf{v}_x,
\end{aligned}
\tag{5.1.2}
$$

we can define $\mathbf{V} = S\mathbf{v}$ and reduce the systems of equations (5.1.1) to the uncoupled system of equations

$$
\mathbf{V}_t = D\mathbf{V}_x,
$$

or

$$
V_{j_t} = \mu_j V_{j_x}, \ j = 1, \cdots, K.
\tag{5.1.3}
$$

Hence, for the introduction to the study of numerical schemes for solving hyperbolic equations, we use a model equation of the form

$$
v_t + av_x = 0.
\tag{5.1.4}
$$

Partial differential equation (5.1.4) is referred to as the **one way wave equation**. The domains that we shall use most often for the hyperbolic partial differential equation (5.1.4) will be $\mathbb{R}$, semi-infinite ($[0, \infty)$) or a bounded interval. The boundary conditions associated with equation (5.1.4) will depend on the domain and the sign of a.

5.2   Initial–Value Problems

As in the parabolic case, we begin our study of numerical methods for solving hyperbolic equations by considering initial–value problems. One of the advantages of this approach is that we can then ignore all the problems associated with boundary conditions. We consider equation (5.1.4) where the domain is the entire real line, $\mathbb{R}$, and

$$
v(x, 0) = f(x)
\tag{5.2.1}
$$

is given.

It is easy to see that the solution with the above initial–value problem is given by

$$
v(x, t) = f(x - at), \ x \in \mathbb{R}, \ t > 0.
\tag{5.2.2}
$$

Thus, not only do we know the solution to the problem, but that the solution is constant along any **characteristic curve**, $x - at =$ constant. The family of characteristics graphed in Figure 5.2.1 completely determines the solution to the problem in that the value of the solution at any point (x, t) can be found by projecting along a characteristic, back to the $t = 0$ axis. It should be noted that the picture is similar for $a < 0$ except that the slope of the lines is in the other direction. See Figure 5.3.1.

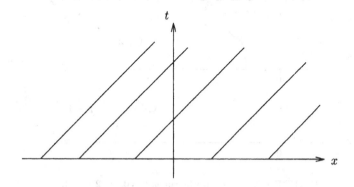

FIGURE 5.2.1. Characteristic curves for $a > 0$.

One of the major properties of solutions of hyperbolic equations is that of wave propagation. That this is the case can be seen easily from the solution (5.2.2). If the initial condition f has a certain form, then the solution for later times looks the same as the original solution except for the fact that the wave form has been translated to the right (left in the case where $a < 0$). The wave form moves to the right with a velocity of a. This wave propagation is illustrated in Figure 5.2.2.

The property described above is one of the properties that makes hyperbolic equations both interesting and difficult to solve. We note that there is no dissipation in the solution. The natural dissipation contained in the parabolic partial differential equations helped make our numerical schemes stable. One of the problems with which we will be constantly confronted will be to develop numerical schemes that do not damp out the solution to the hyperbolic equation and that are sufficiently stable to be useful. Another problem related to the solution given in equation (5.2.2) that we must face while solving hyperbolic equations numerically that we did not have with parabolic equations is the approximation of the **speed of propagation**. When we develop difference schemes for hyperbolic equations, one of the aspects of the approximate solution with which we will be concerned is how well the scheme approximates the speed of propagation of the wave forms in the analytic solution. If a difference scheme did an excellent job of resolving the form of the propagating wave and a terrible job

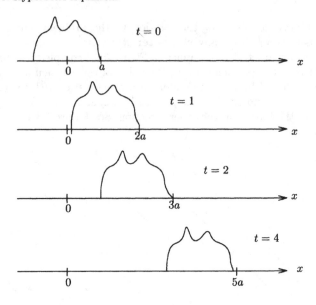

FIGURE 5.2.2. Solutions at times $t = 0$, 1, 2 and 4.

at approximating the speed of propagation, at a given time the solution would appear to be very bad. A discussion of dissipative and dispersive properties of difference schemes is contained in Chapter 7.

One last property of the solution to equation (5.1.4), (5.2.1) that we see immediately from the form of the solution (5.2.2) is that *the solution will only be as nice as the initial condition*. For example, if we are given a discontinuous initial condition, say f has a discontinuity at $x = x_0$, then we must expect the solution to have a discontinuity propagated along the characteristic $x - at = x_0$ (the characteristic that goes through the point $(x_0, 0)$). For example, in Figure 5.2.3 we plot a discontinuous initial function

$$f(x) = \begin{cases} 1 & \text{if } x \le 0 \\ 0 & \text{if } x > 0 \end{cases}$$

along with the solution of equation (5.1.4) with $a = 1$ at times $t = 1.0$ and $t = 2.0$. At any other time, the solution would be identical, except for the location of the discontinuity.

In the above case, and this is common in interesting hyperbolic problems, since the solution is discontinuous and surely cannot satisfy the partial differential equation in the classical sense, we must consider the solution of the hyperbolic partial differential equation in the weak sense. Though we will not dwell on the problems associated with **weak solutions**, we will use them whenever it is convenient. In Chapter 9, Part 2, ref. [13], when we discuss numerical schemes for solving conservation laws, we include

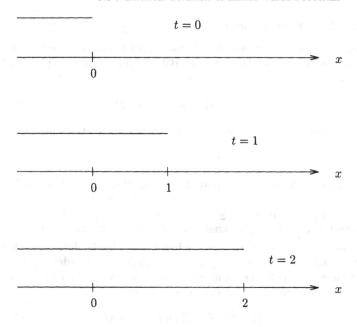

FIGURE 5.2.3. The initial condition and solutions at $t = 1.0$ and $t = 2.0$.

some defintions and results related to weak solutions of partial differential equations.

And finally, before we proceed to begin developing numerical schemes for solving equation (5.1.4), (5.2.1), we emphasize that this is a very simple hyperbolic partial differential equation. But because of the decoupling performed in Section 5.1, many of the difficulties related to more complex hyperbolic problems are shared with this rather simple problem. Also, most difference schemes for solving more complex hyperbolic problems can be seen to be an extension of a scheme for solving problem (5.1.4), (5.2.1).

5.3   Numerical Solution of Initial–Value Problems

We now begin to develop numerical schemes for approximating solutions to problems involving hyperbolic partial differential equations. Of course, we want difference schemes for which the solutions will converge to the solutions of the analogous partial differential equations. As usual, we will obtain convergence by showing the the scheme is consistent and stable, and then use the Lax Theorem. As is so often the case, we will refer to Section 2.3 for the consistency or just claim the the consistency calculation is easy (which most often it is). In this chapter, we will spend most of our time considering the stability of the schemes.

5.3.1   One Sided Schemes

We should recall that we have already considered several schemes for solving problems similar to (5.1.4), (5.2.1). In HW2.3.1(e) we showed that for $R = a\Delta t/\Delta x$ and $a < 0$ the FTFS scheme

$$u_k^{n+1} = u_k^n - R(u_{k+1}^n - u_k^n) \qquad (5.3.1)$$

is a consistent scheme for the hyperbolic partial differential equation

$$v_t + av_x = 0. \qquad (5.3.2)$$

In Example 2.4.2 we showed that difference scheme (5.3.1) is stable for $|R| \leq 1$ (which we showed by another method in Example 3.1.2).

Hence, *if $|R| = |a|\Delta t/\Delta x \leq 1$, difference scheme (5.3.1) is stable and consistent with partial differential equation (5.3.2), and hence, convergent.*

Suppose we now consider the case when $a > 0$. Clearly, difference scheme (5.3.1) is still consistent with equation (5.3.2) (it is first order accurate in both time and space). The same calculation that we did for Example 3.1.2 shows that the symbol of the scheme is given by

$$\rho(\xi) = 1 + R - R\cos\xi - iR\sin\xi. \qquad (5.3.3)$$

Then, again using the calculations from Example 3.1.2, we see that $|\rho(\pm\pi)| = |1+2R|$ and for all $R > 0$ $|\rho(\pm\pi)| = |1+2R| > 1$. Hence, *if $a > 0$, difference scheme (5.3.1) is unstable.*

It is not hard to see why difference scheme (5.3.1) might be a good scheme for partial differential equation (5.3.2) when $a < 0$, yet bad when $a > 0$. We see in Figure 5.3.1 that when $a < 0$ the characteristic for hyperbolic partial differential equation (5.3.2) through the point $(k\Delta x, (n+1)\Delta t)$ runs down and to the right towards the $t = 0$ axis. The stencil for difference scheme (5.3.1) reaches back in that same general direction. If we were lucky (or smart) and chose $R = -1$, then the stencil would reach exactly back onto the characteristic and difference scheme (5.3.1) would give

$$u_k^{n+1} = u_k^n - R(u_{k+1}^n - u_k^n) = u_{k+1}^n,$$

which is on the same characteristic. It is not difficult to see that this procedure would produce the exact solution to our problem (only because we have an easy problem and know exactly what the characteristics are). If we do not choose $R = -1$ but do at least choose R between 0 and -1, we are at least reaching in the correct direction, towards the characteristic.

For the same reason that difference scheme (5.3.1) was especially good for approximating the solution to partial differential equation (5.3.2) when $a < 0$, it is not hard to see that it reaches in the wrong way if $a > 0$. In Figure 5.3.2 we see that when $a > 0$ the characteristic associated with partial differential equation (5.3.2) through the point $(k\Delta x, (n + 1)\Delta t)$

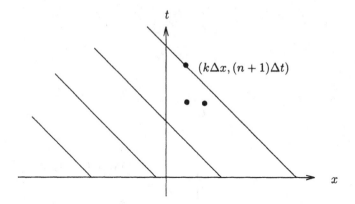

FIGURE 5.3.1. Characteristics for partial differential equation (5.3.2) and stencil for difference scheme (5.3.1) when $a < 0$.

reaches down and to the left toward the $t = 0$ axis, while the difference scheme (5.3.1) reaches down and to the right from the point, away from the characteristic.

When $a < 0$ information related to partial differential equation (5.3.2) flows from right to left. Hence, when we want to know what is going to happen at a given point, it makes sense to reach back to the right to obtain information (upstream). That is precisely what difference scheme (5.3.1) does (with the $k + 1$ index). But if $a > 0$, the information moves from left to right. In this case if you want to know what is going to happen at a given point, you should look to the left (upstream). Obviously, difference scheme (5.3.1) does not do this.

The obvious analog to difference scheme (5.3.1) for solving equation (5.3.2) when $a > 0$ is the FTBS scheme

$$u_k^{n+1} = u_k^n - R(u_k^n - u_{k-1}^n). \tag{5.3.4}$$

It is easy to show that difference scheme (5.3.4) is consistent with partial differential equation (5.3.2) ($\mathcal{O}(\Delta t) + \mathcal{O}(\Delta x)$). If we place the stencil on the characteristics of equation (5.3.2), it is clear that the scheme is trying to reach in the correct direction. (See Figure 5.3.3.) To see that difference scheme (5.3.4) is stable, we take the discrete Fourier transform of equation (5.3.4) to get the symbol

$$\rho(\xi) = (1 - R + R\cos\xi) - iR\sin\xi. \tag{5.3.5}$$

It is then easy to see that $R \le 1$ implies that $|\rho(\xi)| \le 1$. Hence, *if $R \le 1$, difference scheme (5.3.4) is a consistent, stable (and, hence, convergent) scheme for solving partial differential equation (5.3.2) when $a > 0$.*

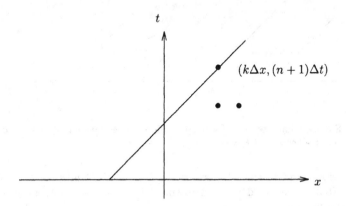

FIGURE 5.3.2. Characteristics for partial differential equation (5.3.2) and stencil for difference scheme (5.3.1) when $a > 0$.

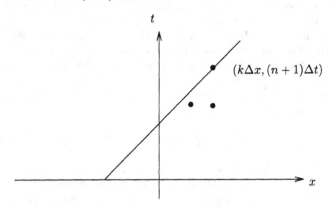

FIGURE 5.3.3. Characteristics for $a > 0$ and stencil for difference scheme (5.3.4).

Thus we see that, depending on the sign of a, we have a scheme that is first order accurate in both time and space for solving

$$v_t + av_x = 0. \tag{5.3.6}$$

We note that when a is negative, we must approximate v_x by $\frac{1}{\Delta x}\delta_+ u_k$ which is called **forward differencing** and when a is positive, we must difference v_x by $\frac{1}{\Delta x}\delta_- u_k$ which is called **backward differencing**.

HW 5.3.1 Perform the calculation that shows that $0 < R \le 1$ implies that $|\rho(\xi)| \le 1$ where ρ is defined by equation (5.3.5).

5.3.2 Centered Scheme

Of course, there are other difference schemes for solving equation (5.3.6). One obvious approach is to try to use a second order difference approximation of v_x and get the FTCS scheme

$$u_k^{n+1} = u_k^n - \frac{R}{2}\delta_0 u_k^n. \tag{5.3.7}$$

It is easy to see that difference scheme (5.3.7) is a $\mathcal{O}(\Delta t) + \mathcal{O}(\Delta x^2)$ approximation of partial differential equation (5.3.6). Taking the discrete Fourier transform of difference scheme (5.3.7) yields the symbol of the operator

$$\rho(\xi) = 1 - iR\sin\xi.$$

If we note that $|\rho(\xi)|^2 = 1 + R^2\sin\xi \ge 1$, we see that *the difference scheme (5.3.7) is unstable for all $R \ne 0$*. This result should not totally surprise us. When we considered the convection–diffusion equation in Example 3.1.3, we found that the stability condition was $R^2/2 < r \le 1/2$. Here we are considering the case when $r = 0$. Thus the necessary condition for stability could never hold and we might have expected to get an unstable scheme.

5.3.3 Lax-Wendroff Scheme

One approach to obtaining a higher order stable scheme is to alter a scheme, such as (5.3.7), to stabilize the scheme. We note that since $v_t = -av_x$,

$$v_{tt} = (-av_x)_t = -av_{xt} = -av_{tx} = -a(v_t)_x = -a(-av_x)_x = a^2 v_{xx}.$$

Then since

$$(v)_k^{n+1} = v_k^n + (v_t)_k^n \, \Delta t + (v_{tt})_k^n \, \frac{\Delta t^2}{2} + \mathcal{O}(\Delta t^3)$$

$$= v_k^n + (-av_x)_k^n \, \Delta t + (a^2 v_{xx})_k^n \, \frac{\Delta t^2}{2} + \mathcal{O}(\Delta t^3)$$

$$= v_k^n - a \left(\frac{v_{k+1}^n - v_{k-1}^n}{2\Delta x} + \mathcal{O}(\Delta x^2) \right) \Delta t$$

$$\quad + a^2 \left(\frac{v_{k+1}^n - 2v_k^n + v_{k-1}^n}{\Delta x^2} + \mathcal{O}(\Delta x^2) \right) \frac{\Delta t^2}{2} + \mathcal{O}(\Delta t^3)$$

$$= v_k^n - \frac{a\Delta t}{2\Delta x} \delta_0 v_k^n + \frac{a^2 \Delta t^2}{2\Delta x^2} \delta^2 v_k^n + \mathcal{O}(\Delta t \Delta x^2) + \mathcal{O}(\Delta t^3)$$

we approximate the partial differential equation $v_t + a v_x = 0$ by the difference scheme

$$u_k^{n+1} = u_k^n - \frac{R}{2} \delta_0 u_k^n + \frac{R^2}{2} \delta^2 u_k^n. \tag{5.3.8}$$

where $R = a\Delta t/\Delta x$.

Difference scheme (5.3.8) is called the **Lax-Wendroff scheme**. The scheme is obviously $\mathcal{O}(\Delta t^2) + \mathcal{O}(\Delta x^2)$. Taking the discrete Fourier transform of equation (5.3.8), we see that the symbol of the Lax-Wendroff scheme is

$$\rho(\xi) = 1 - 2R^2 \sin^2 \frac{\xi}{2} - iR \sin \xi. \tag{5.3.9}$$

Since

$$| \, \rho(\xi) \, |^2 = 1 - 4R^2 \sin^4 \frac{\xi}{2} + 4R^4 \sin^4 \frac{\xi}{2}, \tag{5.3.10}$$

we can differentiate with respect to ξ, set the derivative equal to zero and see that the critical values of the function defined in equation (5.3.10) occur at $\xi = \pm\pi$ and 0. Then we evaluate

$$| \, \rho(0) \, |^2 = 1$$

and

$$| \, \rho(\pm\pi) \, |^2 = | \, \rho(\pi) \, |^2 = (1 - 2R^2)^2. \tag{5.3.11}$$

Because $(1 - 2R^2)^2 \leq 1$ whenever $R^2 \leq 1$, we see that the Lax-Wendroff scheme is conditionally stable with condition $| \, R \, | = | \, a \, | \, \Delta t/\Delta x \leq 1$.

Thus in this situation, we see that *if $| \, a \, | \, \Delta t/\Delta x \leq 1$, the Lax-Wendroff scheme is consistent (second order in both time and space) and stable (and hence, convergent)*. We might note that the last term in difference scheme

Name	Scheme	Stability	Order
FTCS	(5.3.7)	unstable	$\mathcal{O}(\Delta t) + \mathcal{O}(\Delta x^2)$
FTFS	(5.3.1)	stable for $-1 \le R \le 0$	$\mathcal{O}(\Delta t) + \mathcal{O}(\Delta x)$
FTBS	(5.3.4)	stable for $0 \le R \le 1$	$\mathcal{O}(\Delta t) + \mathcal{O}(\Delta x)$
Lax-Wendroff	(5.3.8)	stable for $|R| \le 1$	$\mathcal{O}(\Delta t^2) + \mathcal{O}(\Delta x^2)$
Lax-Friedrichs	$u_k^{n+1} = \frac{1}{2}\left(u_{k+1}^n + u_{k-1}^n\right) - \frac{R}{2}\delta_0 u_k^n$	stable for $|R| \le 1$	conditionally consistent, order $\mathcal{O}(\Delta t) + \mathcal{O}(\Delta x^2/\Delta t)$
Beam-Warming	$u_k^* = u_k^n - R\delta_- u_k^n$ $u_k^{n+1} = \frac{1}{2}\left[u_k^n + u_k^* - R\delta_- u_k^* - R\delta^2 u_{k-1}^n\right]$	stable for $0 \le R \le 2$	$\mathcal{O}(\Delta t^2) + \mathcal{O}(\Delta t \Delta x)$ $+ \mathcal{O}(\Delta x^2)$
MacCormack	$u_k^* = u_k^n - R\delta_+ u_k^n$ $u_k^{n+1} = \frac{1}{2}\left[u_k^n + u_k^* - R\delta_- u_k^*\right]$	stable for $|R| \le 1$	$\mathcal{O}(\Delta t^2) + \mathcal{O}(\Delta x^2)$

TABLE 5.3.1. Explicit difference schemes for initial–value problems.

(5.3.8) can be considered to be added to the unstable scheme (5.3.7) to stabilize it. We should also realize that *the stability condition for the Lax-Wendroff scheme is independent of the sign of a*. This is a property of the Lax-Wendroff scheme that makes it much easier to apply for a large class of problems.

5.3.4 More Explicit Schemes

By this time we must realize that there are a variety of schemes. Most schemes have one or more properties that make them desirable. There are people that claim that for every named difference scheme, there is at least one problem for which that difference scheme is the fastest or best. The purpose of this section is to try to include some more difference schemes. In Table 5.1, we list difference schemes for hyperbolic equations and their properties.

HW 5.3.2 Verify that the order and stability results for the Lax-Friedrichs, Beam-Warming and MacCormack schemes are as given in Table 5.3.1.

HW 5.3.3 Verify that the MacCormack scheme given in Table 5.3.1 is equivalent to the Lax-Wendroff scheme, (5.3.8).

5.4   Implicit Schemes

5.4.1   One Sided Schemes

Just as is the case with parabolic equations, there are also implicit schemes for hyperbolic equations. And, as is usually the case, implicit schemes tend to be more stable than analogous explicit schemes. Many people either state or imply that they feel that implicit schemes are not needed for hyperbolic equations. However, in computational aerodynamics when the calculation involves an extremely high Reynold's number (which would make the time step necessary for stability of an explicit scheme prohibitively small) or when steady state solutions are wanted, implicit schemes for hyperbolic equations have been very useful. Also, implicit schemes have been very useful in the simulation of ground water flows and oil reservoir simulations.

We begin in the obvious way. We consider the initial–value problem for the partial differential equation

$$v_t + av_x = 0, \qquad\qquad (5.4.1)$$

the BTFS scheme analogous to the explicit scheme (5.3.1)

$$(1 - R)u_k^{n+1} + Ru_{k+1}^{n+1} = u_k^n, \qquad\qquad (5.4.2)$$

and the BTBS implicit scheme analogous to explicit scheme (5.3.4)

$$-Ru_{k-1}^{n+1} + (1+R)u_k^{n+1} = u_k^n, \tag{5.4.3}$$

($R = a\Delta t/\Delta x$). Though implicit schemes are most often unconditionally stable, we shall see that *difference schemes (5.4.2) and (5.4.3) are stable if and only if $R \leq 0$ and $R \geq 0$, respectively*. The fact that difference scheme (5.4.2) is stable when $a < 0$ and difference scheme (5.4.3) is stable when $a > 0$ is similar to the stability of the analogous explicit schemes.

To show the stability of difference scheme (5.4.2), we take the discrete Fourier transform and get

$$(1 - R)\hat{u}^{n+1} + Re^{i\xi}\hat{u}^{n+1} = \hat{u}^n.$$

Thus the symbol of the difference scheme (5.4.2) is given by

$$\rho(\xi) = \frac{1}{1 - R + R\cos\xi + iR\sin\xi} \tag{5.4.4}$$

and the magnitude squared of the symbol is given by

$$|\rho(\xi)|^2 = \frac{1}{1 - 4R\sin^2\frac{\xi}{2} + 4R^2\sin^2\frac{\xi}{2}}. \tag{5.4.5}$$

Since $R \leq 0$ ($a < 0$) implies that

$$1 - 4R\sin^2\frac{\xi}{2} + 4R^2\sin^2\frac{\xi}{2} = 1 - 4R\sin^2\frac{\xi}{2}(1 - R) \geq 1,$$

$|\rho(\xi)|^2 \leq 1$. We should also note that for $0 < R < 1$, difference scheme (5.4.2) is unstable. Hence, we see that *difference scheme (5.4.2) is stable if and only if $R \leq 0$ or $R \geq 1$*. Because of the physics involved in the partial differential equation, we will be interested in difference scheme (5.4.2) when $a < 0$. Likewise, *difference scheme (5.4.3) is stable if $R \geq 0$*.

HW 5.4.1 Verify that both difference schemes (5.4.2) and (5.4.3) are consistent with partial differential equation (5.4.1).

HW 5.4.2 Prove that difference scheme (5.4.3) is stable if $R \geq 0$ ($a > 0$).

HW 5.4.3 Show that difference scheme (5.4.2) is not stable when $R > 0$. (A graph of $|\rho(\xi)|^2$ for $\xi \in [-\pi, \pi]$ for several values of R can help.)

5.4.2 Centered Scheme

Next, suppose we would like to obtain a scheme that is second order accurate in space. It would be logical to assume that we should not try to

develop an implicit second order scheme analogous to the explicit scheme
(5.3.7) (which was unstable). However, if we consider the BTCS scheme

$$\frac{-R}{2} u_{k-1}^{n+1} + u_k^{n+1} + \frac{R}{2} u_{k+1}^{n+1} = u_k^n, \tag{5.4.6}$$

we see that the symbol of the difference scheme is

$$\rho(\xi) = \frac{1}{1 + Ri \sin \xi}. \tag{5.4.7}$$

Clearly, difference scheme (5.4.6) is consistent with partial differential equation (5.4.1). Then since

$$|\rho(\xi)|^2 = \frac{1}{1 + R^2 \sin^2 \xi} \leq 1,$$

difference scheme (5.4.6) is unconditionally stable and, hence, convergent
(even though its explicit counterpart is unstable). We also notice that the
stability of difference scheme (5.4.6) is independent of the sign of a. As
usual, having a difference scheme whose stability does not depend on the
sign of a is very convenient.

HW 5.4.4 Verify the consistency and the symbol for the pure implicit
scheme (5.4.6).

5.4.3 Lax-Wendroff Scheme

Another way to obtain a second order accurate implicit scheme (second
order in both time and space) is to develop an implicit version of the Lax-
Wendroff scheme. The implicit version of the Lax-Wendroff scheme is given
by

$$\left(-\frac{R^2}{2} - \frac{R}{2}\right) u_{k-1}^{n+1} + \left(1 + R^2\right) u_k^{n+1} + \left(-\frac{R^2}{2} + \frac{R}{2}\right) u_{k+1}^{n+1} = u_k^n. \tag{5.4.8}$$

Difference scheme (5.4.8) is still consistent with equation (5.4.1). If we
expand the above equation about $(k, n+1)$ we see that the equation is in
fact still second order in both time and space. Taking the discrete Fourier
transform of equation (5.4.8) gives the symbol of the operator

$$\rho(\xi) = \frac{1}{1 + 2R^2 \sin^2 \frac{\xi}{2} + iR \sin \xi}. \tag{5.4.9}$$

Then since

$$|\rho(\xi)|^2 = \frac{1}{(1 + 2R^2 \sin^2 \frac{\xi}{2})^2 + R^2 \sin^2 \xi},$$

it is clear that *the implicit Lax-Wendroff scheme is unconditionally stable
(and hence, convergent).*

HW 5.4.5 Verify the consistency and the symbol for the implicit Lax-Wendroff difference scheme (5.4.8).

5.4.4 Crank-Nicolson Scheme

And finally, another implicit scheme that is an obvious scheme to try and one that has some interesting properties is the Crank-Nicolson scheme for partial differential equation (5.4.1):

$$-\frac{R}{4}u_{k-1}^{n+1} + u_k^{n+1} + \frac{R}{4}u_{k+1}^{n+1} = \frac{R}{4}u_{k-1}^n + u_k^n - \frac{R}{4}u_{k+1}^n.$$
(5.4.10)

We note that as with the Crank-Nicolson scheme for parabolic equations, the spatial difference has been averaged at the nth and $(n+1)$st time levels. Again, this scheme can be considered as a half explicit step (with time step $\Delta t/2$) and a half implicit step (with time step $\Delta t/2$).

If we expand the scheme about the point $(k, n+\frac{1}{2})$, it is easy to see that the scheme is of order $\mathcal{O}(\Delta t^2) + \mathcal{O}(\Delta x^2)$. By taking the discrete Fourier transform, we see that the symbol for the scheme is given by

$$\rho(\xi) = \frac{1 - \frac{iR}{2}\sin\xi}{1 + \frac{iR}{2}\sin\xi},$$
(5.4.11)

and that $\mid \rho(\xi) \mid^2 \equiv 1$. Thus we have an implicit difference scheme where the magnitude of the symbol is identically equal to one. There are advantages and disadvantages of this property. The discussion of dissipation and dispersion given in Chapter 7 will explain most of these advantages and disadvantages. We should at least realize that the stability is close. Any perturbation, due to round-off or due to application of the scheme to a nonlinear problem, could make the magnitude greater than one, and result in instability.

One of the properties of the BTCS scheme, the Lax-Wendroff implicit scheme and the Crank-Nicolson scheme that we should note is that *the stability holds independent of the sign of a*. This is a big advantage when it comes to choosing schemes for problems that the sign of a changes throughout the domain (when a is a function of the independent variables of the problem) or for problems where the sign of a is not known a priori (nonlinear problems where the coefficient depends on the solution).

HW 5.4.6 Verify that ρ as given by equation (5.4.11) is the symbol for the Crank-Nicolson scheme (5.4.10). Also, verify that the Crank-Nicolson scheme is accurate order $\mathcal{O}(\Delta t^2) + \mathcal{O}(\Delta x^2)$.

5.5   Initial–Boundary–Value Problems

The stability results that we obtained for initial–boundary–value problems for parabolic equations do not generally carry over to hyperbolic equations. The reason is clear if we recall that we obtained necessary and sufficient conditions for stability of difference schemes for parabolic initial–boundary–value problems only when we had symmetric matrices. The matrices that we obtain from schemes for hyperbolic equations are almost never symmetric (and maybe never). Thus most often we are faced with obtaining only necessary conditions for stability. As we mentioned earlier, we will develop other methods in Chapter 8 that will give us necessary and sufficient conditions for stability. These methods will be applicable to hyperbolic problems. At this time, we shall be content with necessary conditions. And, as was the case in the parabolic equations, *we can obtain necessary conditions by considering the scheme as an initial–value scheme (calculating the symbol), by using the discrete von Neumann criterion (getting the same results as if we considered the scheme as an initial–value scheme) or by writing the problem as a matrix problem and computing the eigenvalues of the matrix.* Recall that in Examples 3.1.6 and 3.1.8 we found two different necessary conditions for stability of difference scheme (3.1.61)–(3.1.63) by first considering the scheme as an initial–value problem scheme and then by computing the eigenvalues of the matrix operator. The necessary condition found by considering the scheme as an initial–value scheme gave a much better necessary condition. Unless the boundary conditions are very dominant, we shall at this time generally be satisfied with the necessary conditions obtained considering the scheme as an initial–value problem scheme.

Another consideration that we must discuss concerning initial–boundary–value problems for hyperbolic equations is the number and placement of boundary conditions. For parabolic equations, we always had enough boundary conditions and they were always in the right place. This is generally not the case with hyperbolic initial–boundary–value problems. As we shall see, Dirichlet boundary conditions with certain hyperbolic difference schemes cause many problems. Below we consider periodic and Dirichlet boundary conditions. We study the periodic boundary conditions first for several reasons. Surely, periodic boundary conditions are important enough in applications to warrant consideration. More so, periodic boundary conditions are easier. Because they do not cause the problems that Dirichlet boundary conditions cause, they are more attractive both theoretically and numerically. Also, if we want to see numerically how a given difference scheme propagates an initial condition, with any type of Dirichlet boundary conditions we have to make the domain big (as large as it needs to be for the time interval that we want to consider) and carefully plot the regions where the disturbance is nontrivial. With periodic boundary conditions, we can just watch it propagate through the given interval.

5.5.1 Periodic Boundary Conditions

We begin by considering the partial differential equation

$$v_t + av_x = 0, \ x \in (0,1), \tag{5.5.1}$$

along with the initial condition

$$v(x,0) = f(x), \ x \in [0,1], \tag{5.5.2}$$

($f(0) = f(1)$) and the periodic boundary condition

$$v(0,t) = v(1,t). \tag{5.5.3}$$

There are no problems related to the theory for partial differential equation (5.5.1), initial condition (5.5.2) and periodic boundary condition (5.5.3). The approach is to use the periodic boundary conditions to extend the initial value function f and the partial differential equation (5.5.1) periodically to the whole real line. We then consider the initial–boundary–value problem as an initial–value problem defined by partial differential equation (5.5.1) on all of $\mathbb{R}$ with a periodic initial condition.

5.5.2 Dirichlet Boundary Conditions

As we promised earlier, Dirichlet boundary conditions are more difficult to handle than periodic boundary conditions. If we again consider the equation

$$v_t + av_x = 0, \tag{5.5.4}$$

but now on the domain $x > 0$, we recall (as was the case when we considered initial–value problems) that if $a > 0$, the characteristics point upward to the right from the $t = 0$ axis (as in Figure 5.5.1) and if $a < 0$, the characteristics point upward to the left from the $t = 0$ axis (as in Figure 5.5.2).

We see in Figure 5.5.1 that when $a > 0$ the characteristics "come out of" either the t axis, $t > 0$, or the x axis, $x > 0$. These characteristics never cross any other axis. Thus, it is very logical that the partial differential equation have a boundary condition at $x = 0$ in addition to the initial condition.

In the case when $a < 0$, we see in Figure 5.5.2 that if we were to impose a boundary condition at $x = 0$ and an initial condition, we obtain a contradiction. For example, suppose we set

$$v(x,0) = f(x), \ x \geq 0 \tag{5.5.5}$$
$$v(0,t) = g(t), \ t \geq 0 \tag{5.5.6}$$

($f(0) = g(0)$). From our previous discussion, we know that the solution to equation (5.5.4) is of the form

$$v(x,t) = f(x - at). \tag{5.5.7}$$

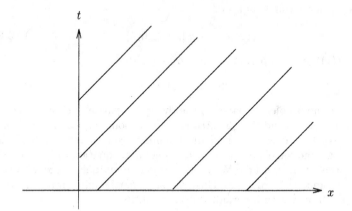

FIGURE 5.5.1. Characteristics for the case when $a > 0$.

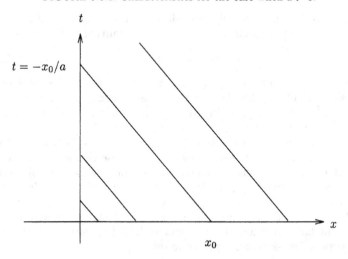

FIGURE 5.5.2. Characteristics for the case when $a < 0$.

We also note that this solution is also constant along characteristics (because $x - at$ is constant along a characteristic).

But, if this solution is also going to satisfy boundary condition (5.5.6), then

$$\lim_{x \to 0+} v(x,t) = g(t).$$

Consider the characteristic emanating from $x = x_0$. Then along the characteristic $x - at = x_0$, the solution is the constant value $f(x_0)$. When $x = 0$ on the above characteristic, $t = -x_0/a$ (which is positive since a is negative). Then, surely along this characteristic

$$\lim_{x \to 0+} v(x,t) = f(x_0).$$

But generally, $f(x_0)$ will not equal $g(-x_0/a)$. Thus, *it is incompatible to assign a boundary condition at $x = 0$ and an initial condition at $t = 0$, when $a < 0$.* Of course, we run into the same problem if $a > 0$ and we treat equation (5.5.4) in the domain $x < 0$.

The above discussion illustrates clearly that we must be very careful about where we assign boundary conditions for hyperbolic problems. For initial–boundary-value hyperbolic problems on $[0, 1]$ (or any other bounded domain), we get only one Dirichlet boundary condition. For example, *if we consider equation (5.5.4) on the domain $[0, 1]$, we get*

(i) a boundary condition at $x = 0$ and none at $x = 1$ if $a > 0$, and

(ii) a boundary condition at $x = 1$ and none at $x = 0$ if $a < 0$.

5.6 Numerical Solution of Initial–Boundary–Value Problems

5.6.1 Periodic Boundary Conditions

We begin our numerical treatment of initial–boundary–value problems by considering periodic boundary conditions. As we saw by Theorem 3.2.3, *we can obtain both necessary and sufficient conditions for stability of difference schemes that approximate initial–value problems with periodic initial conditions.* Specifically, the difference schemes considered in this section are either unconditionally or conditionally convergent where the conditions can be obtained by performing a von Neumann stability analysis on the scheme as if it were an initial–value problem scheme.

To demonstrate how we actually solve a problem such as (5.5.1)–(5.5.3) numerically, we let $a = 1$, $f(x) = \sin 2\pi x$ and consider the one sided difference scheme

$$u_k^{n+1} = u_k^n - R(u_k^n - u_{k-1}^n). \tag{5.6.1}$$

As usual, we choose an integer M and construct a grid on the interval by setting $\Delta x = 1/M$ and $x_k = k\Delta x$, $k = 0, \cdots, M$. We then approximate the solution to problem (5.5.1)–(5.5.3) by using difference scheme (5.6.1) for $k = 1, \cdots, M$, and using the periodic boundary condition to give $u_0^{n+1} = u_M^{n+1}$ for all n. Hence, our difference scheme for numerically solving problem (5.5.1)–(5.5.3) is

$$u_k^{n+1} = u_k^n - R(u_k^n - u_{k-1}^n), \quad k = 1, \cdots, M \tag{5.6.2}$$

$$u_0^{n+1} = u_M^{n+1}. \tag{5.6.3}$$

Below we consider several variations of the problems caused by periodic boundary conditions. Specifically, in Example 5.6.2, we give a treatment of periodic boundary conditions with an implicit scheme. Though periodic boundary conditions cause some difficulties for the solution schemes, we shall see that their advantages far outweigh their disadvantages.

Example 5.6.1 Describe an algorithm to solve problem (5.5.1)–(5.5.3) based on the Lax-Wendroff scheme.

Solution: The major difference between this example and that done immediately before this example is that the Lax-Wendroff scheme reaches in both directions. We use difference scheme (5.3.8) for $k = 1, \cdots, M - 1$. We must use the periodicity to define the difference scheme at $k = 0$. Applying equation (5.3.8) at $k = 0$, we get

$$u_0^{n+1} = u_0^n - \frac{R}{2}(u_1^n - u_{-1}^n) + \frac{R^2}{2}(u_1^n - 2u_0^n + u_{-1}^n). \tag{5.6.4}$$

Because we have extended our problem periodically (and $u_M^n = u_0^n$), we can replace u_{-1}^n in equation (5.6.4) by u_{M-1}^n. We obtain the following Lax-Wendroff scheme for a problem with periodic boundary conditions.

$$u_k^{n+1} = u_k^n - \frac{R}{2}\delta_0 u_k^n + \frac{R^2}{2}\delta^2 u_k^n, \quad k = 1, \cdots, M - 1 \tag{5.6.5}$$

$$u_0^{n+1} = u_0^n - \frac{R}{2}(u_1^n - u_{M-1}^n) + \frac{R^2}{2}(u_1^n - 2u_0^n + u_{M-1}^n) \tag{5.6.6}$$

$$u_M^{n+1} = u_0^{n+1} \tag{5.6.7}$$

We should note that this scheme handles the periodicity in the same way that we handled the periodicity with respect to the θ coordinate in our explicit scheme for solving the two dimensional heat equation given in polar coordinates derived in Section 4.5.

We notice that if we write difference scheme (5.6.5)–(5.6.7) (using (5.6.7) to eliminate u_M^n from our scheme) as $\mathbf{u}^{n+1} = Q\mathbf{u}^n$, the matrix Q will be of the form

$$Q = \begin{pmatrix} 1 - R^2 & -\frac{R}{2} + \frac{R^2}{2} & 0 & \cdots & 0 & \frac{R}{2} + \frac{R^2}{2} \\ \frac{R}{2} + \frac{R^2}{2} & 1 - R^2 & -\frac{R}{2} + \frac{R^2}{2} & 0 & \cdots & \\ & \cdots & & & & \\ & \cdots & 0 & \frac{R}{2} + \frac{R^2}{2} & 1 - R^2 & -\frac{R}{2} + \frac{R^2}{2} \\ -\frac{R}{2} + \frac{R^2}{2} & 0 & \cdots & 0 & \frac{R}{2} + \frac{R^2}{2} & 1 - R^2 \end{pmatrix}. \tag{5.6.8}$$

Matrices of the form Q are referred to as **circulant matrices**.

Example 5.6.2 Describe an algorithm to solve problem (5.5.1)–(5.5.3) based on the Crank-Nicolson scheme.

Solution: If we proceed as we have in the earlier examples, we consider the Crank-Nicolson scheme for equation (5.5.1),

$$-\frac{R}{4}u_{k-1}^{n+1} + u_k^{n+1} + \frac{R}{4}u_{k+1}^{n+1} = \frac{R}{4}u_{k-1}^n + u_k^n - \frac{R}{4}u_{k+1}^n \qquad (5.6.9)$$

at $k = 1, \cdots, M - 2$. Then using the fact that we have extended our initial condition periodically, we use the difference equation

$$-\frac{R}{4}u_{M-1}^{n+1} + u_0^{n+1} + \frac{R}{4}u_1^{n+1} = u_0^n - \frac{R}{4}u_1^n + \frac{R}{4}u_{M-1}^n \qquad (5.6.10)$$

at $k = 0$,

$$-\frac{R}{4}u_{M-2}^{n+1} + u_{M-1}^{n+1} + \frac{R}{4}u_0^{n+1} = -\frac{R}{4}u_0^n + u_{M-1}^n + \frac{R}{4}u_{M-2}^n \qquad (5.6.11)$$

at $k = M - 1$ and set $u_M^{n+1} = u_0^{n+1}$. We should realize that if we write difference scheme (5.6.9)–(5.6.11) as a matrix equation of the form

$$Q_1 \mathbf{u}^{n+1} = \mathbf{r}^n, \qquad (5.6.12)$$

where $\mathbf{u}^{n+1} = (u_0^{n+1}, \cdots, u_{M-1}^{n+1})^T$, the components of $\mathbf{r}^n$ are given by the right hand sides of equations (5.6.9)–(5.6.11) and the matrix Q_1 is of the form

$$Q_1 = \begin{pmatrix} 1 & \frac{R}{4} & 0 & \cdots & 0 & -\frac{R}{4} \\ -\frac{R}{4} & 1 & \frac{R}{4} & 0 & \cdots & \\ & & \cdots & \cdots & & \\ & & \cdots & 0 & -\frac{R}{4} & 1 & \frac{R}{4} \\ \frac{R}{4} & 0 & \cdots & 0 & -\frac{R}{4} & 1 \end{pmatrix}. \qquad (5.6.13)$$

When we look at the matrix equation (5.6.12), we should realize that implicit schemes with periodic boundary conditions have messed up our matrix. Every matrix equation that we have solved so far has been or has been reduced to a tridiagonal matrix equation. A naive approach to the solution of matrix equation (5.6.12) could be very expensive. Fortunately, there is a way of solving equation (5.6.12), and other equations due to periodic boundary conditions, that reduces to the solution of a tridiagonal matrix equation. (A subtitle for this text could be "How to reduce all problems in life to the solution of a tridiagonal matrix equation.") The solution scheme that we introduce is called the **Sherman-Morrison Algorithm**.

We begin by considering two matrices A and B that are related by

$$A = B - \mathbf{w}\mathbf{z}^T, \qquad (5.6.14)$$

where $\mathbf{w}$ and $\mathbf{z}$ are vectors. Note that if A and B are $L \times L$ matrices and $\mathbf{w}$ and $\mathbf{z}$ are L-vectors, then $\mathbf{w}\mathbf{z}^T$ is also an $L \times L$ matrix. We suppose that we wish to solve

$$A\mathbf{x} = \mathbf{b}. \qquad (5.6.15)$$

Generally, $\mathbf{w}$ and $\mathbf{z}$ will be sparse vectors and the term $\mathbf{w}\mathbf{z}^T$ will make the matrix A a slight perturbation of a nicer matrix B. In the example above, we clearly want A to be the matrix Q_1 and B to be some tridiagonal matrix. We then obtain the following proposition.

Proposition 5.6.1 The Sherman-Morrison Formula *If B^{-1} exists and $z^T B^{-1} w \neq 1$, then A^{-1} exists and*

$$A^{-1} = B^{-1} + \alpha B^{-1} w z^T B^{-1}, \qquad (5.6.16)$$

where

$$\alpha = \frac{1}{1 - z^T B^{-1} w}.$$

Proof: The proof follows easily by considering the form of A, (5.6.14), the proposed form of A^{-1}, (5.6.16) and the following calculation.

$$\begin{aligned}
A^{-1} A &= (B^{-1} + \alpha B^{-1} w z^T B^{-1})(B - w z^T) \\
&= I - B^{-1} w z^T + \alpha B^{-1} w z^T - \alpha B^{-1} w z^T B^{-1} w z^T \quad (5.6.17) \\
&= I - B^{-1} w z^T + \alpha B^{-1} w z^T - \alpha (z^T B^{-1} w) B^{-1} w z^T \,(5.6.18) \\
&= I - B^{-1} w z^T \left[1 - \alpha (1 - z^T B^{-1} w) \right] \\
&= I
\end{aligned}$$

We note that (5.6.18) follows from (5.6.17) because the term $z^T B^{-1} w$ in (5.6.17) is a scalar that can be factored out of the rest of the product.

The application of the above result involves using the form of A^{-1} given in equation (5.6.16) to solve equation (5.6.15). Applying formula (5.6.16), we note that

$$\begin{aligned}
x &= A^{-1} b \\
&= B^{-1} b + \alpha B^{-1} w z^T B^{-1} b \\
&= B^{-1} b + \alpha (z^T B^{-1} b) B^{-1} w
\end{aligned}$$

Note that the term $(z^T B^{-1} b)$ is a scalar.

Thus we see that equation (5.6.15) can be solved as follows.

1. Solve

$$B y_1 = b \qquad (5.6.19)$$
$$B y_2 = w \qquad (5.6.20)$$

for y_1 and y_2.

2. Set

$$\beta = \alpha z^T B^{-1} b$$
$$= \frac{z^T y_1}{1 - z^T y_2}. \qquad (5.6.21)$$

3. Then

$$\begin{aligned}
x &= B^{-1} b + \beta B^{-1} w \\
&= y_1 + \beta y_2. \qquad (5.6.22)
\end{aligned}$$

The actual algorithm then consists of steps (5.6.19)–(5.6.22). We *note that, if B is a tridiagonal matrix, equations (5.6.19) and (5.6.20) are both tridiagonal equations.* Moreover, since both of these equations have the same coefficient matrix, *equations (5.6.19) and (5.6.20) can be solved as a single tridiagonal matrix having two right hand sides (*b *and* w*).* The new Trid subroutine that will accomplish this should be a slight perturbation of your previous Trid subroutine.

And, finally, before we leave this topic, we show how matrix Q_1, (5.6.13), will be a perturbation of some tridiagonal matrix. We let

$$B = \begin{pmatrix} 1+\frac{R}{4} & \frac{R}{4} & 0 & \cdots & 0 & 0 \\ -\frac{R}{4} & 1 & \frac{R}{4} & 0 & \cdots & \\ & \cdots & \cdots & & & \\ & \cdots & 0 & -\frac{R}{4} & 1 & \frac{R}{4} \\ 0 & 0 & \cdots & 0 & -\frac{R}{4} & 1-\frac{R}{4} \end{pmatrix}. \tag{5.6.23}$$

To determine w and z, we then set

$$Q_1 = B - \mathbf{w}\mathbf{z}^T. \tag{5.6.24}$$

This will give us L^2 equations in $2L$ unknowns (the components w and z). Of course, in general we would not be able to obtain a solution to this system of equations. But for many simple structures (including our example), it is easy to find a solution.

We see that if we choose

$$\mathbf{w} = \begin{bmatrix} 1 \\ 0 \\ \vdots \\ 0 \\ -1 \end{bmatrix}, \quad \mathbf{z} = \begin{bmatrix} \frac{R}{4} \\ 0 \\ \vdots \\ 0 \\ \frac{R}{4} \end{bmatrix}, \tag{5.6.25}$$

equation (5.6.24) will be satisfied. Hence, to apply the Crank-Nicolson scheme to a problem involving periodic boundary conditions, (5.6.12), we must solve

$$B\begin{bmatrix} \mathbf{y}_1 & \mathbf{y}_2 \end{bmatrix} = \begin{bmatrix} \mathbf{r}^n & \mathbf{w} \end{bmatrix} \tag{5.6.26}$$

at each time step. Generally, the easiest method for finding B, w and z is to write equation (5.6.24) and fiddle with the easiest w and z that will give a tridiagonal matrix B.

HW 5.6.1 (a) Solve

$$v_t + v_x = 0 \; x \in (0,1) \tag{5.6.27}$$
$$v(x,0) = \sin^{80} \pi x \; x \in [0,1] \tag{5.6.28}$$
$$v(0,t) = v(1,t), t \geq 0. \tag{5.6.29}$$

Use the backward difference scheme (5.6.2)–(5.6.3), $M = 20$, and $\Delta t = 0.01$ (so $R = a\Delta t/\Delta x = 0.2$). Plot solutions at times $t = 0.0$, $t = 0.12$, $t = 0.2$ and $t = 0.4$. (b) Repeat part (a) with $M = 100$ and $\Delta t = 0.002$.

HW 5.6.2 Repeat HW5.6.1 using the Crank-Nicolson scheme, (5.6.9)–(5.6.11).

HW 5.6.3 (a) Solve

$$v_t - v_x = 0 \ x \in (0,1) \tag{5.6.30}$$
$$v(x,0) = \sin^{40} \pi x \ x \in [0,1] \tag{5.6.31}$$
$$v(0,t) = v(1,t), t \geq 0. \tag{5.6.32}$$

Use the FTFS difference scheme (5.3.1) along with an appropriate treatment of the periodic boundary conditions. Use $M = 20$, and $\Delta t = 0.04$ (so $R = a\Delta t/\Delta x = -0.8$). Plot solutions at times $t = 0.0$, $t = 0.12$, $t = 0.2$ and $t = 0.8$. (b) Repeat part (a) with $M = 100$ and $\Delta t = 0.008$. (c) Repeat part (b), plotting solutions at times $t = 0.0$, $t = 5.0$, $t = 10.0$ and $t = 20.0$.

HW 5.6.4 Repeat HW5.6.3 using the Crank-Nicolson scheme, (5.6.9)–(5.6.11).

HW 5.6.5 (a) Use the Crank-Nicolson scheme, (5.6.9)–(5.6.11), $M = 20$, and $\Delta t = 0.04$ ($R = 0.8$) to solve problem (5.6.27)–(5.6.29). Plot solutions at times $t = 0.0$, $t = 0.12$, $t = 0.2$ and $t = 0.8$. (b) Repeat part (a) with $M = 100$ and $\Delta t = 0.008$. (c) Repeat part (b), plotting solutions at times $t = 5.0$, $t = 10.0$ and $t = 20.0$.

HW 5.6.6 Solve

$$v_t + v_x = 0 \ x \in (-1,2) \tag{5.6.33}$$
$$v(x,0) = \begin{cases} 0 & -1 \leq x < 0 \\ 1 & 0 \leq x \leq 1 \\ 0 & 1 < x \leq 2 \end{cases} \tag{5.6.34}$$
$$v(-1,t) = v(2,t), \ t \geq 0. \tag{5.6.35}$$

(a) Use the Lax-Friedrichs scheme, Table 5.3.1, with $M = 20$ and $\Delta t = 0.01$. Plot solutions at times $t = 0$, $t = .05$ and $t = 1.0$.
  (b) Repeat part (a) using the Beam-Warming scheme, Table 5.3.1.

HW 5.6.7 Consider the initial–boundary–value problem

$$v_t - v_x = 0 \ x \in (0,1) \tag{5.6.36}$$
$$v(x,0) = f(x) \ x \in [0,1] \tag{5.6.37}$$
$$v(0,t) = v(1,t), t \geq 0, \tag{5.6.38}$$

where f is defined as

$$f(x) = \begin{cases} 1 & \text{if } 0.4 \le x \le 0.6 \\ 0 & \text{otherwise.} \end{cases}$$

Approximate the solution to problem (5.6.36)–(5.6.38) using $R = -0.8$, $M = 100$ and
(a) the FTFS scheme, (5.3.1),
(b) the Lax-Wendroff scheme, (5.3.8), and
(c) the Lax-Friedrichs scheme, Table 5.3.1.
Plot the solutions at times $t = 0.5$, $t = 1.0$ and $t = 2.0$.

5.6.2 Dirichlet Boundary Conditions

Next we consider Dirichlet boundary conditions. Assume that $a > 0$ and consider the hyperbolic partial differential equation (5.5.4) along with the initial and boundary conditions,

$$v(x,0) = f(x), \ x \in [0,1] \tag{5.6.39}$$
$$v(0,t) = g(t), \ t \ge 0. \tag{5.6.40}$$

Recall that when $a > 0$ in the initial–value problem case, we used the explicit, FTBS difference scheme (5.3.4) to approximate the solution. We shall consider using the same scheme for solving partial differential equation (5.5.4) along with initial–boundary values (5.6.39)–(5.6.40). The difference scheme will be consistent (order $\mathcal{O}(\Delta t) + \mathcal{O}(\Delta x)$) with partial differential equation (5.5.4). Also, $0 \le R \le 1$ will be a necessary condition for stability of the scheme. Hence, difference scheme (5.3.4) is surely an excellent candidate for approximating the solution to problem (5.5.4),(5.6.39), (5.6.40).

If we write the scheme as

$$\mathbf{u}^{n+1} = Q\mathbf{u}^n + \mathbf{G}^n, \tag{5.6.41}$$

where

$$Q = \begin{pmatrix} 1-R & 0 & \cdots & & \\ R & 1-R & 0 & \cdots & \\ & & \cdots & & \\ \cdots & R & 1-R & 0 & \\ & \cdots & & R & 1-R \end{pmatrix} \quad \text{and } \mathbf{G}^n = \begin{bmatrix} Rg^n \\ 0 \\ \vdots \\ 0 \end{bmatrix},$$

we see that Q is not symmetric. Hence, at this time, without resorting to using the definitions, we can do no better than a necessary condition for stability. We have no method for actually proving that $0 \le R \le 1$ is also sufficient for stability of difference scheme (5.6.41). As we have done in other instances, we could perform a numerical experiment to convince us that the difference scheme is convergent. We should be careful in such an experiment to both use a number of different Δx's and Δt's and a number of different R's, $R \leq 1$. See HW5.6.12. As promised earlier, we will address the problem of obtaining sufficient conditions for stability in Chapter 8.

One other potential problem we must address is that we have a boundary condition on one side of our domain in this problem. If we look at the situation carefully, we see that since difference scheme (5.5.4) uses the upwinded differencing (backward for $a > 0$) of the v_x term, no boundary condition at $x = 1$ is necessary. The scheme naturally reaches back to $x = 0$, and lets us approximate the solution at $x = 1$ without reaching beyond $x = 1$.

We also note that the placement of the boundary condition is also in the correct place for the implicit scheme for the above problem, (5.4.3) (BTBS). And, likewise, when $a < 0$, the boundary condition at $x = 1$ (and none at $x = 0$) is in the correct place to use either difference scheme (5.3.1) or (5.4.2) to solve the problem.

One special note should be made of the implementation of the implicit schemes (5.4.2) and (5.4.3) for approximating the solution to initial–boundary–value problems. If we write the matrix problem for using difference scheme (5.4.2) for approximating the solution to equation (5.5.4) (with $a < 0$), an initial condition and a Dirichlet boundary condition at $x = 1$, we get

$$
\begin{pmatrix}
1-R & R & \cdots & \\
0 & 1-R & R & \cdots \\
& & \cdots & \\
\cdots & 0 & 1-R & R \\
& \cdots & 0 & 1-R
\end{pmatrix}
\begin{bmatrix}
u_0^{n+1} \\
\vdots \\
\vdots \\
u_{M-1}^{n+1}
\end{bmatrix}
=
\begin{bmatrix}
u_0^n \\
\vdots \\
\vdots \\
u_{M-1}^n
\end{bmatrix}
-
\begin{bmatrix}
0 \\
\vdots \\
0 \\
Rg^{n+1}
\end{bmatrix}.
$$

$$(5.6.42)$$

We note that the matrix is a bi-diagonal matrix. Hence, solving matrix equation (5.6.42) is easier to solve than the analogous matrix equation for parabolic equations. Only the back substitution loop of the solution procedure is needed. Hence, *difference scheme (5.6.42) has all of the usual stability advantages of an implicit scheme, and is computationally as advantageous as an explicit scheme.* If we were to write the matrix equation associated with $a > 0$ and difference scheme (5.4.3), we would see that the matrix is again a bi-diagonal and this time, only the front elimination loop is necessary.

And finally, we ask if we can use the Lax-Wendroff scheme for solving problem (5.5.4),(5.6.39), (5.6.40) to provide a higher order accurate approximation to the solution of the problem. The consistency and necessary condition for stability cause no problem. And, again, we are not able to obtain a sufficient condition for stability at this time. The real problem we encounter is that the difference operator in the Lax-Wendroff scheme is not one sided, the scheme reaches to both $x = 0$ and $x = 1$. Since $a < 0$, we do not have a boundary condition at $x = 1$. Hence, *the Lax-Wendroff scheme needs some sort of boundary condition at $x = 1$.* It should be made very clear that *the analytic problem (5.5.4), (5.6.39), (5.6.40), cannot have a boundary condition at $x = 1$, and if a boundary condition is placed at $x = 1$, the analytic problem would be ill-posed.* The numerical problem (using the Lax-Wendroff scheme) needs a boundary condition at $x = 1$, and if it is done correctly, an extra boundary condition at $x = 1$ can be included with the numerical problem that will make the numerical problem well-posed. This extra boundary condition is referred to as a **numerical boundary condition**. It is, indeed, extra. It must be chosen very carefully. We are not prepared at this time to discuss how to choose this numerical boundary condition. This will be done in Chapter 8. For now, we try some of the possibilities in HW5.6.10.

We should add that the need to include a numerical boundary condition is not unique to the Lax-Wendroff scheme. Any scheme for the one way wave equation that reaches in both directions will need a numerical boundary condition (for example, the Lax-Friedrichs, BTCS, Crank-Nicolson schemes, etc.). Finding acceptable numerical boundary conditions is a difficult problem and has become a very important part of the numerical solution of partial differential equations.

HW 5.6.8 (a) Solve the following problem using both the forward explicit and implicit schemes (FTFS and BTFS).

$$v_t - 2v_x = 0, \ x \in (0, 1), \ t > 0$$
$$v(x, 0) = 1 + \sin 2\pi x, \ x \in [0, 1]$$
$$v(1, t) = 1.0$$

Use $M = 10$, an appropriate Δt and find solutions at $t = 0.06$, $t = 0.1$ and $t = 0.8$.
(b) Repeat part (a) using $M = 100$.

HW 5.6.9 (a) Solve the following problem using both the backward explicit and implicit schemes (FTBS and BTBS).

$$v_t + 2v_x = 0, \ x \in (0, 1), \ t > 0$$
$$v(x, 0) = 1 + \sin 2\pi x, \ x \in [0, 1]$$
$$v(0, t) = 1.0$$

Use $M = 10$, an appropriate Δt and find solutions at $t = 0.06$, $t = 0.1$ and $t = 0.8$.
(b) Repeat part (a) using $M = 20$.

HW 5.6.10 Solve the problem given in HW5.6.8 using the explicit Lax-Wendroff scheme. Try the following *numerical boundary conditions* at $x = 0$.

(a) $u_0^n = 1.0$

(b) $u_0^n = u_1^n$.

(c) $u_0^{n+1} = u_0^n - \frac{a\Delta t}{\Delta x}(u_1^n - u_0^n)$

(d) $u_0^{n+1} + u_1^{n+1} + \frac{a\Delta t}{\Delta x}(u_1^{n+1} - u_0^{n+1}) = u_0^n + u_1^n - \frac{a\Delta t}{\Delta x}(u_1^n - u_0^n)$

HW 5.6.11 Solve the problem given in HW5.6.8 using the Crank-Nicolson scheme with numerical boundary conditions (a)–(d) from HW5.6.10.

HW 5.6.12 Repeat the solution of the problem given in HW5.6.9 using the backward explicit scheme (FTBS) with

(i) $M = 20$ and $\Delta t = 0.02$.

(ii) $M = 40$ and $\Delta t = 0.01$.

(iii) $M = 40$ and $\Delta t = 0.001$.

Compare your solutions with those of HW5.6.9, each other and the exact solution. Do they appear to converge to the exact solution? Is this experiment sufficient to convince you that difference scheme (5.6.41) is convergent? Why or why not? What other tests might you want to perform to better convince yourself (or a colleague) that difference scheme (5.6.41) is convergent?

5.7   The Courant-Friedrichs-Lewy Condition

One of the unique properties of hyperbolic equations is that because of the finite speed of propagation, the solution has a finite domain of dependence. That is, if we consider a problem of the form

$$v_t + av_x = 0 \quad x \in \mathbf{R}, \ t > 0 \tag{5.7.1}$$
$$v(x,0) = f(x), \tag{5.7.2}$$

$(a > 0)$ we know that the solution of the equation at the point (x,t) depends only on the value of f at the point x_0 where $x_0 = x - at$. The point x_0 is called the domain of dependence of the point (x,t). The **domain of dependence** of the point (x,t) is the set of all points that the solution of problem (5.7.1)–(5.7.2) at point (x,t) is dependent upon. Given boundary conditions, nonhomogeneous terms and higher order hyperbolic equations the domain of dependence can also consist of intervals on the $t = 0$ axis and regions in the $x - t$ plane.

Analogous to the domain of dependence for the continuous problem, we wish to define a numerical domain of dependence. Since the numerical domain of dependence seems to be a difficult concept to formally define, we instead define it iteratively in a way we hope is complete and understandable.

We begin by defining the numerical domain of dependence of a point for a particular solution scheme for the problem (5.7.1)–(5.7.2). We consider the difference scheme

$$u_k^{n+1} = u_k^n - R(u_k^n - u_{k-1}^n), \qquad (5.7.3)$$

where $R = a\Delta t/\Delta x$. It is then clear that the solution at a point $(k\Delta x, (n+1)\Delta t)$ depends on the values at the two lattice points below that are illustrated in Figure 5.7.1. If we continue this process down to the zero time level, we see that the solution depends on the points $((k - n - 1)\Delta x, 0), \cdots, (k\Delta x, 0)$. This situation is illustrated in Figure 5.7.2. It is also easy to see (Figure 5.7.3) that if some other δt and δx were used, as long as the ratio $a\delta t/\delta x = R$ and the point $(k\Delta x, (n + 1)\Delta t)$ is still one of the lattice points, the solution at $(k\Delta x, (n + 1)\Delta t)$ will still depend on points starting at $((k - n - 1)\Delta x, 0)$ and ending at $(k\Delta x, 0)$. For this reason, in this situation *we define the* **numerical domain of dependence** *of the point $(k\Delta x, (n + 1)\Delta t)$ for the difference scheme (5.7.3) to be the interval $D_n = [(k - n - 1)\Delta x, k\Delta x]$. It should be emphasized that all of the points that the solution at the point $(k\Delta x, (n + 1)\Delta t)$ depends on for all possible refinements satisfying $a\Delta t/\Delta x = R$ for a fixed R are contained in the interval D_n.*

We note that forcing Δt and Δx to approach zero while satisfying $R = a\Delta t/\Delta x$ is extremely convenient but not absolutely necessary. If we begin with a point such as $(k\Delta x, (n + 1)\Delta t)$ and want to define something like the numerical domain of dependence that will affect the convergence of our scheme, some knowledge of how the grid goes to zero is necessary. If for some strange reason you want to refine your grid by choosing Δt and Δx so that R is changing (say R is decreasing, or R cycles through a finite list of values), it is still possible to define the numerical domain of dependence but more difficult. What is necessary in the definition is that the numerical domain of dependence contains all of the points $(x, 0)$ that the solution at the particular point depends on for sufficiently small Δt and Δx. To make the definition allow for Δt and Δx to approach zero in these erratic ways, makes the definition difficult to formulate and not of much utility.

To illustrate further the numerical domain of dependence, we again consider the partial differential equation (5.7.1) and now consider any difference scheme of the form

$$u_k^{n+1} = \alpha u_k^n + \beta u_{k-1}^n. \qquad (5.7.4)$$

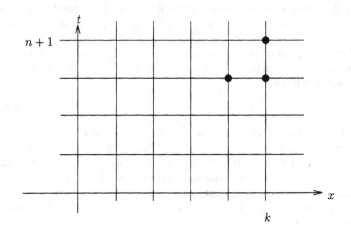

FIGURE 5.7.1. Dependence at the n-th time level of the solution of difference scheme (5.7.3) at point $(k\Delta x, (n+1)\Delta t)$.

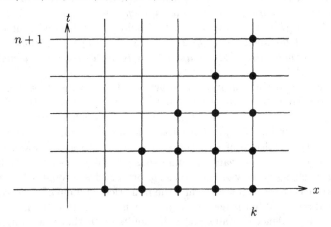

FIGURE 5.7.2. Numerical domain of dependence of point $(k\Delta x, (n+1)\Delta t)$ for difference scheme (5.7.3) where $R = 1\Delta t/\Delta x$.

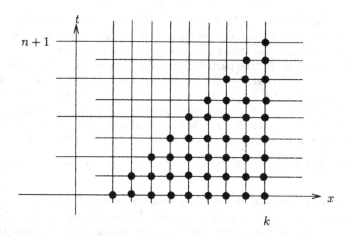

FIGURE 5.7.3. Numerical domain of dependence of point $(k_1 \delta x, (n_1 + 1)\delta t) = (k\Delta x, (n + 1)\Delta t)$ for difference scheme (5.7.3) where $R = a\Delta t / \Delta x$.

After very little thought, we realize that if we consider the point $(k\Delta x, (n+1)\Delta t)$, the numerical domain of dependence of difference scheme (5.7.4) is again the interval $D_n = [(k-n-1)\Delta x, k\Delta x]$. Likewise, it is easy to see that the numerical domain of dependence of the solution at point $(k\Delta x, (n+1)\Delta t)$ for either a difference scheme of the form

$$u_k^{n+1} = \alpha u_{k+1}^n + \beta u_k^n, \tag{5.7.5}$$

or

$$u_k^{n+1} = \alpha u_{k+1}^n + \beta u_k^n + \gamma u_{k-1}^n, \tag{5.7.6}$$

will be the interval $[k\Delta x, (k+n+1)\Delta x]$ or $[(k-n-1)\Delta x, (k+n+1)\Delta x]$, respectively. Thus, we see that *the numerical domain of dependence does not contain much information about the particular difference scheme. The numerical domain of dependence only knows where the difference scheme reaches.*

We now illustrate the utility of the comparison between the analytic domain of dependence and the numerical domain of dependence.

Definition 5.7.1 *A partial differential equation and an associated difference scheme is said to satisfy the Courant–Friedrichs–Lewy (CFL) condition if the analytic domain of dependence, D_a, is contained in the numerical domain of dependence, D_n.*

The importance of the CFL condition is easy to see by the following argument. Suppose that the CFL condition is not satisfied, that is, for some point P_0, $D_a \not\subset D_n$. Let P be a point that is in D_a but is not in D_n. The emphasis here is that when computing the numerical solution at the point P_0, the numerical scheme does not see the point P, but when computing the analytic solution at P_0, the partial differential equation does see the point P. Hence, we let the data be trivial except in a small neighborhood of the point P (small enough so that the entire neighborhood does not intersect the numerical domain of dependence. Then it is clear that as Δt and Δx approach zero, it is impossible for the numerical solution to converge to the analytic solution at the point P_0. We obtain the following theorem.

Theorem 5.7.2 *Satisfying the CFL condition is necessary for convergence.*

We must emphasize that *the above theorem states that the CFL condition provides us with only a necessary condition for convergence and we show below that it is not a sufficient condition.* We emphasize this because the CFL condition is often applied as if it gave both necessary and sufficient conditions for convergence.

We have previously had other conditions that are necessary for convergence. If we compare the CFL necessary condition with other necessary conditions, we have no promise that one is better or worse than the other. But the CFL condition is different from the others. Though the condition is often sharp (it is, in fact, often also sufficient), the CFL condition is a very coarse measure and can be a very bad condition (meaning that it may hint that a totally ridiculous scheme might be convergent). Some of these points are illustrated below. As with other necessary conditions, the CFL condition should be used when we can do no better, using the limit with fingers crossed (and maybe with some numerical experimentation) as a potential stability limit.

We next return to the problem at the beginning of this section and see how the above theorem can be used. The analytic domain of dependence of the point (x, t) for the partial differential equation (5.7.1) is $x_0 = x - at$. If we consider the point $(x, t) = (k\Delta x, (n + 1)\Delta t)$, then

$$x_0 = x - at = k\Delta x - a(n + 1)\Delta t = \{k - R(n + 1)\}\Delta x.$$

Thus we see that it is clear that $x_0 \in [(k - n - 1)\Delta x, k\Delta x]$ (which is the numerical domain of dependence for difference scheme (5.7.3)) if and only if $(k - n - 1)\Delta x \leq \{k - R(n + 1)\}\Delta x \leq k\Delta x$. The following sequence of equivalent inequalities

$$k - n - 1 \leq k - R(n + 1) \leq k$$
$$-(n + 1) \leq -R(n + 1) \leq 0$$
$$1 \qquad \geq R \qquad \geq 0$$

shows that the analytic domain of dependence is contained in the numerical domain of dependence if and only if $0 \leq R \leq 1$. Thus we see that the

CFL condition for partial differential equation (5.7.1) and difference scheme (5.7.3) is equivalent to the inequality $0 \leq R \leq 1$. Thus *by Theorem 5.7.2, $0 \leq R \leq 1$ is a necessary condition for the stability of the FTBS scheme (5.7.3).*

We should note that this necessary condition will imply that difference scheme (5.7.3) **may** be stable if $0 \leq R \leq 1$, and will not be stable if $R < 0$ ($a < 0$) for any Δt or Δx or when $R > 1$. We should also note that this necessary condition is the same as the necessary and sufficient condition derived in Section 5.3. Hence, in this case, the CFL condition provides us with a very good necessary condition.

However, we should note that the CFL condition will be the same for any difference scheme of the form of (5.7.4). So the CFL condition for difference scheme (5.7.4) is $0 \leq R \leq 1$ for any α and β. But the general form of difference scheme (5.7.4) contains a large number of terrible schemes that surely do not converge. In this way, the CFL necessary condition is not a sharp condition. But this should not be a surprise when we remember that the CFL condition only looks where the scheme reaches. It uses no other information about how strongly the difference scheme reaches to these points.

In the same manner, we note that $x_0 \in [k\Delta x, (k+n+1)\Delta x]$ if and only if $-1 \leq R \leq 0$. By Theorem 5.7.2, for any scheme of the form of difference scheme (5.7.5) (including of course, the FTFS scheme (5.3.1)), the CFL condition $-1 \leq R \leq 0$ will give a necessary condition for stability. So we see that if $a < 0$ and $|R| \leq 1$, then difference schemes (5.3.1) and (5.7.5) **may** be stable. We also see that any scheme of the form of (5.7.5) will be unstable if $R > 0$ ($a > 0$). As in the previous case, we see that when the CFL condition is applied to difference scheme (5.3.1), the resulting necessary condition is the same as the necessary and sufficient condition obtained earlier.

And finally, we see that we get the same type of results for difference scheme (5.7.6), i.e. the CFL condition is equivalent to the condition $-1 \leq R \leq 1$, or $|R| \leq 1$. Hence, $|R| \leq 1$ will be a necessary condition for convergence for any scheme of the form (5.7.6) (so this will apply to the explicit Lax-Wendroff difference scheme, (5.3.8)).

We should note that the necessary conditions found above were the same as those found using the von Neumann condition. We emphasize that the von Neumann condition yielded a necessary and sufficient condition for convergence. The CFL condition only provides us with a necessary condition. To see that the CFL condition will not provide a sufficient condition, we consider the following example.

Example 5.7.1 Discuss the CFL condition and the stability results for using the difference scheme (FTCS)

$$u_k^{n+1} = u_k^n - \frac{R}{2}(u_{k+1}^n - u_{k-1}^n) \tag{5.7.7}$$

for solving problem (5.7.1)–(5.7.2).

Solution: Since the difference scheme (5.7.7) is the same form as scheme (5.7.6), we know that the CFL condition for the scheme is $|R| \leq 1$. Hence, $|R| \leq 1$ is a necessary condition for convergence of the difference scheme (5.7.7).

Earlier (in Section 5.3.2), we saw that difference scheme (5.7.7) is unconditionally unstable. Thus, though the CFL condition gives a nontrivial necessary stability condition for difference scheme (5.7.7), the scheme is unstable for all R.

So we see from the above discussion that the CFL condition is truly only a necessary condition while the von Neumann condition provides us with necessary and sufficient conditions. Why then, we should ask, does anyone care about the CFL condition? The best reason is that *the CFL condition is so easy to compute*. We have computed the CFL condition for all schemes that spatially reach one grid point away. And, often, this gives good information. Often, the CFL condition is the same as the stability limit. Hence, it is a good place to begin. The CFL condition is used often in the literature. We should be careful, however, to remember that it only provides a necessary condition for stability. The CFL condition is also often abused.

In the above discussion, we have only considered schemes that spatially reach one grid point to the left or right. The result for a general difference scheme follows in the exactly the same manner. The only new work is to compute the numerical domain of dependence for the more general difference operator. If we consider the general difference scheme

$$u_k^{n+1} = \Big(\sum_{j=-m_1}^{m_2} Q_{2j} S_+^j \Big) u_k^n \tag{5.7.8}$$

(where both $Q_{2\,-m_1}$ and $Q_{2\,m_2}$ are nonzero), it is not hard to see that the numerical domain of dependence of the point $(k\Delta x, (n+1)\Delta t)$ for the difference scheme (5.7.8) is $[(k - m_1(n+1))\Delta x, (k + m_2(n+1))\Delta x]$. And, of course, the CFL condition for scheme (5.7.8) (a necessary condition for stability of the scheme) can be written as

$$-m_1 \leq R \leq m_2.$$

The existence of the CFL condition is useful in giving us some general bad news. This is given in the following theorem.

Theorem 5.7.3 *There is no explicit, consistent, unconditionally stable, difference scheme for solving a hyperbolic partial differential equation.*

The above result follows from the fact that a necessary condition does exist and requires that R be confined to a finite interval. If we assumed that a given explicit difference scheme was unconditionally stable, an R could be chosen sufficiently large to reach beyond the analytic domain of dependence and, hence, contradict Theorem 5.7.2 as Δt and Δx approach zero.

Since earlier we showed that implicit schemes (5.4.2), (5.4.3), (5.4.6), (5.4.8) and (5.4.10) were all unconditionally stable, Theorem 5.7.3 obviously cannot be extended to include implicit schemes.

HW 5.7.1 Use the CFL condition to obtain necessary conditions for the convergence of the Lax-Friedrichs, Beam-Warming and MacCormack difference schemes, Table 5.3.1.

5.8   Two Dimensional Hyperbolic Equations

In this section, We will try to include an introduction to hyperbolic partial differential equations and the numerical solution of hyperbolic equations in more than one dimension (specifically, two). As we did in the one dimensional case, we will work with a model, scalar equation. Specifically, we consider the partial differential equation

$$v_t + av_x + bv_y = 0. \tag{5.8.1}$$

As we shall see, most of the mathematical topics and numerical results will be analogous to the one dimensional case. We begin by noting that if we consider partial differential equation (5.8.1) along with the initial condition

$$v(x, y, 0) = f(x, y), \tag{5.8.2}$$

the solution is given by $v(x, y, t) = f(x - at, y - bt)$. Thus, as in the one dimensional case, the solution consists of translating the initial condition in the appropriate direction. The **speed of propagation** will be a in the x-direction and b in the y-direction. The solution at a given point (x, y) at time t will depend on f at $(x_0, y_0) = (x - at, y - bt)$. Hence, the **domain of dependence** of the point (x, y, t) will be (x_0, y_0). The curve through the point (x, y, t) defined by $x - at = x_0$ and $y - bt = y_0$ is called a **characteristic curve**. And, as was the case in one dimension, the solution of equation (5.8.1) will be constant along a characteristic curve.

5.8.1   Conservation Law Derivation

Solving two dimensional hyperbolic initial–value problems numerically is very similar to solving one dimensional problems. Before we proceed with properties of numerical schemes, we use the conservation law approach to derive several of the numerical schemes. We include another conservation

law derivation of difference schemes because the process is slightly different for hyperbolic equations than it is for parabolic equations. All of our
results will apply to one and three dimensional schemes as well as two dimensional schemes. So that the reader is aware that different approaches
are taken for conservation schemes by different people, we shall approach
the conservation law slightly different from the way we did in Chapters 1
and 4.

We begin by considering the time-space cell shown in Figure 5.8.1. If we
integrate partial differential equation (5.8.1) over this cell (actually performing the three integrations that are possible), we are left with

$$\int_{y_{k-1/2}}^{y_{k+1/2}} \int_{x_{j-1/2}}^{x_{j+1/2}} (v^{n+1} - v^n)dxdy + a \int_{t_n}^{t_{n+1}} \int_{y_{k-1/2}}^{y_{k+1/2}} v(x_{j+1/2}, y, t)dydt$$

$$-a \int_{t_n}^{t_{n+1}} \int_{y_{k-1/2}}^{y_{k+1/2}} v(x_{j-1/2}, y, t)dydt + b \int_{t_n}^{t_{n+1}} \int_{x_{j-1/2}}^{x_{j+1/2}} v(x, y_{k+1/2}, t)dxdt$$

$$-b \int_{t_n}^{t_{n+1}} \int_{x_{j-1/2}}^{x_{j+1/2}} v(x, y_{k-1/2}, t)dxdt = 0. \qquad (5.8.3)$$

We should note that the above integral form of the conservation law is the
same equation that we would either hypothesize or derive as a part of the
derivation of partial differential equation (5.8.1).

As usual with conservation law approaches, we are left with approximating equation (5.8.3). It is easy to see that we can again approximate the
first integral by the midpoint rule. The interesting part of applying the
conservation law approach to deriving difference schemes for hyperbolic
equations is with the remaining integrals. We note, for instance, that the
second integral is of the form

$$\int_{t_n}^{t_{n+1}} \int_{y_{k-1/2}}^{y_{k+1/2}} v(x_{j+1/2}, y, t)dydt. \qquad (5.8.4)$$

We see that the function v has been evaluated at $x_{j+1/2}$, but we must
approximate the integral in terms of functions evaluated at grid points
(x_j, y_k). We do not have the nice first derivative to approximate that we
had for parabolic problems. The integral in expression (5.8.4) does not
cause the problems. It is the x evaluation that causes the problem.

The most obvious approach is to approximate $v(x_{j+1/2}, y, t)$ by the average of $v(x_j, y, t)$ and $v(x_{j+1}, y, t)$. If we do this with the last four integrals
of equation (5.8.3) (and approximate the integrals in the obvious way), we
get

$$\Delta x \Delta y(u_{jk}^{n+1} - u_{jk}^n) + a\Delta y \Delta t \left(\frac{u_{jk}^n + u_{j+1 k}^n}{2} - \frac{u_{j-1 k}^n + u_{jk}^n}{2} \right)$$

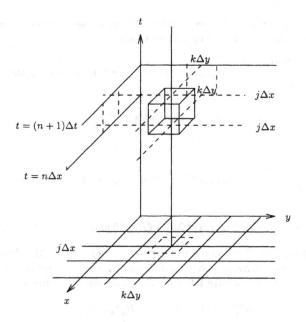

FIGURE 5.8.1. Computational cell associated with the point $(j\Delta x, k\Delta y)$ between times $t_n = n\Delta t$ and $t_{n+1} = (n+1)\Delta t$.

$$+b\Delta x\Delta t\Big(\frac{u^n_{j\,k+1} + u^n_{j\,k}}{2} - \frac{u^n_{j\,k} + u^n_{j\,k-1}}{2}\Big) = 0 \qquad (5.8.5)$$

or

$$u^{n+1}_{j\,k} = u^n_{j\,k} - \frac{a\Delta t}{2\Delta x}\big(u^n_{j+1\,k} - u^n_{j-1\,k}\big)$$

$$-\frac{b\Delta t}{2\Delta y}\big(u^n_{j\,k+1} - u^n_{j\,k-1}\big). \qquad (5.8.6)$$

If we do a consistency argument or approximate the above integrals carefully, we see that difference scheme (5.8.6) is $\mathcal{O}(\Delta t) + \mathcal{O}(\Delta x^2) + \mathcal{O}(\Delta y^2)$. But, we should guess that *the scheme will be unconditionally unstable* (at least the one dimensional analog was unconditionally unstable). Since the symbol of the difference operator is $\rho(\xi, \eta) = 1 - i(R_x \sin\xi + R_y \sin\eta)$, we see that the scheme is unconditionally unstable.

Thus we see that, as was the case with parabolic conservation laws, we can derive unstable difference schemes based on the conservation law as well as stable difference schemes. We shall see that to obtain the stable schemes for hyperbolic partial differential equations, we must make approximations that are somewhat different from the approximations for parabolic problems. For example, let us approximate $v(x_{j+1/2}, y, t)$ and $v(x_{j-1/2}, y, t)$ by $v(x_j, y, t)$ and $v(x_{j-1}, y, t)$; and $v(x, y_{k+1/2}, t)$ and $v(x, y_{k-1/2}, t)$ by

$v(x, y_k, t)$ and $v(x, y_{k-1}, t)$, respectively. Using these approximations and the natural symmetry in our operators yields

$$a \int_{t_n}^{t_{n+1}} \int_{y_{k-1/2}}^{y_{k+1/2}} (v(x_{j+1/2}, y, t) - v(x_{j-1/2}, y, t))dydt = a\Delta y \Delta t \delta_{x-} v_{jk}^n$$

$$+\mathcal{O}(\Delta x^2 \Delta y \Delta t) + \mathcal{O}(\Delta x \Delta y \Delta t^2), \qquad (5.8.7)$$

and

$$b \int_{t_n}^{t_{n+1}} \int_{x_{j-1/2}}^{x_{j+1/2}} (v(x, y_{k+1/2}, t) - v(x, y_{k-1/2}, t))dxdt = b\Delta x \Delta t \delta_{y-} v_{jk}^n$$

$$+\mathcal{O}(\Delta x \Delta y^2 \Delta t) + \mathcal{O}(\Delta x \Delta y \Delta t^2). \qquad (5.8.8)$$

If in equation (5.8.3) we make the obvious integration of the first term, substitute approximations (5.8.7) and (5.8.8) for the last four integrals and drop the order terms, we are left with the following difference scheme

$$u_{jk}^{n+1} = (1 - R_x \delta_{x-} - R_y \delta_{y-})u_{jk}^n, \qquad (5.8.9)$$

where $R_x = a\Delta t/\Delta x$ and $R_y = b\Delta t/\Delta y$. We see that *difference scheme (5.8.9) is a $\mathcal{O}(\Delta t) + \mathcal{O}(\Delta x) + \mathcal{O}(\Delta y)$ order approximation of partial differential equation (5.8.1).*

And, of course, to derive all the different difference schemes for solving initial–value problem (5.8.1)–(5.8.2), we use the different approximate integrations (midpoint rule, rectangular rule, trapezoidal rule) and different approximations to the flux terms.

5.8.2   Initial–Value Problems

Using either the conservation approach as we did in the last section or a Taylor series approach, we should be confident that we can develop a variety of difference schemes for solving partial differential equation (5.8.1). As is so often the case, the method by which we prove convergence is the Lax Theorem. As we did for difference scheme (5.8.6) (in which case we saw that scheme (5.8.6) was unstable), the stability will be analyzed by computing the discrete Fourier transform. There are a large number of difference schemes for solving partial differential equation (5.8.1) discussed in the literature. For many of these it is difficult to perform the stability analysis. For others, the stability conditions become so restrictive as to make the schemes undesirable. In this section we shall analyze the stability (and, hence, convergence) of several difference schemes for approximating the solution of equation (5.8.1). We begin by analyzing the stability of difference scheme (5.8.9).

Example 5.8.1 Assume that a and b are greater than zero. Analyze the stability of difference scheme (5.8.9).

Solution: If we take a two dimensional discrete Fourier transform of equation (5.8.9), we get

$$\hat{u}^{n+1} = \left[1 - R_x(1 - e^{-i\xi}) - R_y(1 - e^{-i\eta})\right]\hat{u}^n. \tag{5.8.10}$$

So the symbol of difference scheme (5.8.9) is given by

$$\rho(\xi, \eta) = 1 - R_x(1 - e^{-i\xi}) - R_y(1 - e^{-i\eta}) \tag{5.8.11}$$

and

$$
\begin{aligned}
|\rho(\xi, \eta)|^2 &= \left[1 - 2R_x \sin^2 \frac{\xi}{2} - 2R_y \sin^2 \frac{\eta}{2}\right]^2 \\
&\quad + \left[R_x \sin \xi + R_y \sin \eta\right]^2.
\end{aligned} \tag{5.8.12}
$$

It is not difficult to differentiate $|\rho|^2$ with respect to ξ and η, set the derivatives equal to zero and find that there are potential maximums at $(\pm\pi, \pm\pi)$, $(\pm\pi, 0)$, $(0, \pm\pi)$ and $(0, 0)$. It is also easy to see that

$$|\rho(0, 0)| = 1, \quad |\rho(\pm\pi, 0)| = (1 - 2R_x)^2, \quad |\rho(0, \pm\pi)| = (1 - 2R_y)^2$$

and

$$|\rho(\pm\pi, \pm\pi)|^2 = (1 - 2R_x - 2R_y)^2.$$

The condition $(1 - 2R_x)^2 \le 1$ requires that R_x satisfies $0 \le R_x \le 1$. Likewise, the condition $(1 - 2R_y)^2 \le 1$ requires that R_y satisfies $0 \le R_y \le 1$. And, the condition $(1 - 2R_x - 2R_y)^2 \le 1$ requires that R_x and R_y satisfy $0 \le R_x + R_y \le 1$. There may be more zeros than those given above. A terrible calculation (and, for that reason, we will not include it here) shows that the conditions imposed because of the other zeros will be less restrictive than the conditions determined above. Therefore, we see that *difference scheme (5.8.9) is first order accurate in time and space, and conditionally stable with condition $0 \le R_x + R_y \le 1$, R_x, $R_y \ge 0$* (and, hence, convergent).

Remark 1: Obviously, the method used above [the statement "A terrible calculation (and, for that reason, we will not include it here)"] is not entirely satisfactory (at least it is not a satisfactory response when students are supposed to prove stability in homework problems). One approach to verifying (not proving) that $|\rho| \le 1$ for all R_x and R_y satisfying $0 \le R_x + R_y \le 1$, R_x, $R_y \ge 0$ is to use graphics. It is reasonably easy to plot $|\rho(\xi, \eta)|$ on $[-\pi, \pi] \times [-\pi, \pi]$ for a carefully chosen set of values of R_x and R_y, strongly indicating that $|\rho| \le 1$ for $0 \le R_x + R_y \le 1$, R_x, $R_y \ge 0$. See HW5.8.3.

Remark 2: Hopefully, it is clear that we can obtain four different one sided schemes by choosing either forward or backward difference approximations for the v_x and v_y terms. Of course, all of them will be first order in both time and space. Only the schemes that use forward differences in the terms where either a and/or b are negative, and backward differences in the terms where either a and/or b are positive will be stable.

HW 5.8.1 Plot $|\rho(\xi, \eta)|$ given by equation (5.8.12) on $[0, \pi] \times [0, \pi]$ for a sequence of R_x's and R_y's satisfying $R_x + R_y \le 1$. Verify that $|\rho(\xi, \eta)| \le 1$.

HW 5.8.2 (a) Give a conservation law derivation of difference scheme

$$u_{jk}^{n+1} = (1 - R_x\delta_{x+} - R_y\delta_{y+})u_{jk}^n$$

for solving partial differential equation (5.8.1).
(b) Derive difference scheme

$$u_{jk}^{n+1} = (1 - R_x\delta_{x+} - R_y\delta_{y-})u_{jk}^n$$

for approximating the solution of equation (5.8.1).

HW 5.8.3 (a) Algebraically verify $\rho(\xi, \eta)| \leq 1$ for R_x and R_y satisfying $0 \leq R_x + R_y \leq 1$, R_x, $R_y \geq 0$ for ρ as in Example 5.8.1.
(b) Verify part (a) numerically. Use a sequence of values R_{xj}, R_{yk} that uniformly cover the region (varying R_x and R_y by 0.1 would be nice, but it might mean too many plots).

Example 5.8.2 Analyze the stability of the two dimensional Lax-Friedrichs scheme

$$u_{jk}^{n+1} = \frac{1}{4}(u_{j+1\,k}^n + u_{j-1\,k}^n + u_{j\,k+1}^n + u_{j\,k-1}^n)$$
$$- \frac{R_x}{2}\delta_{x0}u_{jk}^n - \frac{R_y}{2}\delta_{y0}u_{jk}^n \tag{5.8.13}$$

for approximating the solution of initial–value problem (5.8.1)–(5.8.2).
Solution: We compute the discrete Fourier transform to obtain the symbol for the scheme

$$\rho(\xi, \eta) = \frac{1}{2}(\cos\xi + \cos\eta) - i(R_x\sin\xi + R_y\sin\eta). \tag{5.8.14}$$

The expression $| \rho(\xi, \eta) |^2$ can be written as

$$| \rho(\xi, \eta) |^2 = 1 - (\sin^2\xi + \sin^2\eta)\left[\frac{1}{2} - (R_x^2 + R_y^2)\right]$$
$$- \frac{1}{4}(\cos\xi - \cos\eta)^2 - (R_x\sin\eta - R_y\sin\xi)^2. \tag{5.8.15}$$

Since the last two terms in equation (5.8.15) are negative, we have

$$| \rho(\xi, \eta) |^2 \leq 1 - (\sin^2\xi + \sin^2\eta)\left[\frac{1}{2} - (R_x^2 + R_y^2)\right].$$

If $\frac{1}{2} - (R_x^2 + R_y^2) \geq 0$, then $| \rho(\xi, \eta) |^2 \leq 1$. Hence, if

$$R_x^2 + R_y^2 \leq \frac{1}{2}, \tag{5.8.16}$$

difference scheme (5.8.13) is stable.

We note that stability condition (5.8.16) is very restrictive. In particular, as we shall see in Section 5.8.4 the condition is more restrictive than the CFL condition, $max\{| R_x |, | R_y |\} \leq 1$. It is not obvious that we can always find a scheme with a stability condition the same as the CFL condition, but it is at least what we should try to do (since we know for a given stencil, it is the best that we can do). For the moment consider an equation of the form

$$v_t = Av = (A_1 + A_2)v \qquad (5.8.17)$$

where the A, A_1 and A_2 in equation (5.8.17) are purposely vague to allow for greater flexibility (but surely one example would be to let $Av = -av_x - bv_y$, $A_1v = -av_x$ and $A_2v = -bv_y$). If we use a first order approximation to the time derivative, we can write

$$
\begin{aligned}
v^{n+1} &= v^n + \Delta t A v^n + \mathcal{O}(\Delta t^2) \\
&= (1 + \Delta t A_1 + \Delta t A_2)v^n + \mathcal{O}(\Delta t^2) \\
&= (1 + \Delta t A_1)(1 + \Delta t A_2)v^n - \Delta t^2 A_1 A_2 v^n + \mathcal{O}(\Delta t^2). \quad (5.8.18)
\end{aligned}
$$

Dropping the Δt^2 terms in equation (5.8.18) leaves us with the approximately factored scheme

$$u^{n+1} = (1 + \Delta t A_1)(1 + \Delta t A_2)u^n$$

or

$$u^{n+1/2} = (1 + \Delta t A_2)u^n \qquad (5.8.19)$$

$$u^{n+1} = (1 + \Delta t A_1)u^{n+1/2}. \qquad (5.8.20)$$

Remark 1: We introduced approximately factored schemes in Sections 4.4.2.2 and 4.4.4. The biggest difference between the approximately factored schemes developed in Sections 4.4.2.2 and 4.4.4 and scheme (5.8.19)–(5.8.20) is that in Chapter 4 we developed only implicit approximately factored schemes. Of course, as we see above, approximate factorization can be used equally well to develop explicit schemes.

Remark 2: We must realize that difference scheme (5.8.19)–(5.8.20) is different from other difference schemes we have considered. It would be possible to consider difference scheme (5.8.19)–(5.8.20) as a semi-discrete method (discrete in time and continuous with respect to its spatial variables). However, we will consider difference equation (5.8.19)–(5.8.20) as not yet finished, and will take the opportunity to approximate A_1 and A_2 to our advantage.

We now return to our consideration of the specific hyperbolic partial differential equation

$$v_t + av_x + bv_y = 0,$$

let $A_2v = -av_x$ and $A_1v = -bv_y$, and consider difference scheme (5.8.19)–(5.8.20). If we approximate the A_1 and A_2 by the one dimensional Lax-Wendroff approximation, we get

$$u_{jk}^{n+1/2} = u_{jk}^n - \frac{R_x}{2}\delta_{x0}u_{jk}^n + \frac{R_x^2}{2}\delta_x^2 u_{jk}^n \qquad (5.8.21)$$

$$u_{jk}^{n+1} = u_{jk}^{n+1/2} - \frac{R_y}{2}\delta_{y0}u_{jk}^{n+1/2} + \frac{R_y^2}{2}\delta_y^2 u_{jk}^{n+1/2}. \qquad (5.8.22)$$

Example 5.8.3 Analyze the consistency and stability of difference scheme (5.8.21)–(5.8.22).

Solution: By substituting $u_{jk}^{n+1/2}$ from equation (5.8.21) into equation (5.8.22), it is easy to see that difference scheme (5.8.21)–(5.8.22) is second order accurate in Δt. A more difficult calculation (remember, you do not have to consider the terms that are order Δt^3 and Δt^4) shows that the scheme is also accurate $\mathcal{O}(\Delta x^2) + \mathcal{O}(\Delta y^2)$. See HW5.8.5. (We should remember that the computer should do these calculations, not us.)

To analyze the stability of difference scheme (5.8.21)–(5.8.22), we take the discrete Fourier transforms of both equations, note that

$$\hat{u}^{n+1} = \left(1 - iR_y \sin\eta - 2R_y^2 \sin^2 \frac{\eta}{2}\right)\hat{u}^{n+1/2} \tag{5.8.23}$$

$$= \left(1 - iR_y \sin\eta - 2R_y^2 \sin^2 \frac{\eta}{2}\right)\left(1 - iR_x \sin\xi - 2R_x^2 \sin^2 \frac{\xi}{2}\right)\hat{u}^n, \tag{5.8.24}$$

and use the analysis for the one dimensional Lax-Wendroff scheme to see that $\mid \rho(\xi, \eta) \mid^2$ will be less than or equal to one if and only if $max\{\mid R_x \mid, \mid R_y \mid\} \leq 1$. Hence, difference scheme (5.8.21)–(5.8.22) is accurate order $\mathcal{O}(\Delta t^2) + (\Delta x^2) + \mathcal{O}(\Delta y^2)$ and conditionally stable, so difference scheme (5.8.21)–(5.8.22) is conditionally convergent order $\mathcal{O}(\Delta t^2) + \mathcal{O}(\Delta x^2) + \mathcal{O}(\Delta y^2)$. .

We conclude this section by including the analysis of two implicit schemes for approximating the solution of initial–value problem (5.8.1)–(5.8.2).

Example 5.8.4 Analyze the stability of the following BTCS implicit scheme.

$$u_{jk}^{n+1} + \frac{R_x}{2}\delta_{x0}u_{jk}^{n+1} + \frac{R_y}{2}\delta_{y0}u_{jk}^{n+1} = u_{jk}^n \tag{5.8.25}$$

Solution: Since the symbol for difference scheme (5.8.25) is given by

$$\rho(\xi, \eta) = \frac{1}{1 + iR_x \sin\xi + iR_y \sin\eta},$$

it is easy to see that

$$|\rho(\xi, \eta)|^2 = \frac{1}{1 + (R_x \sin\xi + R_y \sin\eta)^2} \leq 1.$$

Hence, difference scheme (5.8.25) is unconditionally stable.

Example 5.8.5 Analyze the Crank-Nicolson scheme

$$u_{jk}^{n+1} + \frac{R_x}{4}\delta_{x0}u_{jk}^{n+1} + \frac{R_y}{4}\delta_{y0}u_{jk}^{n+1} = u_{jk}^n - \frac{R_x}{4}\delta_{x0}u_{jk}^{n+1} - \frac{R_y}{4}\delta_{y0}u_{jk}^{n+1} \tag{5.8.26}$$

for approximating the solution to initial–value problem (5.8.1)–(5.8.2).
Solution: Again, since the symbol for difference scheme (5.8.26) is

$$\rho(\xi, \eta) = \frac{1 - i\frac{R_x}{2}\sin\xi - i\frac{R_y}{2}\sin\eta}{1 + i\frac{R_x}{2}\sin\xi + i\frac{R_y}{2}\sin\eta},$$

clearly $|\rho(\xi, \eta)| = 1$ and difference scheme (5.8.26) is unconditionally stable.

HW 5.8.4 Show that the difference scheme (not the two dimensional Lax-Wendroff scheme)

$$u_{jk}^{n+1} = u_{jk}^n - \frac{R_x}{2}\delta_{x0}u_j^n k + \frac{R_x^2}{2}\delta_x^2 u_{jk}^n - \frac{R_y}{2}\delta_{y0}u_j^n k + \frac{R_y^2}{2}\delta_y^2 u_{jk}^n$$

is not second order accurate in time.

HW 5.8.5 Analyze the order of accuracy of difference scheme (5.8.21)–(5.8.22).

HW 5.8.6 Analyze the stability of the scheme

$$u_{jk}^{n+1} = (1 - R_x \delta_{x\pm} - R_y \delta_{y\pm})u_{jk}^n,$$

where all different combinations of forward and backward differences are considered.

5.8.3   ADI Schemes

If we consider the implementation of implicit schemes for approximating the solution of partial differential equation (5.8.1) (along with some initial and boundary conditions), we run into the same problem that we faced with using implicit schemes for solving two dimensional parabolic problems: the matrices are broadly banded. Again, as we did for parabolic problems, the approach is to use alternating direction implicit (ADI) schemes.

We begin by considering a locally one dimensional (lod) scheme (analogous to the scheme considered for parabolic equations in HW4.4.6) for solving initial–value problem (5.8.1)–(5.8.2) given by

$$\left(1 + \frac{R_x}{2}\delta_{x0}\right)u_{jk}^{n+1/2} = u_{jk}^n \qquad (5.8.27)$$

$$\left(1 + \frac{R_y}{2}\delta_{y0}\right)u_{jk}^{n+1} = u_{jk}^{n+1/2}. \qquad (5.8.28)$$

Example 5.8.6 Analyze the consistency and stability (and, hence, convergence) of difference scheme (5.8.27)–(5.8.28).

Solution: By eliminating $u^{n+1/2}$ from equations (5.8.27)–(5.8.28), we see that difference equations (5.8.27)–(5.8.28) are equivalent to

$$0 = \frac{u_{jk}^{n+1} - u_{jk}^n}{\Delta t} + \frac{a}{2\Delta x}\delta_{x0}u_{jk}^{n+1} + \frac{b}{2\Delta y}\delta_{y0}u_{jk}^{n+1} + ab\Delta t \frac{\delta_{x0}}{\Delta x}\frac{\delta_{y0}}{\Delta y}u_{jk}^{n+1}.$$

Since the last term is order Δt, it is clear that difference scheme (5.8.27)–(5.8.28) is order $\mathcal{O}(\Delta t) + \mathcal{O}(\Delta x^2) + \mathcal{O}(\Delta y^2)$.

If we take the discrete Fourier transform of equations (5.8.27)–(5.8.28) and eliminate the $\hat{u}^{n+1/2}$ term, we get that the symbol for the system of difference equations (5.8.27)–(5.8.28) is

$$\rho(\xi, \eta) = \frac{1}{(1 + iR_x \sin \xi)(1 + iR_y \sin \eta)}. \qquad (5.8.29)$$

Then since

$$|\rho(\xi, \eta)|^2 = \frac{1}{(1 + R_x^2 \sin^2 \xi)(1 + R_y^2 \sin^2 \eta)},$$

it is clear that $0 \le |\rho(\xi, \eta)| \le 1$ and difference scheme (5.8.27)–(5.8.28) is unconditionally stable. Therefore, *difference scheme (5.8.27)–(5.8.28) is a $\mathcal{O}(\Delta t) + \mathcal{O}(\Delta x^2) + \mathcal{O}(\Delta y^2)$ order accurate, unconditionally stable scheme for solving initial–value problem (5.8.1)–(5.8.2)* (and, hence, convergent).

Remark: We must realize that because we have central differences in difference scheme (5.8.27)–(5.8.28), when we use this scheme for solving an initial–boundary–value problem, if we have either Dirichlet boundary conditions, we will have to provide a numerical boundary condition at one of the sides in each direction. As usual, this numerical boundary condition must be chosen very carefully. If we have periodic boundary conditions, as was the case with one dimensional problems with periodic boundary conditions, the matrices will have terms in the opposite corners and the Sherman-Morrison Algorithm, Section 5.6.1, will have to be used to solve the matrix equations.

Before we leave this section, we include another ADI scheme for solving partial differential equation (5.8.1) that has some special properties. We begin with the centered, Crank-Nicolson scheme

$$\frac{u_{jk}^{n+1} - u_{jk}^n}{\Delta t} + \frac{a}{4\Delta x}\delta_{x0}(u_{jk}^n + u_{jk}^{n+1}) + \frac{b}{4\Delta y}\delta_{y0}(u_{jk}^n + u_{jk}^{n+1}) = 0,$$

(5.8.30)

or

$$\left(1 + \frac{R_x}{4}\delta_{x0} + \frac{R_y}{4}\delta_{y0}\right)u_{jk}^{n+1} = \left(1 - \frac{R_x}{4}\delta_{x0} - \frac{R_y}{4}\delta_{y0}\right)u_{jk}^n.$$

(5.8.31)

As usual, the Crank-Nicolson scheme is accurate order $\mathcal{O}(\Delta t^2) + \mathcal{O}(\Delta x^2) + \mathcal{O}(\Delta y^2)$. If we factor both sides of equation (5.8.31) (adding and subtracting the necessary terms on both sides), we get

$$\left(1 + \frac{R_x}{4}\delta_{x0}\right)\left(1 + \frac{R_y}{4}\delta_{y0}\right)u_{jk}^{n+1} = \left(1 - \frac{R_x}{4}\delta_{x0}\right)\left(1 - \frac{R_y}{4}\delta_{y0}\right)u_{jk}^n$$

$$+ \frac{R_x R_y}{16}\delta_{x0}\delta_{y0}(u_{jk}^{n+1} - u_{jk}^n).$$

(5.8.32)

The last term is clearly order Δt^3 (remembering that we still must divide by Δt). Hence, dropping the last term will leave us with a difference scheme that is second order with respect to both time and space of the form

$$\left(1 + \frac{R_x}{4}\delta_{x0}\right)\left(1 + \frac{R_y}{4}\delta_{y0}\right)u_{jk}^{n+1} = \left(1 - \frac{R_x}{4}\delta_{x0}\right)\left(1 - \frac{R_y}{4}\delta_{y0}\right)u_{jk}^n.$$

(5.8.33)

The above scheme is referred to as the **Beam-Warming scheme** and is most often written as

$$\left(1 + \frac{R_x}{4}\delta_{x0}\right)u_{jk}^* = \left(1 - \frac{R_x}{4}\delta_{x0}\right)\left(1 - \frac{R_y}{4}\delta_{y0}\right)u_{jk}^n$$

(5.8.34)

$$\left(1 + \frac{R_y}{4}\delta_{y0}\right)u_{jk}^{n+1} = u_{jk}^*.$$

(5.8.35)

It is not hard to see that the symbol for the Beam-Warming scheme is

$$\rho(\xi, \eta) = \frac{(1 - i\frac{R_x}{2}\sin\xi)(1 - i\frac{R_y}{2}\sin\eta)}{(1 + i\frac{R_x}{2}\sin\xi)(1 + i\frac{R_y}{2}\sin\eta)}.$$

(5.8.36)

Thus we see that $|\rho(\xi, \eta)|^2 = 1$ for all ξ, $\eta \in [-\pi, \pi]$. Therefore, *the Beam-Warming scheme is a second order, unconditionally stable scheme* (and, hence, convergent).

Finally, by subtracting

$$\left(1 + \frac{R_x}{4} \delta_{x0}\right) \left(1 + \frac{R_y}{4} \delta_{y0}\right) u_{jk}^n$$

from both sides of equation (5.8.33) we rewrite difference scheme (5.8.34)–(5.8.35) in the common computational form, called the **delta formulation**,

$$\left(1 + \frac{R_x}{4} \delta_{x0}\right) \Delta u_{jk}^* = \left(-\frac{R_x}{2} \delta_{x0} - \frac{R_y}{2} \delta_{y0}\right) u_{jk}^n \qquad (5.8.37)$$

$$\left(1 + \frac{R_y}{4} \delta_{y0}\right) \Delta u_{jk} = \Delta u_{jk}^*. \qquad (5.8.38)$$

where $\Delta u_{jk} = u_{jk}^{n+1} - u_{jk}^n$.

5.8.4 Courant–Friedrichs–Lewy Condition for Two Dimensional Problems

Just as we had a CFL condition and results using the CFL condition to give us a necessary condition for stability for one dimensional schemes, we also have analogous results for multi-dimensional schemes. As with the one dimensional CFL condition, the multi-dimensional CFL condition is a coarse measure of stability and convergence. However, the CFL condition is always easy to compute and serves as an excellent upper bound for the stability condition. If for our computational requirements this limit is too severe, then we know that we must change either the stencil or go to an implicit scheme. For a given equation and stencil, we will often try to find an explicit difference scheme that has a stability condition that matches the CFL condition. We know that we should be able to find a scheme that does as well as the CFL condition. In this section we introduce the CFL condition for two dimensional equations. As we shall see, the definitions and results related to the two dimensional CFL condition are fairly obvious extensions for those for the one dimensional CFL condition. Results concerning the CFL condition for dimensions greater than two are easy extensions of the two dimensional results.

In Section 5.8 we defined the analytic domain of dependence for partial differential equation (5.8.1). As we did in the one dimensional case, we define the two dimensional numerical domain of dependence assuming that $R_x = a\Delta t/\Delta x$ and $R_y = b\Delta t/\Delta y$ are fixed. Under this assumption, *the numerical domain of dependence of a difference scheme at the point $(j\Delta x, k\Delta y, (n+1)\Delta t)$ is defined to be the smallest rectangle in $\mathbb{R}^2$ that contains all the points on which the solution of the difference scheme depends as Δt and Δx approach zero.*

Using the same argument as we did in Section 5.7, we see that if we consider difference schemes of the form

$$u_{jk}^{n+1} = a_1 u_{j-1\,k}^n + a_2 u_{j\,k-1}^n + a_3 u_{j\,k}^n + a_4 u_{j+1\,k}^n + a_5 u_{j\,k+1}^n, \qquad (5.8.39)$$

$$u_{jk}^{n+1} = a_3 u_{j\,k}^n + a_4 u_{j+1\,k}^n + a_5 u_{j\,k+1}^n, \qquad (5.8.40)$$

and

$$u_{jk}^{n+1} = a_1 u_{j-1\,k}^n + a_2 u_{j\,k-1}^n + a_3 u_{j\,k}^n, \qquad (5.8.41)$$

we obtain the numerical domains of dependence at the point $(j\Delta x, k\Delta y, (n+1)\Delta t)$

$$[(j-n-1)\Delta x, (j+n+1)\Delta x] \times [(k-n-1)\Delta y, (k+n+1)\Delta y], \qquad (5.8.42)$$

$$[j\Delta x, (j+n+1)\Delta x] \times [k\Delta y, (k+n+1)\Delta y] \qquad (5.8.43)$$

and

$$[(j-n-1)\Delta x, j\Delta x] \times [(k-n-1)\Delta y, k\Delta y], \qquad (5.8.44)$$

respectively.

As in the one dimensional case, for a given partial differential equation and an associated difference scheme, *the CFL condition is satisfied if the analytic domain of dependence is contained in the numerical domain of dependence.* The very important result that *satisfying the CFL condition is a necessary condition for convergence* follows from the same argument used in Section 5.7 to prove Theorem 5.7.2. And again as we did in Section 5.7, it is not difficult to see that difference schemes (5.8.39), (5.8.40) and (5.8.41) satisfy the CFL condition if

$$-1 \leq R_x \leq 1 \text{ and } -1 \leq R_y \leq 1, \qquad (5.8.45)$$

$$-1 \leq R_x \leq 0 \text{ and } -1 \leq R_y \leq 0 \qquad (5.8.46)$$

and

$$0 \leq R_x \leq 1 \text{ and } 0 \leq R_y \leq 1, \qquad (5.8.47)$$

respectively.

Thus we see that the two dimensional CFL conditions and the associated necessary conditions are very similar to the one dimensional results. We see that we immediately obtain the following results.

- The CFL condition for difference scheme (5.8.9) is satisfied if $0 \leq R_x \leq 1$ and $0 \leq R_y \leq 1$. Hence, $0 \leq R_x \leq 1$ and $0 \leq R_y \leq 1$ $(\max\{|R_x|, |R_y|\} \leq 1)$ is a necessary condition for convergence for difference scheme (5.8.9). Thus we see that the necessary and sufficient stability condition for difference scheme (5.8.9) found in Example 5.8.1, $0 \leq R_x + R_y \leq 1$, R_x, $R_y \geq 0$ is more restrictive than the CFL necessary condition.

- The CFL condition for the Lax-Friedrichs scheme (5.8.13) is satisfied if $\max\{|R_x|, |R_y|\} \leq 1$. Hence, this is also a necessary condition for convergence of difference scheme (5.8.13). Again the necessary and sufficient condition obtained in Example 5.8.2 is more restrictive than the necessary condition found using the CFL condition.

- The CFL condition for the time split scheme (5.8.21)–(5.8.22) is satisfied if $\max\{|R_x|, |R_y|\} \leq 1$. Hence, this is also a necessary condition for convergence of difference scheme (5.8.21)–(5.8.22). As we saw in Example 5.8.3, this necessary condition is the same as the necessary and sufficient condition for convergence of difference scheme (5.8.21)–(5.8.22).

HW 5.8.7 (a) Find the numerical domains of dependence at the point $(j\Delta x, k\Delta y, (n+1)\Delta t)$ for the following difference schemes.

1. $u_{j\,k}^{n+1} = a_2 u_{j\,k-1}^n + a_3 u_{j\,k}^n + a_4 u_{j+1\,k}^n$

2. $u_{j\,k}^{n+1} = a_1 u_{j-1\,k}^n + a_3 u_{j\,k}^n + a_4 u_{j+1\,k}^n + a_5 u_{j\,k+1}^n$

3. $u_{j\,k}^{n+1} = a_1 u_{j-1\,k}^n + a_3 u_{j\,k}^n + a_5 u_{j\,k+1}^n$

4. $u_{j\,k}^{n+1} = a_6 u_{j+1\,k+1}^n + a_7 u_{j+1\,k-1}^n + 8_4 u_{j-1\,k-1}^n + a_9 u_{j-1\,k+1}^n$

(b) Determine the CFL condition for the difference schemes given in part (a).

5.8.5 Two Dimensional Initial–Boundary–Value Problems

As is usually the case, when we proceed to consider initial–boundary–value problems for partial differential equation (5.8.1), things become more difficult. For example, consider equation (5.8.1) and initial condition (5.8.2) on $\overline{R} = [0, 1] \times [0, 1]$. We know that we will not generally have boundary conditions on all of the sides of the boundary. If we have periodic boundary conditions, we will have to wrap around the domain (now in two directions) as we did for one dimensional problems with periodic boundary conditions. With Dirichlet boundary conditions, we again must be aware of how many boundary conditions we have and where they are applied. For example, **if a and b are positive,** then we have boundary conditions of the form

$$v(0, y, t) = g_1(y, t), \ y \in [0, 1], \ t > 0 \qquad (5.8.48)$$

$$v(x, 0, t) = g_2(x, t), \ x \in [0, 1], \ t > 0. \qquad (5.8.49)$$

The decision on where the boundary conditions may be placed is the same as it was in the one dimensional case. Because a and b are both positive, the characteristic curves will be leaving the domain along the $x = 1$ and $y = 1$ boundaries. If we placed a boundary condition on either of these boundaries, the solution would have one value along the characteristic due to the point where the characteristic intersects the $t = 0$, $x = 0$ or $y = 0$ plane and another value (probably different) due to the prescribed boundary condition.

Hence, as was the case for the one dimensional schemes, if we use the appropriate one sided scheme, we have enough boundary conditions and they are in the correct place. If we use a centered scheme (such as Lax-Wendroff, Lax-Friedrichs, BTCS or Crank-Nicolson), we must provide a numerical boundary condition on the boundaries that do not have analytic boundary conditions. (See Section 5.6.2.) We shall return to the choice of these numerical boundary conditions in Chapter 8.

The implementation of two dimensional difference schemes for solving hyperbolic partial differential equations is not very much different from the implementation of difference schemes for parabolic equations. Except for the problems involving boundary conditions discussed above (whether we have enough or do we have to find a numerical boundary condition to use), implementation of two dimensional, explicit schemes for hyperbolic problems is the same as that for parabolic equations. (In fact, you can often use the same code, slightly altering the appropriate parts.) Also, as was the case with parabolic equations, implementation of a pure implicit or pure Crank-Nicolson scheme is usually not smart (the big matrix is just to large to solve).

When implementing ADI schemes for hyperbolic equations, the only difference with the parabolic counterpart is when we have periodic boundary conditions (which, for parabolic equations, we had only when we considered polar coordinates). If we return to the discussion given in Section 4.4.3.4, we recall that we can consider an ADI scheme such as (5.8.37)–(5.8.38) as either two big $(M_x - 1)(M_y - 1) \times (M_x - 1)(M_y - 1)$ matrices or one group of $M_y - 1$ $M_x - 1 \times M_x - 1$ matrices and one group of $M_x - 1$ $M_y - 1 \times M_y - 1$ matrices. If we briefly consider what the two big matrices will look like, it makes it clear that we should use the second approach.

Consider approximating the solution to problem (5.8.1)–(5.8.2) with periodic boundary conditions in both directions on $\bar{R} = [0, 1] \times [0, 1]$ using the Beam-Warming scheme, (5.8.34)–(5.8.35). If we consider applying this scheme for a grid with $M_x = 5$ and $M_y = 3$ (assuming the generalization to

$$
\left[
\begin{array}{ccccc|ccccc|ccccc}
1 & R/4 & 0 & 0 & -R/4 & 0 & 0 & 0 & 0 & 0 & 0 & 0 & 0 & 0 & 0 \\
-R/4 & 1 & R/4 & 0 & 0 & 0 & 0 & 0 & 0 & 0 & 0 & 0 & 0 & 0 & 0 \\
0 & -R/4 & 1 & R/4 & 0 & 0 & 0 & 0 & 0 & 0 & 0 & 0 & 0 & 0 & 0 \\
0 & 0 & -R/4 & 1 & R/4 & 0 & 0 & 0 & 0 & 0 & 0 & 0 & 0 & 0 & 0 \\
R/4 & 0 & 0 & -R/4 & 1 & 0 & 0 & 0 & 0 & 0 & 0 & 0 & 0 & 0 & 0 \\
\hline
0 & 0 & 0 & 0 & 0 & 1 & R/4 & 0 & 0 & -R/4 & 0 & 0 & 0 & 0 & 0 \\
0 & 0 & 0 & 0 & 0 & -R/4 & 1 & R/4 & 0 & 0 & 0 & 0 & 0 & 0 & 0 \\
0 & 0 & 0 & 0 & 0 & 0 & -R/4 & 1 & R/4 & 0 & 0 & 0 & 0 & 0 & 0 \\
0 & 0 & 0 & 0 & 0 & 0 & 0 & -R/4 & 1 & R/4 & 0 & 0 & 0 & 0 & 0 \\
0 & 0 & 0 & 0 & 0 & R/4 & 0 & 0 & -R/4 & 1 & 0 & 0 & 0 & 0 & 0 \\
\hline
0 & 0 & 0 & 0 & 0 & 0 & 0 & 0 & 0 & 0 & 1 & R/4 & 0 & 0 & -R/4 \\
0 & 0 & 0 & 0 & 0 & 0 & 0 & 0 & 0 & 0 & -R/4 & 1 & R/4 & 0 & 0 \\
0 & 0 & 0 & 0 & 0 & 0 & 0 & 0 & 0 & 0 & 0 & -R/4 & 1 & R/4 & 0 \\
0 & 0 & 0 & 0 & 0 & 0 & 0 & 0 & 0 & 0 & 0 & 0 & -R/4 & 1 & R/4 \\
0 & 0 & 0 & 0 & 0 & 0 & 0 & 0 & 0 & 0 & R/4 & 0 & 0 & -R/4 & 1 \\
\end{array}
\right]
$$

FIGURE 5.8.2. Coefficient matrix associated the Beam-Warming scheme, solving in the x-direction with $M_x = 5$ and $M_y = 3$. The partitioning included separates the matrix according to blocks associated with the k-lines, $k = 0$, 1 and 2.

large values of M_x and M_y is clear) using the "large matrix approach", we must solve two matrix equations involving a 15×15 matrix. The coefficient matrix for solving in the x-direction is given in Figure 5.8.2. The matrix is not tridiagonal (but we should not have expected it to be tridiagonal for periodic boundary conditions) and the structure does not appear to be especially nice.

Using the "small matrix approach" will involve solving three matrix equations in the x-direction involving a 5×5 matrix of the form

$$
\begin{pmatrix}
1 & R/4 & 0 & 0 & -\frac{R}{4} \\
-R/4 & 1 & R/4 & 0 & 0 \\
0 & -R/4 & 1 & R/4 & 0 \\
0 & 0 & -R/4 & 1 & R/4 \\
\frac{R}{4} & 0 & 0 & -\frac{R}{4} & 1
\end{pmatrix}
\tag{5.8.50}
$$

and five matrix equations in the y-direction involving 3×3 matrices similar to matrix (5.8.50) (except smaller). The form of matrix (5.8.50) is made rather obvious by the partitions included in the matrix given in Figure 5.8.2. Of course, each of the eight matrix equations that must be solved when we use the "small matrix approach" is solved using the Sherman-Morrison Algorithm as described in Section 5.6.1.

One last warning should be included when we use either the "large" or "small" matrix approach. Filling the right hand side of the matrix equations is reasonably clear and easy except when we are filling the right hand side associated with solving in the x-direction for the $j = 0$ terms and the $j = M_x - 1$ terms and on all terms associated with $k = 0$ (assuming that we solve on $j = 0$ and $k = 0$ and not on $j = M_x$ and $k = M_y$). When we use the right hand side of equation (5.8.34) and reach to $j = -1$, $j = M_y$ or $k = -1$, we must remember to use the periodicity and replace that u^n value by the one associated with $j = M_x - 1$, $j = 0$ and $k = M_y - 1$, respectively.

5.9 Computational Interlude III

5.9.1 Review of Computational Results

Hopefully by this time, you have completed computations with the parabolic equations. After doing the work necessary to program both the explicit schemes and the ADI schemes, surely you realize one of the advantages of explicit schemes over implicit schemes.

Also, hopefully, you have seen that there are advantages in using the ADI schemes rather than the explicit schemes. The major advantage of the ADI scheme is the lack of stability limitations compared to their explicit counterparts. Of course, *we must always be aware that when the larger time steps are taken, the temporal accuracy is decreased.* This was made very clear in the results of HW4.4.8 where we solved a parabolic partial differential equation using the Douglas-Rachford scheme with a time step of $\Delta t = 0.01$. Given that the Douglas-Rachford scheme is only first order accurate in time, we cannot expect the solution to be highly accurate. If we compare the results from HW4.4.8 to the exact answers and those from HW4.4.3 using the Peaceman-Rachford scheme, we see that the answers from the Douglas-Rachford scheme barely give two places of accuracy. This should not be surprising considering the order of accuracy of the two schemes.

One other consideration of whether an ADI scheme or explicit scheme should be used is the amount of computation time needed for each scheme. Of course, if the solution desired is to be done just once, then the computational cost may not be such a great consideration. Clearly, if the scheme is intended for frequent use, such as a part of a production code, then the cost of the computation is very important.

If you compare the explicit scheme with any of the ADI schemes, it is clear that the ADI schemes require many more operations. The amount of extra computational cost is not trivial. In the past, it would be assumed that all of the operations were done in a scalar mode and an operation count would be used. Since the relative speeds of certain operations differ greatly, care must be taken when interpreting the operation count. Ignoring the differing speeds of different operations and considering only an operation count, an implicit scheme generally contains approximately twelve times as many operations than an analogous explicit scheme. In the case of vector or parallel machines, the situation is much different. The multiple can vary from 12 to almost any large number, depending on the number of points in each direction (and hence, specifically on the amount of parallelism available).

In any case, it is safe to assume that the cost of an ADI scheme is in the neighborhood of 10–20 times as expensive as the explicit scheme. Hence, *to justify using an ADI scheme, the user must be able to use a time step that is 10–20 times the allowable time step of the explicit scheme. The determining factor on whether we are able to use the larger time step is the accuracy that is necessary (or desired) in the calculation.* We saw in HW4.2.1 that for stability, the time step would have to be nearly 0.0001 (really, 0.000625 would also work). In HW4.4.3, using the Peaceman-Rachford scheme, the results using $\Delta t = 0.01$ are actually better than those found in HW4.2.1 (Recall that the explicit scheme is first order in time while the Peaceman-Rachford scheme is second order in time.) The fact that with the implicit scheme we have been able to use many fewer time steps (one one-hundredth of the number used with the explicit scheme), indicates that computing costs have been lowered. These are among the considerations that must be made when choosing a difference scheme for solving a particular problem.

5.9.2   Convection–Diffusion Equations

In Example 3.1.3, we saw that if we tried to solve partial differential equation

$$v_t + av_x = \nu v_{xx} \tag{5.9.1}$$

by the difference scheme

$$u_k^{n+1} = u_k^n - \frac{R}{2}\delta_0 u_k^n + r\delta^2 u_k^n, \tag{5.9.2}$$

we must satisfy the stability condition $R^2/2 \leq r \leq 1/2$. We pointed out that if a was significantly larger than ν, this condition could be difficult to satisfy. In Section 4.3.1, we showed that this was also the case for the analogous two dimensional problem. And, finally, we saw in Section 2.6.2 that when we considered HW0.0.1 for small ν, it was impossible to make Δt small enough to stabilize the nonlinear analog to difference scheme (5.9.2).

The difficulty encountered in the above cases is that when the convective term dominates the diffusive term, we must use a difference equation that is designed for a hyperbolic equation. For example, when we want to approximate the solution to a problem involving partial differential equation (5.9.1) where the convective term dominates the diffusive term (say for reasonable sizes of Δt and Δx, $R^2/2 > r$), it might be logical to try a difference scheme of the form

$$u_k^{n+1} = u_k^n - R\delta_+ u_k^n + r\delta^2 u_k^n \text{ if } a < 0, \tag{5.9.3}$$

$$u_k^{n+1} = u_k^n - R\delta_- u_k^n + r\delta^2 u_k^n \text{ if } a > 0, \tag{5.9.4}$$

or

$$u_k^{n+1} = u_k^n - \frac{R}{2}\delta_0 u_k^n + \frac{R^2}{2}\delta^2 u_k^n + r\delta^2 u_k^n \tag{5.9.5}$$

for either sign of a. Hopefully it is clear that difference schemes (5.9.3)–(5.9.5) were obtained by treating the av_x term in partial differential equation (5.9.1) as we did in the BTFS, BTBS and Lax-Wendroff schemes for hyperbolic partial differential equations. It should also be clear that there are other "hyperbolic like" schemes that can be developed for approximating the solutions to partial differential equation (5.9.1).

The above schemes can be used, they can be helpful in certain situations, but they must still be applied carefully. We see in HW5.9.1 that the stability limits have been improved just as we had hoped. When ν is small, the problem in the past was that r was small (and it was difficult to satisfy $R^2/2 \leq r$). Now when r is small, it is very easy to satisfy the stability conditions for difference schemes (5.9.3)–(5.9.5).

The potential problem with applying difference schemes (5.9.3)–(5.9.5) is a problem that we have with many of our solution schemes and is one that we have not yet addressed (but we will consider these problems completely in Chapter 7). However, here the problem seems more pronounced. We started looking for a new scheme for approximating the solutions to partial differential equation (5.9.1) because the νv_{xx} term was not dominant (r was not large enough). However, as we shall see in Chapter 7, in approximating the av_x term by $\delta_+ u_k^n/\Delta x$ or $\delta_- u_k^n/\Delta x$, we are adding a term like $\nu_1 v_{xx}$ to our equation (and this is what gives us the stability). The bad part of this additional dissipation is that the $\nu_1 v_{xx}$ term will swamp the νv_{xx} term. Hence, using either difference scheme (5.9.3) or (5.9.4) is very much like saying, "because ν is not big enough to use difference scheme (5.9.2), we should make ν bigger."

Difference scheme (5.9.5) appears to have the same problem. The stabilizing term added to the Lax-Wendroff scheme is and looks like an additional dissipative term. However, as we saw in our derivation of the Lax-Wendroff scheme, the effect of the added $(R^2/2)\delta^2 u_k^n$ term is to eliminate the v_{tt} term in the Taylor series expansion. Hence, the added dissipative term in difference scheme (5.9.5) should not swamp the νv_{xx} term in partial differential equation (5.9.1) and difference scheme (5.9.5) should do a good job in modeling the dissipation of equation (5.9.1). (See HW5.9.2).

HW 5.9.1 (a) Show that difference scheme (5.9.3) is stable if and only if $R \le 2r \le 1 + R$. (Hint: Use the technique introduced in Remark 2, page 109 to show the the above condition is a necessary condition. Then show that if this condition is violated, the scheme will be unstable.)
(b) Show that difference scheme (5.9.4) is stable if and only if $-2r \le R \le 1 - 2r$.
(b) Show that difference scheme (5.9.5) is stable if and only if $r + R^2/2 \le 1/2$.

HW 5.9.2 Use difference schemes (5.9.4) and (5.9.5) to approximate the solutions to the following initial–boundary–value problem.

$$v_t + v_x = \nu v_{xx}, \ t > 0, \ x \in (0,1)$$
$$v(0,t) = v(1,t) = 0, \ t > 0$$
$$v(x,0) = 3\sin 4\pi x, \ x \in [0,1]$$

Let $\nu = 0.00001$ and use $M = 100$, $\Delta t = 0.005$ and plot solutions at times $t = 0.1$, $t = 0.5$, $t = 1.0$ and $t = 2.0$.

5.9.3 HW0.0.1

If we review the approaches that we have taken to solve HW0.0.1 to this time in light of what we have learned in this chapter, hopefully, we have

some idea what we have been doing wrong. In Section 3.4.2, we discussed
the fact that for small values of ν, the explicit scheme was not good because of the stability condition necessary on the analogous scheme for the
linear model problem, (3.1.31). We then reasoned that this was a stability problem and tried to solve the problem using implicit schemes. As we
have probably seen by now, unless a very fine grid was used, the implicit
schemes do not perform much better than the explicit scheme. From this
chapter and specifically from the discussions in Section 5.9.2, we see that
the model problem for the first order derivative term in Burgers' equation
is the one way wave equation, equation (5.1.4). When ν is very small, Burgers' equation (and even more so, any discrete approximation to Burgers'
equation) acts like a hyperbolic equation. Hence, we must try to use hyperbolic schemes to approximate the solution to Burgers' equation. From
our discussion in Section 5.9.2, it seems clear that it is logical to try using
the nonlinear analog to difference scheme (5.9.5),

$$u_k^{n+1} = u_k^n - \frac{R}{2} u_k^n \delta_0 u_k^n + \left(r + \frac{R^2}{2} (u_k^n)^2 \right) \delta^2 u_k^n, \qquad (5.9.6)$$

to solve HW0.0.1.

The reasoning used above is completely based on using the one way wave
equation as a model equation for the first two terms of the viscous Burgers'
equation. Obviously, this reasoning is directly applicable to the inviscid
Burgers' equation. For this reason, and because any success we have in
approximating the solution to the inviscid Burgers' equation should carry
over to approximating solutions to the viscous Burgers' equation for very
small values of ν, we suggest that the reader leave HW0.0.1 temporarily
and proceed to work on HW0.0.2 (Section 5.9.4).

5.9.4   HW0.0.2

As we stated in the previous section, the one way wave equation is a very
logical model equation for the inviscid Burgers' equation. Hence, we suggest
that you approximate the inviscid Burgers' equation by a discrete equation
that is analogous to one of the schemes discussed and developed in this
chapter. Since v in the inviscid Burgers' equation takes the place of the
a in equation (5.1.4), we know that we cannot use either of the one sided
schemes (since v will change sign in the interval). We suggest that you try
the following analog to the Lax-Wendroff scheme (the inviscid analog to
difference scheme (5.9.6)),

$$u_k^{n+1} = u_k^n - \frac{R}{2} u_k^n \delta_0 u_k^n + \frac{R^2}{2} (u_k^n)^2 \delta^2 u_k^n. \qquad (5.9.7)$$

Of course, there are other schemes that we could try and other approaches that we could use to derive a Burgers' equation analog to one

of our hyperbolic difference schemes. For example, we could try any of the implicit schemes given in Section 5.2.2, including using Newton's method to solve the nonlinear problem as we did in Section 3.4.2. However, difference scheme (5.9.7) is an excellent scheme to try first in that it is clearly a good candidate, and it is easy to implement.

Before we leave this subject, we should at least mention boundary conditions. Since the inviscid Burgers' equation is a first order hyperbolic partial differential equation, we might expect that we would get only one boundary condition. However, difference scheme (5.9.7) requires more than one boundary condition and in the problem statement, we are given two boundary conditions. As we shall see in Chapter 9, both $x = 0$ and $x = 1$ will be characteristic curves for Burgers' equation with the given initial condition (of course, for this we need to define characteristic curves for nonlinear partial differential equations). Though it appears dangerous from what we have seen for linear hyperbolic equations, it is correct to use zero Dirichlet boundary conditions at both ends of your interval.

6

Systems of Partial Differential Equations

6.1 Introduction

As we discussed earlier, many applications of partial differential equations involve systems of partial differential equations. The approach that we used in the introduction to Chapter 5 was to uncouple a system to motivate the importance of the model equation

$$v_t + av_x = 0.$$

Most often when we face systems of partial differential equations, we do not want to or are unable to uncouple the system. The uncoupling is more of a theoretical tool than it is a practical tool. Hence, we must be able to solve systems as a system of equations and be able to analyze the difference schemes for these systems of equations.

We begin by considering an equation of the form

$$\mathbf{v}_t = \left\{ \sum_{j=0}^{J} B_j \frac{\partial^j}{\partial x^j} \right\} \mathbf{v} + \mathbf{F}, \qquad (6.1.1)$$

where $\mathbf{v} = \mathbf{v}(x, t) = [v_1, \cdots, v_K]^T$, $\mathbf{F} = \mathbf{F}(x, t) = [F_1, \cdots, F_K]^T$ and B_j is a $K \times K$ matrix. (It is possible to obtain some results for and solve non-constant matrices, but we shall not do so here.) For example, if $J = 2$, $K = 3$ and B_2 is positive definite, then

$$\mathbf{v}_t = B_2 \mathbf{v}_{xx} + B_1 \mathbf{v}_x + B_0 \mathbf{v} + \mathbf{F} \qquad (6.1.2)$$

denotes a nonhomogeneous, second order parabolic system of three equations in one spatial unknown. Likewise, if A is a diagonalizable, constant matrix, then for $J = 1$ we get the first order hyperbolic system of partial differential equations of the form

$$\mathbf{v}_t = A\mathbf{v}_x + B_0\mathbf{v}.$$

As we did in the case with scalar equations, we shall generally restrict our work to first and second order equations. The solutions to an equation of the form of (6.1.2) behave much like their scalar counterparts. The solutions are generally dominated by the diffusive term (the second derivative with respect to x term). If, in some fashion the B_1 term is larger than the B_2 term, then the solutions of the equation behave more like solutions to hyperbolic equations and we would have to treat them as if they were hyperbolic equations.

The solutions of hyperbolic systems of partial differential equations also behave similar to their scalar counterparts. If we consider a hyperbolic system of the form

$$\mathbf{v}_t = A\mathbf{v}_x, \ x \in \mathbb{R}, \ t > 0, \tag{6.1.3}$$

and return to the analysis performed in Section 5.1, we see that equation (6.1.3) is equivalent to solving the uncoupled system of equations

$$V_{j_t} - \mu_j V_{j_x} = 0, \ x \in \mathbb{R}, \ t > 0, \ j = 1, \cdots, K. \tag{6.1.4}$$

We know that equations of the form of (6.1.4) have a solution of the form

$$V_j(x,t) = V_{0_j}(x + \mu_j t) \tag{6.1.5}$$

where $\mathbf{V}_0 = [V_{0_1} \ \cdots \ V_{0_K}]^T$ denotes the initial condition for the system (6.1.4) (and is given in terms of the initial condition for the primitive variables $\mathbf{v}_0$ by $\mathbf{V}_0 = S\mathbf{v}_0$). The solutions (6.1.5) can then be recoupled to provide a solution to the original initial–value problem (6.1.3) along with $\mathbf{v}(x,0) = \mathbf{v}_0(x)$ as

$$\mathbf{v}(x,t) = S^{-1}\mathbf{V} = S^{-1} \begin{bmatrix} V_{0_1}(x + \mu_1 t) \\ \vdots \\ V_{0_K}(x + \mu_K t) \end{bmatrix}. \tag{6.1.6}$$

We note in solution (6.1.6) that though they are coupled, the solution consists of K waves, propagating with speed $-\mu_j$, $j = 1, \cdots, K$. As in the case of scalar hyperbolic partial differential equations, the phase velocities for the solutions of systems of hyperbolic partial differential equations must also be approximated. The difference is that we know have K phase velocities to approximate and that each component of the solution will involve sums of these propagating waves. It should be clear that it will be much more difficult to recognize or see these propagating waves when they are coupled together as in solution (6.1.6).

For a complete theoretical discussion of initial–value problems and initial–boundary–value problems for systems of partial differential equations, see ref. [5].

6.2 Initial–Value Difference Schemes

We next consider equation (6.1.1) on all of $\mathbb{R}$ along with the initial condition

$$\mathbf{v}(x,0) = \mathbf{f}(x) = (f_1(x), \cdots, f_K(x))^T. \qquad (6.2.1)$$

As usual, we place a grid on $\mathbb{R}$ with grid spacing Δx and consider functions defined at the n–th time level on the grid of the form $\mathbf{u}_k^n$, $-\infty < k < \infty$ where $\mathbf{u}_k^n = [u_{1_k}^n, \cdots, u_{K_k}^n]^T$. We define $\mathbf{u}^n$ to be the vector (of vectors) $[\cdots, \mathbf{u}_{-1}^n, \mathbf{u}_0^n, \mathbf{u}_1^n, \cdots]^T$. We can then let $\mathbf{u}_k^0 = \mathbf{f}(k\Delta x)$, $-\infty < k < \infty$ and write a general two level difference scheme in the form

$$Q_1 \mathbf{u}_k^{n+1} = Q_2 \mathbf{u}_k^n + \Delta t \mathbf{G}_k^n, \qquad (6.2.2)$$

where

$$Q_1 = \sum_{j=-m_3}^{m_4} Q_{1j} S_+^j, \qquad (6.2.3)$$

$$Q_2 = \sum_{j=-m_1}^{m_2} Q_{2j} S_+^j, \qquad (6.2.4)$$

and Q_{1j}, Q_{2j} are $K \times K$ matrices. We note that when we defined the shift operator $S_\pm$ in Section 3.1.1, they were defined to work on the element in $\ell_{2,\Delta x}$ which is an infinitely long vector. However, in equations (6.2.3) and (6.2.4) a slight variation of the notation is used in that though $S_\pm$ is still applied to the entire vector, the vector in question is now a K-vector. By $S_\pm \mathbf{u}_k$ we mean $\mathbf{u}_{k\pm 1}$. Also, we use the notation that $S_+^{-j} = S_-^j$, where j is a positive integer. At certain times, it will be convenient to difference scheme (6.2.2)–(6.2.4) as an operator equation on $\ell_{2\,\Delta x}$. We can then express difference scheme (6.2.2) as

$$\hat{Q} \mathbf{u}^{n+1} = Q \mathbf{u}^n + \Delta t \mathbf{G}^n, \qquad (6.2.5)$$

or

$$\mathbf{u}^{n+1} = Q \mathbf{u}^n + \Delta t \mathbf{G}^n. \qquad (6.2.6)$$

For example, consider the potential difference scheme for solving partial differential equation (6.1.2),

$$\frac{\mathbf{u}_k^{n+1} - \mathbf{u}_k^n}{\Delta t} = \frac{1}{2\Delta x^2} B_2(\mathbf{u}_k^{n+1} - 2\mathbf{u}_{k+1}^{n+1} + \mathbf{u}_{k+2}^{n+1}) + \frac{1}{2\Delta x^2} B_2(\mathbf{u}_k^n$$
$$- 2\mathbf{u}_{k+1}^n + \mathbf{u}_{k+2}^n) + \frac{1}{4\Delta x} B_1(\mathbf{u}_{k+1}^{n+1} - \mathbf{u}_{k-1}^{n+1})$$
$$+ \frac{1}{4\Delta x} B_1(\mathbf{u}_{k+1}^n - \mathbf{u}_{k-1}^n) + \Delta t B_0 \mathbf{u}_k^n$$

(obviously not a good scheme for solving (6.1.2)). If we rewrite the above difference scheme as

$$\frac{R}{4} B_1 \mathbf{u}_{k-1}^{n+1} + \left(I - \frac{r}{2} B_2\right) \mathbf{u}_k^{n+1} + \left(r B_2 - \frac{R}{4} B_1\right) \mathbf{u}_{k+1}^{n+1} + \frac{-r}{2} B_2 \mathbf{u}_{k+2}^{n+1} =$$
$$\frac{-R}{4} B_1 \mathbf{u}_{k-1}^n + \left(I + \frac{r}{2} B_2 + \Delta t B_0\right) \mathbf{u}_k^n + \left(-r B_2 + \frac{R}{4} B_1\right) \mathbf{u}_{k+1}^n + \frac{r}{2} B_2 \mathbf{u}_{k+2}^n,$$

then the $Q_{\ell j}$'s can be copied off as (with $m_2 = m_4 = 2$ and $m_1 = m_3 = 1$)

$$Q_{1-1} = \frac{R}{4} B_1, \ Q_{10} = I - \frac{r}{2} B_2, \ Q_{11} = r B_2 - \frac{R}{4} B_1, \text{ and } Q_{12} = \frac{-r}{2} B_2,$$

and

$$Q_{2-1} = \frac{-R}{4} B_1, \ Q_{20} = I + \frac{r}{2} B_2 + \Delta t B_0, \ Q_{21} = -r B_2 + \frac{R}{4} B_1,$$

and

$$Q_{22} = \frac{r}{2} B_2.$$

The definitions of consistency, stability and convergence are the same as they were in the scalar case. The only modification that must be made is that we must use a different norm. Definitions 2.2.2 and 2.3.2–2.4.1 were all given in terms of a general norm. Most often, when we worked with initial–value problems, we used the energy norm, (2.2.28). For systems, the norms given in Definitions 2.2.2 and 2.3.2–2.4.1 are replaced by the vector version of the energy norm, (2.2.37). Then, using the vector norm, we see that *the Lax Theorem (Theorem 2.5.2) also applies to systems of equations*. Thus, as before, *the job of proving convergence is replaced by showing consistency and stability*. Again, the consistency is relatively easy to show and we proceed to discuss methods for showing that difference schemes for systems of partial differential equations are stable.

The basic tool for proving a difference scheme is stable is to use the discrete Fourier transform. As is the case in Chapter 3, we define the discrete Fourier transform of $\mathbf{u}$ (for $\mathbf{u} = [\cdots, \mathbf{u}_{-1}, \mathbf{u}_0, \mathbf{u}_1, \cdots]^T$ and $\mathbf{u}_k = [u_{1_k}, \cdots, u_{K_k}]^T$) as

$$\mathcal{F}(\mathbf{u})(\xi) = \hat{\mathbf{u}}(\xi) = \frac{1}{\sqrt{2\pi}} \sum_{k=-\infty}^{\infty} e^{-ik\xi} \mathbf{u}_k \tag{6.2.7}$$

for $\xi \in [-\pi, \pi]$, and the inverse discrete Fourier transform as

$$\mathbf{u}_k = \frac{1}{\sqrt{2\pi}} \int_{-\pi}^{\pi} e^{ik\xi} \hat{\mathbf{u}}(\xi) d\xi \tag{6.2.8}$$

for $-\infty < k < \infty$. Then Parseval's Identity ($\| \mathbf{u} \|_2 = \| \hat{\mathbf{u}} \|_2$ where the first two-norm is the vector ℓ_2 norm and the second two-norm is the vector L_2 norm on $[-\pi, \pi]$) still holds and we can transfer the job of proving stability of our difference schemes from the sequence space to the transform space. Also, as before, we still have

$$\mathcal{F}(S_{\pm}\mathbf{u}) = e^{\pm i\xi} \mathcal{F}(\mathbf{u}). \tag{6.2.9}$$

The approach that we use to analyze the stability is to take the discrete Fourier transform of equation (6.2.2) (where, as usual, for the stability argument we eliminate the nonhomogeneous term) to get

$$\sum_{j=-m_3}^{m_4} Q_{1j} e^{ij\xi} \hat{\mathbf{u}}^{n+1} = \sum_{j=-m_1}^{m_2} Q_{2j} e^{ij\xi} \hat{\mathbf{u}}^n \tag{6.2.10}$$

or

$$\hat{\mathbf{u}}^{n+1}(\xi) = \left(\sum_{j=-m_3}^{m_4} Q_{1j} e^{ij\xi} \right)^{-1} \left(\sum_{j=-m_1}^{m_2} Q_{2j} e^{ij\xi} \right) \hat{\mathbf{u}}^n(\xi). \tag{6.2.11}$$

We define

$$G(\xi) = \left(\sum_{j=-m_3}^{m_4} Q_{1j} e^{ij\xi} \right)^{-1} \left(\sum_{j=-m_1}^{m_2} Q_{2j} e^{ij\xi} \right) \tag{6.2.12}$$

to be the **symbol** or **amplification matrix** of the difference scheme. We see that

$$\hat{\mathbf{u}}^{n+1}(\xi) = G(\xi) \hat{\mathbf{u}}^n \tag{6.2.13}$$
$$= G^{n+1}(\xi) \hat{\mathbf{u}}^0. \tag{6.2.14}$$

So, the growth of $\hat{\mathbf{u}}^n$ (and $\mathbf{u}^n$) and, hence, the stability of the scheme, depends on the growth of the amplification matrix raised to the n-th power, G^n.

Remark 1: In taking the discrete Fourier transform, we have transformed our problem from a problem in a vector $\ell_{2,\Delta x}$ space (which is the space in which we are doing our stability analysis) containing numerical spatial derivatives to a problem (with no spatial variations) in a vector L_2 space. As is the case when we transform partial differential equations, our problem has been made easier because we have transformed away the spatial derivatives.

Remark 2: Our problem has been made easier because we do not have to work with operators defined on the vector $\ell_{2,\Delta x}$ space. The operators with which we must work have been transformed from infinite dimensional operators defined on $\ell_{2,\Delta x}$ to a matrix operator that depends on the parameter $\xi \in [-\pi, \pi]$. Since the vector L_2 space is also a difficult space, it is not obvious that the problem has been made easier. But, as we will see below, the problem is made easier because we are able to work in this latter space pointwise in ξ.

If we then apply Parseval's Identity to the definition of stability, equation (6.2.13)–(6.2.14) gives us the following theorem which is analogous to Proposition 3.1.6 for scalar equations.

Theorem 6.2.1 *Difference scheme (6.2.2) is stable with respect to the $\ell_{2,\Delta x}$ norm if and only if there exists positive constants Δx_0 and Δt_0 and non-negative constants K and β so that*

$$\| G^n(\xi) \|_2 \leq K e^{\beta t}, \tag{6.2.15}$$

for $0 < \Delta x \leq \Delta x_0$, $0 < \Delta t \leq \Delta t_0$, $t = n\Delta t$ and all $\xi \in [-\pi, \pi]$.

We should specifically note that the norm used on G in (6.2.15) is the matrix 2-norm, based on the fact that G operates on K-vectors. The variable ξ in the matrix is treated as a parameter and is not integrated out. The proof of the above theorem is analogous to the proof of Proposition 3.1.6 and depends heavily on the assumption that $G(\xi)$ is a smooth function of ξ (at least continuous). Theorem 6.2.1 is also analogous to Proposition 3.1.6 in that it relates stability to the growth of the symbol of the difference operator. However, the setting in Chapter 3 was much easier since then the symbol was a scalar function of ξ. The fact that our symbol is now a matrix requires us to be very careful and, ultimately, work much harder for stability. The result that is similar to Theorem 3.1.7 that we are able to prove is as follows.

Theorem 6.2.2 *If the difference scheme (6.2.2) is stable in the $\ell_{2,\Delta x}$ norm, then there exists positive constants Δx_0, Δt_0 and C, independent of Δt, Δx and ξ, so that*

$$\sigma(G(\xi)) \leq 1 + C\Delta t \tag{6.2.16}$$

for $0 < \Delta t \leq \Delta t_0$, $0 < \Delta x \leq \Delta x_0$ and all $\xi \in [-\pi, \pi]$.

We note that $\sigma(G(\xi))$ denotes the spectral radius of the amplification matrix G. The proof of Theorem 6.2.2 follows readily from Theorem 6.2.1 if we use the fact that

$$\sigma(G(\xi)) \leq \| G(\xi) \|_2 \tag{6.2.17}$$

and the same kind of contradiction argument used in the proof of Proposition 3.1.7. The fact that (6.2.17) is an inequality instead of an equality is what stops us from using the other direction of Theorem 6.2.1 ((6.2.15) implies stability) to prove the converse of Theorem 6.2.2. We should realize that when G is normal ($G^*G = GG^*$), then $\sigma(G(\xi)) = \parallel G(\xi) \parallel_2$ and we are able to prove the converse of Theorem 6.2.2.

Theorem 6.2.3 *Suppose G is normal and there exist positive constants Δx_0, Δt_0 and C, independent of Δx, Δt and ξ, so that $\sigma(G(\xi)) \leq 1 + C\Delta t$ for $0 < \Delta t \leq \Delta t_0$, $0 < \Delta x \leq \Delta x_0$ and all $\xi \in [-\pi, \pi]$. Then difference scheme (6.2.2) is stable.*

And finally, we say that *when G satisfies (6.2.16), G satisfies the* **von Neumann condition**. We must be careful to realize that *we obtain a weaker result involving the von Neumann condition for systems than we do when using the von Neumann condition for scalar equations. Satisfying the von Neumann condition for systems is generally a necessary condition (not sufficient) for stability.*

We next provide an example that shows that the von Neumann condition is not sufficient for stability.

Example 6.2.1 Consider the difference scheme

$$u_{1_k}^{n+1} = u_{1_k}^n + \delta^2 u_{2_k}^n \tag{6.2.18}$$

$$u_{2_k}^{n+1} = u_{2_k}^n, \tag{6.2.19}$$

or

$$\mathbf{u}_k^{n+1} = \mathbf{u}_k^n + \begin{pmatrix} 0 & 1 \\ 0 & 0 \end{pmatrix} \delta^2 \mathbf{u}_k^n. \tag{6.2.20}$$

Show that the difference scheme (6.2.20) satisfies the von Neumann condition, yet is not stable.

Solution: If we take the discrete Fourier transform of difference scheme (6.2.20), we see that the amplification matrix is

$$G(\xi) = \begin{pmatrix} 1 & -4\sin^2 \frac{\xi}{2} \\ 0 & 1 \end{pmatrix}. \tag{6.2.21}$$

Then since the eigenvalues of G are obviously both 1, it is easy to see that G *satisfies the von Neumann condition*. It is also easy to see inductively that

$$G^n(\xi) = \begin{pmatrix} 1 & -4n\sin^2 \frac{\xi}{2} \\ 0 & 1 \end{pmatrix}.$$

If $\parallel G^n(\xi) \parallel_2$ is to be bounded, it must be bounded for all n and ξ (including $\xi = \pi$) and satisfy

$$\parallel G^n(\pi)\hat{\mathbf{u}} \parallel_2 \leq \parallel G^n(\pi) \parallel_2 \parallel \hat{\mathbf{u}} \parallel$$

for any $\hat{\mathbf{u}}$ (including $\hat{\mathbf{u}} = [0\ 1]^T$). Clearly, if we let $n \to \infty$, we see that

$$\parallel G^n(\pi) \begin{bmatrix} 0 \\ 1 \end{bmatrix} \parallel_2 = \parallel \begin{bmatrix} -4n \\ 1 \end{bmatrix} \parallel_2 = \sqrt{16n^2 + 1} \to \infty,$$

while $\| \ [0 \quad 1]^T \ \|_2 = 1$. Thus, clearly, $\| \ G^n(\pi) \ \|_2$ is not bounded. Hence, $\| \ G^n(\xi) \ \|_2$ is surely not bounded and the scheme is unstable. We have *an example of an unstable difference scheme that satisfies the von Neumann condition.*

Remark 1: We also should note that if we find the eigenvectors associated with the eigenvalue 1, we see that there is only one eigenvector. Thus the matrix has a nontrivial Jordan canonical form and can not be diagonalized. In Theorem 6.2.5 we shall see that since G satisfies the von Neumann condition, if G were similar to the diagonal matrix with eigenvalues on the diagonal, then the scheme would be stable. *Amplification matrix (6.2.21) is not similar to a diagonal matrix with the eigenvalues on the diagonal.*

Remark 2: And finally, we note that this example is especially nice in that the instability can also be analyzed in sequence space. The reason that we take the discrete Fourier transform is that we do not want to work in the sequence space. We examine it here because it is one of the few examples that can be treated in sequence space.

We begin by noting that the difference scheme is described by the operator Q defined on the space $\ell_{2\,\Delta x}$. We can represent this operator as an infinite matrix where a general row is given by

$$\cdots \quad \begin{pmatrix} 0 & 1 \\ 0 & 0 \end{pmatrix} \quad \begin{pmatrix} 1 & -2 \\ 0 & 1 \end{pmatrix} \quad \begin{pmatrix} 0 & 1 \\ 0 & 0 \end{pmatrix} \quad \cdots \tag{6.2.22}$$

A careful calculation will then show that the general row of Q^n will be

$$\cdots \quad n\begin{pmatrix} 0 & 1 \\ 0 & 0 \end{pmatrix} \quad \begin{pmatrix} 1 & -2n \\ 0 & 1 \end{pmatrix} \quad n\begin{pmatrix} 0 & 1 \\ 0 & 0 \end{pmatrix} \quad \cdots \tag{6.2.23}$$

Then choosing $\mathbf{u}$ in $\ell_{2\,\Delta x}$ so that it has the vector $[0\,1]$ in the kth component and zeros elsewhere, we see that we get $\| \ Q^n \mathbf{u} \ \|_{2,\Delta x} = \sqrt{(1 + 6n^2)\Delta x}$. And finally, using the same reasoning used in the transform space, we can show that $\| \ Q^n \ \|_{2,\Delta x}$ is not bounded.

Theorem 6.2.1 in an excellent result and Theorem 6.2.2 can be useful to show that a scheme is not stable. However, at this time we have only the restrictive Theorem 6.2.3 to show that a scheme is stable. We recall that in the scalar case, we did have trouble showing stability for initial–boundary–value problems. We could almost always get sufficient conditions for stability for initial–value problems. We next state and prove several results that will give us sufficient conditions for stability.

Theorem 6.2.4 *Suppose that the amplification matrix G associated with difference scheme (6.2.6) satisfies the von Neumann condition. Then*

(i) if Q is self adjoint, the scheme is stable.

(ii) if there exists an operator S such that $\| \ S \ \|_{2,\Delta x} \le C_1$, $\| \ S^{-1} \ \|_{2,\Delta x} \le C_2$, ($C_1$ and C_2 constant) and SQS^{-1} is self adjoint, the scheme is stable.

(iii) if G is Hermitian, the scheme is stable.

(iv) if there exists a matrix S such that $\| \ S \ \|_2 \le C_1$, $\| \ S^{-1} \ \|_2 \le C_2$ (C_1 and C_2 constant), and SGS^{-1} is Hermitian, the scheme is stable.

(v) if the elements of $G(\xi)$ are bounded and if all the eigenvalues of G, with the possible exception of one, lies in a circle inside the unit circle, the scheme is stable.

Proof: The proofs of (i) and (ii) will follow from the proofs of (iii) and (iv), respectively, using the fact that the amplification matrix and the difference operator will have the same symmetry.

(iii) Since G is Hermitian, $\sigma(G(\xi)) = \parallel G(\xi) \parallel_2$. Then since $\sigma(G(\xi)) \leq 1 + C\Delta t$,

$$\parallel G(\xi) \parallel_2 \leq 1 + C\Delta t \leq e^{C\Delta t}.$$

Then $\parallel G^n(\xi) \parallel_2 \leq \parallel G \parallel_2^n \leq e^{Cn\Delta t}$ so the scheme is stable by Theorem 6.2.1.

(iv) Let $H = SGS^{-1}$. Then

$$
\begin{aligned}
\parallel G^n(\xi) \parallel_2 &= \parallel S^{-1} H^n S \parallel_2 \\
&\leq \parallel S^{-1} \parallel_2 \parallel H^n \parallel_2 \parallel S \parallel_2 \\
&\leq C_1 C_2 \parallel H^n \parallel_2 \\
&\leq C_1 C_2 \parallel H \parallel_2^n \\
&= C_1 C_2 [\sigma(H)]^n \\
&= C_1 C_2 [\sigma(G)]^n \\
&\leq C_1 C_2 (1 + C\Delta t)^n \\
&\leq C_1 C_2 e^{Cn\Delta t}.
\end{aligned}
$$

So again, by Theorem 6.2.1, the scheme is stable.

(v) Since the proof of (v) involves results that we do not wish to introduce, we refer the reader to ref. [9], page 86 or ref. [10], page 78 for the proof.

Theorem 6.2.5 *(i) If there exists a non-negative C such that $G^*G \leq (1 + C\Delta t)^2 I$, then the scheme is stable.*

(ii) If there exists a matrix S such that $\parallel S \parallel_2 \leq C_1$, $\parallel S^{-1} \parallel_2 \leq C_2$ (C_1 and C_2 constant) and non-negative C such that $H = SGS^{-1}$ satisfies $H^*H \leq (1 + C\Delta t)^2 I$, then the scheme is stable.

Proof: Before we begin the proof of the above theorem, we explain that the inequalities between matrices given above should be interpreted as $A \leq B$ if $u * (B - A)u \geq 0$ for all $\mathbf{u}$.

(i) Since

$$\parallel G\hat{u} \parallel_2^2 = \hat{u}^* G^* G\hat{u} \leq (1 + C\Delta t)^2 \hat{u}^* \hat{u} = (1 + C\Delta t)^2 \parallel \hat{u} \parallel_2^2,$$

$\parallel G \parallel_2 \leq 1 + C\Delta t \leq e^{C\Delta t}$, $\parallel G^n \parallel_2 \leq \parallel G \parallel_2^n \leq e^{Cn\Delta t}$ and the stability follows from Theorem 6.2.1.

(ii) Since $G = S^{-1}HS$, $\| G^n \|_2 = \| S^{-1}H^n S \|_2 \leq \| S^{-1} \|_2 \| S \|_2 \| H^n \|_2$. Then since H satisfies $H^* H \leq (1 + C\Delta t)^2 I$, as in the proof of (i) we see that $\| H^n \|_2 \leq e^{C\Delta t}$, which completes the proof.

The hypotheses that must be checked carefully (and that are sometimes difficult to check) in the above theorems are the bounds on $\| S \|_2$ and $\| S^{-1} \|_2$. In some of the cases (for example, in the cases below where we apply Theorem 6.2.5-(ii) to hyperbolic equations and S comes from the original partial differential operator), it is clear that the bounds will be constant. But in other cases (such as when we attempt to apply the results to systems related to multilevel schemes), these bounds are not obvious and can be difficult to establish.

Before we go to examples illustrating the use of the above theorem, we simplify our work somewhat by stating and proving the following theorem. We notice that while the hypothesis of being stable via Definition 2.4.1–(2.4.3) is not the strongest possible definition, it is not overly restrictive. The utility of Theorem 6.2.6 is that we are able to omit the zeroth order term (zeroth order derivative) and analyze difference schemes without such a term (which will generally be stable according to Definition 2.4.1–(2.4.3)). And finally, we notice that this theorem is analogous to Proposition 3.1.8 for scalar equations. Also note that the proofs are not analogous and that Theorem 6.2.6 can not be proved in the same manner as was Proposition 3.1.8.

Theorem 6.2.6 *Suppose we have a difference scheme*

$$\mathbf{u}^{n+1} = Q\mathbf{u}^n \tag{6.2.24}$$

which is stable by Definition 2.4.1–(2.4.3) and a bounded operator B. Then

$$\mathbf{u}^{n+1} = (Q + \Delta t B)\mathbf{u}^n \tag{6.2.25}$$

is a stable difference scheme.

Proof: We begin by noting that if difference scheme (6.2.24) is stable by Definition 2.4.1–(2.4.3), there exists a constant K so that $\| Q^m \| \leq K$. We also note that we can assume that $K > 1$ (otherwise, we could define a new K_1 to be $\max\{K, 2\}$, which would satisfy Definition 2.4.1–(2.4.3)). We consider the binomial expansion of $(Q + \Delta t B)^n$ and emphasize that it is not in the usual form of the binomial expansion. However, it does have a Q^n term, n terms containing one $\Delta t B$ and $n - 1$ Q's, $\binom{n}{2}$ terms containing two $\Delta t B$'s and $n - 2$ Q's, $\cdots$, and finally, a $\Delta t^n B^n$ term. We cannot rearrange these terms to look like the usual binomial expansion because the B and Q do not commute, so the order of the Q's and B's in the expansion is important.

But, if we look at this expansion carefully (write out about five terms), we will see that the Q's appear in groups. How many groups of Q there are depends on how many B's we have in the particular term.

$j = 0$:   For example, the first term in the expansion is

$$Q^n. \tag{6.2.26}$$

Clearly, this term satisfies

$$\| Q^n \| \le K. \tag{6.2.27}$$

$j = 1$:   The next group of terms contains n terms with $n - 1$ Q's and one B. Each of these terms will satisfy

$$\| Q \cdots Q \Delta tBQ \cdots Q \| \le \| Q \cdots Q \| \| \Delta tB \| \| Q \cdots Q \| \tag{6.2.28}$$
$$\le K \Delta t \| B \| K = K^2 \Delta t \| B \| \tag{6.2.29}$$

(The two terms in this group that do not look like the left hand side of inequality (6.2.28) will still satisfy inequality (6.2.29) because $K > 1$,i.e. $\| \Delta tBQ \cdots Q \| \le \Delta t \| B \| K \le K^2 \Delta t \| B \|$).

$j = 2$:   The third group is the one of $\binom{n}{2}$ terms containing $n - 2$ Q's and two ΔtB's. Most of the terms in this group satisfy

$$\| Q \cdots Q \Delta tBQ \cdots Q \Delta tBQ \cdots Q \| \le K^3 \Delta t^2 \| B \|^2 . \tag{6.2.30}$$

(Again, the terms not in the above form will also satisfy the inequality because $K > 1$.)

$j = m - 1$:   The mth group of terms contains $\binom{n}{m-1}$ terms containing $n - (m - 1)$ Q's and $m - 1$ ΔtB's. These terms satisfy

$$\| Q \cdots Q \| \le K^m \Delta t^{m-1} \| B \|^{m-1} . \tag{6.2.31}$$

(As usual, the terms where the ΔtB terms are not separated or come first or last still satisfy inequality (6.2.31) because $K > 1$.)

$j = n$:   And, finally, the $n + 1$-st group has $\binom{n}{n}$ terms containing 0 Q's and n ΔtB's. This term satisfies

$$\| (\Delta tB)^n \| \le \Delta t^n \| B \|^n \le K^{n+1} \Delta t^n \| B \|^n \tag{6.2.32}$$

(since $K > 1$).

Using inequalities (6.2.27)–(6.2.32), we get

$$\| (Q + \Delta tB)^n \| \le \sum_{j=0}^{n} \binom{n}{j} K^{j+1} \Delta t^j \| B \|^j$$
$$\le K \sum_{j=0}^{n} \frac{(nK \Delta t \| B \|)^j}{j!}$$
$$\le K e^{n \Delta t K \| B \|}. \tag{6.2.33}$$

Hence, since $\| B \|$ is a constant, difference scheme (6.2.25) is stable by Definition 2.4.1.

Remark: We note that most of the results proved above refer to stability with respect to the vector $\ell_{2,\Delta x}$ norm. However, Theorem 6.2.6 is true with respect to a general norm (though we will most often apply it with respect to the $\ell_{2,\Delta x}$ norm).

We now proceed to see what the above theorems will allow us to say about the stability of several obvious schemes.

Example 6.2.2 Let $\mathbf{v} = \mathbf{v}(x,t) = [v_1 \cdots v_K]^T$ and consider the parabolic partial differential equation

$$\mathbf{v}_t = B_2\mathbf{v}_{xx} + B_0\mathbf{v}, \quad x \in \mathbb{R}, \quad t > 0 \tag{6.2.34}$$

where B_2 is a real, positive definite, $K \times K$, symmetric, constant matrix and B_0 is bounded. Discuss the consistency, stability and convergence of the difference scheme

$$\mathbf{u}_k^{n+1} = \mathbf{u}_k^n + \frac{\Delta t}{\Delta x^2}B_2\delta^2\mathbf{u}_k^n + \Delta t B_0\mathbf{u}_k^n \tag{6.2.35}$$

for solving the partial differential equation (6.2.34) along with initial condition $\mathbf{v}(x,0) = \mathbf{f}(x)$.

Solution: We begin by noting that due to Theorem 6.2.6 we can ignore the B_0 term. Any stability result that we get without the B_0 term will also be true for a scheme that includes the B_0 term. Also, we realize that the scheme is clearly consistent with the partial differential equation so our approach will be to study stability and get convergence by the Lax Theorem.

To investigate the stability of difference scheme (6.2.35), we take the discrete Fourier transform of equation (6.2.35) (without the B_0 term) and see that the transformed equation can be written as

$$\hat{\mathbf{u}}^{n+1} = \left(I - 4r\sin^2\frac{\xi}{2}B_2\right)\hat{\mathbf{u}}^n. \tag{6.2.36}$$

The amplification matrix, G, is given by

$$G(\xi) = I - 4r\sin^2\frac{\xi}{2}B_2. \tag{6.2.37}$$

Since B_2 is symmetric, G is Hermitian. We can then apply Theorem 6.2.4-(iii). But *we must remember that to apply Theorem 6.2.4, we must still satisfy the von Neumann condition.* Theorem 6.2.4-(iii) states that if the scheme satisfies the von Neumann condition and G is Hermitian, then the scheme is stable. In this case, we can see that the eigenvalues of G will be

$$\lambda_j = 1 - 4r\mu_j\sin^2\frac{\xi}{2}, \quad j = 1, \cdots, K,$$

where μ_j, $j = 1, \cdots, K$ are the eigenvalues of B_2. Hence, it is possible, based on knowledge of the eigenvalues of B_2, to decide whether G satisfies the von Neumann condition. We see that

$$\left|1 - 4r\mu_j\sin^2\frac{\xi}{2}\right| \leq 1$$

is the same as

$$-1 \leq 1 - 4r\mu_j\sin^2\frac{\xi}{2} \leq 1.$$

The right inequality is always true. As in the scalar case, the left most inequality is true if $r\mu_j \leq 1/2$. Hence, *if B_2 is a symmetric, positive definite matrix, then difference scheme (6.2.35) is stable if and only if $r\mu_j \leq 1/2$ for $j = 1, \cdots, K$.*

And, of course, since the scheme is consistent, then difference scheme (6.2.35) will be a convergent scheme for approximating the solution to the system of partial differential equations (6.2.34) if and only if $r\mu_j \le 1/2$ for all $j = 1, \cdots, K$.

Remark: To prove that difference scheme (6.2.35) is stable, we could also use Theorem 6.2.4-(i). However, to apply Theorem 6.2.4-(i) we must verify that Q is self adjoint. The concept of an operator being self adjoint is much more difficult than that of a matrix being Hermitian. Hence, it is easier to apply Theorem 6.2.4-(iii) instead of Theorem 6.2.4-(i).

Example 6.2.3 Let v be as in Example 6.2.2 and consider the hyperbolic partial differential equation

$$\mathbf{v}_t = A\mathbf{v}_x + B_0\mathbf{v} \quad x \in \mathbb{R}, \quad t > 0 \tag{6.2.38}$$

where A and B_0 are real, $K \times K$ constant matrices and A is diagonalizable (so that the system is hyperbolic). Discuss the consistency, stability and convergence of the forward difference scheme

$$\mathbf{u}_k^{n+1} = \frac{\Delta t}{\Delta x}A\mathbf{u}_{k+1}^n + \left(I - \frac{\Delta t}{\Delta x}A\right)\mathbf{u}_k^n + \Delta t B_0\mathbf{u}_k^n \tag{6.2.39}$$

for approximating the solution to partial differential equation (6.2.38) along with initial condition $\mathbf{v}(x, t) = \mathbf{f}(x)$.

Solution: We begin by using Theorem 6.2.6 to eliminate the B_0 term. Thus we consider the stability of difference scheme (6.2.39) without the last term. As in Example 6.2.2, we investigate the stability by taking the discrete Fourier transform of the difference scheme (6.2.39) (without the B_0 term). We get

$$\hat{\mathbf{u}}^{n+1}(\xi) = \left(R\cos\xi A + I - RA + iR\sin\xi A\right)\hat{\mathbf{u}}^n, \tag{6.2.40}$$

and

$$G(\xi) = I + R\cos\xi A - RA + iR\sin\xi A. \tag{6.2.41}$$

It is clear that G is not Hermitian. Matrix G can be diagonalized but it seems likely (and it is the case) that the eigenvalues will be complex. In this case, then, the diagonal matrix would not be Hermitian so we could not apply part (iv) of Theorem 6.2.4.

We diagonalize G by using the fact that A was assumed to be diagonalizable. Hence, there exists a matrix S such that $D = SAS^{-1}$ is a diagonal matrix with the eigenvalues of A on the diagonal. If we then multiply G on the left by S and on the right by S^{-1}, we get

$$\begin{aligned} H &= SGS^{-1} \\ &= S[I + R\cos\xi A - RA + iR\sin\xi A]S^{-1} \\ &= I + R\cos\xi D - RD + iR\sin\xi D. \end{aligned}$$

The general term on the diagonal of H is

$$\lambda_j(\xi) = 1 - R\mu_j + R\mu_j\cos\xi + iR\mu_j\sin\xi, \quad j = 1, \cdots, K \tag{6.2.42}$$

(which are also the eigenvalues of G). We note that if we uncouple the system (6.2.38) (without the B_0 term) as we did in the beginning of Chapter 5, we get

$$V_{jt} - \mu_j V_{jx} = 0 \quad j = 1, \cdots, K. \tag{6.2.43}$$

If we solve the K uncoupled equations using difference scheme (5.3.1), the symbol of the difference scheme for solving the j-th equation associated with eigenvalue μ_j would be

$\rho(\xi) = \lambda_j(\xi)$. Hence, we know that if $\mu_j \geq 0$ and $\mu_j \Delta t/\Delta x \leq 1$, then $|\lambda_j(\xi)| \leq 1$. Since the matrix H^*H is a diagonal matrix with $|\lambda_j(\xi)|^2$ on the diagonal, $H^*H \leq I$ and we can apply Theorem 6.2.5-(ii) to show that *if $\mu_j \geq 0$ and $\mu_j \Delta t/\Delta x \leq 1$ $j = 1, \cdots, K$, then difference scheme (6.2.39) is a stable scheme for solving the system of partial differential equations (6.2.38)*. And, of course, since difference scheme (6.2.39) is a consistent scheme for the system of partial differential equations (6.2.38), then difference scheme (6.2.39) will be a convergent scheme for the solution of the system of partial differential equations (6.2.38).

We note that before we can apply Theorem 6.2.5 above, we must show that the matrix S satisfies $\| S \| \leq C_1$ and $\| S^{-1} \| \leq C_2$. These conditions are clearly satisfied since the matrix S is the matrix that diagonalizes the coefficient matrix A (the columns of S^{-1} are the eigenvectors of A). Since A is a constant matrix, both $\| S \|$ and $\| S^{-1} \|$ will be bounded.

And finally, we note above that if $\mu_j < 0$ for some j, then at least one of the eigenvalues of G would be greater than one in magnitude and the difference scheme would not satisfy the von Neumann condition. In this case, the scheme would be unstable. Thus, the condition $\mu_j \geq 0$ and $\mu_j \Delta t/\Delta x \leq 1$, $j = 1, \cdots, K$ is both necessary and sufficient for stability, and hence, convergence.

Remark: We emphasize that there will be many applications to Theorem 6.2.5-(ii) similar to the one used in Example 6.2.3. For example, we immediately realize that *the scheme*

$$\mathbf{u}_k^{n+1} = \mathbf{u}_k^n + RA(\mathbf{u}_k^n - \mathbf{u}_{k-1}^n) + \Delta t B_0 \mathbf{u}_k^n \qquad (6.2.44)$$

will be a stable difference scheme for solving the system of partial differential equations (6.2.38) if and only if $\mu_j \leq 0$ and $|\mu_j|\Delta t/\Delta x \leq 1$ for $j = 1, \cdots, K$.

Another very important difference scheme for hyperbolic systems of partial differential equations is the Lax-Wendroff scheme. We consider the stability of this scheme in the following example.

Example 6.2.4 Discuss the consistency, stability and convergence of the Lax-Wendroff scheme for solving system (6.2.38).

Solution: If we consider the vector version of the Lax-Wendroff scheme for solving the system of partial differential equations (6.2.38),

$$\mathbf{u}_k^{n+1} = \mathbf{u}_k^n + \frac{\Delta t}{2\Delta x} A \delta_0 \mathbf{u}_k^n + \frac{\Delta t^2}{2\Delta x^2} A^2 \delta^2 \mathbf{u}_k^n + \Delta t B_0 \mathbf{u}_k^n, \qquad (6.2.45)$$

as in the scalar case the scheme will be accurate of $\mathcal{O}(\Delta t^2) + \mathcal{O}(\Delta x^2)$. If we again use Theorem 6.2.6 to eliminate the B_0 term and take the discrete Fourier transform of equation (6.2.45) (without the B_0 term), we see that

$$G(\xi) = I - 2R^2 A^2 \sin^2 \frac{\xi}{2} + iR \sin \xi A. \qquad (6.2.46)$$

Since we know that there exists a matrix S such that $A = S^{-1}DS$ where D is a diagonal matrix with the eigenvalues of A on the diagonal (and note that $A^2 = S^{-1}D^2S$), we see that we can write

$$\begin{aligned} H &= SGS^{-1} \\ &= S[I - 2R^2 A^2 \sin^2 \frac{\xi}{2} + iR \sin \xi A]S^{-1} \\ &= I - 2R^2 D^2 \sin^2 \frac{\xi}{2} + iR \sin \xi D. \end{aligned}$$

The general term on the diagonal is the same as the symbol for the scalar Lax-Wendroff scheme (5.3.9). Thus, we again apply Theorem 6.2.5-(ii) and get that *the Lax-Wendroff scheme, (6.2.45), is a stable difference scheme for solving the system of partial differential equations (6.2.38) if and only if* $|\mu_j|\Delta t/\Delta x \leq 1$, $j = 1, \cdots, K$.

Again the fact that S satisfies the hypotheses of Theorem 6.2.5-(ii) follows from the fact that S diagonalizes the constant matrix A.

Remark: We note that the Lax-Wendroff scheme is very important for systems because *the Lax-Wendroff scheme does not require that all of the eigenvalues have the same sign.* Schemes (6.2.39) and (6.2.44) are not as useful because they only apply when all of the eigenvalues are either negative or positive, respectively.

We next consider a system of partial differential equations of the form

$$\mathbf{v}_t = \mathbf{v}_{xx} + B_1\mathbf{v}_x + B_0\mathbf{v}, \quad x \in \mathbb{R}, \quad t > 0, \tag{6.2.47}$$

where B_j, $j = 0, 1$ are real, $K \times K$ constant matrices and B_1 has K distinct eigenvalues. An obvious and often used scheme for solving system (6.2.47) is

$$\mathbf{u}_k^{n+1} = \mathbf{u}_k^n + \frac{\Delta t}{\Delta x^2}\left[\mathbf{u}_{k-1}^n - 2\mathbf{u}_k^n + \mathbf{u}_{k+1}^n\right]$$
$$+ \frac{\Delta t}{2\Delta x}B_1\left[\mathbf{u}_{k+1}^n - \mathbf{u}_{k-1}^n\right] + \Delta t B_0\mathbf{u}_k^n. \tag{6.2.48}$$

Recall that the scalar analog of difference equation (6.2.48) was discussed in Example 3.1.3.

Example 6.2.5 Discuss the consistency, stability and convergence of the difference scheme (6.2.48) for solving the partial differential equation (6.2.47) along with initial condition $\mathbf{v}(x, 0) = \mathbf{f}(x)$.

Solution: As usual, we begin by realizing that due to Theorem 6.2.6 we can ignore the B_0 term. Any result that we get without the B_0 term, will also be true with that term. Also, we realize that the scheme is clearly consistent with the partial differential equation so that our approach will be to study stability and get convergence by the Lax Theorem.

To investigate the stability of the scheme we take the discrete Fourier transform of equation (6.2.48) (without the B_0 term), see that the transformed equation can be written as

$$\hat{\mathbf{u}}^{n+1} = \left[\left(1 - 4r\sin^2\frac{\xi}{2}\right)I + iR\sin\xi B_1\right]\hat{\mathbf{u}}^n \tag{6.2.49}$$

and the amplification matrix, G, is given by

$$G(\xi) = \left(1 - 4r\sin^2\frac{\xi}{2}\right)I + iR\sin\xi B_1. \tag{6.2.50}$$

It is easy to see that the eigenvalues of G will be

$$\lambda_j(\xi) = 1 - 4r\sin^2\frac{\xi}{2} + i\mu_j R\sin\xi, \quad j = 1, \cdots, K \tag{6.2.51}$$

where μ_j, $j = 1, \cdots, K$ are the eigenvalues of B_1.

We can now use the results from Example 3.1.3 and state

- if $r^2 - \mu_j^2 R^2/4 \geq 0$ and $r \leq 1/2$, then $|\lambda_j| \leq 1$.
- if $r^2 - \mu_j^2 R^2/4 < 0$, $r \geq \mu_j^2 R^2/2$ and $r \leq 1/2$, then $|\lambda_j| \leq 1$.
- if $r^2 - \mu_j^2 R^2/4 < 0$ and $r < \mu_j^2 R^2/2$, then $|\lambda_j| > 1$.

Thus we see that to satisfy the von Neumann condition, it is necessary that for all j, $\mu_j^2 R^2/2 \leq r \leq 1/2$.

If S is the matrix such that $B_1 = S^{-1} D S$ where D is the diagonal matrix with the eigenvalues of B_1 on the diagonal,

$$H = SGS^{-1} = \left(1 - 4r\sin^2 \frac{\xi}{2}\right) I + iR \sin \xi D$$

will be the diagonal matrix with the eigenvalues of G, λ_j, $j = 1, \cdots, K$ on the diagonal. Then using the results given above and Theorem 6.2.5-(ii) (S will satisfy the hypotheses of Theorem 6.2.5-(ii) because A is a constant matrix), we see that *the condition $\mu_j^2 R^2/2 \leq r \leq 1/2$ for $j = 1, \cdots, K$ is both necessary and sufficient for stability* (and, hence, convergence).

Remark: We should note that if we are considering a partial differential equation of the form (6.2.47) where the convective terms dominate the flow ($r < \mu_j^2 R^2/2$ for some j), then difference scheme (6.2.48) is not a very good scheme to use. From our previous experience, it should be clear that we would prefer to use a scheme that treats the first order term by a scheme designed for a hyperbolic equation.

Before we leave this section, we add that we might want to consider a system of partial differential equations as general as

$$\mathbf{v}_t = B_2 \mathbf{v}_{xx} + B_1 \mathbf{v}_x + B_0 \mathbf{v} \quad -\infty < x < \infty, \quad t > 0 \qquad (6.2.52)$$

where B_j, $j = 0, \cdots, 2$ are real, $K \times K$ constant matrices and B_2 is positive definite and the obvious explicit difference scheme

$$\mathbf{u}_k^{n+1} = \mathbf{u}_k^n + \frac{\Delta t}{\Delta x^2} B_2 \left[\mathbf{u}_{k-1}^n - 2\mathbf{u}_k^n + \mathbf{u}_{k+1}^n\right]$$
$$+ \frac{\Delta t}{2\Delta x} B_1 \left[\mathbf{u}_{k+1}^n - \mathbf{u}_{k-1}^n\right] + \Delta t B_0 \mathbf{u}_k^n. \qquad (6.2.53)$$

As usual, its clear that the difference scheme consistent with the partial differential equation. Also, as usual we can ignore the B_0 term.

To investigate the stability of the scheme we take the discrete Fourier transform of equation (6.2.53) (without the B_0 term) and find that the amplification matrix, G, is given by

$$G(\xi) = I - 4r\sin^2 \frac{\xi}{2} B_2 + iR \sin \xi B_1. \qquad (6.2.54)$$

It is easy to see that the eigenvalues of G will depend on B_2 and B_1 in such a way that it will be difficult apriori to decide whether G satisfies the von Neumann condition or if any parts of Theorems 6.2.4 or 6.2.5 can be used to obtain sufficient conditions for stability. Most often the systems of equations that must be solved will not be as general as equation (6.2.52). However, when confronted with the situation where we must solve a very general system of partial differential equations, we must again resort to numerical experimentation. The scheme is analyzed for model equations similar to the given equation (probably including those in Examples 6.2.2 and 6.2.5). From these analyses, potential stability limits are proposed. And, finally, the difference scheme is tested, using a series of different initial conditions and boundary conditions.

When we consider approximate solutions to partial differential equations such as (6.2.52), we should also realize that the relative sizes of the eigenvalues of B_1 and B_2 will effect what type of scheme we should use. It should not surprise us that a scheme such as (6.2.53) will work well when the eigenvalues of B_2 are larger than those of B_1, but does not work when the eigenvalues of B_2 are smaller than those of B_1. Also, it should not surprise us that in the latter case, some scheme using a "hyperbolic type" differencing of the first order term is necessary. The most difficult case is when we have a system of partial differential equations where some of the eigenvalues of B_2 are larger than the eigenvalues of B_1 and some of the eigenvalues of B_2 are smaller than the eigenvalues of B_1. In this situation, one must proceed very carefully, and often must uncouple the problem enough to group the equations that are alike together.

HW 6.2.1 Verify the stability result given for difference scheme (6.2.44).

HW 6.2.2 Discuss the consistency, stability and convergence of difference scheme

$$\mathbf{u}_k^{n+1} - \frac{R}{2} A \delta_0 \mathbf{u}_k^{n+1} = \mathbf{u}_k^n + \Delta t B_0 \mathbf{u}_k^n$$

as a solution scheme for the initial–value problem given by partial differential equations (6.2.38) along with initial condition $\mathbf{v}(x,0) = \mathbf{f}(x)$, $x \in \mathbb{R}$.

HW 6.2.3 Discuss the consistency, stability and convergence of difference scheme

$$\mathbf{u}_k^{n+1} = \mathbf{u}_k^n + \frac{R}{2} A \delta_0 \mathbf{u}_k^n + \Delta t B_0 \mathbf{u}_k^n$$

as a solution scheme for the initial–value problem given by partial differential equations (6.2.38) along with initial condition $\mathbf{v}(x,0) = \mathbf{f}(x)$, $x \in \mathbb{R}$.

HW 6.2.4 Discuss the consistency, stability and convergence of difference scheme

$$\mathbf{u}_k^{n+1} = \mathbf{u}_k^n + r \delta^2 \mathbf{u}_k^n + R A \delta_+ \mathbf{u}_k^n + \Delta t B_0 \mathbf{u}_k^n$$

as a solution scheme for the initial–value problem given by partial differential equations (6.2.47) along with initial condition $\mathbf{v}(x,0) = \mathbf{f}(x)$, $x \in \mathbb{R}$.

HW 6.2.5 Discuss the consistency, stability and convergence of difference scheme

$$\mathbf{u}_k^{n+1} - \frac{R}{2} A \delta_0 \mathbf{u}_k^{n+1} = \mathbf{u}_k^n + r \delta^2 \mathbf{u}_k^n + \Delta t B_0 \mathbf{u}_k^n$$

as a solution scheme for the initial–value problem given by partial differential equations (6.2.47) along with initial condition $\mathbf{v}(x,0) = \mathbf{f}(x)$, $x \in \mathbb{R}$.

6.2.1 Flux Splitting

Most often when we are dealing with systems of hyperbolic equations, we are not able to assume that all of the eigenvalues of A are either all positive or all negative. Of course, if a scheme is used that does not require all of the eigenvalues to be the same sign (such as the Lax-Wendroff scheme), then this is no problem. Another approach to eliminate the assumption that all of the eigenvalues are of the same sign is to use **flux splitting**. The major application of flux splitting is to nonlinear equations and there are several different forms of flux splitting. However, we shall introduce a very basic flux splitting in the linear case, more as a preparation for the more difficult nonlinear case than as a scheme to be used in the form presented.

We again consider the system of partial differential equations (6.2.38) and ignore the $B_0\mathbf{v}$ term. Our goal is to devise a scheme that uses one sided differences but allows for A to have both positive and negative eigenvalues. If we let S be such that $D = SAS^{-1}$ is the diagonal matrix with the eigenvalues of A on the diagonal, then equation (6.2.38) can be decoupled by multiplying on the left by S to get

$$
\begin{aligned}
S\mathbf{v}_t &= (S\mathbf{v})_t \\
&= SA\mathbf{v}_x \\
&= SAS^{-1}S\mathbf{v}_x \\
&= D(S\mathbf{v})_x.
\end{aligned}
$$

Hence, if we let $\mathbf{V} = S\mathbf{v}$, system (6.2.38) $(B_0 = \Theta)$ is equivalent to the uncoupled system

$$
\mathbf{V}_t = D\mathbf{V}_x, \quad x \in \mathbb{R}, \quad t > 0. \tag{6.2.55}
$$

We will refer to the variables $\mathbf{v}$ as the **primitive variables** and the variables $\mathbf{V}$ as the **characteristic variables**.

We first write D as $D = D_+ + D_-$ where D_+ and D_- are the $K \times K$ diagonal matrices containing the positive and negative eigenvalues of D, respectively, with zeros elsewhere. Suppose the matrix A has K_+ positive eigenvalues and $K_- = K - K_+$ negative eigenvalues. It is generally convenient to suppose that the original equations in system (6.2.38) have been ordered so that the first K_+ eigenvalues are positive and the last K_- eigenvalues are negative. We next let D_{1+} and D_{2-} denote the $K_+ \times K_+$ and $K_- \times K_-$ diagonal matrices containing the positive and negative terms in D, and let $\mathbf{V}_1$ and $\mathbf{V}_2$ K_+ and K_--vectors, respectively. Then the system of partial differential equations (6.2.55) can be written as the pair of equations

$$\mathbf{V}_{1t} = D_{1+}\mathbf{V}_{1x} \tag{6.2.56}$$
$$\mathbf{V}_{2t} = D_{2-}\mathbf{V}_{2x}. \tag{6.2.57}$$

But we know that these two equations can be solved using one sided differencing by the pair of difference equations

$$\mathbf{U}_{1_k}^{n+1} = \mathbf{U}_{1_k}^n + RD_{1+}(\mathbf{U}_{1_{k+1}}^n - \mathbf{U}_{1_k}^n) \tag{6.2.58}$$
$$\mathbf{U}_{2_k}^{n+1} = \mathbf{U}_{2_k}^n + RD_{2-}(\mathbf{U}_{2_k}^n - \mathbf{U}_{2_{k-1}}^n). \tag{6.2.59}$$

The pair of equations given in (6.2.58)–(6.2.59) can be written as

$$\mathbf{U}_k^{n+1} = \mathbf{U}_k^n + RD_+(\mathbf{U}_{k+1}^n - \mathbf{U}_k^n) + RD_-(\mathbf{U}_k^n - \mathbf{U}_{k-1}^n) \tag{6.2.60}$$

where $\mathbf{U}$ is the K vector made of the appropriate composition of vectors $\mathbf{U}_1$ and $\mathbf{U}_2$. We then recouple the system of difference equations (6.2.60) by multiplying through on the left by S^{-1} to get

$$S^{-1}\mathbf{U}_k^{n+1} = S^{-1}\mathbf{U}_k^n + RS^{-1}D_+S(S^{-1}\mathbf{U}_{k+1}^n - S^{-1}\mathbf{U}_k^n)$$
$$+ RS^{-1}D_-S(S^{-1}\mathbf{U}_k^n - S^{-1}\mathbf{U}_{k-1}^n)$$

or

$$\mathbf{u}_k^{n+1} = \mathbf{u}_k^n + RA_+\delta_+\mathbf{u}_k^n + RA_-\delta_-\mathbf{u}_k^n \tag{6.2.61}$$

where $A_+ = S^{-1}D_+S$ and $A_- = S^{-1}D_-S$.

Remark 1: We must realize that it is fairly obvious that if you are going to go through all of the work to form A_+ and A_- to enable you to solve your problem, you might consider solving the uncoupled system (6.2.58)–(6.2.59). As we said in the introduction to the topic, we have not introduced the method for its use on linear equations. It is a common technique used with nonlinear equations.

Remark 2: It should be fairly obvious that the difference scheme given by equation (6.2.61) is stable. This follows from the fact that when we constructed the scheme for the uncoupled systems (6.2.58)–(6.2.59), we chose stable schemes for each piece. The recoupling procedure will not make the scheme unstable. However, it is also quite easy to take the finite Fourier transform of difference scheme (6.2.61) to obtain G, use S to diagonalize the matrix and apply Theorem 6.2.5(ii) as we have done so often before. See HW6.2.6.

Remark 3: Above we have assumed that the eigenvalues of A are either positive or negative. If there are one or more zero eigenvalues (with a full complement of eigenvectors), everything works equally well. The partial differential equation associated with a zero eigenvalue analogous to partial differential equations (6.2.56)–(6.2.57) is $\mathbf{V}_t = \boldsymbol{\theta}$. The difference equation approximation to this partial differential equation is $\mathbf{U}_k^{n+1} = \mathbf{U}_k^n$. Hence, if there are zero eigenvalues, difference equation (6.2.61) will still give a stable, flux split scheme for approximating the solution to the system of partial differential equations (6.2.38). To accommodate the zero eigenvalues, D_+ and D_- will have to be redefined as

D_+ will be the $K \times K$ diagonal matrix with the K_+ positive eigenvalues as the first K_+ diagonal elements and zeros elsewhere,

D_- will be the $K \times K$ diagonal matrix with the K_- negative eigenvalues as the $(K_+ + 1)$st through the $(K_+ + K_-)$–st diagonal elements and zeros elsewhere.

We have assumed that the equations have been ordered so that the first K_+ eigenvalues are positive, the next K_- eigenvalues are negative and the last K_0 eigenvalues are zero $(K_+ + K_- + K_0 = K)$.

HW 6.2.6 Analyze the stability of difference scheme (6.2.61).

HW 6.2.7 Analyze the consistency and stability of difference scheme

$$\mathbf{u}_k^{n+1} = \mathbf{u}_k^n + r\delta^2\mathbf{u}_k^n + RB_{1+}\delta_+\mathbf{u}_k^n + RB_{1-}\delta_-\mathbf{u}_k^n + \Delta t B_0\mathbf{u}_k^n,$$

for solving the system of partial differential equations (6.2.47), where B_{1+} and B_{1-} are defined analogous to A_+ and A_-.

6.2.2   Implicit Schemes

As was the case with scalar equations, we also have the possibility of developing implicit difference schemes for solving systems of partial differential equations. We have already introduced the notation for implicit schemes in that the general difference scheme given by equation (6.2.2) includes the Q_1 matrix to account for implicit terms and the Q_2 matrix to account for explicit terms. To this point, we have always considered the case when $Q_1 = I$. We introduce implicit schemes and the convergence analysis of these schemes through a series of examples.

Example 6.2.6 Consider the parabolic system of partial differential equations

$$\mathbf{v}_t = B_2\mathbf{v}_{xx}   -\infty < x < \infty, \quad t > 0 \tag{6.2.62}$$

where again $\mathbf{v}$ is a K-vector and B_2 is a $K \times K$, positive definite, symmetric, constant matrix. Discuss the consistency, stability and convergence of the difference scheme

$$\mathbf{u}_k^{n+1} - rB_2\delta^2\mathbf{u}_k^{n+1} = \mathbf{u}_k^n \tag{6.2.63}$$

for solving the system of partial differential equations (6.2.62) along with the initial condition $\mathbf{v}(x,0) = \mathbf{f}(x)$.

Solution: We should begin by noting that we could just as well add a term $B_0 \mathbf{v}$ to the system of equations (6.2.62) and account for this term in (6.2.63) by adding a term $\Delta t B_0 \mathbf{u}_k^n$. Any result that we obtain for this example would also apply for the perturbed system. Also, we again note that we obtain convergence by proving consistency and stability. And, as usual, the consistency is clear. Hence, we spend our time considering the stability off difference scheme (6.2.63).

We begin by taking the discrete Fourier transform of difference scheme (6.2.63) to get

$$\left(I + 4r \sin^2 \frac{\xi}{2} B_2 \right) \hat{\mathbf{u}}^{n+1} = \hat{\mathbf{u}}^n. \tag{6.2.64}$$

Hence, the amplification matrix G is given by

$$G(\xi) = \left(I + 4r \sin^2 \frac{\xi}{2} B_2 \right)^{-1}. \tag{6.2.65}$$

To determine the eigenvalues of G, λ_j, $j = 1, \cdots, K$, we again let μ_j, $\hat{\mathbf{u}}_j$, $j = 1, \cdots, K$ denote the eigenvalues and eigenvectors of B_2, respectively, and consider for each $j = 1, \cdots, K$

$$G \hat{\mathbf{u}}_j = \lambda_j \hat{\mathbf{u}}_j,$$

or

$$\begin{aligned}
\hat{\mathbf{u}}_j &= G^{-1}(\lambda_j \hat{\mathbf{u}}_j) \\
&= \lambda_j G^{-1}(\hat{\mathbf{u}}_j) \\
&= \lambda_j \left(I + 4r \sin^2 \frac{\xi}{2} B_2 \right) \hat{\mathbf{u}}_j \\
&= \lambda_j \left(1 + 4r \sin^2 \frac{\xi}{2} \mu_j \right) \hat{\mathbf{u}}_j.
\end{aligned}$$

Thus we see that if $1 + 4r \sin^2 \frac{\xi}{2} \mu_j > 0$ (which is the case since r, $\sin^2 \frac{\xi}{2}$ and μ_j are all greater or equal to zero), then the eigenvalues of G are given by

$$\lambda_j = \frac{1}{1 + 4r \sin^2 \frac{\xi}{2} \mu_j} \quad j = 1, \cdots, K. \tag{6.2.66}$$

The calculation to see that $0 < \lambda_j \le 1$ is exactly the same as it was in the scalar case for the analogous implicit scheme. Hence, the scheme satisfies the von Neumann condition. Since G is Hermitian, by Theorem 6.2.4(iii) difference scheme (6.2.63) is unconditionally stable, and hence, convergent.

Example 6.2.7 Consider the linear hyperbolic system of partial differential equations

$$\mathbf{v}_t = A \mathbf{v}_x \quad -\infty < x < \infty, \quad t > 0 \tag{6.2.67}$$

where $\mathbf{v}$ is a K-vector and A is a $K \times K$, constant diagonalizable matrix. Discuss the consistency, stability and convergence of the implicit Lax-Wendroff difference scheme

$$\mathbf{u}_k^{n+1} - \frac{R}{2} A \delta_0 \mathbf{u}_k^{n+1} - \frac{R^2}{2} A^2 \delta^2 \mathbf{u}_k^{n+1} = \mathbf{u}_k^n \tag{6.2.68}$$

for solving the system of partial differential equations (6.2.67) along with the initial condition $\mathbf{v}(x, 0) = \mathbf{f}(x)$.

Solution: We begin by taking the discrete Fourier transform of equation (6.2.68) to get

$$\left(I - iR \sin \xi A + 2R^2 \sin^2 \frac{\xi}{2} A^2 \right) \hat{\mathbf{u}}^{n+1} = \hat{\mathbf{u}}^n \tag{6.2.69}$$

or that

$$G(\xi) = \left(I + 2R^2 \sin^2 \frac{\xi}{2} A^2 - iR \sin \xi A\right)^{-1}. \qquad (6.2.70)$$

Using the fact that there exists a matrix S such that $A = S^{-1}DS$ where D is a diagonal matrix with the eigenvalues of A on the diagonal, it is not hard to see that G can be written as $G = S^{-1}HS$ where

$$H = \left(I + 2R^2 \sin^2 \frac{\xi}{2} D^2 - i \sin \xi D\right)^{-1}. \qquad (6.2.71)$$

Then since H^*H is the diagonal matrix with

$$\frac{1}{(1 + 2R^2 \sin^2 \frac{\xi}{2}\mu_j^2)^2 + \sin^2 \xi \mu_j^2} \qquad j = 1, \cdots, K \qquad (6.2.72)$$

on the diagonal (where μ_j, $j = 1, \cdots, K$ are the eigenvalues of A), it is easy to see that H satisfies $H^*H \leq I$. Since the elements on the diagonal of H^*H are the magnitude squared of the eigenvalues of G, it is easy to see that the von Neumann condition is satisfied and that we can use Theorem 6.2.5(ii) (since S is again the constant matrix associated with the partial differential equation) to show that difference scheme (6.2.68) is unconditionally stable. Because difference scheme (6.2.68) is consistent and unconditionally stable, it is convergent.

Example 6.2.8 Again consider partial differential equation (6.2.67) along with the difference scheme (Crank-Nicolson)

$$\mathbf{u}_k^{n+1} - \frac{R}{4} A \delta_0 \mathbf{u}_k^{n+1} = \mathbf{u}_k^n + \frac{R}{4} A \delta_0 \mathbf{u}_k^n. \qquad (6.2.73)$$

Discuss the consistency, stability and convergence of the above scheme.

Solution: It is fairly obvious that the above difference scheme is consistent with partial differential equation (6.2.67). If we repeat what we have done earlier for Crank-Nicolson schemes and expand about the point $(k, n + \frac{1}{2})$, we will see that the scheme is $\mathcal{O}(\Delta t^2) + \mathcal{O}(\Delta x^2)$.

To investigate the stability of difference scheme (6.2.73), we take the discrete Fourier transform and get

$$\left(I - \frac{iR}{2} \sin \xi A\right) \hat{\mathbf{u}}^{n+1} = \left(I + \frac{iR}{2} \sin \xi A\right) \hat{\mathbf{u}}^n.$$

Hence, the amplification matrix is given by

$$G(\xi) = \left(I - \frac{iR}{2} \sin \xi A\right)^{-1} \left(I + \frac{iR}{2} \sin \xi A\right). \qquad (6.2.74)$$

If we note that

$$0 = \det(G - \lambda I)$$

$$= \det\left(\left(I - \frac{iR}{2} \sin \xi A\right)^{-1}\right) \det\left(\left(I + \frac{iR}{2} \sin \xi A\right) - \lambda\left(I - \frac{iR}{2} \sin \xi A\right)\right)$$

$$= \det\left(\left(I - \frac{iR}{2} \sin \xi A\right)^{-1}\right) \det\left((1 - \lambda)I - [-(1 + \lambda)]\frac{iR}{2} \sin \xi A\right)$$

$$= \left[-(1 + \lambda)\frac{iR}{2} \sin \xi\right]^K \det\left(\left(I - \frac{iR}{2} \sin \xi A\right)^{-1}\right) \det\left[\frac{1 - \lambda}{-(1 + \lambda)\frac{iR}{2} \sin \xi} I - A\right].$$

So

$$\mu_j = \frac{-(1 - \lambda)}{(1 + \lambda)\frac{iR}{2} \sin \xi},$$

where μ_j, $j = 1, \cdots, K$ are the eigenvalues of the matrix A. Thus, the eigenvalues of G are given by

$$\lambda_j = \frac{1 + \frac{iR}{2}\mu_j \sin \xi}{1 - \frac{iR}{2}\mu_j \sin \xi}, \quad j = 1, \cdots, K. \tag{6.2.75}$$

Since $|\lambda_j| = 1$ for all $j = 1, \cdots, K$, it is clear that the Crank-Nicolson scheme satisfies the von Neumann condition. Thus, the Crank-Nicolson **may be** stable (with no restrictions on R).

If we proceed as we have for several of the explicit difference schemes and multiply G on the left by S and on the right by S^{-1} (where S^{-1} is the matrix made up of the eigenvectors of A and the matrix $D = SAS^{-1}$ is a diagonal matrix with the eigenvalues of A on the diagonal), we get

$$
\begin{aligned}
H &= SGS^{-1} \\
&= S\left(I - \frac{iR}{2}\sin \xi A\right)^{-1}\left(I + \frac{iR}{2}\sin \xi A\right)S^{-1} \\
&= S\left(I - \frac{iR}{2}\sin \xi A\right)^{-1}S^{-1}S\left(I + \frac{iR}{2}\sin \xi A\right)S^{-1} \\
&= \left[S\left(I - \frac{iR}{2}\sin \xi A\right)S^{-1}\right]^{-1}\left(I + \frac{iR}{2}\sin \xi D\right) \\
&= \left(I - \frac{iR}{2}\sin \xi D\right)^{-1}\left(I + \frac{iR}{2}\sin \xi D\right).
\end{aligned}
$$

Thus H is a diagonal matrix with the eigenvalues of G on the diagonal. Then it is clear that $H^*H = I \leq I$, so by Theorem 6.2.5-(ii) (S is again a constant, invertible matrix) we see that the Crank-Nicolson scheme is unconditionally stable (and hence, convergent).

Remark: We should note that the method that we have used so often to apply Theorem 6.2.5-(ii) will apply any time the scheme G can be represented as a polynomial or rational function of the matrix A. In all of these cases, the stability results reduces to stability results for the difference scheme for the analogous scalar partial differential equation where the eigenvalue of the matrix A replaces the matrix A.

Example 6.2.9 Develop a flux splitting implicit finite difference scheme.

Solution: We again consider the hyperbolic partial differential equation (6.2.67). We let S, D, D_+, D_-, A_+ and A_- be as in Section 6.2.1 and rewrite partial differential equation (6.2.67) as

$$\mathbf{v}_t = A_+\mathbf{v}_x + A_-\mathbf{v}_x. \tag{6.2.76}$$

This is surely equivalent to the original system, and in this form the "positive" and "negative" directions are clear. By considering our approach for an explicit flux splitting scheme (specifically, difference scheme (6.2.61)), we consider the difference scheme

$$\mathbf{u}_k^{n+1} - RA_+\delta_+\mathbf{u}_k^{n+1} - RA_-\,\delta_-\mathbf{u}_k^{n+1} = \mathbf{u}_k^n. \tag{6.2.77}$$

To analyze the stability of the above scheme, we take the discrete Fourier transform to see that the amplification matrix is given by

$$G(\xi) = \left[I - R(\cos \xi - 1)A_+ - R(1 - \cos \xi)A_- + i\left[-R\sin \xi A_+ - R\sin \xi A_-\right]\right]^{-1}.$$

It is not hard to see that the eigenvalues of G, $\lambda_j(\xi)$, $j = 1, \cdots, K$ are

$$\lambda_j(\xi) = \begin{cases} \frac{1}{1 - R(\cos \xi - 1)\mu_j - iR\sin \xi \mu_j} & \text{when } \mu_j \geq 0 \\ \frac{1}{1 - R(1 - \cos \xi)\mu_j - iR\sin \xi \mu_j} & \text{when } \mu_j < 0 \end{cases} \tag{6.2.78}$$

By the work done in Section 5.3.1, we see that $|\lambda_j| \leq 1$, $j = 1, \cdots, K$. Hence, difference scheme (6.2.77) satisfies the von Neumann condition.

If, as we have done often before, we let

$$H = SGS^{-1}$$
$$= \left[I - R(\cos\xi - 1)D_+ - R(1 - \cos\xi)D_- + i\left[- R\sin\xi D_+ - R\sin\xi D_- \right] \right]^{-1},$$

we can use Theorem 6.2.5(ii) to see that difference scheme (6.2.77) is unconditionally stable, and, hence, convergent (as before the hypotheses on S are satisfied because S is the constant matrix associated with the partial differential equation).

Remark: As with the explicit, flux split scheme, zero eigenvalues (having an appropriate number of associated eigenvectors) are also permissible by altering the definition of D_+ and D_- as we did in Section 6.2.1.

Thus we see that as in the scalar case, implicit schemes are generally unconditionally stable. Of course, there are many other schemes that can and should be considered.

HW 6.2.8 Analyze the consistency, stability and convergence of the following difference schemes for solving the system of partial differential equations given in (6.2.67):

$$(i) \qquad \mathbf{u}_k^{n+1} - \frac{R}{2} A\delta_0 \mathbf{u}_k^{n+1} = \mathbf{u}_k^n \qquad\qquad (6.2.79)$$

$$(ii) \qquad \mathbf{u}_k^{n+1} - RA\delta_+ \mathbf{u}_k^{n+1} = \mathbf{u}_k^n. \qquad\qquad (6.2.80)$$

HW 6.2.9 Discuss the consistency, stability and convergence of the following flux split difference scheme

$$\left(I - r\delta^2 - RB_{1_+}\delta_+ - RB_{1_-}\delta_- \right)\mathbf{u}_k^{n+1} = (I + B_0)\mathbf{u}_k^n \qquad (6.2.81)$$

for approximating the solution to system (6.2.47).

6.3   Initial–Boundary–Value Problems

We next wish to consider initial–boundary–value problems for systems of equations. We will concentrate most of our work on hyperbolic systems. This is not because parabolic systems are unimportant, but because parabolic systems are more nicely behaved. As with scalar equations, we can easily obtain necessary conditions for stability of numerical schemes for systems by using the fact that *the von Neumann criteria for the analogous initial–value problem will be necessary for stability for an initial–boundary–value problem*. Also, as was the case for scalar equations, symmetry will be needed to give a sufficient condition for stability (and, hence, convergence). For an equation like partial differential equation (6.1.2), this will generally mean requiring at least that the matrix B_2 be symmetric and $B_1 = \Theta$. For initial–boundary–value problems for systems, except for very nice parabolic systems and hyperbolic equations with periodic boundary conditions, we have almost no chance of obtaining a sufficient condition for stability. Hence, we must be content with the necessary conditions that we inherit from the associated initial–value problem and proceed carefully with numerical experimentation.

Other than the difficulty of obtaining sufficient conditions for convergence, the numerical treatment of systems of equations should not cause many new problems. For parabolic systems, we will generally have enough boundary conditions that are in the right places. This is not the case for hyperbolic systems (which by now, we should expect). In Section 6.3.1 we discuss boundary conditions for hyperbolic systems of partial differential equations.

The implementation of implicit schemes for systems of partial differential equations will be more difficult than their scalar analog. We discuss the implementation of numerical schemes for systems of equations in Section 6.3.2.

6.3.1   Boundary Conditions

6.3.1.1   Periodic Boundary Conditions

We first mention periodic boundary conditions. It is not difficult to see that systems of hyperbolic equations having periodic boundary conditions are as nicely behaved as their scalar counterpart. The partial differential equations will have solutions and numerical schemes can be used (obtaining sufficient conditions for convergence based on the analog to Theorem 3.2.3 for systems of equations) to approximate these solutions. The only difficulty that we face related to periodic boundary conditions for systems of equations is the part of the implementation of implicit schemes when we must solve block circulant matrices. This topic will be discussed as a part of the implementation discussion given in Section 6.3.2.3.

6.3.1.2   Dirichlet Boundary Conditions

The difficulty that occurs when we consider Dirichlet boundary conditions is how many boundary conditions we can apply and where can we apply them. This problem should be predictable. Just as was the case with scalar hyperbolic initial–boundary–value problems, we cannot arbitrarily assign boundary conditions for systems of hyperbolic equations. To see how many boundary conditions we can assign and where we can assign them, we begin by considering the system of hyperbolic partial differential equations

$$\mathbf{V}_t = D\mathbf{V}_x, \ x \in (0,1), \ t > 0, \tag{6.3.1}$$

where D is a $K \times K$ diagonal matrix with $\mu_1, \cdots, \mu_K$ on the diagonal. Suppose that $\mu_1 > 0, \cdots, \mu_{K_+} > 0$ and $\mu_{K_++1} < 0, \cdots, \mu_K < 0$. The system of equations (6.3.1) can be written as

$$V_{j_t} = \mu_j V_{j_x}, \ j = 1, \cdots, K_+ \tag{6.3.2}$$

$$V_{j_t} = \mu_j V_{j_x}, \ j = K_+ + 1, \cdots, K \tag{6.3.3}$$

and from Section 5.5.2 we know that

- each of the equations in (6.3.2) gets a boundary condition at $x = 1$ (and none at $x = 0$), and

- each of the equations in (6.3.3) gets a boundary condition at $x = 0$ (and none at $x = 1$.)

Thus, it is clear that *for system (6.3.1), we can assign boundary conditions at $x = 0$ to $V_{K_++1}, \cdots, V_K$ and boundary conditions at $x = 1$ to $V_1, \cdots, V_{K_+}$. Any other boundary conditions needed must be* **numerical boundary conditions** *discussed briefly in Section 5.6.2 and in-depth in Chapter 8.*

We should also realize that since the solution at $x = 0$ and $t > 0$ for $V_1, \cdots, V_{K_+}$ is completely determined by the initial conditions and boundary conditions at $x = 1$, they can be considered as known functions and we can assign the boundary condition

$$\mathbf{V}_2(0, t) = \mathbf{g}_2(t) + S_2 \mathbf{V}_1(0, t) \tag{6.3.4}$$

at $x = 0$, where $\mathbf{V}_1 = [V_1, \cdots, V_{K_+}]^T$, $\mathbf{V}_2 = [V_{K_++1}, \cdots, V_K]^T$, S_2 is a $K_- \times K_+$ ($K_- = K - K_+$) matrix and $\mathbf{g}_2$ is a known function. Likewise, we can assign the boundary condition

$$\mathbf{V}_1(1, t) = \mathbf{g}_1(t) + S_1 \mathbf{V}_2(1, t) \tag{6.3.5}$$

at $x = 1$ where S_1 is a $K_+ \times K_-$ matrix and $\mathbf{g}_2$ is a known function.

We next consider a more general hyperbolic system of partial differential equations

$$\mathbf{v}_t = A\mathbf{v}_x, \ x \in (0, 1), \ t > 0. \tag{6.3.6}$$

As in Section 6.2.1, we have a matrix S such that $D = SAS^{-1}$ is the diagonal matrix with the eigenvalues of A on the diagonal (we will assume that the diagonalization was done in such a way that the first K_+ eigenvalues are positive and the last $K_- = K - K_+$ eigenvalues are negative), $\mathbf{V} = [\mathbf{V}_1^T \ \mathbf{V}_2^T]^T = S\mathbf{v}$ and D_{1+} and D_{2-} the diagonal matrices containing the positive and negative eigenvalues of A, respectively. Using the uncoupling procedure used in Section 6.2.1, we are left with the pair of systems of equations in characteristic variables

$$\mathbf{V}_{1_t} = D_{1+}\mathbf{V}_{1_x} \tag{6.3.7}$$

$$\mathbf{V}_{2_t} = D_{2-}\mathbf{V}_{2_x}, \tag{6.3.8}$$

which are equivalent to system (6.3.6). Since both systems (6.3.7) and (6.3.8) have diagonal matrices and are uncoupled, we know (and have seen from the discussion above)

- each of the equations in system (6.3.7) can have a boundary condition at $x = 1$ (and none at $x = 0$), and

- each of the equations in system (6.3.8) can have a boundary condition at $x = 0$ (and none at $x = 1$).

We see clearly that we can make the problem well defined by assigning boundary conditions to V_1 at $x = 1$ and to V_2 at $x = 0$. Moreover, we should realize that the general form of the boundary conditions that we can assign are given by (6.3.4) and (6.3.5). Assigning boundary conditions to V_1 and V_2 is equivalent to assigning boundary conditions to certain linear combinations of the v_j's. Since $V = Sv$, we can write $V_1 = S_T v$ and $V_2 = S_B v$, where S_T and S_B are the top K_+ and bottom K_- rows of S, respectively. Then it is clear exactly which linear combination of the v_j's to which we are assigning boundary conditions when we assign boundary conditions to V_1 and V_2.

Often, it is difficult or impossible to assign boundary conditions in terms of the characteristic variables. For physical reasons, it is often advantageous to assign boundary conditions directly to some of the v's rather than the given linear combinations of the v's. Clearly, we will not be able to assign boundary conditions at $x = 0$ to more than K_- of the v's and at $x = 1$ to more than K_+ of the v's. Also, it should be fairly clear that the decision on which of the v's to assign boundary conditions cannot be made arbitrarily. However, we would like to have more flexibility on how we choose which v's or which linear combination of the v's get assigned boundary conditions.

Consider the boundary condition at $x = 1$. At $x = 1$ we know that we can assign a general boundary condition of the form

$$V_1(1,t) = g_1(t) + S_1 V_2(1,t) \tag{6.3.9}$$

where both g_1 and S_1 are free for us to choose. Also, we know that $V(1,t) = S_T v(1,t)$ and $V_2(1,t) = S_B v(1,t)$. Hence, the most general boundary condition that we can apply to the primitive variables can be written as

$$S_T v(1,t) = g_1(t) + S_1 S_B v(1,t) \tag{6.3.10}$$

or

$$(S_T - S_1 S_B)\, v(1,t) = g_1(t). \tag{6.3.11}$$

Thus, we can assign the K_+ boundary conditions at $x = 1$ given by the linear combinations described by equation (6.3.11) *where the $K_+ K_-$ values* *of S_1 are free to be chosen and may be chosen so as to simplify the form of the boundary conditions.* Generally, relationship (6.3.11) and the choice of the terms of S_1 enable us to assign reasonable boundary conditions to the primitive variables. Likewise, the general boundary condition at $x = 0$ can be written in terms of the characteristic variables as

$$\mathbf{V}_2(0, t) = \mathbf{g}_2(t) + S_2 \mathbf{V}_1(0, t) \qquad (6.3.12)$$

and in terms of the primitive variables as

$$S_B \mathbf{v}(0, t) = \mathbf{g}_2(t) + S_2 S_T \mathbf{v}(0, t) \qquad (6.3.13)$$

or

$$(S_B - S_2 S_T) \mathbf{v}(0, t) = \mathbf{g}_2(t). \qquad (6.3.14)$$

As with the case of assigning boundary conditions at $x = 1$, the choice of the $K_+ K_-$ elements of S_2 will often allow us to assign reasonably nice boundary conditions at $x = 0$. If we are lucky, we may be able to assign boundary conditions to the primitive variables of our choice. Consider the following example.

Example 6.3.1 Consider the system of equations

$$\mathbf{v}_t = A \mathbf{v}_x \ x \in (0, 1) \qquad (6.3.15)$$

where the matrix A is given by

$$A = \begin{pmatrix} 0 & 4/3 & -2/3 \\ 1 & -2/3 & -5/3 \\ -1 & -4/3 & -1/3 \end{pmatrix}.$$

Discuss some of the possible boundary conditions for partial differential equation (6.3.15).

Solution: Be begin by noting that the eigenvalues of A are 2, -1 and -2, $K_+ = 1$, $K_- = 2$, the eigenvectors are given by

$$\mathbf{u}_1 = \begin{bmatrix} 1 \\ 1 \\ -1 \end{bmatrix}, \ \mathbf{u}_2 = \begin{bmatrix} 2 \\ -1 \\ 1 \end{bmatrix} \text{ and } \mathbf{u}_3 = \begin{bmatrix} -1 \\ 2 \\ 1 \end{bmatrix},$$

and the matrices S and S^{-1} that diagonalize A ($D = SAS^{-1}$ where D is the diagonal matrix with 2, -1 and -2 on the diagonal) are given by

$$S = \begin{pmatrix} 1/3 & 1/3 & -1/3 \\ 1/3 & 0 & 1/3 \\ 0 & 1/3 & 1/3 \end{pmatrix} \text{ and } S^{-1} = \begin{pmatrix} 1 & 2 & -1 \\ 1 & -1 & 2 \\ -1 & 1 & 1 \end{pmatrix}.$$

These calculations are not difficult, but they are even easier if the machine does them for you (Maple did mine). Thus we see that we get to assign two boundary conditions at $x = 0$ and one boundary condition at $x = 1$. The system of equations (6.3.15) is equivalent to the system

$$\mathbf{V}_t = D \mathbf{V}_x \ x \in (0, 1) \qquad (6.3.16)$$

where the matrix D is the diagonal matrix with 2, -1 and -2 on the diagonal, $\mathbf{V} = S\mathbf{v}$ and $\mathbf{V}$ are the characteristic variables associated with system (6.3.15). We get to assign boundary conditions to V_1 at $x = 1$ and V_2 and V_3 at $x = 0$. In the notations of this section, $\mathbf{V}_1$ is the 1-vector $\mathbf{V}_1 = [V_1]$ and $\mathbf{V}_2$ is the 2-vector $\mathbf{V}_2 = [V_2 \quad V_3]^T$. As in equations (6.3.9) and (6.3.12), we can assign boundary conditions

$$\mathbf{V}_1(1, t) = \mathbf{g}_1 + S_1\mathbf{V}_2(1, t) \tag{6.3.17}$$

where $\mathbf{g}_1$ is a 1-vector and S_1 is a 1×2 matrix, and

$$\mathbf{V}_2(0, t) = \mathbf{g}_2(t) + S_2\mathbf{V}_1(0, t) \tag{6.3.18}$$

where $\mathbf{g}_2$ is a 2-vector and S_2 is a 2×1 matrix. Since $\mathbf{V} = S\mathbf{v}$,

$$\mathbf{V}_1 = S_T\mathbf{v} = \left(\begin{array}{ccc} 1/3 & 1/3 & -1/3 \end{array}\right)\mathbf{v} = \left(\begin{array}{c} \frac{1}{3}v_1 + \frac{1}{3}v_2 - \frac{1}{3}v_3 \end{array}\right) \tag{6.3.19}$$

and

$$\mathbf{V}_2 = S_B\mathbf{v} = \left(\begin{array}{ccc} 1/3 & 0 & 1/3 \\ 0 & 1/3 & 1/3 \end{array}\right)\mathbf{v} = \left(\begin{array}{c} \frac{1}{3}v_1 + \frac{1}{3}v_3 \\ \frac{1}{3}v_2 + \frac{1}{3}v_3 \end{array}\right). \tag{6.3.20}$$

Then assigning boundary conditions (6.3.17) and (6.3.18) to the characteristic variables is equivalent to assigning boundary conditions to the following combinations of the primitive variables.

$$\frac{1}{3}v_1(1, t) + \frac{1}{3}v_2(1, t) - \frac{1}{3}v_3(1, t) = g_{1_1}(t) + s^1_{1\,1}\left(\frac{1}{3}v_1(1, t) + \frac{1}{3}v_3(1, t)\right)$$
$$+ s^1_{1\,2}\left(\frac{1}{3}v_2(1, t) + \frac{1}{3}v_3(1, t)\right) \tag{6.3.21}$$

$$\frac{1}{3}v_1(0, t) + \frac{1}{3}v_3(0, t) = g_{2_1}(t) + s^2_{1\,1}\left(\frac{1}{3}v_1(0, t) + \frac{1}{3}v_2(0, t) - \frac{1}{3}v_3(0, t)\right) \tag{6.3.22}$$

$$\frac{1}{3}v_2(0, t) + \frac{1}{3}v_3(0, t) = g_{2_2}(t) + s^2_{2\,1}\left(\frac{1}{3}v_1(0, t) + \frac{1}{3}v_2(0, t) - \frac{1}{3}v_3(0, t)\right) \tag{6.3.23}$$

where

$$S_1 = \left(\begin{array}{cc} s^1_{1\,1} & s^1_{1\,2} \end{array}\right) \text{ and } S_2 = \left(\begin{array}{c} s^2_{1\,1} \\ s^2_{2\,1} \end{array}\right).$$

We should realize that equation (6.3.21) is the same as equation (6.3.10) and equations (6.3.22) and (6.3.23) are the same as equation (6.3.13). We can rewrite equations (6.3.21)–(6.3.23) as

$$(1 - s^1_{11})v_1(1, t) + (1 - s^1_{12})v_2(1, t) + (-1 - s^1_{11} - s^1_{12})v_3(1, t) = 3g_{1_1}(t) \tag{6.3.24}$$
$$(1 - s^2_{11})v_1(0, t) - s^2_{11}v_2(0, t) + (1 + s^2_{11})v_3(0, t) = 3g_{2_1}(t) \tag{6.3.25}$$
$$-s^2_{21}v_1(0, t) + (1 - s^2_{21})v_2(0, t) + (1 + s^2_{21})v_3(0, t) = 3g_{2_2}(t). \tag{6.3.26}$$

We want to emphasize that equations (6.3.24)–(6.3.26) represent the most general forms of boundary conditions allowable for partial differential equation (6.3.15). We should also point out that equation (6.3.24) is equivalent to equation (6.3.11) and equations (6.3.25) and (6.3.26) are equivalent to equation (6.3.14) (in each case multiplied by three).

Thus we see that if we choose $s^1_{12} = 1$, $s^1_{11} = -2$, $s^2_{11} = 0$ and $s^2_{21} = 1$, we are left with boundary conditions

$$v_1(1, t) = g_{1_1}(t)$$
$$v_1(0, t) + v_3(0, t) = 3g_{2_1}(t) \tag{6.3.27}$$
$$-v_1(0, t) + 2v_3(0, t) = 3g_{2_2}(t). \tag{6.3.28}$$

Equations (6.3.27) and (6.3.28) can be solved for $v_1(0, t)$ and $v_3(0, t)$. Then, we find that it is permissible to assign the following boundary conditions to the primitive variables.

$$v_1(1, t) = g_{1_1}(t) \tag{6.3.29}$$
$$v_1(0, t) = 2g_{2_1}(t) - g_{2_2}(t) \tag{6.3.30}$$
$$v_3(0, t) = g_{2_1}(t) + g_{2_2}(t) \tag{6.3.31}$$

Since g_{1_1}, g_{2_1} and g_{2_2} are arbitrary, we see that we can assign arbitrary boundary conditions to these variables at the designated points.

Remark 1: We should note that the boundary conditions found above are not the only permissible boundary conditions. If we choose $s^1_{11} = 1$ and $s^1_{12} = -2$, we see that boundary condition (6.3.24) becomes $v_2(1, t) = g_{1_1}(t)$. If we choose $s^1_{11} = 1$ and $s^1_{12} = 1$, boundary condition (6.3.24) becomes $v_3(1, t) = -g_{1_1}(t)$. Thus, though we can assign only one boundary condition at $x = 1$, we see that we can assign a condition to any one of the three primitive variables at $x = 1$. Likewise, if we let $s^2_{11} = 1$ and $s^2_{21} = 0$, we get to assign boundary conditions to $v_2(0, t)$ and $v_3(0, t)$. And, if we let $s^2_{11} = -1$ and $s^2_{21} = -1$, we get to assign boundary conditions to $v_1(0, t)$ and $v_2(0, t)$. Thus, though we can assign only two boundary conditions at $x = 0$, we can assign boundary conditions to any pair of the variables at $x = 0$.

Remark 2: The result of the above example showed that at least some time we can assign boundary conditions to whichever of the variables that we would like (as long as we assign the correct number at each end). If we consider the trivial system of partial differential equations

$$\mathbf{v}_t = A\mathbf{v}_x$$

where

$$A = \begin{pmatrix} 2 & 0 & 0 \\ 0 & -1 & 0 \\ 0 & 0 & -2 \end{pmatrix}$$

we see that in terms of the primitive variables we can only assign boundary conditions to v_1 at $x = 1$ and v_2 and v_3 at $x = 0$. Hence, the result is not always as nice as that demonstrated in Example 6.3.1.

Remark 3: It should be clear that it is also possible to have zero eigenvalues (as long as the zero eigenvalue is associated with a full complement of eigenvectors which will be the case for hyperbolic systems). The partial differential equation analogous to partial differential equations (6.3.2) and (6.3.3) associated with a zero eigenvalue is of the form $V_{j_t} = 0$. Hence, the characteristic variable associated with a zero eigenvalue is completely deter-mined by the initial condition and will not be assigned boundary conditions at either boundary (or can be assigned the constant boundary condition consistent with the given initial condition).

HW 6.3.1 Consider the system of equations

$$\mathbf{v}_t = A\mathbf{v}_x \ x \in (0,1) \tag{6.3.32}$$

where the matrix A is given by

$$A = \begin{pmatrix} 5/3 & 2/3 \\ 1/3 & 4/3 \end{pmatrix}.$$

(a) For the above system, show that we can assign two boundary conditions at $x = 1$ and none at $x = 0$.
(b) Compute the characteristic variables V_1 and V_2 for the above system in terms of the primitive variables v_1 and v_2.
(c) Determine boundary conditions that can be assigned to the charac-teristic variables that will be equivalent to assigning Dirichlet boundary conditions $v_1(1,t) = 2.0$ and $v_2(1,t) = \sin 2t$.

HW 6.3.2 Consider the system of equations (6.3.32) where the matrix A is given by

$$A = \begin{pmatrix} -5/3 & -2/3 \\ -1/3 & -4/3 \end{pmatrix}.$$

(a) For the above system, show that we can assign two boundary conditions at $x = 0$ and none at $x = 1$.
(b) Compute the characteristic variables V_1 and V_2 for the above system in terms of the primitive variables v_1 and v_2.
(c) Determine boundary conditions that can be assigned to the charac-teristic variables that will be equivalent to assigning Dirichlet boundary conditions $v_1(0,t) = \sin 2t$ and $v_2(0,t) = \sin 3t$.

HW 6.3.3 Consider the system of equations (6.3.32) where the matrix A is given by

$$A = \begin{pmatrix} -1 & -2 \\ -1 & 0 \end{pmatrix}.$$

(a) For the above system, show that we can assign one boundary condition at $x = 0$ and one boundary condition at $x = 1$.
(b) Compute the characteristic variables V_1 and V_2 for the above system in terms of the primitive variables v_1 and v_2.
(c) Determine boundary conditions that can be assigned to the charac-teristic variables that will be equivalent to assigning Dirichlet boundary conditions $v_1(0,t) = \sin 2t$ and $v_1(1,t) = 0$.

HW 6.3.4 Consider the system of equations (6.3.32) where the matrix A is given by

$$A = \begin{pmatrix} 5/3 & 1 & 5/3 \\ -1/3 & -1 & -7/3 \\ 1/3 & -1 & 1/3 \end{pmatrix}.$$

(a) For the above system, show that we can assign one boundary condition at $x = 0$ and two boundary conditions at $x = 1$.

(b) Compute the characteristic variables V_1 and V_2 for the above system in terms of the primitive variables v_1 and v_2.

(c) Determine boundary conditions that can be assigned to the characteristic variables that will be equivalent to assigning Dirichlet boundary conditions $v_1(1,t) = 1$ $v_2(1,t) = \sin 2t$ and $v_3(0,t) = \cos 2t$.

(d) Determine boundary conditions that can be assigned to the characteristic variables that will be equivalent to assigning Dirichlet boundary conditions $v_1(1,t) = 1$ $v_3(1,t) = \cos 2t$ and $v_1(0,t) = \sin 2t$.

6.3.2   Implementation

6.3.2.1   Explicit Schemes

We next proceed with the implementation of difference schemes for systems of partial differential equations. We first state that the implementation of an explicit scheme is not very different from the analogous scalar implementation. Using the notation introduced in Section 6.1, we know that a general explicit scheme for an initial–boundary–value problem can be written as

$$\mathbf{u}_k^{n+1} = Q_2 \mathbf{u}_k^n + \Delta t \mathbf{G}_k^n, \quad k = 1, \cdots, M-1, \qquad (6.3.33)$$

where $Q_2 = \sum_{j=-m_1}^{m_2} Q_{2j} S_+^j$, the vectors $\mathbf{u}_k^{n+1}$, $\mathbf{u}_k^n$ and $\mathbf{G}_k^n$ are K vectors and the matrices Q_{2j} are $K \times K$ matrices. Equation (6.3.33) can be considered as K separate equations and each of the K equations can be implemented just as the scalar explicit schemes were implemented. It is also possible, and sometimes preferable, to implement equation (6.3.33) directly using subroutines that generate the matrices and do the appropriate matrix-vector multiplies and vector adds. Generally, difference schemes such as equation (6.3.33) are not much more difficult to implement than an explicit difference scheme for a scalar equation.

HW 6.3.5 Use the FTFS scheme to approximate the solution to the problem given in HW6.3.1. Use the boundary conditions given in part (c) on the primitive variables, the initial condition

$$\mathbf{v}(x,0) = \begin{bmatrix} 2e^{x-1} \\ \sin \pi x \end{bmatrix}$$

and $M = 100$.

HW 6.3.6 Use the FTBS scheme to approximate the solution to the problem given in HW6.3.2. Use the boundary conditions given in part (c) on the primitive variables, the initial condition

$$\mathbf{v}(x,0) = \begin{bmatrix} 2\sin 2\pi x \\ \sin \pi x \end{bmatrix}$$

and $M = 100$.

HW 6.3.7 (a) Use the Lax-Wendroff scheme to approximate the solution to the problem given in HW6.3.1. Use the boundary conditions given in part (c) on the primitive variables, $M = 100$ and the initial condition given in HW6.3.5. Try numerical boundary conditions $u_{1_0}^n = u_{1_1}^n$ and $u_{2_0}^n = u_{2_1}^n$.
(b) Show that the numerical boundary conditions $u_{1_0}^n = u_{1_1}^n$ and $u_{2_0}^n = u_{2_1}^n$ are equivalent to the numerical boundary conditions on the characteristic variables, $U_{1_0}^n = U_{1_1}^n$ and $U_{2_0}^n = U_{2_1}^n$.

HW 6.3.8 Use the Lax-Wendroff scheme to approximate the solution to the problem given in HW6.3.3. Use the boundary conditions given in part (c) on the primitive variables, the initial condition

$$\mathbf{v}(x,0) = \begin{bmatrix} 2\sin 4\pi x \\ \sin 2\pi x \end{bmatrix}$$

and $M = 100$. Try numerical boundary conditions $u_{1_0}^n = u_{1_1}^n$ and $u_{2_0}^n = u_{2_1}^n$.

HW 6.3.9 Use the Lax-Wendroff scheme to approximate the solution to the problem given in HW6.3.4. Use the boundary conditions given in part (c) on the primitive variables, the initial condition

$$\mathbf{v}(x,0) = \begin{bmatrix} \cos 4\pi x \\ (1-x)\cos \pi x \\ e^{-x} \end{bmatrix}$$

and $M = 100$. Try numerical boundary conditions $u_{3_M}^n = u_{3_{M-1}}^n$, $u_{1_0}^n = u_{1_1}^n$ and $u_{2_0}^n = u_{2_1}^n$.

HW 6.3.10 Use the Lax-Wendroff scheme to approximate the solution to the problem given in HW6.3.4. Use the boundary conditions given in part (d) on the primitive variables, $M = 100$ and the initial condition

$$\mathbf{v}(x,0) = \begin{bmatrix} \sin \frac{\pi}{2} x \\ \cos \pi x \\ xe^{1-x} \end{bmatrix}.$$

Try numerical boundary conditions $u_{2_M}^n = u_{2_{M-1}}^n$, $u_{2_0}^n = u_{2_1}^n$ and $u_{3_0}^n = u_{3_1}^n$.

6.3.2.2 Implicit Schemes

We next consider the implementation of implicit schemes. Instead of the general implicit scheme given by equation (6.2.2), we consider the case that is most common where $m_3 = m_4 = 1$. If we consider an implicit scheme for solving an initial–boundary–value problem assuming some combination of appropriate boundary conditions, we face solving a block matrix equation of the form

$$\hat{Q}\mathbf{u}^{n+1} = Q\mathbf{u}^n + \Delta t\mathbf{G}^n, \qquad (6.3.34)$$

where

$$\mathbf{u}^{n+1} = [\mathbf{u}_1^{n+1} \cdots \mathbf{u}_{M-1}^{n+1}]^T, \quad \mathbf{u}^n = [\mathbf{u}_1^n \cdots \mathbf{u}_{M-1}^n]^T, \quad \mathbf{G}^n = [\mathbf{G}_1^n \cdots \mathbf{G}_{M-1}^n]^T,$$

and Q and $\hat{Q}$ are $K(M-1) \times K(M-1)$ matrices ($(M-1)\times(M-1)$ matrices with elements that are $K \times K$ matrices). For example, consider the system of partial differential equations (6.3.6) on $(0,1)$ (where we assume that A has K positive eigenvalues) along with Dirichlet boundary conditions at $x = 1$ and the following Crank-Nicolson difference scheme

$$\mathbf{u}_k^{n+1} - \frac{R}{4}A\delta_0\mathbf{u}_k^{n+1} = \mathbf{u}_k^n + \frac{R}{4}A\delta_0\mathbf{u}_k^n. \qquad (6.3.35)$$

Since we assumed that all of the eigenvalues of A were positive, it is appropriate that we have K Dirichlet boundary conditions given at $x = 1$. To apply difference scheme (6.3.35), we must provide K numerical boundary conditions at $x = 0$. At this time, let us use

$$u_{j_0}^{n+1} = u_{j_0}^n - R\left(u_{j_1}^n - u_{j_0}^n\right), \quad j = 1,\cdots,K \qquad (6.3.36)$$

as our numerical boundary conditions at $x = 0$. Of course, *we do not know that difference scheme (6.3.35) along with Dirichlet boundary conditions at $x = 1$ and numerical boundary conditions (6.3.36) at $x = 0$ will be stable.* The choice of numerical boundary conditions (6.3.36) are very convenient because they are implemented much like Dirichlet boundary conditions (i.e. they are easy to implement). As we said earlier, the Crank-Nicolson scheme (6.3.35) along with the Dirichlet and numerical boundary conditions will require that we solve an equation of the form (6.3.34). In this case the coefficient matrix $\hat{Q}$ is the block tridiagonal matrix

$$\hat{Q} = T\left[\frac{R}{4}A, I, -\frac{R}{4}A\right] \qquad (6.3.37)$$

or

$$\hat{Q} = \begin{pmatrix} I & -\frac{R}{4}A & \Theta & \cdots & \\ \frac{R}{4}A & I & -\frac{R}{4}A & \Theta & \cdots \\ & & \cdots & & \\ \cdots & \Theta & \frac{R}{4}A & I & -\frac{R}{4}A \\ & & \Theta & \frac{R}{4}A & I \end{pmatrix}. \qquad (6.3.38)$$

One approach to solve equation (6.3.34) is to solve it as a banded matrix. However, even when K is reasonably small, the band width of matrix (6.3.38) is reasonably large. Also, using this approach, we are not using the information that we know $\hat{Q}$ is a block tridiagonal matrix. A more common approach is to solve equation (6.3.34) where $\hat{Q}$ is given by (6.3.38) as a block tridiagonal matrix, using the essentially the same algorithm used earlier for tridiagonal matrices. If we extend the Thomas Algorithm, introduced in Section 2.6.3 to the case for block tridiagonal matrices, $T[A, B, C] = \mathbf{R}$, we obtain the following algorithm.

$$C_1' = B_1^{-1} C_1$$
$$\mathbf{R}_1'' = B_1^{-1} \mathbf{R}_1$$

For $j = 2, \cdots, m - 1$
$$B_j' = B_j - A_j C_{j-1}'$$
$$\mathbf{R}_j' = \mathbf{R}_j - A_j \mathbf{R}_{j-1}''$$
$$C_j' = B_j'^{-1} C_j$$
$$\mathbf{R}_j'' = B_j'^{-1} \mathbf{R}_j'$$
Next j

$$B_m' = B_m - A_m C_{m-1}'$$
$$\mathbf{R}_m' = \mathbf{R}_m - A_m \mathbf{R}_{m-1}''$$
$$\mathbf{R}_m''' = B_m'^{-1} \mathbf{R}_m'$$

For $j = m - 1, \cdots, 1$
$$\mathbf{R}_j''' = \mathbf{R}_j'' - C_j' \mathbf{R}_{j+1}'''$$
Next j

Before the above algorithm is implemented, it is best to build a system of subroutines designed to do the necessary matrix multiplies, matrix-vector multiplies, vector adds, and matrix inverses. Generally, unless K is very small, it is advantageous to use an LU decomposition to calculate the B'^{-1}'s. Hence, we see that except for the inconvenience of replacing the TRID subroutine by a BLKTRID subroutine, the implementation of most implicit schemes of the form (6.3.34) is similar to that for the scalar analog.

One last comment that we make concerns the form of matrix (6.3.38). Two very big assumptions that we made to make this example look nice was that (1) all of the eigenvalues of A were positive (so that we could have K Dirichlet boundary conditions given at $x = 1$ and the K numerical boundary conditions that are necessary are all given at $x = 0$), and (2) that the numerical boundary conditions are of the form (6.3.36). Both of these assumptions made the coefficient matrix $\hat{Q}$ nicer and hid all of the affects of the numerical boundary condition in the right hand side (which, since we never wrote out the right hand side, we never saw). We must realize that if we have a mixture of analytic and numerical boundary conditions at both ends of the interval and if instead of numerical boundary condition (6.3.36) we use some implicit condition (say

$$u_{j_0}^{n+1} + R\left(u_{j_1}^{n+1} - u_{j_0}^{n+1}\right) = u_{j_0}^n, \; j = 1, \cdots, K)$$

the first and last rows of matrix (6.3.38) could be very different (and not necessarily easy to find). An implicit implementation of a mixture of analytic and numerical boundary conditions can be difficult. Consider the following example.

Example 6.3.2 Discuss the implementation of the BTCS scheme to approximate the solution to the system of partial differential equations (6.3.15) along with boundary conditions $v_1(1,t) = \sin t$, $v_1(0,t) = \sin 4t$ and $v_3(0,t) = \cos 2t$.

Solution: We saw in Example 6.3.1 that the above boundary conditions are acceptable boundary conditions to be assigned along with partial differential equation (6.3.15). Since the BTCS scheme is a centered scheme, we know that we will have to have one numerical boundary condition at $x = 0$ and two numerical boundary conditions at $x = 1$. As we have done in the past, we will suggest some numerical boundary conditions to try and return to show that they are acceptable (or unacceptable) in Chapter 8. Suppose we try numerical boundary conditions $u_{2_0}^n = u_{2_1}^n$, $u_{2_M}^n = u_{2_{M-1}}^n$ and $u_{3_M}^n = u_{3_{M-1}}^n$. Of course, the BTCS scheme will be applied at the interior points of the grid, i.e.

$$RA\mathbf{u}_{k-1}^{n+1} + \mathbf{u}_k^{n+1} - RA\mathbf{u}_{k+1}^{n+1} = \mathbf{u}_k^n, \quad k = 1, \cdots, M - 1. \tag{6.3.39}$$

When this scheme is applied to $k = 1$, the left-most term is given by

$$RA\mathbf{u}_0^{n+1} = RA \begin{bmatrix} u_{1_0}^{n+1} \\ u_{2_0}^{n+1} \\ u_{3_0}^{n+1} \end{bmatrix} = RA \left(\begin{bmatrix} \sin 4(n+1)\Delta t \\ 0 \\ \cos 2(n+1)\Delta t \end{bmatrix} + \begin{bmatrix} 0 \\ u_{2_1}^{n+1} \\ 0 \end{bmatrix} \right). \tag{6.3.40}$$

Note that we have used both the boundary conditions at $x = 0$ and the numerical boundary condition at $x = 0$ to obtain the above expression. Obviously, the first term of the right hand side of equation (6.3.40) will go to the right hand side of our difference equation and the second term of the right hand side of equation (6.3.40) will stay on the left hand side of our difference equation. Since

$$RA \begin{bmatrix} 0 \\ u_{2_1}^{n+1} \\ 0 \end{bmatrix} = R \begin{bmatrix} \frac{4}{3}u_{2_1}^{n+1} \\ -\frac{2}{3}u_{2_1}^{n+1} \\ -\frac{4}{3}u_{2_1}^{n+1} \end{bmatrix}$$

and

$$\mathbf{u}_1^{n+1} + RA \begin{bmatrix} 0 \\ u_{2_1}^{n+1} \\ 0 \end{bmatrix} = \begin{pmatrix} 1 & \frac{4}{3}R & 0 \\ 0 & 1 - \frac{2}{3}R & 0 \\ 0 & -\frac{4}{3}R & 1 \end{pmatrix} \mathbf{u}_1^{n+1},$$

the difference equation associated with $k = 1$ becomes

$$\begin{pmatrix} 1 & \frac{4}{3}R & 0 \\ 0 & 1 - \frac{2}{3}R & 0 \\ 0 & -\frac{4}{3}R & 1 \end{pmatrix} \mathbf{u}_1^{n+1} - RA\mathbf{u}_2^{n+1} = \mathbf{u}_1^n - RA \begin{bmatrix} \sin 4(n+1)\Delta t \\ 0 \\ \cos 2(n+1)\Delta t \end{bmatrix}. \tag{6.3.41}$$

Likewise, at $k = M - 1$ the term $-RA\mathbf{u}_M^{n+1}$ becomes

$$-RA\mathbf{u}_M^{n+1} = -RA\left(\begin{bmatrix} \sin t \\ 0 \\ 0 \end{bmatrix} + \begin{bmatrix} 0 \\ u_{2_{M-1}}^{n+1} \\ u_{3_{M-1}}^{n+1} \end{bmatrix}\right)$$

(using both the boundary condition and the two numerical boundary conditions at $x = 1$) and the difference equation associated with $k = M - 1$ becomes

$$RA\mathbf{u}_{M-2}^{n+1} + \begin{pmatrix} 1 & -\frac{4}{3}R & \frac{2}{3}R \\ 0 & 1 + \frac{2}{3}R & \frac{5}{3}R \\ 0 & \frac{4}{3}R & 1 + \frac{1}{3}R \end{pmatrix} \mathbf{u}_{M-1}^{n+1} = \mathbf{u}_{M-1}^n + RA\begin{bmatrix} \sin(n+1)\Delta t \\ 0 \\ 0 \end{bmatrix}.$$

$$(6.3.42)$$

Therefore, we combine equations (6.3.39) at $k = 2, \cdots, M - 2$ along with equations (6.3.41) and (6.3.42) to get a system of equations of the form of (6.3.34) where

$$\hat{Q} = \begin{pmatrix} A_1 & -RA & \Theta & \cdots \\ RA & I & -RA & \Theta & \cdots \\ & & \cdots & & \\ \cdots & \Theta & RA & I & -RA \\ & \cdots & \Theta & RA & A_2 \end{pmatrix},$$

$$A_1 = \begin{pmatrix} 1 & \frac{4}{3}R & 0 \\ 0 & 1 - \frac{2}{3}R & 0 \\ 0 & -\frac{4}{3}R & 1 \end{pmatrix}, \quad A_2 = \begin{pmatrix} 1 & -\frac{4}{3}R & \frac{2}{3}R \\ 0 & 1 + \frac{2}{3}R & \frac{5}{3}R \\ 0 & \frac{4}{3}R & 1 + \frac{1}{3}R \end{pmatrix},$$

$Q = I$ and

$$\Delta t\mathbf{G}^n = \begin{bmatrix} -RA\begin{bmatrix} \sin 4(n+1)\Delta t \\ 0 \\ \cos 2(n+1)\Delta t \end{bmatrix} \\ \boldsymbol{\theta} \\ \vdots \\ \boldsymbol{\theta} \\ RA\begin{bmatrix} \sin(n+1)\Delta t \\ 0 \\ 0 \end{bmatrix} \end{bmatrix}.$$

As usual with implicit schemes for systems of equations, these equations must (or probably should) be solved using a block tridiagonal solver.

HW 6.3.11 Solve the problem described in HW6.3.9 (including the numerical boundary conditions) using the BTCS scheme.

HW 6.3.12 Solve the problem described in HW6.3.10 (including the numerical boundary conditions) using the Crank-Nicolson scheme.

6.3.2.3 Periodic Boundary Conditions

And finally, there are times when we wish to solve a system of partial differential equations of the form (6.3.6) with periodic boundary conditions by an implicit scheme. For example, if we again wish to use the Crank-Nicolson scheme (6.3.35), we must consider equation (6.3.35) for

$k = 0, \cdots, M - 1$ and use the periodicity to given us that $\mathbf{u}_M = \mathbf{u}_0$ and $\mathbf{u}_{-1} = \mathbf{u}_{M-1}$. At each time step, we are again left with solving an equation of the form (6.3.34) where $\mathbf{G}^n$ would not appear unless we had a nonhomogeneous term present and the vectors $\mathbf{u}^{n+1}$ and $\mathbf{u}^n$ would be M-vectors (for example $\mathbf{u}^n = [\mathbf{u}_0^n, \cdots, \mathbf{u}_{M-1}^n]^T$). Similar to the case for scalar schemes with periodic boundary conditions, the coefficient matrix $\hat{Q}$ will now be an $M \times M$ block circulant matrix of the form

$$
\hat{Q} = \begin{pmatrix}
I & -\frac{R}{4}A & \Theta & \cdots & \Theta & \frac{R}{4}A \\
\frac{R}{4}A & I & -\frac{R}{4}A & \Theta & \cdots & \\
& \cdots & \cdots & & & \\
& \cdots & \Theta & \frac{R}{4}A & I & -\frac{R}{4}A \\
-\frac{R}{4}A & \Theta & \cdots & & \frac{R}{4}A & I
\end{pmatrix}. \tag{6.3.43}
$$

It is fairly clear that to solve an equation such as (6.3.34) where $\hat{Q}$ is given by (6.3.43) a block Sherman-Morrison Algorithm is needed. Similar to the situation that we considered in Proposition 5.6.1, we suppose that we wish to solve a matrix equation of the form

$$
A\mathbf{x} = \mathbf{b}, \tag{6.3.44}
$$

where A is a $km \times km$ block matrix (an $m \times m$ matrix of $k \times k$ blocks), and $\mathbf{x}$ and $\mathbf{b}$ are km-vectors (m-vectors of k blocks). In addition, we suppose that there is a "nice" matrix B such that A and B are related by

$$
A = B - WZ^T, \tag{6.3.45}
$$

where W and Z are $km \times k$ matrices. If we assume that B and $I - Z^T B^{-1} W$ are nonsingular, then

$$
A^{-1} = B^{-1} + B^{-1}W(I - Z^T B^{-1}W)^{-1}Z^T B^{-1}. \tag{6.3.46}
$$

The proof of the above statement is easy to see by performing the computation AA^{-1}, see HW6.3.13. The implementation of using equation (6.3.46) to solve equation (6.3.44) is as follows.

1. Solve

$$
\begin{aligned}
B\mathbf{y}_1 &= \mathbf{b} \tag{6.3.47} \\
BY_2 &= W \tag{6.3.48}
\end{aligned}
$$

Note that $\mathbf{y}_1$ will be a km-vector and Y_2 wil be a $km \times k$ matrix.

2. Solve

$$
(I - Z^T Y_2)\mathbf{z} = Z^T \mathbf{y}_1. \tag{6.3.49}
$$

The solution $\mathbf{z}$ will be a k-vector.

3. Set

$$\mathbf{x} = A^{-1}\mathbf{b} = \mathbf{y}_1 + Y_2\mathbf{z}. \tag{6.3.50}$$

As in Section 5.6.1, the solutions to equations (6.3.47) and (6.3.48) can be done together. We see that since $\mathbf{b}$ is a km-vector and W is a $km \times k$ matrix, the solution of (6.3.47)–(6.3.48) will consist of solving a system of equations with $k + 1$ right hand sides. When k is relatively small, equation (6.3.49) is a small system to invert. The matrix $I - Z^T Y_2$ is a $k \times k$ matrix. In our applications, $k = K$ will generally be small.

Since the block cyclic matrix (an $m \times m$ matrix with $k \times k$ blocks)

$$\begin{pmatrix} B & C & \Theta & \cdots & \Theta & A \\ A & B & C & \Theta & \cdots & \\ & & \cdots & \cdots & & \\ & \cdots & \Theta & A & B & C \\ C & \Theta & \cdots & \Theta & A & B \end{pmatrix}, \tag{6.3.51}$$

can be written as

$$\begin{pmatrix} B_1 & C & \Theta & \cdots & \\ A & B & C & \Theta & \cdots \\ & & \cdots & & \\ \cdots & \Theta & A & B & C \\ & \cdots & \Theta & A & B_2 \end{pmatrix} + \begin{bmatrix} I \\ \Theta \\ \vdots \\ \Theta \\ I \end{bmatrix} \begin{bmatrix} C & \Theta & \cdots & \Theta & A \end{bmatrix}, \tag{6.3.52}$$

where $B_1 = B - C$ and $B_2 = B - A$, it is clear that the most common implicit schemes (pure implicit, Crank-Nicolson, etc) can be applied to problems involving periodic boundary conditions where the solution scheme involves the rank k Sherman-Morrison update algorithm described above.

HW 6.3.13 Verify the formula given for A^{-1} in equation (6.3.46).

HW 6.3.14 Consider the system of partial differential equations given in HW6.3.4 with initial condition

$$\mathbf{v}(x,0) = \begin{bmatrix} \sin 4\pi x \\ x \sin \pi x \\ x(1-x)e^{-x} \end{bmatrix}$$

and boundary condition $\mathbf{v}(0,t) = \mathbf{v}(1,t)$ for all $t > 0$.
(a) Use the Lax-Wendroff scheme to approximate the solution to the problem described above.
(b) Use the flux split scheme, (6.2.61), to approximate the solution to the problem described above.
(c) Use the Crank-Nicolson scheme to approximate the solution to the problem described above.

In each of the above cases, plot one or two of the components of the solution at enough times so that you can view the interaction of the propagating waves. If possible, use some animation software to view one of the interesting components.

HW 6.3.15 Use the Lax-Wendroff scheme to approximate the solution to the following initial–boundary–value problem

$$\mathbf{v}_t = \begin{pmatrix} -1/3 & -4/3 \\ -2/3 & 1/3 \end{pmatrix} \mathbf{v}_x$$
$$\mathbf{v}(x,0) = [v_{1_0} \quad v_{2_0}]^T$$
$$\mathbf{v}(0,t) = \mathbf{v}(1,t),$$

where

$$v_{1_0}(x) = \begin{cases} 1 & \text{if } x < 0 \\ 2 & \text{if } x \ge 0 \end{cases} \quad \text{and} \quad v_{2_0}(x) = \begin{cases} -1 & \text{if } x < 0 \\ 1 & \text{if } x \ge 0 \end{cases}.$$

6.4 Multilevel Schemes

6.4.1 Scalar Multilevel Schemes

The schemes that have considered so far, for both scalar equations and systems of equations, have generally involved only two time levels (n and $n+1$). The only multilevel schemes, schemes involving more than two time levels, that we have considered are in Chapter 1 where we introduced the leapfrog scheme for solving the one dimensional heat equation (and showed experimentally in HW1.3.1 that the scheme was unstable) and in Chapter 2 where we showed in HW2.3.1-(d) that the Dufort-Frankel scheme was inconsistent. The reason for postponing the discussion of scalar multilevel schemes is that a natural way to analyze the convergence of such schemes is to pose them as a system and then use the results given in this chapter to discuss the convergence.

We begin by considering the leapfrog difference scheme considered in Chapter 1 for solving the one dimensional heat equation, i.e.

$$u_k^{n+1} = u_k^{n-1} + 2r\delta^2 u_k^n. \tag{6.4.1}$$

Example 6.4.1 Discuss the consistency, stability and convergence of the leapfrog difference scheme for the one dimensional heat equation.

Solution: We saw in Chapter 1 that difference scheme (6.4.1) was a consistent scheme for solving the one dimensional heat equation of order $\mathcal{O}(\Delta t^2) + \mathcal{O}(\Delta x^2)$. Also, if the code written for HW1.3.1 was correct, we saw that the scheme was unstable. To consider the stability of such a scheme, we reduce the three level scheme (6.4.1) to a two level system by setting

$$U_{1_k}^{n+1} = u_k^{n+1} \tag{6.4.2}$$

$$U_{2_k}^{n+1} = u_k^n. \tag{6.4.3}$$

We note that

$$U_{1_k}^n = u_k^n \tag{6.4.4}$$
$$U_{2_k}^n = u_k^{n-1} \tag{6.4.5}$$

and that we can rewrite equation (6.4.1) as the system

$$U_{1_k}^{n+1} = U_{2_k}^n + 2r\delta^2 U_{1_k}^n \tag{6.4.6}$$
$$U_{2_k}^{n+1} = U_{1_k}^n, \tag{6.4.7}$$

or

$$\mathbf{U}_k^{n+1} = \begin{pmatrix} 2r\delta^2 & 1 \\ 1 & 0 \end{pmatrix} \mathbf{U}_k^n \tag{6.4.8}$$

where

$$\mathbf{U}_k^n = \begin{pmatrix} U_{1_k}^n \\ U_{2_k}^n \end{pmatrix} \quad \text{and} \quad \mathbf{U}_k^{n+1} = \begin{pmatrix} U_{1_k}^{n+1} \\ U_{2_k}^{n+1} \end{pmatrix}. \tag{6.4.9}$$

It should be noted that the method for rewriting the three level difference scheme as a two level system is analogous to how higher order differential equations are rewritten as a system of first order equations.

As in most of our earlier analyses, the approach we will use to study convergence of the solution of the difference scheme is to apply either the Lax or the Lax Equivalence Theorem and replace the study of convergence by a study of the stability of the scheme. We proceed by taking the finite Fourier transform of equation (6.4.8) to get

$$\hat{U}^{n+1} = \begin{pmatrix} -8r\sin^2\frac{\xi}{2} & 1 \\ 1 & 0 \end{pmatrix} \hat{U}^n. \tag{6.4.10}$$

Thus, we see that the amplification matrix of system (6.4.8) is given by

$$G(\xi) = \begin{pmatrix} -8r\sin^2\frac{\xi}{2} & 1 \\ 1 & 0 \end{pmatrix}. \tag{6.4.11}$$

Noting that matrix G is Hermitian, we know that if the scheme satisfies the von Neumann condition, the scheme will be stable by Theorem 6.2.4-(iii). (We should also note that since Q is self adjoint, Theorem 6.2.4-(i) can also be used if the scheme satisfies the von Neumann condition.) Evaluating

$$0 = \det(G - \lambda I) = \det\begin{pmatrix} -8r\sin^2\frac{\xi}{2} - \lambda & 1 \\ 1 & -\lambda \end{pmatrix} = \lambda^2 + 8r\sin^2\frac{\xi}{2}\lambda - 1,$$

we see that the eigenvalues of G are

$$\lambda_\pm = -4r\sin^2\frac{\xi}{2} \pm \sqrt{1 + 16r^2\sin^4\frac{\xi}{2}}. \tag{6.4.12}$$

Then since $1 + 16r^2\sin^4\frac{\xi}{2} \geq 1$, we see that the eigenvalue

$$\lambda_- = -4r\sin^2\frac{\xi}{2} - \sqrt{1 + 16r^2\sin^4\frac{\xi}{2}} < -1$$

for some $\xi \in [-\pi, \pi]$ (for example, $\xi = \pi$). And it is clear that $|\lambda_-| \not\leq 1 + C\Delta t$ for any C. Hence, the scheme will not satisfy the von Neumann condition, so *by Theorem 6.2.2 the scheme will be unstable.*

Though the difference scheme given in (6.4.1) is unstable, it can be stabilized by expanding the term

$$\delta^2 u_k^n = u_{k+1}^n - 2u_k^n + u_{k-1}^n$$

and replacing the u_k^n term by the average $(u_k^{n+1} + u_k^{n-1})/2$. The new difference scheme, written as

$$u_k^{n+1} = \frac{2r}{1+2r}(u_{k+1}^n + u_{k-1}^n) + \frac{1-2r}{1+2r}u_k^{n-1}, \qquad (6.4.13)$$

is called the **Dufort-Frankel scheme**. In Chapter 2, HW2.3.1-(d), we saw that the Dufort-Frankel scheme was inconsistent with the one dimensional heat equation. However, we notice that if we restrict the relationship between Δx and Δt so that $\Delta t/\Delta x \to 0$ as Δx, $\Delta t \to 0$, then the scheme will be consistent. For example, if we require that $r = \Delta t/\Delta x^2$ remain constant, then the scheme will be **conditionally consistent**. It is in this setting that we consider the stability (and, hence, convergence) of the Dufort-Frankel scheme. It should be noted that if we use conditional consistency in the Lax Theorem in place of consistency, the result is still true with the appropriate condition imposed on the convergence.

Example 6.4.2 Analyze the stability of the Dufort-Frankel difference scheme.

Solution: We begin our analysis by using (6.4.2)–(6.4.5) and (6.4.9) to rewrite our scheme as

$$\mathbf{U}_k^{n+1} = \begin{pmatrix} \frac{2r}{1+2r}(S_+ + S_-) & \frac{1-2r}{1+2r} \\ 1 & 0 \end{pmatrix} \mathbf{U}_k^n. \qquad (6.4.14)$$

If we take the discrete Fourier transform of equation (6.4.14), we get

$$G(\xi) = \begin{pmatrix} \frac{4r}{1+2r}\cos\xi & \frac{1-2r}{1+2r} \\ 1 & 0 \end{pmatrix}. \qquad (6.4.15)$$

The eigenvalues of G are

$$\lambda_\pm = \frac{2r\cos\xi \pm \sqrt{1 - 4r^2\sin^2\xi}}{1 + 2r}. \qquad (6.4.16)$$

When $1 - 4r^2\sin^2\xi < 0$, $\lambda_\pm$ can be written as

$$\lambda_\pm = \frac{1}{1+2r}\left(2r\cos\xi \pm i\sqrt{4r^2\sin^2\xi - 1}\right),$$

so

$$|\lambda_\pm| = \sqrt{\frac{2r-1}{2r+1}} < 1$$

for all r.

When $1 - 4r^2\sin^2\xi \geq 0$, then $\sqrt{1 - 4r^2\sin^2\xi} \leq 1$ and

$$|\lambda_\pm| \leq \frac{1}{1+2r}\left(2r|\cos\xi| + \sqrt{1 - 4r^2\sin^2\xi}\right) \leq \frac{1}{1+2r}(2r+1) = 1$$

for any r.

Thus, in either case, G satisfies the von Neumann condition and there is no condition to prevent the stability of the Dufort-Frankel scheme. To show that the scheme is, in fact, stable, we apply Theorem 6.2.4(v). To apply Theorem 6.2.4(v), we must show that one of the eigenvalues is strictly less than one in magnitude. Below in Figures 6.4.1 and 6.4.2 we include the plots of the magnitudes of the eigenvalues as a function of $\cos \xi$ for a representative value of $r > 1/2$ and $r \leq 1/2$. (It is not difficult to see that plots for all other values of r will be similar.) We see that in each case ($r > 1/2$ and $r \leq 1/2$), for all ξ, at least one of the eigenvalues is strictly less than one in magnitude. Hence, by Theorem 6.2.4(v) the Dufort-Frankel scheme is unconditionally stable, and, since, the scheme is conditionally consistent, the Dufort-Frankel scheme is conditionally convergent with a conditions such as $r = constant$.

Obviously, since difference scheme (6.4.1) has a name, there must be some place that leapfrog schemes are usable. We next consider a **leapfrog scheme** for solving the hyperbolic partial differential equation

$$v_t + av_x = 0 \tag{6.4.17}$$

of the form

$$u_k^{n+1} = u_k^{n-1} - R(u_{k+1}^n - u_{k-1}^n). \tag{6.4.18}$$

Example 6.4.3 Analyze the stability of leapfrog difference scheme, (6.4.18).

Solution: If we rewrite the three level difference scheme (6.4.18) as a two level system as we did in the previous example (again using (6.4.2)–(6.4.5) and (6.4.9)), we obtain the difference scheme

$$\mathbf{U}_k^{n+1} = \begin{pmatrix} -R\delta_0 & 1 \\ 1 & 0 \end{pmatrix} \mathbf{U}_k^n. \tag{6.4.19}$$

We note that because of the central difference operator δ_0, the difference operator Q will not be self adjoint.

If we take the discrete Fourier transform of equation (6.4.19), we get

$$\hat{\mathbf{U}}^{n+1} = \begin{pmatrix} -2iR\sin \xi & 1 \\ 1 & 0 \end{pmatrix} \hat{\mathbf{U}}^n \tag{6.4.20}$$

so that the amplification matrix is given by

$$G(\xi) = \begin{pmatrix} -2iR\sin \xi & 1 \\ 1 & 0 \end{pmatrix}. \tag{6.4.21}$$

It is clear that G is not Hermitian and reasonably certain that it would be difficult to find a matrix S so that SGS^{-1} will be Hermitian. There is hope that we may be able to use Theorem 6.2.5-(ii). We begin by calculating the eigenvalues of G so that we can check to see when the von Neumann condition is satisfied and if we can find matrices S and S^{-1} that will diagonalize G.

We evaluate

$$0 = \det(G - \lambda I) = \det \begin{pmatrix} -2iR\sin \xi - \lambda & 1 \\ 1 & -\lambda \end{pmatrix} = \lambda^2 + 2iR\sin \xi \lambda - 1$$

to see that the eigenvalues of G are given by

$$\lambda_\pm = \pm\sqrt{1 - R^2 \sin^2 \xi} - iR\sin \xi. \tag{6.4.22}$$

It is easy to see that if $R^2 > 1$, then for some $\xi \in [-\pi, \pi]$ (especially near $\xi = \pi/2$),

$$\sqrt{1 - R^2 \sin^2 \xi} = i\sqrt{R^2 \sin^2 \xi - 1}$$

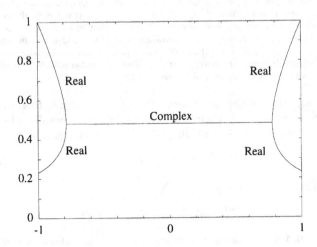

FIGURE 6.4.1. Plots of the magnitude of the eigenvalues $\lambda_\pm$ for a representative value of $r > 1/2$, $r = 0.8$.

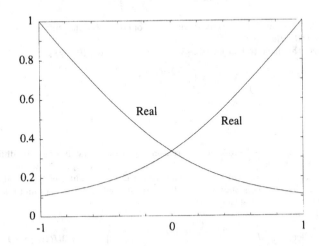

FIGURE 6.4.2. Plots of the magnitude of the eigenvalues $\lambda_\pm$ for a representative value of $r \leq 1/2$, $r = 0.4$.

and

$$|\lambda_-|(\pi/2) = R + \sqrt{R^2 - 1} > 1 \quad a > 0$$
$$|\lambda_+|(\pi/2) = |R| + \sqrt{R^2 - 1} > 1 \quad a < 0.$$

In either case, the difference scheme will not satisfy the von Neumann condition if $R^2 > 1$.

In the case where $R^2 \leq 1$, then $1 - R^2 \sin^2 \xi \geq 0$ and $|\lambda_\pm| = 1$. Hence, *if $R^2 \leq 1$, difference scheme (6.4.18) satisfies the von Neumann condition.*

We next try to apply Theorem 6.2.5-(ii). As usual, we find S and S^{-1} so that $H = SGS^{-1}$ is a diagonal matrix with the eigenvalues of G on the diagonal of H. This is done by letting S^{-1} be the matrix whose columns are the eigenvectors of G. If we compute the eigenvectors of G, we see that they are given by

$$\hat{u}_1 = \begin{pmatrix} \bar{z} \\ 1 \end{pmatrix} \quad \text{and} \quad \hat{u}_2 = \begin{pmatrix} -z \\ 1 \end{pmatrix}$$

where $z = \alpha + i\beta$, $\alpha = \sqrt{1 - R^2 \sin^2 \xi}$ and $\beta = R \sin \xi$. (The eigenvalues of G are $\bar{z}$ and $-z$.) Hence, we see that

$$S^{-1} = \begin{pmatrix} \bar{z} & -z \\ 1 & 1 \end{pmatrix} \tag{6.4.23}$$

and

$$S = \frac{1}{2\alpha} \begin{pmatrix} 1 & z \\ -1 & \bar{z} \end{pmatrix}. \tag{6.4.24}$$

We first note that when $R^2 = 1$ and $\xi = \pi/2$, then the two eigenvectors are identical (and S is not defined). Hence, the transformation will not be invertible and can not be used in this case. For this reason we restrict the analysis to the case when $|R| < 1$ and consider equality later.

Secondly, we must be careful of assuming that the norms of S and S^{-1} will be bounded by constants. When we used the analogous transformation earlier, the matrix S always came from the fact that we knew that the matrix associated with our hyperbolic system was diagonalizable. Since S and S^{-1} came from the matrix A, we knew that it did not depend on ξ, Δx, etc. In this case, that is not necessarily so. Hence, we must be careful to show that $\| S \|_2 \leq C_1$ and $\| S^{-1} \|_2 \leq C_2$ where C_1 and C_2 are constants. This computation can be done using the facts that $\| S \|_2 = \sqrt{\sigma(S^*S)}$ and $\| S^{-1} \|_2 = \sqrt{\sigma((S^{-1})^*S^{-1})}$ where S^* and $(S^{-1})^*$ denote the conjugate transpose of S and S^{-1}, respectively. Thus the norms and S and S^{-1} are given by the square roots of the largest eigenvalues of S^*S and $(S^{-1})^*S^{-1}$, respectively.

If we compute $(S^{-1})^*S^{-1}$, it is not hard to show that the eigenvalues of $(S^{-1})^*S^{-1}$ are given by

$$2 \pm 2|R||\sin \xi|.$$

Hence, for $|R| < 1$, $\| S^{-1} \|_2 \leq 2$. Likewise, it is not hard to show that the eigenvalues of S^*S are given by

$$\frac{1}{1 \pm |R||\sin \xi|}.$$

Clearly, we must bound $|R|$ away from one. Thus we see that for $|R| \leq R_0$ for any $R_0 < 1$,

$$\| S \|_2 \leq \frac{1}{\sqrt{1 - R_0}}.$$

Since $H = SGS^{-1}$ will be the matrix

$$H = \begin{pmatrix} \bar{z} & 0 \\ 0 & -z \end{pmatrix},$$

it is easy to see that $H^*H = I$ so that condition (ii) of Theorem 6.2.5 is trivially satisfied. Hence, *the leapfrog scheme, (6.4.18), is stable if $|R| \leq R_0 < 1$.*

We must now consider the case when $R^2 = 1$. If we specifically consider the case when $R = 1$ (the case when $R = -1$ is similar) and evaluate $G^n(\pi/2)$, we see that

$$G^n(\pi/2) = (-i)^{n-1} \begin{pmatrix} -i(n+1) & n \\ n & i(n-1) \end{pmatrix}.$$

Hence, when $R = 1$ and $\xi = \pi/2$ (and also near there), G^n grows linearly. Hence, the scheme can not be stable for $R = 1$.

Therefore, *the leapfrog scheme (6.4.18) is a stable scheme (and hence, convergent) for solving the partial differential equation (6.4.17), if $|R| \leq R_0$ for any $R_0 < 1$.*

Remark: We should be careful to understand the difference between the stability level calculated above and the condition $|R| < 1$. The latter condition can be satisfied by a sequence of Δx's and Δt's so that $R_j = \Delta t_j / \Delta x_j$, $R_j < 1$ but $R_j \to 1$ as $j \to \infty$. Application of Theorem 6.2.5-(ii) is not allowable in this situation since $\| S \|_2$ will not be bounded by a constant.

HW 6.4.1 Analyze the stability of the following difference schemes.

(a)

$$u_k^{n+1} = u_k^{n-1} - R\delta_0 u_k^n + \frac{R}{6}\delta^2\delta_0 u_k^n$$

(b)

$$u_k^{n+1} = u_k^{n-1} - R\delta_0 u_k^n + \frac{R}{6}\delta^2\delta_0 u_k^n - \frac{R}{30}\delta^4\delta_0 u_k^n$$

(c)

$$u_k^{n+2} = u_k^{n-2} - \frac{2R}{3}\left(1 - \frac{1}{6}\delta^2\right)\delta_0\left(2u_k^{n+1} - u_k^n + 2u_k^{n-1}\right)$$

See Hw2.3.3.

6.4.2 Implementation of Scalar Multilevel Schemes

We discuss briefly some of the differences that we encounter when we try to implement a multilevel scheme. Of course, the first problem we encounter is that we now must save at least one more level of data than we had to do for the two level schemes. Other than choosing a better name for the array than *uoldold* and the extra storage necessary, this doesn't really present a problem.

The most difficult task that we must face is that of getting our scheme started. For a first order equation, we are given an initial condition. But to get started computing with a three level scheme, we need function values at time levels $n = 0$ and $n = 1$, and we start our computation at level $n = 2$. As a part of any multilevel scheme, we must provide an **initialization scheme** to make the difference scheme well defined. The problem we encounter for a three level scheme is to obtain function values at $n = 1$ that will not lower the order of our computational accuracy. For example, consider the leapfrog scheme, (6.4.18) (or any other three level scheme that is second order in both space and time). We must be able to obtain values for u_k^1 in such a way so as to not lower the order of the scheme below second order in time and space.

Two rather obvious approaches to solving this problem are to

(i) use a two level scheme that is first order in time and second order in space with $\Delta t / \delta t$ time steps (where Δt is the time step to be used with the three level scheme and $\delta t = \Delta t^2$) to approximate u_k^1

or

(ii) use a two level scheme that is second order in both time and space to approximate u_k^1 (such as Lax-Wendroff).

And, of course, analogous approaches will work with schemes having more than three time levels.

Both of the methods described will work. However, the situation is nicer that is assumed above (where we assumed that to insure that we kept our second order accuracy, we must compute u_k^1 to second order). *We can use a scheme that is lower order than our three level scheme to start our calculation (even an unstable scheme) and still retain the order of accuracy of the three level scheme.* The scheme used for the first few time steps to start our calculation introduces a small growth in the solution. The growth is small because of the consistency of the initialization scheme. The stability of our multilevel scheme keeps this small initial growth from being amplified.

To see why either (i) or (ii) above work, and also to explain the claims made in the last paragraph, we return to Section 2.5.1 to the proof of the Lax Theorem, Theorem 2.5.2. We notice that in the calculation performed in (2.5.4), we reduced the term $\mathbf{w}^{n+1}$ to a term involving $\mathbf{w}^0$ and a sum of $n + 1$ terms involving the truncation errors. Since the error term $\mathbf{W}^n$ for two level schemes is given by

$$
\mathbf{W}^n = \begin{bmatrix} \vdots \\ \mathbf{W}_1^n \\ \mathbf{W}_0^n \\ \mathbf{W}_{-1}^n \\ \vdots \end{bmatrix}, \text{ where } \mathbf{W}_k^n = \begin{bmatrix} w_k^n \\ w_k^{n-1} \end{bmatrix},
$$

we see that $\mathbf{W}^0$ is not defined (and general multilevel schemes will have a similar problem). Thus, we must consider one step less than the reduction done in calculation (2.5.4) and get

$$\mathbf{W}^{n+1} = Q^n \mathbf{W}^1 + \Delta t \sum_{j=0}^{n-1} Q^j \boldsymbol{\tau}^{n-j}. \tag{6.4.25}$$

We note that if $\mathbf{W}^1 = \boldsymbol{\theta}$, then *the proof of Theorem 2.5.2 applies to multilevel schemes just as it did for two level schemes. But, for multilevel schemes,* $\mathbf{W}^1$ *will not generally be zero* (unless we do as we did in HW1.3.1 and use exact values for u_k^1). Thus, we must adjust the proof of Theorem 2.5.2 slightly so that it will apply to multilevel schemes. If we write our initialization scheme as

$$u_k^1 = \sum_{j=-m_3}^{j=m_4} q_{1\,j} S_+^j u_k^0$$

and let τ_k^0 denote the truncation error associated with this initialization scheme, then

$$\mathbf{W}_k^1 = \begin{bmatrix} w_k^1 \\ w_k^0 \end{bmatrix},$$

$w_k^0 = \boldsymbol{\theta}$ and

$$w_k^1 = \sum_{j=-m_3}^{j=m_4} q_{1\,j} S_+^j w_k^0 + \Delta t \tau_k^0 = \Delta t \tau_k^0.$$

Thus,

$$\mathbf{W}^1 = \Delta t \boldsymbol{\tau}^0 = \Delta t \begin{bmatrix} \vdots \\ \begin{bmatrix} \tau_1^0 \\ 0 \end{bmatrix} \\ \begin{bmatrix} \tau_0^0 \\ 0 \end{bmatrix} \\ \begin{bmatrix} \tau_{-1}^0 \\ 0 \end{bmatrix} \\ \vdots \end{bmatrix}. \tag{6.4.26}$$

It must be made very clear that $\boldsymbol{\tau}^0$ *is the truncation error due to the initialization scheme used to get the multilevel difference scheme running, whereas* $\boldsymbol{\tau}^j$, $j \geq 1$ *is the truncation error due to the multilevel scheme.*

Using expression (6.4.26), equation (6.4.25) can be written as

$$\mathbf{W}^{n+1} = \Delta t Q^n \boldsymbol{\tau}^0 + \Delta t \sum_{j=0}^{n-1} Q^j \boldsymbol{\tau}^{n-j} \tag{6.4.27}$$

where τ^0 is as defined in equation (6.4.26). In both situations considered below, just as we did in the proof of Theorem 2.5.2, we use the stability and the order of accuracy of the multilevel scheme to write the summation term of equation (6.4.27) as

$$\| \Delta t \sum_{j=0}^{n-1} Q^j \tau^{n-j} \| \le n\Delta t K e^{\beta n \Delta t} C^*(t)(\Delta x^p + \Delta t^q). \qquad (6.4.28)$$

It is then easy to see that if the truncation error of the initialization scheme is such that $\| \tau^0 \| = \mathcal{O}(\Delta x^p) + \mathcal{O}(\Delta t^q)$, the first term on the right hand side of equation (6.4.27) can be treated as

$$\| \Delta t Q^n \tau^0 \| \le \Delta t K e^{n\Delta t} C(t)(\Delta x^p + \Delta t^q). \qquad (6.4.29)$$

Equations (6.4.29) and (6.4.28) can be combined with equation (6.4.27) to give

$$\| \mathbf{W}^{n+1} \| \le (n+1)\Delta t K e^{\beta n \Delta t} C^{**}(t)(\Delta x^p + \Delta t^q). \qquad (6.4.30)$$

Then by the same discussion of convergence used in Theorem 2.5.2, we have: *if the truncation error of the scheme used to start the multilevel scheme is the same as that of the multilevel scheme, then the difference scheme converges to the same order as the multilevel scheme.*

It is also easy to see that we can treat equation (6.4.27) above differently and obtain a better result. We treat the summation term as we did above to get inequality (6.4.28). If we then require that

$$\Delta t \| \tau^0 \| = \mathcal{O}(\Delta x^p) + \mathcal{O}(\Delta t^q), \qquad (6.4.31)$$

equation (6.4.27) along with inequalities (6.4.28) and (6.4.31) gives

$$\begin{aligned} \| \mathbf{W}^{n+1} \| \quad \le \quad & K e^{\beta n \Delta t} C_1(t)(\Delta x^p + \Delta t^q) \\ & + n\Delta t K e^{\beta n \Delta t} C^*(t)(\Delta x^p + \Delta t^q). \end{aligned} \qquad (6.4.32)$$

In the same manner as inequality (2.5.8) in Section 2.5.2 or inequality (6.4.30) above, inequality (6.4.32) implies that the difference scheme converges order (p, q). Thus, *if the initialization scheme used to start a multilevel computation (where the multilevel scheme is order (p,q)) satisfies*

$$\Delta t \| \tau^0 \| = \mathcal{O}(\Delta x^p) + \mathcal{O}(\Delta t^q),$$

then the multilevel computation converges order (p, q).

Remark 1: A fact that should be mentioned is that there was never anything said about the stability of the initialization scheme. This is because *it is not necessary that the scheme used to start the multilevel computation be stable.* The initialization scheme never gets iterated. Hence, it can't cause an instability. The order of the truncation error is the only thing about the initialization scheme that is important.

Remark 2: One last comment that can be made about the above computations is that the same results will hold for systems (not related to multilevel schemes) and scalar equations where we assume that the initial conditions are not given exactly. Hence, if $\mathbf{W}^0$ (or w^0 in the scalar case) is $\mathcal{O}(\Delta x^p) + \mathcal{O}(\Delta t^q)$ (where we are assuming that the difference scheme is order (p, q)), then the difference scheme will converge order (p, q).

HW 6.4.2 Solve the problems given in both HW5.6.8 and HW5.6.9 using the leapfrog scheme, (6.4.18), with numerical boundary conditions (a)–(d) from HW5.6.10 and (e) $u_0^n = 2u_1^n - u_2^n$.

HW 6.4.3 (a) Repeat HW5.6.5 (a), (b) and (c) using the leapfrog scheme, (6.4.18).

HW 6.4.4 Consider the Dufort-Frankel scheme (6.4.13) along with the condition that $r = \nu \Delta t / \Delta x^2$ is held constant (to make the Dufort-Frankel scheme conditionally consistent). Discuss the order of convergence of the Dufort-Frankel scheme using the following initialization scheme.

$$u^1 = u^0 + r\delta^2 u_k^0$$

HW 6.4.5 Consider the use of a leapfrog difference scheme

$$u_k^{n+1} = u_k^{n-1} - R\delta_0 u_k^n + r\delta^2 u_k^n \tag{6.4.33}$$

as an approximation of the convection–diffusion equation

$$v_t + av_x = \nu v_{xx}.$$

(a) Analyze the stability of difference scheme (6.4.33).
(b) Resolve the problem given in HW5.9.2 using difference scheme (6.4.33). Compare your results with those from HW5.9.2.

6.4.3 Multilevel Systems

Just as we introduced multilevel difference schemes for solving scalar one dimensional equations, multilevel schemes can be very useful for solving systems in one or more dimensions. In this section, we use the example of the leapfrog scheme for a one dimensional system to illustrate the minor differences between analyzing multilevel scalar schemes and multilevel schemes for systems.

We begin by considering the initial–value problem

$$\mathbf{v}_t = A\mathbf{v}_x, \quad x \in \mathbb{R},\ t > 0, \tag{6.4.34}$$
$$\mathbf{v}(x, 0) = \mathbf{f}(x), \quad x \in \mathbb{R} \tag{6.4.35}$$

along with the leapfrog difference scheme

$$\mathbf{u}_k^{n+1} = \mathbf{u}_k^{n-1} + RA\delta_0\mathbf{u}_k^n. \tag{6.4.36}$$

Example 6.4.4 Analyze the stability of difference scheme (6.4.36).

Solution: We proceed as we did with the scalar leapfrog scheme in Section 6.4.1, let

$$\mathbf{U}_k^{n+1} = \begin{bmatrix} \mathbf{u}_k^{n+1} \\ \mathbf{u}_k^n \end{bmatrix}$$

and rewrite difference scheme (6.4.36) as

$$\mathbf{U}_k^{n+1} = \begin{pmatrix} RA\delta_0 & I \\ I & \Theta \end{pmatrix}\mathbf{U}_k^n. \tag{6.4.37}$$

If we take the discrete Fourier transform of equation (6.4.37), we see that the amplification matrix for difference scheme (6.4.37) is given by

$$G(\xi) = \begin{pmatrix} 2iR\sin\xi A & I \\ I & \Theta \end{pmatrix}. \tag{6.4.38}$$

Obviously, amplification matrix (6.4.38) looks very much like that given in (6.4.20). The difference is that the matrix G given in (6.4.38) is not really a 2×2 matrix. It is, instead, a $2K \times 2K$ matrix. However, we can combine techniques that we have used to analyze hyperbolic systems with those that we used to analyze the scalar leapfrog scheme to determine stability conditions for difference scheme (6.4.36).

We begin by letting S be the matrix that diagonalizes matrix A, i.e. $D = SAS^{-1}$, where D is a diagonal matrix with the eigenvalues of A on its diagonal. An easy calculation shows that if we let

$$S_1 = \begin{pmatrix} S & \Theta \\ \Theta & S \end{pmatrix},$$

then

$$G_1 = S_1 G S_1^{-1} = \begin{pmatrix} 2iR\sin\xi D & I \\ I & \Theta \end{pmatrix}.$$

If we then let P_1 denote the permutation matrix obtained by interchanging the kth and the jth rows of $I_{2K \times 2K}$, then P_1^{-1} will be the matrix obtained by interchanging the kth and the jth columns of $I_{2K \times 2K}$ and $\| P_1 \| = \| P_1^{-1} \| = 1$. It is not hard to see that we can use a series of K permutations $P_1, \cdots, P_K$ to reduce G_1 to the form

$$G_2 = P_K \cdots P_1 G_1 P_1^{-1} \cdots P_K^{-1} = \begin{pmatrix} B_1 & \Theta & \cdots \\ & \ddots & \\ \cdots & \Theta & B_K \end{pmatrix}$$

where

$$B_j = \begin{pmatrix} 2iR\lambda_j\sin\xi & 1 \\ 1 & 0 \end{pmatrix}, \ j = 1, \cdots, K$$

and $\lambda_j, \ j = 1, \cdots, K$ are the eigenvalues of the matrix A. (The easiest way to see this is to take the cases $K = 2$ and $K = 3$ and interchange the appropriate rows and columns until you get it in the correct form.) We note that we have now reduced our problem to K small problems that each look like the matrix associated with the scalar leapfrog scheme. The eigenvalues of G will be the the same as those of G_2 and the eigenvalues of G_2 will be the same as those of B_j, $j = 1, \cdots, K$. And, finally, the eigenvalues of B_j will be the same as those given in Example 6.4.3, equation (6.4.22), with R replaced by $\lambda_j R$.

Hence, the eigenvalues of G are given by

$$\lambda_{j\pm} = \pm\sqrt{1 - \lambda_j^2 R^2 \sin^2\xi} - i\lambda_j R\sin\xi, \ j = 1, \cdots, K.$$

As in Example 6.4.3, if $\mid \lambda_j R \mid > 1$ for any j, then either $\mid \lambda_+(\pi/2) \mid$ or $\mid \lambda_-(\pi/2) \mid$ will be greater than one, depending on the sign of λ_j. When $\mid \lambda_j R \mid \leq 1$, $\mid \lambda_{j_\pm} \mid = 1$. Thus for $\mid \lambda_j R \mid \leq 1$, $j = 1, \cdots, K$, difference scheme (6.4.36) satisfies the von Neumann condition.

As with the scalar leapfrog scheme, we obtain sufficient conditions for stability by applying Theorem 6.2.5-(ii). If we let

$$
S = \begin{pmatrix} S_1 & \Theta & \cdots \\ & \ddots & \\ \cdots & \Theta & S_K \end{pmatrix}
$$

where S_j, $j = 1, \cdots, K$ are the 2×2 matrices analogous to matrix (6.4.24) (where in S_j, $R = \lambda_j \Delta t / \Delta x$), then

$$
SG_2 S^{-1}
$$

will be the diagonal matrix with the eigenvalues of G its diagonal. Hence, we use Theorem 6.2.5-(ii) to see that *if* $\mid \lambda_j R \mid \leq R_0 < 1$, $j = 1, \cdots, K$, *then difference scheme (6.4.36) is stable*. Using the same example that we used in Example 6.4.3, we see that if $\mid \lambda_j R \mid = 1$ for any $j = 1, \cdots, K$, then difference scheme (6.4.36) will be unstable.

Remark: Of course, to apply the leapfrog scheme or any other multi-level schemes to approximate the solution of a hyperbolic system of partial differential equations, we have all of the usual problems associated with schemes for systems of hyperbolic equations (boundary conditions) and multilevel schemes (finding an initialization scheme). Boundary conditions for the leapfrog scheme do not cause any new problems and the initialization scheme can be developed and analyzed in exactly the same way we did it for the scalar leapfrog scheme.

HW 6.4.6 Consider a stable, three level difference scheme for solving initial–value problem (6.4.34)–(6.4.35) (accurate order $\mathcal{O}(\Delta x^p) + \mathcal{O}(\Delta t^q)$) along with an initialization scheme of the form

$$
\mathbf{u}_k^1 = \sum_{j=-m_1}^{m_2} Q_{2j} S_+^j \mathbf{u}_k^0.
$$

Show that if the initialization scheme satisfies

$$
\mathbf{v}_k^1 = \sum_{j=-m_1}^{m_2} Q_{2j} S_+^j \mathbf{v}_k^0 + \Delta t \tau_k^0
$$

where $\mathbf{v}$ is the solution to initial–value problem (6.4.34)–(6.4.35) and $\Delta t \parallel \tau_k^0 \parallel = \mathcal{O}(\Delta x^p) + \mathcal{O}(\Delta t^q)$, then the difference scheme converges $\mathcal{O}(\Delta x^p) + \mathcal{O}(\Delta t^q)$.

6.5   Higher Order Hyperbolic Equations

Until this time, all of the hyperbolic partial differential equations that we have studied have been first order equations. There are higher order equations that are important. The most obvious of these is the wave equation

$$v_{tt} = c^2 v_{xx}, \tag{6.5.1}$$

Though we obviously felt that the first order, one way wave equation was the logical model to use for the introduction to hyperbolic partial differential equations, the wave equation has long been the traditional model used for this purpose. We shall use the wave equation to illustrate how we reduce the numerical treatment of higher order equations back to first order systems.

We should first state that it would be logical to treat equation (6.5.1) much the same way we treated the heat equation and write a difference scheme

$$\frac{u_k^{n+1} - 2u_k^n + u_k^{n-1}}{\Delta t^2} = c^2 \frac{u_{k+1}^n - 2u_k^n + u_{k-1}^n}{\Delta x^2} \tag{6.5.2}$$

or

$$u_k^{n+1} = R^2(u_{k+1}^n + u_{k-1}^n) + 2(1 - R^2)u_k^n - u_k^{n-1} \tag{6.5.3}$$

where $R = c\Delta t/\Delta x$ (c is assumed positive). If we are given initial conditions

$$v(x,0) = f(x) \tag{6.5.4}$$
$$v_t(x,0) = f_1(x) \tag{6.5.5}$$

and boundary conditions

$$v(0,t) = a(t), \quad v(1,t) = b(t), \tag{6.5.6}$$

using difference scheme (6.5.3) to approximate the solution to initial–boundary–value problem (6.5.1), (6.5.4)–(6.5.6) would be about the same as any explicit, multilevel difference schemes that we have used. As with multilevel schemes for first order equations, we must have an initialization scheme to help start the calculation. Explicitly, for difference scheme (6.5.3), we need an initialization scheme to give u_k^1. Fortunately, for equations second order in time, we get two initial conditions. One of the ways that initial condition (6.5.5) can be used to obtain u_k^1 is by setting

$$v_t(x,0) = f_1(x) \approx \frac{u_k^1 - u_k^0}{\Delta t}, \tag{6.5.7}$$

or

$$u_k^1 = u_k^0 + \Delta t f_1(k\Delta x)$$
$$= f(k\Delta x) + \Delta t f_1(k\Delta x), \quad k = 0, \cdots, M. \tag{6.5.8}$$

Then, implementing difference scheme (6.5.3) is easy. For example, see HW6.5.1.

HW 6.5.1 Use difference scheme (6.5.3) to solve the initial–boundary-value problem (6.5.1), (6.5.4)–(6.5.6) with $c = 1$, $f(x) = \sin 2\pi x$, $f_1(x) = \sin 3\pi x$ and $a(t) = b(t) = 0$.

(a) Use $M = 10$, $\Delta t = 0.05$ and compute solutions at times $t = 0.5$, $t = 1.0$, $t = 3.0$ and $t = 10.0$.

(b) Repeat (a) using $M = 40$ and $\Delta t = 0.01$.

6.5.1   Initial–Value Problems

To see what we should expect from a difference scheme such as (6.5.3) or any other difference scheme for partial differential equation (6.5.1), we must consider convergence of the scheme. As usual, one way (maybe the best way) to consider convergence is to consider consistency and stability. We begin by using the approach we used for multilevel schemes and rewrite difference scheme (6.5.3) as a two level system of equations

$$\mathbf{U}_k^{n+1} = \begin{bmatrix} u_k^{n+1} \\ u_k^n \end{bmatrix} = \begin{pmatrix} R^2(S^+ + S^-) + 2(1 - R^2) & -1 \\ 1 & 0 \end{pmatrix} \begin{bmatrix} u_k^n \\ u_k^{n-1} \end{bmatrix}$$

$$= Q\mathbf{U}_k^n. \tag{6.5.9}$$

At this time, it would be very easy to consider difference equation (6.5.9) as a system and proceed as we have in the rest of this chapter. If we were to do so (and we will in Example 6.5.1), we would decide that difference scheme (6.5.9) is unstable. However, *difference equation (6.5.9) is fundamentally different from the difference equations that we have considered previously.* Because of the Δt^2 in the denominator of the first term in equation (6.5.2), if we substitute v_k^n (where v is a solution the partial differential equation (6.5.1)) into difference equation (6.5.3), we get

$$v_k^{n+1} = R^2(v_{k+1}^n + v_{k-1}^n) + 2(1 - R^2)v_k^n - v_k^{n-1} + \Delta t^2 \tau_k^n, \tag{6.5.10}$$

where τ_k^n is the truncation error, $\tau_k^n = \mathcal{O}(\Delta x^2) + \mathcal{O}(\Delta t^2)$. Hence, instead of the error satisfying equation (2.5.3) as in the proof of Theorem 2.5.2, the error associated with difference scheme (6.5.3) will satisfy

$$w_k^{n+1} = R^2(w_{k+1}^n + w_{k-1}^n) + 2(1 - R^2)w_k^n - w_k^{n-1} + \Delta t^2 \tau_k^n. \tag{6.5.11}$$

or

$$\mathbf{W}^{n+1} = Q\mathbf{W}^n + \Delta t^2 \tau^n, \tag{6.5.12}$$

where $\| \tau^n \| = \mathcal{O}(\Delta t^2) + \mathcal{O}(\Delta x^2)$. If we then perform the calculation to obtain the expression analogous to equation (6.4.25) (which is the system version of equation (2.5.5)), we get

$$\mathbf{W}^{n+1} = Q^n \mathbf{W}^1 + \Delta t^2 \sum_{j=0}^{n-1} Q^j \tau^{n-j}. \tag{6.5.13}$$

We should realize that equations (6.5.12) and (6.5.13) are valid for more than just difference scheme (6.5.3). These expressions will be true for any three level difference scheme for partial differential equations that are second order in time. Because of equation (6.5.13), it is easy to see that *if* $\mathbf{W}^1$ *is zero or sufficiently small and we have a consistent scheme (equation (6.5.12) is satisfied where* $\tau_k^n \to 0$ *as* Δt, $\Delta x \to 0$), *our previous definition of stability (the system version of Definition 2.4.1 will imply convergence.* In fact, we see that the Δt^2 in expression (6.5.13) would give an order of convergence that is Δt higher than the order of accuracy.

If we consider equation (6.5.13) carefully (along with the proof of Theorem 2.5.2), we see that we can allow for more growth in $\| Q^n \|$ than we allowed in Proposition 2.4.2 (really Definition 2.4.1) and still get convergence. We state the following alternative to Definition 2.4.1.

Definition 6.5.1 *Difference scheme*

$$\mathbf{u}^{n+1} = Q\mathbf{u}^n \tag{6.5.14}$$

is said to be stable order n with respect to the norm $\|\cdot\|$ *if there exists positive constants* Δx_0 *and* Δt_0, *and non-negative constants* K_1, K_2 *and* β *so that*

$$\| \mathbf{u}^{n+1} \| \leq (K_1 + nK_2)e^{\beta t} \| \mathbf{u}^0 \|, \tag{6.5.15}$$

for $0 \leq t = (n+1)\Delta t$, $0 < \Delta x \leq \Delta x_0$, *and* $0 < \Delta t \leq \Delta t_0$.

Obviously, *if a scheme is stable—(Definition 2.4.1), it will be stable order n*. We should also realize that the above definition is equivalent to requiring that Q satisfy

$$\| Q^n \| \leq (K_1 + nK_2)e^{\beta t}. \tag{6.5.16}$$

We next carefully analyze equation (6.5.13) along with Definition 6.5.1 to see what is needed for convergence. We assume that the difference scheme is accurate order (p, q). Following the proof of Theorem 2.5.2, it is easy to see that equation (6.5.13) gives

$$\| \mathbf{W}^{n+1} \| = \| Q^n \mathbf{W}^1 + \Delta t^2 \sum_{j=0}^{n-1} Q^j \tau^{n-j} \|$$

$$\leq \| Q^n \mathbf{W}^1 \|$$
$$+ \Delta t^2 (K_1 + nK_2)e^{\beta n\Delta t} nC^*(t)(\Delta x^p + \Delta t^q). \tag{6.5.17}$$

We see that the second term in equation (6.5.17) is in a form that will give us convergence order (p, q). We have an n in the expression from eliminating the summation and an n from the stability order n. These n's are eliminated by the Δt^2 term because $n\Delta t \to t$. Especially with the "new" definition of stability, we must be careful about the first term. Assuming that $w_k^0 = 0$ (we take $u_k^0 = v_k^0$),

$$\mathbf{W}_k^1 = \begin{bmatrix} w_k^1 \\ 0 \end{bmatrix}. \tag{6.5.18}$$

Then,

$$\| Q^n \mathbf{W}^1 \| \leq \| Q^n \| \| \mathbf{W}^1 \|$$
$$\leq (K_1 + nK_2)e^{\beta n \Delta t} \| \mathbf{W}^1 \|. \tag{6.5.19}$$

If

$$w_k^1 = \mathcal{O}(\Delta t \Delta x^p) + \mathcal{O}(\Delta t^{q+1}), \tag{6.5.20}$$

then

$$\| Q^n \mathbf{W}^1 \| \leq (K_1 + nK_2)e^{\beta n \Delta t}C(\Delta t \Delta x^p + \Delta t^{q+1}), \tag{6.5.21}$$

and the difference scheme will be convergent order (p, q). (The "extra" Δt is necessary to account for the n term in the definition of stable order n.)

We summarize the above Initialization Condition as follows.

IC: Suppose a three level difference scheme is consistent order (p, q) with a partial differential equation that is second order in time (the error vector satisfies equation (6.5.12) where $\| \tau^n \| = \mathcal{O}(\Delta x^p) + \mathcal{O}(\Delta t^q)$) and is stable order n. Then if the initial conditions approximate w^0 and w^1 are such that $w_k^0 = 0$ and $w_k^1 = \mathcal{O}(\Delta t \Delta x^p) + \mathcal{O}(\Delta t^{q+1})$, the difference scheme is convergent order (p, q).

We note that the above result is a combination of a convergence result (the Lax Theorem) and an initialization condition. Instead of treating this result as a theorem, we have listed it as an initialization condition (IC) and will refer to it as IC.

Example 6.5.1 Analyze the stability of difference scheme (6.5.3).

Solution: To analyze the stability of difference scheme (6.5.3), we take the discrete Fourier transform of the system of equations (6.5.9) to find that the amplification matrix, G, is given by

$$G = \begin{pmatrix} 2R^2 \cos \xi + 2(1 - R^2) & -1 \\ 1 & 0 \end{pmatrix} \tag{6.5.22}$$

$$= \begin{pmatrix} 2(1 - 2R^2 \sin^2 \frac{\xi}{2}) & -1 \\ 1 & 0 \end{pmatrix}. \tag{6.5.23}$$

The eigenvalues of G are solutions to the following equation

$$\lambda^2 - 2(1 - 2R^2 \sin^2 \frac{\xi}{2})\lambda + 1 = 0, \tag{6.5.24}$$

and are given by

$$\lambda_{\pm} = 1 - 2R^2 \sin^2 \frac{\xi}{2} \pm 2R \left| \sin \frac{\xi}{2} \right| \left[R^2 \sin^2 \frac{\xi}{2} - 1 \right]^{1/2}. \qquad (6.5.25)$$

If $R > 1$, then we can take $\xi = \pi$ and get $\lambda_- = 1 - 2R^2 - 2R\sqrt{R^2 - 1}$. It is not difficult to see that $\lambda_- < -1$ (and, less than -1 in a way that there is no multiple of Δt to help bound its growth). Note that $\| G^n \| \geq (\lambda_-)^n$ and $(\lambda_-)^n$ grows faster than $K_1 + nK_2$ for any K_1 and K_2. As in the case with stability, Definition 2.4.1, $\lambda < -1$ implies that difference scheme (6.5.3) is unstable order n. Hence, *$R \leq 1$ is a necessary condition for difference scheme (6.5.3) to be stable order n.* Note also that when $R \leq 1$, $|\lambda_{\pm}| = 1$.

Stability via Definition 2.4.1

If we were using Definition 2.4.1 for our stability definition, the approach we have used in the past was to use Theorems 6.2.4 and 6.2.5 (often, Theorem 6.2.5-(ii)). Since stability via Definition 2.4.1 implies stability via Definition 6.5.1, this approach is permissible. If we let $\alpha = 1 - 2R^2 \sin^2 \frac{\xi}{2}$ and $\beta = 2R|\sin \frac{\xi}{2}|[1 - R^2 \sin^2 \frac{\xi}{2}]^{1/2}$, then we can write $\lambda_+ = \alpha + i\beta = z$ and $\lambda_- = \bar{z}$. It is then not hard to see that the eigenvectors associated with eigenvalues $\lambda_+ = z$ and $\lambda_- = \bar{z}$ are

$$\mathbf{x}_1 = \begin{bmatrix} z \\ 1 \end{bmatrix} \qquad (6.5.26)$$

and

$$\mathbf{x}_2 = \begin{bmatrix} \bar{z} \\ 1 \end{bmatrix}, \qquad (6.5.27)$$

respectively. If we let

$$S^{-1} = \begin{bmatrix} \mathbf{x}_1 & \mathbf{x}_2 \end{bmatrix} = \begin{pmatrix} z & \bar{z} \\ 1 & 1 \end{pmatrix}, \qquad (6.5.28)$$

$$S = \frac{1}{2i\beta} \begin{pmatrix} 1 & -\bar{z} \\ -1 & z \end{pmatrix}. \qquad (6.5.29)$$

Then

$$H = SGS^{-1} = \begin{pmatrix} \lambda_+ & 0 \\ 0 & \lambda_- \end{pmatrix} = \begin{pmatrix} z & 0 \\ 0 & \bar{z} \end{pmatrix}. \qquad (6.5.30)$$

Therefore, if S is such that $\| S \|$ and $\| S^{-1} \|$ are bounded, then Theorem 6.2.5-(ii) can be used to show that the difference scheme is stable—(Definition 2.4.1) (and, hence stable—(Definition 6.5.1)). But, *since $\beta = 0$ when $\xi = 0$, this is clearly not the case.*

It is not at all difficult to see why the above approach does not work. If we look back at the form of the eigenvectors, we see that when $\xi = 0$, the eigenvalues and eigenvectors are equal. The repeated eigenvalue (for $\xi = 0$) $\lambda_{\pm} = 1$ is associated with only one eigenvector. Hence, *for any R the matrix G is not diagonalizable for all ξ.*

Stability Order n

An approach that may be used to analyze the above situation is to use a different special transformed form of the matrix, the Schur decomposition, ref. [3], page 79. We begin with a normalized version of the eigenvector $\mathbf{x}_1$,

$$\mathbf{y}_1 = \frac{1}{\sqrt{2}} \begin{bmatrix} z \\ 1 \end{bmatrix}. \qquad (6.5.31)$$

Then choose any vector $\mathbf{z}_2$ orthogonal to $\mathbf{y}_1$ so that $\{\mathbf{y}_1, \mathbf{z}_2\}$ will form a basis for $\mathbb{C}^2$. We then apply the Gram-Schmidt orthogonalization process, ref. [3], page 15, to $\{\mathbf{y}_1, \mathbf{z}_2\}$ to get

$$\mathbf{y} = \mathbf{z}_2 - (\mathbf{y}_1^* \cdot \mathbf{z}_2)\mathbf{y}_1 \qquad (6.5.32)$$

(where $\mathbf{y}_1^*$ is the conjugate transpose of $\mathbf{y}_1$) and set

$$\mathbf{y}_2 = \frac{\mathbf{y}}{\|\mathbf{y}\|}. \tag{6.5.33}$$

In our case, if we choose

$$\mathbf{z}_2 = \begin{bmatrix} 1 \\ 0 \end{bmatrix},$$

then

$$\mathbf{y} = \begin{bmatrix} 1 \\ 0 \end{bmatrix} - \left(\frac{1}{\sqrt{2}} \begin{bmatrix} \bar{z} & 1 \end{bmatrix} \begin{bmatrix} 1 \\ 0 \end{bmatrix} \right) \frac{1}{\sqrt{2}} \begin{bmatrix} z \\ 1 \end{bmatrix}$$

$$= \begin{bmatrix} \frac{1}{2} \\ -\frac{1}{2}\bar{z} \end{bmatrix},$$

and

$$\mathbf{y}_2 = \frac{1}{\sqrt{2}} \begin{bmatrix} 1 \\ -\bar{z} \end{bmatrix}. \tag{6.5.34}$$

Note that when $\xi = 0$,

$$\mathbf{y}_1 = \frac{1}{\sqrt{2}} \begin{bmatrix} 1 \\ 1 \end{bmatrix} \quad \text{and} \quad \mathbf{y}_2 = \frac{1}{\sqrt{2}} \begin{bmatrix} 1 \\ -1 \end{bmatrix}.$$

So, $\mathbf{y}_1$ and $\mathbf{y}_2$ are still orthonormal when $\xi = 0$. Also note that $G\mathbf{y}_1 = z\mathbf{y}_1$ but $G\mathbf{y}_2 \neq \bar{z}\mathbf{y}_2$ ($\mathbf{y}_2$ is not an eigenvector).

We then set

$$U = [\mathbf{y}_1 \ \mathbf{y}_2] \tag{6.5.35}$$

and note that

$$U^*GU = \begin{pmatrix} z & a\bar{z} + \frac{1}{2}(\bar{z}^2 + 1) \\ 0 & \bar{z} \end{pmatrix} \tag{6.5.36}$$

(where U^* is the conjugate transpose of U). Let $b = a\bar{z} + \frac{1}{2}(\bar{z}^2 + 1)$ and

$$U^*GU = T = \begin{pmatrix} z & b \\ 0 & \bar{z} \end{pmatrix}. \tag{6.5.37}$$

We should realize that U is unitary ($U^{-1} = U^*$) and $G^n = UT^nU^*$. Since U is unitary $\| G^n \|_2 = \| T^n \|_2$ for the l_2 norm. Thus, to prove stability order n, it suffices to satisfy inequality (6.5.16) for T instead of G. By inspection, we note that

$$T^n = \begin{pmatrix} z^n & b\sum_{j=0}^{n-1} z^{n-1-j}\bar{z}^j \\ 0 & \bar{z}^n \end{pmatrix}. \tag{6.5.38}$$

We use the fact that $\| T^n \|_2 = \sqrt{\sigma(T^{n*}T^n)}$ and a routine but ugly calculation to see that the eigenvalues of $T^{n*}T^n$ are given by

$$\lambda = \frac{1}{2}\left[2 + a\bar{a} \pm \sqrt{(a\bar{a})^2 + 4a\bar{a}}\right], \tag{6.5.39}$$

where $a = b\sum_{j=0}^{n-1} z^{n-1-j}\bar{z}$. Then another rather nasty calculation gives us that

$$\| T^n \|_2 \leq \begin{cases} 2 + 10n & \text{if } |a| \geq 1 \\ 4 & \text{if } |a| < 1 \end{cases}. \tag{6.5.40}$$

As usual, using Parseval's Identity, the fact that $\| G^n \|_2$ is bounded is enough to imply that $\| Q^n \|_2$ is bounded. Hence, we see that if $R \leq 1$, difference scheme (6.5.3) is stable order n, and, since the scheme is consistent with partial differential equation (6.5.1), the scheme is convergent if $R \leq 1$.

Remark 1: We see that in this situation where we are able to have more growth in $\| G^n \|$, *the growth is due to the fact that the eigenvalues coalesce when $\xi = 0$.* This will always be the case when for some values of ξ, the matrix has a nontrivial Jordan canonical form (is not diagonalizable).

Remark 2: To see why the factor n is needed (other than to make difference scheme (6.5.3) stable order n), we consider partial differential equation (6.5.1) along with initial conditions $v(x,0) = 0$ and $v_t(x,0) = 1$. The solution to this problem is $v(x,t) = t$. This is clearly a problem that we would like to able to solve. Since the approximate solution using difference scheme (6.5.3) is $u_k^{n+1} = (n + 1)\Delta t$, it is clear that stability order n is needed to approximately solve this problem.

Remark 3: We note the similarities with the analysis of the leapfrog scheme done in Section 6.4. We found in Section 6.4 that the leapfrog scheme was stable if $R \leq R_0$ for any $R_0 < 1$. The problem that occurred at $R = 1$ was that for $\xi = \frac{\pi}{2}$, the eigenvalues became equal. In that case, since we could not get convergence for that scheme using stability order n (because the leapfrog scheme did not satisfy an equation of the form (6.5.12)), it was necessary to restrict R so that the eigenvalues were never equal.

We next include two examples that demonstrate how we should choose a starting scheme to be used in conjunction with difference scheme (6.5.3).

Example 6.5.2 Analyze equation (6.5.8) as an initialization scheme for difference scheme (6.5.3).

Solution: Returning to equation (6.5.8), our first impression should probably be that initialization scheme (6.5.8) will not maintain the accuracy of difference scheme (6.5.3). If we return to equation (6.5.7), we see that equation (6.5.8) was derived using a first order approximation of the derivative $v_t(x,0)$ while difference scheme (6.5.3) was derived using second order differences. But, it does not look quite as bad when we let v denote the solution to initial–value problem (6.5.1), (6.5.4)–(6.5.5) and calculate the error as

$$
\begin{aligned}
w_k^1 &= v_k^1 - u_k^1 \\
&= v_k^0 + (v_t)_k^0 \Delta t + \mathcal{O}(\Delta t^2) - [u_k^0 + \Delta t f_{1k}] \\
&= \mathcal{O}(\Delta t^2).
\end{aligned}
$$

However, if we return to the Initialization Condition (**IC**) derived earlier, and. specifically to expression (6.5.20), we see that the above error is not sufficient to ensure us that we maintain the accuracy of difference scheme (6.5.3). The convergence of difference scheme (6.5.3) and (6.5.8) will be convergent order $\mathcal{O}(\Delta t) + \mathcal{O}(\Delta x^2)$. Clearly, *the choice of the initialization scheme can make the scheme lose accuracy.*

Example 6.5.3 Analyze

$$
u_k^1 = \frac{1}{2} R^2 (f_{k+1} + f_{k-1}) + (1 - R^2) f_k + \Delta t f_{1k} \tag{6.5.41}
$$

as an initialization scheme for difference scheme (6.5.3).

Solution: Before we perform this analysis, we note that equation (6.5.41) can be derived by approximating initial condition (6.5.5) by

$$
\frac{u_k^1 - u_k^{-1}}{2\Delta t} = f_{1k} \tag{6.5.42}
$$

and assume that difference scheme (6.5.3) is valid at $n = 0$,

$$u_k^1 = R^2(u_{k+1}^0 + u_{k-1}^0) + 2(1 - R^2)u_k^0 - u_k^{-1}. \qquad (6.5.43)$$

If we then use $u_k^0 = f_k$ and equation (6.5.42) along with equation (6.5.43), we arrive at our proposed initialization scheme (6.5.41).

As in Example 6.5.2, we begin by calculating the error associated with initialization scheme (6.5.41). We let v denote a function satisfying (6.5.1), (6.5.4)–(6.5.5) and note that

$$
\begin{aligned}
w_k^1 &= v_k^1 - u_k^1 \\
&= v_k^0 + (v_t)_k^0 \Delta t + (v_{tt})_k^0 \frac{\Delta t^2}{2} + \mathcal{O}(\Delta t^3) \\
&\quad - \frac{1}{2}R^2(f_{k+1} + f_{k-1}) - (1 - R^2)f_k - \Delta t f_{1\,k} \\
&= c^2(v_{xx})_k^0 \frac{\Delta t^2}{2} + \mathcal{O}(\Delta t^3) - \frac{1}{2}R^2(f_{k+1} + f_{k-1}) + R^2 f_k \\
&= \frac{c^2}{\Delta x^2}(\delta^2 v_k^0)\frac{\Delta t^2}{2} + c^2\mathcal{O}(\Delta x^2)\frac{\Delta t^2}{2} + \mathcal{O}(\Delta t^3) \\
&\quad - \frac{1}{2}R^2(f_{k+1} + f_{k-1}) + R^2 f_k \\
&= \mathcal{O}(\Delta x^2 \Delta t^2) + \mathcal{O}(\Delta t^3).
\end{aligned}
$$

Thus, again using the IC criteria (and equation (6.5.20)), we see that the initialization scheme (6.5.41) used along with difference scheme (6.5.3) and $u_k^0 = f_k$ is sufficient to imply convergence of order $(2, 2)$.

HW 6.5.2 Redo HW6.5.1 using (6.5.41) as the initialization scheme.

HW 6.5.3 (a) Use the substitution $w_1 = v_x$, $w_2 = v_t$ to reduce the second order initial–value problem (6.5.1), (6.5.4)–(6.5.5) to a first order initial–value problem (assuming that the function f is a differentiable function).
(b) Analyze the stability of the following difference schemes to be used for solving the first order system found in part (a).
(1) forward time, backward space (6.2.44)
(2) Lax-Wendroff, (6.2.45)
(c) Use the difference schemes found in (b)-(1) and (b)-(2) above (if they are stable) to solve the problem given in HW6.5.1. Compare your results with both the exact solution and the numerical solutions found in HW6.5.1 and HW6.5.2.

6.5.2   More

All of the work done in the previous section was done for the wave equation, (6.5.1). Of course there are other equations that are higher order in time. The results for these equations will follow in a similar manner as those for the wave equations.

Also, all of the analysis done in Section 6.5.1 was for initial–value problems (and, of course, the computations done in HW6.5.1 and HW6.5.2 was done for initial–boundary–value problems). As is usually the case, due to the lack of symmetry in our problems, it is very difficult to obtain sufficient conditions for stability of schemes for initial–boundary–value problems. Thus, we will be content with using the conditions obtained for the schemes for the initial–value problems as necessary conditions for the stability of schemes for the initial–boundary–value problems.

And, finally, before we leave this section, we discuss briefly the CFL condition for the one dimensional wave equation. If we consider partial differential equation (6.5.1) along with initial conditions (6.5.4) and (6.5.5) on $\mathbb{R}$, we see the the solution is given by

$$v(x,t) = \frac{1}{2}\left[f(x+ct) + f(x-ct)\right] + \frac{1}{2c}\int_{x-ct}^{x+ct} f_1(s)ds.$$

The above solution makes it clear that the analytic domain of dependence of the point (x,t) will include the points $x+ct$ and $x-ct$ due to initial condition (6.5.4) and the interval $[x-ct, x+ct]$ due to the initial condition (6.5.5). And, if we were to consider the nonhomogeneous analog of equation (6.5.1), the analytic domain of dependence would be a triangle in (x,t) space with vertices (x,t), $(x-ct,0)$ and $(x+ct,0)$. (Of course, all of these facts and much more can be found in any elementary textbook on partial differential equations.)

Also, as we did for scalar equations, we can *define the numerical domain of dependence of a difference scheme for higher order hyperbolic partial differential equations at the point $(k\Delta x, (n+1)\Delta t)$ as the smallest interval containing the set points in $\mathbb{R}$ on which the solution of the difference scheme depends as Δx and Δt approach zero and $c\Delta t/\Delta x$ is held constant.* For example, the numerical domain of dependence of difference scheme (6.5.3) for the point $(k\Delta x, (n+1)\Delta t)$ is the interval $[(k-n-1)\Delta x, (k+n+1)\Delta x]$.

Again, using exactly the same reasoning we did for scalar equations, we find that *for a difference scheme for approximating the solution of partial differential equation (6.5.1) to be convergent, it is necessary that the difference scheme satisfy the CFL condition.* Of course, the CFL condition is again defined to be the requirement that the analytic domain of dependence be contained in the numerical domain of dependence. And as we have done before, it is easy to see that difference scheme (6.5.3) and partial differential equation (6.5.1) will satisfy the CFL condition if

$$[k\Delta x - c(n+1)\Delta t, k\Delta x + c(n+1)\Delta t] \subset [(k-n-1)\Delta x, (k+n+1)\Delta x],$$

or

$$R = c\frac{\Delta t}{\Delta x} \leq 1.$$

Thus, we see that as is so often the case, the CFL necessary condition for convergence is the same as the necessary and sufficient condition derived in Section 6.5.1. And, as was the case with the CFL condition for one and two dimensional scalar equations for the one way wave equation, as we choose which explicit difference scheme to use for a given stencil, the CFL condition gives an easily calculated lower limit for the stability condition. Most often, we try to choose a scheme having a stability condition as close to the CFL condition as we can find. We next consider a CFL theory for general systems of hyperbolic partial differential equations.

HW 6.5.4 Consider the partial differential equation

$$v_{tt} = c^2 v_{xx} + F(x, t) \tag{6.5.44}$$

and the finite difference scheme

$$u_k^{n+1} = R^2(u_{k+1}^n + u_{k-1}^n) + 2(1 - R^2)u_k^n - u_k^{n-1} + \Delta t^2 F_k^n. \tag{6.5.45}$$

Show that the numerical domain of dependence of difference scheme (6.5.45) at point $(k\Delta x, (n+1)\Delta t)$ is given by the triangle with vertices $(k\Delta x, (n+1)\Delta t)$, $((k - n - 1)\Delta x, 0)$ and $((k + n + 1)\Delta x, 0)$. Show also that if the analytic domain of dependence of partial differential equation (6.5.44) is not contained in the numerical domain of dependence of difference scheme (6.5.45), the solution of the difference scheme will not converge to the solution of the partial differential equation.

6.6 Courant-Friedrichs-Lewy Condition for Systems

Just as we were able to define a numerical domain of dependence for explicit difference schemes for first and second order hyperbolic equations and obtain a CFL condition which was necessary for stability, the same program is possible for hyperbolic systems. *Because stability analyses become more difficult with systems, the CFL necessary condition becomes more valuable.*

We begin by considering a first order, constant coefficient hyperbolic system of partial differential equations of the form

$$\mathbf{v}_t = A\mathbf{v}_x, \quad x \in \mathbb{R}, \quad t > 0, \tag{6.6.1}$$

where as before, we assume that A is diagonalizable. We should recall that any results we obtain will also hold true for the case where we add a term of the form $B\mathbf{v}$ to the above partial differential equation. We approximate the systems of partial differential equations in (6.6.1) by the general explicit difference equation

$$\mathbf{u}_k^{n+1} = Q_2\mathbf{u}_k^n, \tag{6.6.2}$$

where Q_2 is given by

$$Q_2 = \sum_{j=-m_1}^{m_2} Q_{2j} S_+^j, \tag{6.6.3}$$

and Q_{2j}, $j = -m_1, \cdots, j = m_2$ are $K \times K$ matrices.

As in the scalar case, the system of partial differential equations (6.6.1) has a domain of dependence. Since A is diagonalizable, there exists a matrix S such that $D = SAS^{-1}$ is a diagonal matrix with the eigenvalues of A on the diagonal. Then using the decomposition used in Section 6.2.1, we see that the systems of partial differential equations (6.6.1) can be rewritten as

$$\mathbf{V}_t = D\mathbf{V}_x, \qquad (6.6.4)$$

where $\mathbf{V} = S\mathbf{v}$. If the initial condition for system (6.6.1) is $\mathbf{v}(x,0) = \mathbf{v}_0$, then the initial condition for system (6.6.4) is given by $\mathbf{V}(x,0) = \mathbf{V}_0 = S\mathbf{v}_0$, and the solution to the initial–value problem is given by

$$V_j(x,t) = V_{0j}(x + \mu_j t), \quad j = 1, \cdots, K,$$

where μ_j, $j = 1, \cdots, K$ are the eigenvalues of the matrix A. Then the solution to the original system, (6.6.1), can be written as

$$\mathbf{v}(x,t) = S^{-1}\mathbf{V}(x,t).$$

We see that the solution to the system of partial differential equations (6.6.1) at the point (x,t) depends on the points

$$(x + \mu_j t, 0): \ j = 1, \cdots, K.$$

Thus *the analytic domain of dependence of the point (x,t) for the system of partial differential equations (6.6.1) is the set of points*

$$(x + \mu_j t, 0): \ j = 1, \cdots, K.$$

As we did in the scalar case, we define the numerical domain of dependence of the point $(k\Delta x, (n+1)\Delta t)$ for difference scheme (6.6.2) to be the interval $D_n = [(k - m_1(n+1))\Delta x, (k + m_2(n+1))\Delta x]$. The rationale for this definition is the same as for the scalar case. If we consider convergence for $R = \Delta t/\Delta x = constant$, then the interval D_n contains all of the points at $t = 0$ to which the solution will reach through convergence. This can be easily seen by applying the same approach we used in the scalar case to equations (6.6.2)–(6.6.3).

The definition of when a system of partial differential equations and an associated difference scheme is said to satisfy the Courant-Friedrichs-Lewy (CFL) condition is again given by Definition 5.7.1, when the analytic domain of dependence of a given point, D_a, is contained in the numerical domain of dependence of that point, D_n. So the system of partial differential equations (6.6.1) and the difference scheme (6.6.2)–(6.6.3) satisfy the CFL condition if for the point $(x,t) = (k\Delta x, (n+1)\Delta t)$

$$x + \mu_j t = k\Delta x + \mu_j(n+1)\Delta t \in [(k - m_1(n+1))\Delta x, (k + m_2(n+1))\Delta x]$$
$$\text{for } j = 1, \cdots, K$$

where $R = \Delta t/\Delta x$. It is easy to see that the difference scheme will satisfy the CFL condition if

$$- m_1 \leq \mu_j R \leq m_2. \tag{6.6.5}$$

As with everything else concerning the CFL condition for systems, Theorems 5.7.2 and 5.7.3 are also true for systems. Then, for example, if we consider the forward difference scheme given in (6.2.39), we see that the CFL condition is satisfied for this scheme if $0 \leq \mu_j R \leq 1$, $j = 1, \cdots, K$ ($m_1 = 0$ and $m_2 = 1$). Thus, $0 \leq \mu_j R \leq 1$, $j = 1, \cdots, K$ is a necessary condition for convergence of difference scheme (6.2.39). Again we see that this is the same as the necessary and sufficient stability condition derived for difference scheme (6.2.39) in Section 6.1. As was the situation in the scalar case, *the CFL condition gives only a necessary condition while the stability analysis* may *give a necessary and sufficient condition.*

Clearly, satisfying the CFL condition gives only a necessary condition and not a sufficient condition since that was the case for the 1×1 example considered in Example 5.7.1. However, as we have mentioned before, when we wish to solve a problem using an explicit scheme, the CFL condition gives a lower limit on the stability condition, a condition that we *may* want to strive to equal.

6.7   Two Dimensional Systems

To solve systems of equations with two or more dimensions, we get all of the difficulties caused by the facts that it is a system and that we have more than one dimension. We will concentrate our efforts on two dimensional problems. Other than size, three or more dimensional problems do not add any new difficulties to the solution process.

Before we begin our work, we emphasize that though the numerical solution of multi-dimensional systems of partial differential equations is difficult, it is done often. Most often the multi-dimensional systems that are approximately solved are difficult nonlinear systems. Most of the numerical weather prediction, computational aerodynamics, computational fluid dynamics, simulation of groundwater flows and oil reservoir simulation involve solving time dependent, nonlinear systems (parabolic and/or hyperbolic) of partial differential equations. Much of the work is based on schemes developed for linear model equations and careful numerical experiments. Most of the numerical techniques that have been developed for two and three dimensional systems were driven by these application areas.

To consider general two dimensional systems of partial differential equations, we would like to consider an equation of the form

$$\mathbf{v}_t = B_0 \mathbf{v}_{xx} + B_1 \mathbf{v}_{yy} + A_1 \mathbf{v}_x + A_2 \mathbf{v}_y + C_0 \mathbf{v}. \tag{6.7.1}$$

As we shall see as we proceed, it is virtually impossible to obtain results for an equation as general as equation (6.7.1). We will include difference schemes that will (should?) work for obtaining approximate solutions to equation (6.7.1). However, we will present these results for special sub-cases of equation (6.7.1) and do all of our stability analysis on these easier model equations. In addition, when we do some of the analysis for our schemes, we will see that it is necessary to include strong assumptions that will most often not be satisfied for most applied problems. *The approach taken must be to obtain the best analytic results possible and then proceed carefully with the appropriate numerical experiments.* The methods that we include for two dimensional systems are the linear analogs of some successful schemes used in the area of computational fluid dynamics.

We begin this section by discussing difference schemes for initial–value problems. Because of the difficulties involved with stability analyses for difference schemes for multi-dimensional systems, we will confine our analyses to schemes for initial–value problems. As was the case for implicit schemes for two dimensional scalar equations, implicit schemes for initial boundary–value problems involving systems leave us with a broadly banded matrix equation to solve (worse in the case of systems). This difficulty is overcome by introducing alternating direction implicit (ADI) schemes to obtain efficient, implicit schemes.

And, finally, we close the section by discussing boundary conditions for hyperbolic systems. The problems caused by the boundary conditions for two dimensional systems are similar to those for one dimensional systems and two dimensional scalar equations. In Section 6.7.2, we give a brief discussion about how to determine how many mathematical boundary conditions we can apply, and where we can apply them. For a more complete discussion, the reader should consult a textbook such as ref. [5].

6.7.1 Initial–Value Problems

In this section we will investigate several schemes for solving special cases of equation (6.7.1). We will only do analysis for these schemes as initial–value schemes. The approach we shall use is to prove consistency and stability and obtain convergence by the Lax Theorem. We should be aware that for two dimensional systems, the $\ell_{2,\Delta x}$ norm used with regard to consistency, stability and convergence is the vector version of the norm given in (2.2.35) (it will be a combination of norms (2.2.35) and (2.2.37)). And, as usual, stability will be obtaining by using the discrete Fourier transform.

6.7.1.1 Hyperbolic Systems

We begin by considering partial differential equation (6.7.1) for the case when $B_0 = B_1 = \Theta$,

$$\mathbf{v}_t = A_1 \mathbf{v}_x + A_2 \mathbf{v}_y + C_0 \mathbf{v}. \tag{6.7.2}$$

Partial differential equation (6.7.2) is **hyperbolic** if (i) the eigenvalues of $\omega_1 A_1 + \omega_2 A_2$ are real for all ω_1, ω_2, and (ii) the matrix $\omega_1 A_1 + \omega_2 A_2$ is diagonalizable by a matrix $S = S(\omega_1, \omega_2)$ where norms of S and S^{-1} are bounded independent of ω_1, ω_2, $|\omega_1|^2 + |\omega_2|^2 = 1$. We note that by the above definition, A_1 and A_2 are both diagonalizable, and each of them have real eigenvalues.

Consider the forward difference scheme for approximating the solution to partial differential equation (6.7.2),

$$\mathbf{u}_{jk}^{n+1} = \mathbf{u}_{jk}^{n} + R_x A_1 \delta_{x+} \mathbf{u}_{jk}^{n} + R_y A_2 \delta_{y+} \mathbf{u}_{jk}^{n} + \Delta t C_0 \mathbf{u}_{jk}^{n} \qquad (6.7.3)$$

Example 6.7.1 Analyze the consistency and stability of difference scheme (6.7.3).

Solution: If we consider briefly what must be done to analyze the consistency of the scheme, it becomes clear that the analysis in this case is the same as that for the scalar equation. Thus, difference scheme (6.7.3) is consistent order $\mathcal{O}(\Delta t) + \mathcal{O}(\Delta x)$.

In order to study the stability of difference scheme (6.7.3) (without the C_0 term) we take the discrete Fourier transform and obtain the amplification matrix

$$G(\xi, \eta) = I + R_x(\cos\xi - 1)A_1 + R_y(\cos\eta - 1)A_2 + i(R_x \sin\xi A_1 + R_y \sin\eta A_2).$$
$$(6.7.4)$$

Any result we obtain using amplification matrix (6.7.4) without the C_0 term will apply also for the appropriate scheme including the C_0 term by Theorem 6.2.6.

One might expect that it might now be possible to use the definition of hyperbolicity to obtain sufficient conditions for stability in much the same way we did so often with one dimensional systems. We can use the definition of hyperbolicity to obtain a matrix S that diagonalizes G. The difference is that the values on the diagonal are generally not given in terms of the eigenvalues of A_1 and A_2. We have no way to relate the values that would appear on the diagonal of G directly to the matrices A_1 and A_2 (i.e. we cannot relate these diagonal elements to the eigenvalues of A_1 and A_2).

Hence, we will most often make the stronger *assumption that the matrices A_1 and A_2 are simultaneously diagonalizable.* The matrices A_1 and A_2 are said to be simultaneously diagonalizable if there exists a matrix S such that $D_1 = SA_1S^{-1}$ and $D_2 = SA_2S^{-1}$ are both diagonal matrices. Clearly, *if A_1 and A_2 are simultaneously diagonalizable, then the system will be hyperbolic.*

We emphasize that this is a strong assumption. If A_1 and A_2 are simultaneously diagonalizable, then both matrices have the same eigenvectors and the values on the diagonals of D_1 and D_2 will the the eigenvalues of A_1 and A_2, respectively. Of course, these eigenvalues need not be the same. Another indication that this assumption is strong is the following result: *A_1 and A_2 are simultaneously diagonalizable if and only if A_1 and A_2 commute,* ref. [3], page 52.

With this assumption, we see that using S we can easily write SGS^{-1} as

$$SGS^{-1} = I + R_x(\cos\xi - 1)D_1 + R_y(\cos\eta - 1)D_2 + i(R_x \sin\xi D_1 + R_y \sin\eta D_2)$$
$$(6.7.5)$$

where D_1 and D_2 are the diagonal matrices and the diagonal elements of D_1 and D_2 are the eigenvalues of A_1 and A_2, respectively. Thus we see that the eigenvalues of G will be given by

$$\lambda_j = 1 + R_x(\cos\xi - 1)\mu_j + R_y(\cos\eta - 1)\nu_j + i(R_x\mu_j \sin\xi + R_y\nu_j \sin\eta),$$
$$j = 1, \cdots, K, \qquad (6.7.6)$$

where μ_j and $\nu_j, j = 1, \cdots, K$ are the eigenvalues of A_1 and A_2, respectively. Then using an analysis similar to that used in Example 5.8.1, we see that μ_j, $\nu_j \geq 0$, $0 \leq$

$\mu_j R_x + \nu_j R_y \leq 1$, $j = 1, \cdots, K$ is a necessary condition for convergence (difference scheme (6.7.3) satisfies the von Neumann condition).

To see that the above condition is also a sufficient condition, we need two dimensional analogs of Theorems 6.2.4 and 6.2.5. However, we see that this is no problem. For example, for this problem we specifically need a two dimensional version of Theorem 6.2.5(ii). But, as long as

$$H = SGS^{-1}$$
$$= I + R_x(\cos \xi - 1)D_1 + R_y(\cos \eta - 1)D_2 + i(R_x \sin \xi D_1 + R_y \sin \eta D_2),$$

it is easy to see that

$$\| G^n \| \leq \| S \| \| S^{-1} \| \| H^n \|.$$

Hence, as long as we can restrict H^n from growing (by restricting R_x and R_y so that the diagonal elements of H will be less than or equal to one), it makes no difference whether G originally came from a one or two dimensional problem. Hence, we see that μ_j, $\nu_j \geq 0$, $0 \leq \mu_j R_x + \nu_j R_y \leq 1$, $j = 1, \cdots, K$ is both a necessary and sufficient condition for stability (and, hence, convergence) of difference scheme (6.7.3).

We next consider the analog to difference scheme (5.8.21)–(5.8.22) for solving the system of partial differential equations (6.7.2).

Example 6.7.2 Analyze the consistency and stability of difference scheme

$$\mathbf{u}_{jk}^{n+1/2} = \mathbf{u}_{jk}^n + \frac{R_x}{2} A_1 \delta_{x0} \mathbf{u}_{jk}^n + \frac{R_x^2}{2} A_1^2 \delta_x^2 \mathbf{u}_{jk}^n \qquad (6.7.7)$$

$$\mathbf{u}_{jk}^{n+1} = \mathbf{u}_{jk}^{n+1/2} + \frac{R_y}{2} A_2 \delta_{y0} \mathbf{u}_{jk}^{n+1/2} + \frac{R_y^2}{2} A_2^2 \delta_y^2 \mathbf{u}_{jk}^{n+1/2} + \Delta t C_0 \mathbf{u}_{jk}^n. \qquad (6.7.8)$$

Solution: If we proceed as usual, omit the C_0 term and take the discrete Fourier transform of equations (6.7.7)–(6.7.8), we find that the amplification matrix is given by

$$G(\xi, \eta) = \left(I + iR_y \sin \eta A_2 - 2R_y^2 \sin^2 \frac{\eta}{2} A_2^2 \right)\left(I + iR_x \sin \xi A_1 - 2R_x^2 \sin^2 \frac{\xi}{2} A_1^2 \right).$$
$$(6.7.9)$$

If we again *assume that A_1 and A_2 are simultaneously diagonalizable*, we see that we can write SGS^{-1} as

$$SGS^{-1} = \left(I + iR_y \sin \eta D_2 - 2R_y^2 \sin^2 \frac{\eta^2}{2} D_2^2 \right)\left(I + iR_x \sin \xi D_1 - 2R_x^2 \sin^2 \frac{\xi}{2} D_1^2 \right).$$
$$(6.7.10)$$

Then applying the two dimensional version of Theorem 6.2.5(ii) discussed in Example 6.7.1 along with the analysis done in Example 5.8.3 (done for the scalar version of difference scheme (6.7.7)–(6.7.8)), we find that *difference scheme (6.7.7)–(6.7.8) is stable if and only if*

$$\max\{| \mu_1 | R_x, \cdots, | \mu_K | R_x, | \nu_1 | R_y, \cdots, | \nu_k | R_y\} \leq 1.$$

And since it is not difficult to show that difference scheme (6.7.7)–(6.7.8) approximates the system of hyperbolic partial differential equations (6.7.2) to order $\mathcal{O}(\Delta t^2) + \mathcal{O}(\Delta x^2) + \mathcal{O}(\Delta y^2)$, condition

$$\max\{| \mu_1 | R_x, \cdots, | \mu_K | R_x, | \nu_1 | R_y, \cdots, | \nu_k | R_y\} \leq 1$$

will be necessary and sufficient for convergence of difference scheme (6.7.7)–(6.7.8).

Now that we have seen how to obtain stability conditions for difference schemes for two dimensional systems, we include the following examples for the implicit scheme (BTCS) and the Crank-Nicolson scheme.

Example 6.7.3 Assume that the matrices A_1 and A_2 are simultaneously diagonalizable. Discuss the stability of the difference scheme

$$\mathbf{u}_{jk}^{n+1} + \frac{R_x}{2} A_1 \delta_{x0} \mathbf{u}_{jk}^{n+1} + \frac{R_y}{2} A_2 \delta_{y0} \mathbf{u}_{jk}^{n+1} = \mathbf{u}_{jk}^n. \tag{6.7.11}$$

Solution: Taking the discrete Fourier transform of equation (6.7.11) gives us the amplification matrix

$$G(\xi, \eta) = (I + iR_x \sin \xi A_1 + iR_y \sin \eta A_2)^{-1}. \tag{6.7.12}$$

Since

$$SGS^{-1} = (I + iR_x \sin \xi D_1 + iR_y \sin \eta D_2)^{-1},$$

the two dimensional analog of Theorem 6.2.5(ii) will imply that difference scheme (6.7.11) is unconditionally stable.

Example 6.7.4 Assume that the matrices A_1 and A_2 are simultaneously diagonalizable. Analyze the stability of the Crank-Nicolson scheme

$$\mathbf{u}_{jk}^{n+1} - \frac{R_x}{4} A_1 \delta_{x0} \mathbf{u}_{jk}^{n+1} - \frac{R_y}{4} A_2 \delta_{y0} \mathbf{u}_{jk}^{n+1} = \mathbf{u}_{jk}^n + \frac{R_x}{4} A_1 \delta_{x0} \mathbf{u}_{jk}^n + \frac{R_y}{4} A_2 \delta_{y0} \mathbf{u}_{jk}^n. \tag{6.7.13}$$

Solution: The amplification matrix for the Crank-Nicolson scheme is given by

$$G(\xi, \eta) = \left(I + i\frac{R_x}{2} \sin \xi A_1 + i\frac{R_y}{2} \sin \eta A_2\right)^{-1} \left(I - i\frac{R_x}{2} \sin \xi A_1 - i\frac{R_y}{2} \sin \eta A_2\right).$$

We note that

$$SGS^{-1} = \left(I + i\frac{R_x}{2} \sin \xi D_1 + i\frac{R_y}{2} \sin \eta D_2\right)^{-1} \left(I - i\frac{R_x}{2} \sin \xi D_1 - i\frac{R_y}{2} \sin \eta D_2\right).$$

If we let $H = SGS^{-1}$, it is easy to see that $H^*H = I$ and that by the two dimensional analog of Theorem 6.2.5(ii) difference scheme (6.7.13) is unconditionally stable.

6.7.1.2 Parabolic Systems

We next consider systems of partial differential equations that are parabolic. A system of equations of form (6.7.1) is said the be **parabolic** if for all $\omega = [\omega_1, \omega_2]^T \in \mathbb{R}^2$, the eigenvalues $\kappa_j(\omega)$, $j = 1, \cdots, K$ of $-\omega_1^2 B_0 - \omega_2^2 B_1$ satisfy $\mathcal{R}e\ \kappa_j(\omega) \leq \delta \mid \omega \mid^2, j = 1, \cdots, K$ for some $\delta > 0$ independent of ω. Specifically, we consider the parabolic system of partial differential equations of the form

$$\mathbf{v}_t = B_0 \mathbf{v}_{xx} + B_1 \mathbf{v}_{yy}, \tag{6.7.14}$$

where we assume that B_0 and B_1 are positive definite and simultaneously diagonalizable. Below we give two difference schemes for solving partial differential equation (6.7.14) and give stability results for these schemes.

Example 6.7.5 Discuss the stability of the difference scheme

$$\mathbf{u}_{j\,k}^{n+1} = \mathbf{u}_{j\,k}^{n} + r_x B_0 \delta_x^2 \mathbf{u}_{j\,k}^{n} + r_y B_1 \delta_y \mathbf{u}_{j\,k}^{n}. \tag{6.7.15}$$

Solution: If we take the discrete Fourier transform of equation (6.7.15), we find that the amplification matrix is given by

$$G(\xi, \eta) = I - 4r_x \sin^2 \frac{\xi}{2} B_0 - 4r_y \sin^2 \frac{\eta}{2} B_1. \tag{6.7.16}$$

If B_0 and B_1 are simultaneously diagonalizable, it is not hard to see that

$$SGS^{-1} = I - 4r_x \sin^2 \frac{\xi}{2} D_0 - 4r_y \sin^2 \frac{\eta}{2} D_1,$$

where SGS^{-1} has the eigenvalues of G on its diagonal. Using the analysis for the analogous scalar scheme, Section 4.3.1, we see that $\mu_j r_x + \nu_j r_y \leq \frac{1}{2}$ (where μ_j and ν_j, $j = 1, \cdots, K$ are the eigenvalues of B_0 and B_1, respectively) is necessary for convergence. Then, applying Theorem 6.2.5(ii), we see that $\mu_j r_x + \nu_j r_y \leq \frac{1}{2}$, $j = 1, \cdots, K$ is both necessary and sufficient for convergence of difference scheme (6.7.15).

Example 6.7.6 Discuss the stability for the Crank-Nicolson scheme for solving the system of partial differential equations (6.7.14).

$$\mathbf{u}_{j\,k}^{n+1} - \frac{r_x}{2} B_0 \delta_x^2 \mathbf{u}_{j\,k}^{n+1} - \frac{r_y}{2} B_1 \delta_y \mathbf{u}_{j\,k}^{n+1} = \mathbf{u}_{j\,k}^{n} + \frac{r_x}{2} B_0 \delta_x^2 \mathbf{u}_{j\,k}^{n} + \frac{r_y}{2} B_1 \delta_y \mathbf{u}_{j\,k}^{n} \tag{6.7.17}$$

Solution: The amplification matrix for difference scheme (6.7.17) is given by

$$G(\xi, \eta) = \left(I + 2r_x \sin^2 \frac{\xi}{2} B_0 + 2r_y \sin^2 \frac{\eta}{2} B_1 \right)^{-1} \left(I - 2r_x \sin^2 \frac{\xi}{2} B_0 - 2r_y \sin^2 \frac{\eta}{2} B_1 \right). \tag{6.7.18}$$

Using the fact that B_0 and B_1 are simultaneously diagonalizable, we see that SGS^{-1} is a diagonal matrix with the eigenvalues of G,

$$\lambda_j = \frac{1 - 2r_x \sin^2 \frac{\xi}{2} \mu_j - 2r_y \sin^2 \frac{\eta}{2} \nu_j}{1 + 2r_x \sin^2 \frac{\xi}{2} \mu_j + 2r_y \sin^2 \frac{\eta}{2} \nu_j}, \quad j = 1, \cdots, K$$

on the diagonal. Then by the calculations done in Example 4.3.3 and the two dimensional version of Theorem 6.2.5(ii), we see that the Crank-Nicolson scheme is unconditionally stable.

We next consider a parabolic system of partial differential equations that is an excellent model equation for the compressible Navier-Stokes equations. We consider the equation

$$\mathbf{v}_t = \mathbf{v}_{xx} + \mathbf{v}_{yy} + A_1 \mathbf{v}_x + A_2 \mathbf{v}_y + C_0 \mathbf{v} \tag{6.7.19}$$

where we assume that A_1 and A_2 are simultaneously diagonalizable. One obvious way to try to solve equation (6.7.19) is by using the following difference equation.

$$\mathbf{u}_{j\,k}^{n+1} = \mathbf{u}_{j\,k}^{n} + r_x \delta_x^2 \mathbf{u}_{j\,k}^{n} + r_y \delta_y^2 \mathbf{u}_{j\,k}^{n} + \frac{R_x}{2} A_1 \delta_{x0} \mathbf{u}_{j\,k}^{n}$$

$$+ \frac{R_y}{2} A_2 \delta_{y0} \mathbf{u}_{j\,k}^{n} + \Delta t C_0 \mathbf{u}_{j\,k}^{n} \tag{6.7.20}$$

Example 6.7.7 Analyze the stability of difference scheme (6.7.20).

Solution: As usual, we begin by omitting the C_0 term. We then note that the amplification matrix is given by

$$G(\xi, \eta) = \left(1 - 4r_x \sin^2 \frac{\xi}{2} - 4r_y \sin^2 \frac{\eta}{2}\right) I + iR_x \sin \xi A_1 + iR_y \sin \eta A_2 \qquad (6.7.21)$$

and use the fact that A_1 *and* A_2 *are simultaneously diagonalizable* to write SGS^{-1} as

$$SGS^{-1} = \left(1 - 4r_x \sin^2 \frac{\xi}{2} - 4r_y \sin^2 \frac{\eta}{2}\right) I + iR_x \sin \xi D_1 + iR_y \sin \eta D_2. \qquad (6.7.22)$$

We then return to the stability analysis performed on the scalar analog to difference scheme (6.7.20), Example 4.3.2, to see that if we fix r_x and r_y to be constants satisfying $r_x + r_y \leq 1/2$, the difference scheme (6.7.20) will be stable. Again we emphasize that the stability is not as nice as we have for most other schemes. Specifically, for certain values of Δt, Δx and Δy satisfying $r_x + r_y \leq 1/2$, computations may appear to be going unstable. What we know is that if we make Δt sufficiently small, the computations will behave nicely.

HW 6.7.1 Assuming that A_1 and A_2 are simultaneously diagonalizable, analyze the stability of the following Crank-Nicolson scheme for solving the system of partial differential equations (6.7.19) (without the C_0 term).

$$\mathbf{u}_{jk}^{n+1} - \frac{r_x}{2} \delta_x^2 \mathbf{u}_{jk}^{n+1} - \frac{r_y}{2} \delta_y^2 \mathbf{u}_{jk}^{n+1} - \frac{R_x}{4} A_1 \delta_{x0} \mathbf{u}_{jk}^{n+1} - \frac{R_y}{4} A_2 \delta_{y0} \mathbf{u}_{jk}^{n+1} \mathbf{u}_{jk}^{n+1}$$

$$= \mathbf{u}_{jk}^{n} + \frac{r_x}{2} \delta_x^2 \mathbf{u}_{jk}^{n} + \frac{r_y}{2} \delta_y^2 \mathbf{u}_{jk}^{n} + \frac{R_x}{4} A_1 \delta_{x0} \mathbf{u}_{jk}^{n} + \frac{R_y}{4} A_2 \delta_{y0} \mathbf{u}_{jk}^{n}$$

$$\qquad (6.7.23)$$

6.7.1.3   ADI Schemes

Clearly, there are many other schemes for either system (6.7.14) or (6.7.19) that we could consider and analyze. Hopefully, by now the reader is very capable of performing these stability analyses. We note that as we consider more dimensions, the stability constraint for explicit schemes generally becomes more severe. The situation is worse for systems in that the few results that do not require that the coefficient matrices be simultaneously diagonalizable have even more restrictive stability constraints. And, of course, we must realize by now that using implicit schemes in more than one dimension is too expensive. Hence, as we did with scalar, two dimensional parabolic and hyperbolic equations, we develop ADI schemes for solving systems of partial differential equations.

We begin our discussion by considering the difference scheme analogous to the locally one dimensional (lod) scheme discussed in Section 5.8.3, (5.8.27)–(5.8.28). We consider the hyperbolic system of equations (6.7.2) (without the C_0 term) and the difference scheme

$$\left(I - \frac{R_x}{2} A_1 \delta_{x0}\right) \mathbf{u}_{jk}^{n+1/2} = \mathbf{u}_{jk}^{n} \qquad (6.7.24)$$

$$\left(I - \frac{R_y}{2} A_2 \delta_{y0}\right) \mathbf{u}_{jk}^{n+1} = \mathbf{u}_{jk}^{n+1/2}. \qquad (6.7.25)$$

Example 6.7.8 Discuss the consistency and stability of difference scheme (6.7.24)–(6.7.25).

Solution: We first consider the consistency of difference scheme (6.7.24)–(6.7.25). Eliminating $\mathbf{u}^{n+1/2}$ from equations (6.7.24) and (6.7.25) leaves us with

$$\left(I - \frac{R_x}{2} A_1 \delta_{x0}\right)\left(I - \frac{R_y}{2} A_2 \delta_{y0}\right)\mathbf{u}^{n+1}_{j\,k} = \mathbf{u}^n_{j\,k}.$$

This is the same as

$$\left(I - \frac{R_x}{2} A_1 \delta_{x0} - \frac{R_y}{2} A_2 \delta_{y0} + \frac{R_x R_y}{4} A_1 A_2 \delta_{x0}\delta_{y0}\right)\mathbf{u}^{n+1}_{j\,k} = \mathbf{u}^n_{j\,k},$$

or

$$\frac{\mathbf{u}^{n+1}_{j\,k} - \mathbf{u}^n_{j\,k}}{\Delta t} - A_1 \frac{1}{2\Delta x}\delta_{x0}\mathbf{u}^{n+1}_{j\,k} - A_2\frac{1}{2\Delta x}\delta_{y0}\mathbf{u}^{n+1}_{j\,k}$$
$$+ \Delta t A_1 A_2 \frac{1}{2\Delta x}\delta_{x0}\frac{1}{2\Delta y}\delta_{y0}\mathbf{u}^{n+1}_{j\,k} = \boldsymbol{\theta}.$$

Since

$$\Delta t A_1 A_2 \frac{1}{2\Delta x}\delta_{x0}\frac{1}{2\Delta y}\delta_{y0}\mathbf{u}^{n+1}_{j\,k}$$

is a $\mathcal{O}(\Delta t) + \mathcal{O}(\Delta t \Delta x^2) + \mathcal{O}(\Delta t \Delta y^2)$ term and the fact that the difference equation without this term is accurate $\mathcal{O}(\Delta t) + \mathcal{O}(\Delta x^2) + \mathcal{O}(\Delta y^2)$, difference scheme (6.7.24)–(6.7.25) will be a $\mathcal{O}(\Delta t) + \mathcal{O}(\Delta x^2) + \mathcal{O}(\Delta y^2)$ accurate scheme.

To perform a stability analysis for this scheme, we take the discrete Fourier transform of equations (6.7.24) and (6.7.25) to get

$$(I - iR_x \sin\xi A_1)\hat{\mathbf{u}}^{n+1/2} = \hat{\mathbf{u}}^n$$

and

$$(I - iR_y \sin\eta A_2)\hat{\mathbf{u}}^{n+1} = \hat{\mathbf{u}}^{n+1/2}.$$

Solving for $\hat{\mathbf{u}}^{n+1}$, we see that

$$G(\xi,\eta) = (I - iR_y \sin\eta A_2)^{-1}(I - iR_x \sin\xi A_1)^{-1}. \tag{6.7.26}$$

As is so often the case with systems, *we assume that A_1 and A_2 are simultaneously diagonalizable.* Then

$$SGS^{-1} = (I - iR_y \sin\eta D_2)^{-1}(I - iR_x \sin\xi D_1)^{-1}, \tag{6.7.27}$$

where the matrices D_1 and D_2 are the diagonal matrices with the eigenvalues of A_1 and A_2 on the diagonal. We see that the eigenvalues of G are given by

$$\lambda_j = \frac{1}{(1 - iR_y \nu_j \sin\eta)(1 - iR_x \mu_j \sin\xi)}, \quad j = 1,\cdots,K$$

where μ_j and ν_j, $j = 1,\cdots,K$ are the eigenvalues of A_1 and A_2, respectively. Since

$$|\lambda_j|^2 = \frac{1}{(1 + R_y^2 \nu_j^2 \sin^2\eta)(1 + R_x^2 \mu_j^2 \sin^2\xi)} \le 1$$

for all $j = 1,\cdots,K$, difference scheme (6.7.24)–(6.7.25) satisfies the von Neumann condition.

If we let $H = SGS^{-1}$, we see that $H^* H$ is a diagonal matrix with

$$\frac{1}{(1 + R_y^2 \nu_j^2 \sin^2\eta)(1 + R_x^2 \mu_j^2 \sin^2\xi)}$$

on the diagonal. Therefore, we can use the two dimensional version of Theorem 6.2.5-(ii) to see that difference scheme (6.7.24)–(6.7.25) is unconditionally stable.

We next consider the the scheme obtained by the approximate factorization of the Crank-Nicolson scheme, (6.7.13), in a manner analogous to the way we obtained difference scheme (5.8.34)–(5.8.35) from the scalar Crank-Nicolson scheme. We begin by factoring difference equation (6.7.13) to get

$$\left(I - \frac{R_x}{4} A_1 \delta_{x0}\right)\left(I - \frac{R_y}{4} A_2 \delta_{y0}\right) \mathbf{u}_{jk}^{n+1} = \left(I + \frac{R_x}{4} A_1 \delta_{x0}\right)\left(I + \frac{R_y}{4} A_2 \delta_{y0}\right) \mathbf{u}_{jk}^n$$

$$+ \frac{R_x R_y}{16} A_1 A_2 \delta_{x0} \delta_{y0} (\mathbf{u}_{jk}^{n+1} - \mathbf{u}_{jk}^n). \tag{6.7.28}$$

Since the last term is order $\mathcal{O}(\Delta t^3)$, we see that it can be omitted. We then rewrite the scheme as

$$\left(I - \frac{R_x}{4} A_1 \delta_{x0}\right) \mathbf{u}_{jk}^* = \left(I + \frac{R_x}{4} A_1 \delta_{x0}\right)\left(I + \frac{R_y}{4} A_2 \delta_{y0}\right) \mathbf{u}_{jk}^n \tag{6.7.29}$$

$$\left(I - \frac{R_y}{4} A_2 \delta_{y0}\right) \mathbf{u}_{jk}^{n+1} = \mathbf{u}_{jk}^* \tag{6.7.30}$$

Difference scheme (6.7.29)–(6.7.30) is referred to as the **Beam-Warming scheme**.

Example 6.7.9 Analyze the consistency and stablilty (and, hence, convergence) of the Beam-Warming difference scheme.

Solution: To see that difference scheme (6.7.29)–(6.7.30) is consistent with the system of partial differential equations (6.7.2) it is only necessary to use the accuracy of the Crank-Nicolson Scheme and the fact that the term from equation (6.7.28) was a $\mathcal{O}(\Delta t^3)$ term to see that difference scheme (6.7.29)–(6.7.30) is a $\mathcal{O}(\Delta t^2) + \mathcal{O}(\Delta x^2) + \mathcal{O}(\Delta y^2)$ approximation of equation (6.7.2).

To analyze the stability of difference scheme (6.7.29)–(6.7.30), we take the discrete Fourier transform of equations (6.7.29) and (6.7.30), eliminate $\hat{\mathbf{u}}^*$ and solve for $\hat{\mathbf{u}}^{n+1}$ to get the amplification matrix

$$G = \left(I - i\frac{R_y}{2} \sin \eta A_2\right)^{-1}\left(I - i\frac{R_x}{2} \sin \xi A_1\right)^{-1}\left(I + i\frac{R_x}{2} \sin \xi A_1\right)\left(I + i\frac{R_y}{2} \sin \eta A_2\right).$$
$$\tag{6.7.31}$$

As we have had to do so often before, we assume that A_1 and A_2 are simultaneously diagonalizable. If we let $H = SGS^{-1}$, then

$$H = SGS^{-1}$$
$$= \left(I - i\frac{R_y}{2} \sin \eta D_2\right)^{-1}\left(I - i\frac{R_x}{2} \sin \xi D_1\right)^{-1}\left(I + i\frac{R_x}{2} \sin \xi D_1\right)\left(I + i\frac{R_y}{2} \sin \eta D_2\right).$$
$$\tag{6.7.32}$$

Thus the eigenvalues of G are

$$\lambda_j = \frac{(1 + i\frac{R_x}{2}\mu_j \sin \xi)(1 + i\frac{R_y}{2}\nu_j \sin \eta)}{1 - i\frac{R_x}{2}\mu_j \sin \xi)(1 - i\frac{R_y}{2}\nu_j \sin \eta)}$$

for $j = 1, \cdots, K$ (and since $|\lambda_j| = 1$, $j = 1, \cdots, K$, we see that the Beam-Warming scheme satisfies the von Neumann condition) and H^*H is the diagonal matrix with the square of magnitude of these eigenvalues on the diagonal. Then since $|\lambda_j| = 1$, $j = 1, \cdots, K$, we apply the two dimensional version of Theorem 6.2.5-(ii) to see that difference scheme (6.7.29)–(6.7.30) is unconditionally stable.

HW 6.7.2 Rewrite difference scheme (6.7.29)–(6.7.30) in delta formulation (analogous to the scalar form given in (5.8.37)–(5.8.38)).

6.7.2 Boundary Conditions

The situation concerning boundary conditions for equation (6.7.1) is easy for the cases when equation (6.7.1) is parabolic (in which case, we "usually" have the correct number of boundary conditions in the correct place) or when equation (6.7.1) is hyperbolic and we have periodic boundary conditions (in which case, we extend the problem periodically and treat the problem as an initial–value problem). Hence, we assume that $B_0 = B_1 = \Theta$ and that system (6.7.1) is hyperbolic, i.e. consider the equation

$$\mathbf{v}_t = A_1\mathbf{v}_x + A_2\mathbf{v}_y, \ (x,y) \in (0,1) \times (0,1), \ t > 0. \qquad (6.7.33)$$

For the special case when A_1 and A_2 are simultaneously diagonalizable, it is not too difficult to determine how to assign the boundary conditions. Just as in the one dimensional case, we multiply equation (6.7.33) on the left by S (where S is the matrix for which SA_1S^{-1} and SA_2S^{-1} are diagonal matrices) and rewrite equation (6.7.33) in an uncoupled form as

$$V_{jt} = \mu_j V_{jx} + \nu_j V_{jy}, \ j = 1,\cdots,K, \qquad (6.7.34)$$

where $\mathbf{V} = S\mathbf{v}$; and μ_j, $j = 1,\cdots,K$ and ν_j, $j = 1,\cdots,K$ are the eigenvalues of A_1 and A_2, respectively. Thus, just as was the case for one dimensional systems, the problem of assigning the boundary conditions to a simultaneously diagonalizable, two dimensional, hyperbolic system reduces back to assigning boundary conditions to K two dimensional, scalar hyperbolic equations in terms of the characteristic variables. We know that if $K_{\mu+}$ and $K_{\mu-}$ of the eigenvalues μ_j are positive and negative, respectively, and $K_{\nu+}$ and $K_{\nu-}$ of the eigenvalues ν_j are positive and negative, respectively, then

1. we can assign $K_{\mu-}$ boundary conditions at $x = 0$ and $K_{\mu+}$ boundary conditions at $x = 1$, and

2. we can assign $K_{\nu-}$ boundary conditions at $y = 0$ and $K_{\nu+}$ boundary conditions at $y = 1$.

Remember that because the spatial derivatives in equation (6.7.34) are on the right hand side of the equality, the assignment of boundary conditions based on the sign of μ_j and ν_j is opposite of what it was in Section 5.8.5. Of course, we know more than this. We know that

1. a boundary condition at $x = 0$ can be assigned to each characteristic variable V_j that is associated with a negative μ_j,

2. a boundary condition at $x = 1$ can be assigned to each characteristic variable V_j that is associated with a positive μ_j,

3. a boundary condition at $y = 0$ can be assigned to each characteristic variable V_j that is associated with a negative ν_j and

4. a boundary condition at $y = 1$ can be assigned to each characteristic variable V_j that is associated with a positive ν_j.

The boundary conditions described above can be thought of as either boundary conditions on the characteristic variables V_j (which they clearly are) or as boundary conditions on combinations of the primitive variables through the relationship $\mathbf{V} = S\mathbf{v}$. Also, we can proceed as described in Section 6.3.1 to use the information of which characteristic variables can be assigned a boundary condition at a given boundary to indicate which primitive variables can be assigned boundary conditions at that same boundary.

When we cannot make the assumption that A_1 and A_2 are simultaneously diagonalizable, the situation gets more difficult. We will give an *approximate version* of the full result that will cover most situations that are faced (we omit a technically difficult hypothesis). For a complete discussion of boundary conditions for system (6.7.33), again see ref. [5]. Approximately speaking, just as we had for the simultaneously diagonalizable case, the number of boundary conditions that we can assign at $x = 0$ and $x = 1$ depends on the sign of the eigenvalues of A_1, and the number of boundary conditions we can assign at $y = 0$ and $y = 1$ depends on the signs of the eigenvalues of A_2. Or,

1. if A_1 has K_+ positive eigenvalues and K_- negative eigenvalues, then we can assign boundary conditions at $x = 1$ to the K_+ characteristic variables associated with the positive eigenvalues and boundary conditions at $x = 0$ to the K_- characteristic variables associated with the negative eigenvalues. And,

2. if A_2 has K_+ positive eigenvalues and K_- negative eigenvalues, then we can assign boundary conditions at $y = 1$ to the K_+ characteristic variables associated with the positive eigenvalues and boundary conditions at $y = 0$ to the K_- characteristic variables associated with the negative eigenvalues.

The above description does not eliminate all of the problems and is not especially well defined. At this time, we do not really know how to determine the characteristic variables of a two dimensional system that is not simultaneously diagonalizable. However, the characteristic variables referred to above are not the characteristic variables of the two dimensional system. We treat the determination of the boundary conditions as a local phenomena and base the placement of boundary conditions at $x = 0$ and $x = 1$ on the one dimensional equation $\mathbf{v}_t = A_1 \mathbf{v}_x$ (omit the $A_2 \mathbf{v}_y$ term) and the placement of boundary conditions at $y = 0$ and $y = 1$ on the one dimensional equation $\mathbf{v}_t = A_2\mathbf{v}_y$. Hence, *the one-dimensional results derived in Section 6.3.1 can also be used to choose primitive variables on which to prescribe the boundary conditions.* Thus, the approach that we suggest is to

1. determine the boundary conditions at $x = 0$ and $x = 1$ by considering the one dimensional system of equations $\mathbf{v}_t = A_1\mathbf{v}_x$ (where we can use either of two approaches described in Section 6.3.1, assigning boundary conditions combinations of primitive variables based on their relationship to the appropriate characteristic variables (one dimensional characteristic variables) or assigning boundary conditions to primitive variables determined by the procedure in Section 6.3.1, and

2. determine the boundary conditions at $y = 0$ and $y = 1$ by considering the one dimensional system of equations $\mathbf{v}_t = A_2\mathbf{v}_y$ (where again either of the two approaches described in Section 6.3.1 can be used to assign the boundary conditions).

Remark 1: For the non-simultaneously diagonalizable case, it is impossible to consider any approach above that would treat the boundary conditions as a boundary condition for the characteristic variables. We must remember that the characteristic variables referred to above are local characteristic variables and are not necessarily the characteristic variables of the two dimensional system under consideration.

Remark 2: If we use this procedure described above for general A_1 and A_2 and A_1 and A_2 are simultaneously diagonalizable, we will obtain the same results as we would using the method described in the beginning of this section where we assumed that A_1 and A_2 were simultaneously diagonalizable.

Remark 3: If some of the eigenvalues are zero (and this case is not uncommon when the variables represent the velocities for either an inviscid fluid (the normal component) or a viscous fluid (all components)), then the situation for either of the cases considered above is essentially the same as in the one dimensional case discussed in Section 6.3.1. Since the characteristic curve is normal to the boundary, the associated characteristic variable is completely determined by the initial condition. Hence, either no boundary condition is prescribed for this characteristic variable or a boundary condition is prescribed that is completely consistent with the given initial condition.

We should be aware that in order to implement the ADI schemes developed in Section 6.7.1.3, it will be necessary to have half-step boundary conditions. Just as was the case with scalar ADI schemes, a bad choice of half-step boundary conditions can lower the order of the scheme. The method for choosing half-step boundary conditions for systems is the same as for scalar equations. For example, when choosing half-step boundary conditions for u^* (in order to solve equation (6.7.29)), the method is to use equation (6.7.30) to see that

$$u_{j\,k}^* = \left(I - \frac{R_y}{4} A_2 \delta_{y0}\right) u_{j\,k}^{n+1} \tag{6.7.35}$$

and then use the boundary conditions given for u^{n+1} to define boundary conditions for u^*.

And, finally, we remind the reader that as usual with hyperbolic equations there may not be enough boundary conditions for all schemes. It will be necessary to use numerical boundary conditions that we have referred to so often (and that will be discussed in-depth in Chapter 8). Of course, since systems are so much more difficult than scalar equations, it becomes more difficult to choose the numerical boundary conditions. The most common approach is to use a successful numerical boundary condition for a scalar model problem and then proceed with care (and/or with numerical experimentation). We must realize that for ADI schemes for hyperbolic systems, we will most likely have to obtain numerical boundary conditions for the half-steps, as well as for the primitive variables.

HW 6.7.3 Consider the system of equations

$$v_t = A_1 v_x + A_2 v_y \quad (x, y) \in (0, 1) \times (0, 1) \tag{6.7.36}$$

where the matrix A_1 and A_2 are given by

$$A_1 = \begin{pmatrix} 5/3 & 2/3 \\ 1/3 & 4/3 \end{pmatrix} \text{ and } A_2 = \begin{pmatrix} -1 & -2 \\ -1 & 0 \end{pmatrix}.$$

(a) For the above system, show that we can assign two boundary conditions at $x = 1$, none at $x = 0$ and one boundary condition at each of $y = 0$ and $y = 1$.

(b) Show that boundary conditions $v_1(1, y, t) = \sin 2t$, $v_2(1, y, t) = \cos 2t$, $v_1(x, 0, t) = \sin 2t$ and $v_1(x, 1, t) = \sin 5t$ are acceptable boundary conditions for partial differential equation (6.7.36).

(c) Use the Lax-Wendroff split scheme

$$u_{j\,k}^{n+1/2} = u_{j\,k}^n + \frac{R_x}{2} A_1 \delta_{x0} u_{j\,k}^n + \frac{R_x^2}{2} A_1^2 \delta_x^2 u_{j\,k}^n \tag{6.7.37}$$

$$u_{j\,k}^{n+1} = u_{j\,k}^{n+1/2} + \frac{R_y}{2} A_2 \delta_{y0} u_{j\,k}^{n+1/2} + \frac{R_y^2}{2} A_2^2 \delta_y^2 u_{j\,k}^{n+1/2} \tag{6.7.38}$$

(for the scalar analog, see (5.8.21)–(5.8.22)) to solve the initial–boundary–value problem given by partial differential equation (6.7.36), initial condition

$$v(x, y, 0) = \begin{bmatrix} \sin \pi x \sin 2\pi y \\ \cos 2\pi x \end{bmatrix},$$

boundary conditions given above in part (b) and numerical boundary conditions $u_{1_{0k}}^n = u_{1_{1k}}^n$, $u_{2_{0k}}^n = u_{2_{1k}}^n$, $u_{2_{j0}}^n = u_{2_{j1}}^n$ and $u_{1_{jM}}^n = u_{1_{jM-1}}^n$.
(c) Analyze the consistency and stability of difference scheme (6.7.37)–(6.7.38) as an initial–value problem scheme.

HW 6.7.4 (a) Solve the problem described in HW6.7.3-(c) using the Lax-Friedrichs scheme

$$\mathbf{u}_{jk}^{n+1} = \frac{1}{4}\left(\mathbf{u}_{jk+1}^n + \mathbf{u}_{jk-1}^n + \mathbf{u}_{j-1k}^n + \mathbf{u}_{j+1k}^n\right)$$
$$+ \frac{R_x}{2}A_1\delta_{x0}\mathbf{u}_{jk}^n + \frac{R_y}{2}A_2\delta_{y0}\mathbf{u}_{jk}^n. \tag{6.7.39}$$

(b) Analyze the consistency and stability of difference scheme (6.7.39) as an initial–value problem scheme.

HW 6.7.5 Consider the system of equations

$$\mathbf{v}_t = A_1\mathbf{v}_x + A_2\mathbf{v}_y \quad (x,y) \in (0,1) \times (0,1) \tag{6.7.40}$$

where the matrix A_1 and A_2 are given by

$$A_1 = \begin{pmatrix} 0 & 4/3 & -2/3 \\ 1 & -2/3 & -5/3 \\ -1 & -4/3 & -1/3 \end{pmatrix} \text{ and } A_2 = \begin{pmatrix} 5/3 & 1 & 5/3 \\ -1/3 & -1 & -7/3 \\ 1/3 & -1 & 1/3 \end{pmatrix}.$$

(a) For the above system, show that we can assign two boundary conditions at $x = 0$, one boundary condition at $x = 1$, one boundary condition at $y = 0$ and two boundary conditions at $y = 1$.
(b) Show that boundary conditions $v_1(1,y,t) = \sin t$, $v_2(0,y,t) = \sin 4t$, $v_3(0,y,t) = \cos 2\pi y \cos 2t$, $v_1(x,0,t) = \sin 2t$, $v_1(x,1,t) = \sin \pi x$ and $v_3(x,1,t) = \cos 2\pi x \cos 2t$ are acceptable boundary conditions for partial differential equation (6.7.40).
(c) Use the Lax-Wendroff split scheme (6.7.37)–(6.7.38) to solve the initial–boundary–value problem given by partial differential equation (6.7.36) (with A_1 and A_2 as defined above), initial condition

$$\mathbf{v}(x,y,0) = \begin{bmatrix} \sin \pi x \sin \frac{\pi}{2}y \\ \sin 2\pi x \\ \cos 2\pi x \cos 2\pi y \end{bmatrix},$$

boundary conditions given in part (b) and numerical boundary conditions $u_{1_{0k}}^n = u_{1_{1k}}^n$, $u_{2_{Mk}}^n = u_{2_{M-1k}}^n$, $u_{3_{Mk}}^n = u_{3_{M-1k}}^n$, $u_{2_{j0}}^n = u_{2_{j1}}^n$, $u_{3_{j0}}^n = u_{3_{j1}}^n$ and $u_{2_{jM}}^n = u_{2_{jM-1}}^n$.

HW 6.7.6 Solve both problems given in HW6.7.3 and HW6.7.5) (same initial condition, boundary conditions and numerical boundary conditions) using the Beam-Warming scheme, (6.7.29)–(6.7.30).

6.7.3   Two Dimensional Multilevel Schemes

Earlier we considered multilevel schemes for one dimensional scalar and systems of partial differential equations. It should not be surprising that multilevel schemes can also be useful for solving problems involving two or more dimensions. In this section we introduce multilevel schemes for two dimensional partial differential equations. As we did in Section 6.4.3, we introduce two dimensional, multilevel schemes by developing the leapfrog scheme for two dimensional hyperbolic partial differential equations. We do so by first considering a scalar equation and then extending these results to to include two dimensional systems of hyperbolic partial differential equations.

We consider the hyperbolic partial differential equation

$$v_t + av_x + bv_y = 0 \tag{6.7.41}$$

and the leapfrog scheme for approximating the solution to equation (6.7.41),

$$u_{jk}^{n+1} = u_{jk}^{n-1} - R_x\delta_{x0}u_{jk}^n - R_y\delta_{y0}u_{jk}^n, \tag{6.7.42}$$

where $R_x = a\Delta t/\Delta x$ and $R_y = b\Delta t/\Delta y$.

Example 6.7.10 Analyze the stability of difference scheme (6.7.42).

Solution: If we proceed as we did in Section 6.4 and let

$$U_{1jk}^n = u_{jk}^n$$
$$U_{2jk}^n = u_{jk}^{n-1}$$

and

$$\mathbf{U}_{jk}^n = \begin{bmatrix} u_{1jk}^n \\ u_{2jk}^n \end{bmatrix},$$

difference scheme (6.7.42) can be written as

$$\mathbf{U}_{jk}^{n+1} = Q\mathbf{U}_{jk}^n = \begin{pmatrix} -R_x\delta_{x0} - R_y\delta_{y0} & 1 \\ 1 & 0 \end{pmatrix} \mathbf{U}_{jk}^n. \tag{6.7.43}$$

Taking the discrete Fourier transform of difference scheme (6.7.43) gives us the amplification matrix

$$G = \begin{pmatrix} -2i(R_x\sin\xi + R_y\sin\eta) & 1 \\ 1 & 0 \end{pmatrix}. \tag{6.7.44}$$

To determine a stability limit for difference scheme (6.7.42), we analyze the amplification matrix (6.7.44) in the same manner that we did for the scalar, one dimensional scheme in Example 6.4.3. We begin by noting that the eigenvalues of G are given by

$$\lambda_\pm = -i(R_x\sin\xi + R_y\sin\eta) \pm \sqrt{1 - (R_x\sin\xi + R_y\sin\eta)^2}. \tag{6.7.45}$$

By considering $|\lambda_\pm|^2$ $(\pm\pi/2, \pm\pi/2)$, we see that *if $|R_x| + |R_y| > 1$, then the von Neumann condition is not satisfied and the difference scheme is unstable.*

We next note that when $|R_x| + |R_y| \leq 1$, $|\lambda_\pm|^2 = 1$. We also note that the rest of the stability analysis for difference scheme (6.7.42) follows exactly from the analysis for the one dimensional case in Example 6.4.3 (with $R\sin\xi$ replaced by $R_x\sin\xi + R_y\sin\eta$).

Likewise, if $\mid R_x \mid + \mid R_y \mid = 1$, we choose ξ and η so that $R_x \sin \xi + R_y \sin \eta = 1$ (ξ and η must be chosen to be $\pm \pi/2$, where the sign depends on the sign of R_x and R_y.) Then, as in Example 6.4.3, G^n will grow linearly with n. Hence, difference scheme (6.7.42) cannot be stable if $\mid R_x \mid + \mid R_y \mid = 1$.

We are left with the result that *difference scheme (6.7.42) is stable if and only if* $\mid R_x \mid + \mid R_y \mid \leq R_0$ *for any* $R_0 < 1$.

Thus we see that the stability analysis for the two dimensional, scalar leapfrog difference scheme is very similar to that for the one dimensional, scalar leapfrog difference scheme. Below, we shall see that the stability analysis for the two dimensional leapfrog difference scheme for systems of hyperbolic equations will follow analogous to the one dimensional leapfrog scheme for systems.

We consider the two dimensional hyperbolic system of partial differential equations

$$\mathbf{v}_t = A_1 \mathbf{v}_x + A_2 \mathbf{v}_y, \ x, y \in \mathbb{R}, \ t > 0, \tag{6.7.46}$$

the initial condition $\mathbf{v}(x, y, 0) = \mathbf{f}(x, y)$ and the leapfrog scheme for numerically solving equation (6.7.46)

$$\mathbf{u}_{jk}^{n+1} = \mathbf{u}_{jk}^{n-1} + R_x A_1 \delta_{x0} \mathbf{u}_{jk}^n + R_y A_2 \delta_{y0} \mathbf{u}_{jk}^n \tag{6.7.47}$$

where $R_x = \Delta t / \Delta x$ and $R_y = \Delta t / \Delta y$.

Example 6.7.11 Analyze the stability of difference scheme (6.7.47).

Solution: If we again define U as

$$\mathbf{U}_{jk}^{n+1} = \left[\begin{array}{c} \mathbf{u}_{jk}^{n+1} \\ \mathbf{u}_{jk}^n \end{array} \right],$$

difference scheme (6.7.47) can be written as

$$\mathbf{U}_{jk}^{n+1} = Q \mathbf{U}_{jk}^n = \left(\begin{array}{cc} R_x A_1 \delta_{x0} + R_y A_2 \delta_{y0} & I \\ I & \Theta \end{array} \right) \mathbf{U}_{jk}^n, \tag{6.7.48}$$

and the amplification matrix for difference scheme (6.7.48) is given by

$$G(\xi, \eta) = \left(\begin{array}{cc} 2i(\sin \xi A_1 + \sin \eta A_2) & I \\ I & \Theta \end{array} \right). \tag{6.7.49}$$

If we try to follow the analysis done for the leapfrog scheme for one dimensional systems of hyperbolic equations, we next diagonalize the $(1, 1)$ term of the matrix G. *To be able to perform this diagonalization, we must resort to the method and the assumption that has been necessary so often for two dimensional systems of hyperbolic equations, that the matrices A_1 and A_2 be simultaneously diagonalizable.* With this assumption in place, we can replace the matrix S used in Example 6.4.4 by the matrix that performs that simultaneous diagonalization of A_1 and A_2, and proceed as we did in Example 6.4.4 (with any reference to the results of Example 6.4.3 replaced by those of Example 6.7.10). We obtain the following: *difference scheme (6.7.46) is stable if* $\mid \mu_j R_x \mid + \mid \nu_j R_y \mid \leq R_0$, $j = 1, \cdots, K$ *for any* $R_0 < 1$ *where* μ_j *and* ν_j, $j = 1, \cdots, K$ *are the eigenvalues of* A_1 *and* A_2, *respectively.*

6.8   A Consistent, Convergent, Unstable Difference Scheme?

Obviously from the title of this section, we introduce a difference scheme that appears to be consistent (actually conditionally consistent), convergent and unstable. Since we do not want a counterexample to the Lax Equivalence Theorem, we show why this scheme is not as it appears. The example shows how we must be careful when we do these consistency and stability analyses. We note that this example is the same as was used in Example 6.2.1 to show that the von Neumann condition is not sufficient for stability.

Example 6.8.1 Consider the difference scheme

$$u_{k\,1}^{n+1} = u_{k\,1}^{n} + \delta^2 u_{k\,2}^{n} \tag{6.8.1}$$

$$u_{k\,2}^{n+1} = u_{k\,2}^{n}, \tag{6.8.2}$$

or

$$\mathbf{u}_k^{n+1} = \mathbf{u}_k^n + \begin{pmatrix} 0 & 1 \\ 0 & 0 \end{pmatrix} \delta^2 \mathbf{u}_k^n. \tag{6.8.3}$$

Show that the difference scheme (6.8.3) is both consistent and convergent, yet is not stable.

Solution: If we return to Example 6.2.1, we first see that the scheme is not stable. We consider the consistency of difference scheme (6.8.3) with the system of partial differential equations

$$v_{1\,t} = 0 \tag{6.8.4}$$

$$v_{2\,t} = 0 \tag{6.8.5}$$

along with the initial condition

$$\mathbf{v}(x,0) = \mathbf{f}(x) = [f^1(x)\ f^2(x)]^T. \tag{6.8.6}$$

It is not difficult to see that difference scheme (6.8.1)–(6.8.2) is not consistent with this system of partial differential equations. However, if we rewrite equation (6.8.1) as

$$\frac{u_{k\,1}^{n+1} - u_{k\,1}^{n}}{\Delta t} = \frac{\Delta x^2}{\Delta t}\frac{1}{\Delta x^2}\delta^2 u_{k\,2}^{n}, \tag{6.8.7}$$

it is easy to see that we can restrict the way that Δx and Δt go to zero by requiring that $R = \Delta t/\Delta x$ be a fixed constant and get that difference equations (6.8.2), (6.8.7) are conditionally consistent with the system of partial differential equations (6.8.4)–(6.8.5).

And finally, we let $\mathbf{f} = [\cdots \mathbf{f}_{-1}\mathbf{f}_0\mathbf{f}_1 \cdots]^T$ where $\mathbf{f}_k = [f_k^1 \quad f_k^2]^T$ and return to the form of the operator Q^n given in (6.2.23) to see that the kth component of a general row of the vector $Q^n\mathbf{f}$ is given by

$$\begin{pmatrix} f_k^1 + n\delta^2 f_k^2 \\ f_k^2 \end{pmatrix} = \begin{pmatrix} f_k^1 + n\Delta x^2 \frac{\delta^2 f_k^2}{\Delta x^2} \\ f_k^2 \end{pmatrix}. \tag{6.8.8}$$

Since our condition on the consistency is that $R = \Delta t/\Delta x$ is held fixed (as Δx and Δt go to zero), we can replace the $n\Delta x^2$ term in equation (6.8.8) by $n\Delta x\Delta t/R$. But now since $\frac{\delta^2}{\Delta x^2}f_k^2 \to f^{2\prime\prime}$ and $n\Delta t \to t$, we see that as Δx and Δt approach zero, the vector given in (6.8.8) approachs

$$\begin{pmatrix} f^1 \\ f^2 \end{pmatrix}$$

(which is the exact solution to system (6.8.4)–(6.8.5) with initial condition (6.8.6)). Thus difference scheme (6.8.3) is convergent.

Remark: It appears as if we have a conditionally consistent scheme that is unstable and convergent. From the Lax Equivalence Theorem (even though we did not prove it), we know that this cannot be the case. The problem with the above case is that the function $\mathbf{f}$ needs no smoothness for $\mathbf{v} = \mathbf{f}$ to be a solution to system (6.8.4)–(6.8.5). But the convergence argument above requires that f^2 be twice differentiable. The instability argument used a vector $[\cdots \boldsymbol{\theta}^T \quad [0\,1]^T \boldsymbol{\theta}^T \cdots]^T$. As the grid is refined (Δx goes to zero), this vector is not smooth enough to converge to a differentiable function. Hence, we see that an important part of the instability argument is the space of functions for which it is unstable. If we restrict Δx and Δt so that $R = \Delta t / \Delta x$ is constant and work on a space of initial conditions that the second component is at least twice differentiable (meaning that for any vector $\mathbf{f}$ in the space, $\mathbf{f} = [\cdots, \boldsymbol{\theta}^T \,[f_1 \; f_2]^T \,\boldsymbol{\theta}^T \cdots]^T$, $\delta^2 f_k^2 / \Delta x^2$ converges as $\Delta x \to 0$ for any k), then difference scheme (6.8.3) is stable.

6.9   Computational Interlude IV

6.9.1   HW0.0.1 and HW0.0.2

By this time we will have applied the Lax-Wendroff scheme to the linearized equations associated with HW0.0.2 (Section 5.9.4) and the appropriate convection–diffusion form of the Lax-Wendroff scheme to the linearized equation associated with HW0.0.1 (Section 5.9.3). If the implementation has been done properly, your results should be pretty good—by far the best results yet for small values of ν. For certain grids or if you ran the calculation out sufficiently far in time, you probably obtain small oscillations near $x = 1/2$. As we shall see in Chapter 7, this should not surprise us and is one of the problems that we must expect with the Lax-Wendroff scheme. As we shall see later, the oscillations in these examples are really much better than we might expect. Because of these oscillations—and because we will have more to learn—in this section, we introduce another method that might be effective for solving HW0.0.1 and HW0.0.2. We introduce the method for solving the problem given in HW0.0.1. The corresponding scheme for the problem given in HW0.0.2 will be obvious.

We temporarily return to our consideration of implicit schemes for solving HW0.0.1. Though for small values of ν it was sometimes difficult to see, when implicit schemes were used to solve the viscous Burgers' equation, the results were more stable than for the explicit schemes. All that a stable scheme promises is that the solutions will remain bounded. And bounded, in this case, means bounded mathematically with infinite precision arithmetic, not on a computer. If you examine your solutions carefully (and a lot of graphics makes this much easier), you would see that the schemes were performing reasonably well until a steep front formed. (This was also the case with the explicit schemes. When the steep front formed,

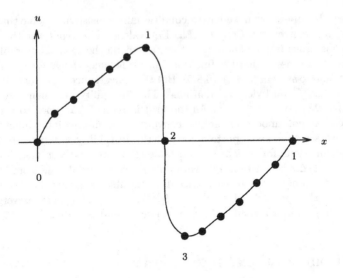

FIGURE 6.9.1. Example of a function with a steep front.

the explicit scheme became unstable.) If you recall the order arguments that were used earlier, all of the accuracy with which we have approximated the partial differential equations by difference equations depends on truncated Taylor series. The coefficients of the lowest ordered terms are then the appropriate derivatives evaluated in the interval over which the function is expanded (generally from $(k-1)\Delta x$ to $(k+1)\Delta x$). When the front becomes steep, these derivatives are quite large. Hence, sometimes it is not necessarily an instability that is occurring, but, instead, that you have calculated a very inaccurate approximation to the solution.

In light of the argument given above, we consider a function such as the one graphed in Figure 6.9.1 (which is much like the solution to HW0.0.1 for small ν and the solution to HW0.0.2). We see that for the centered schemes that we have been using to solve the problems in HW0.0.1 and HW0.0.2, approximations of the derivative at points 1 and 3 are very bad. Also, we note that the derivative at point 2 is such that the error may be very large (because the error depends on the derivatives of the solution near the point). If we could use a backward difference scheme for $x < 1/2$, we would never reach across the steep front. Likewise, if we could use a forward differencing scheme for $x > 1/2$, we would never reach across the front. If we were to use this approach, it would be another case of differencing the hyperbolic terms in Burgers' equation like hyperbolic equations should be differenced (which is what we did in Sections 5.9.3 and 5.9.4).

The difficulty to the method to which we are alluding is that the direction of the differencing that we used would have to change during the calculation. When we considered using forward and backward differencing for the one way wave equation, we found that the scheme was stable using forward differencing if a was less than zero and the scheme was stable using backward differencing if a was greater than zero. In our case, the a term is replaced by v, and v is greater than zero for $x > 1/2$ and less than zero for $x < 1/2$.

The difficulty that we are facing is very similar to the difficulty faced with using one sided difference schemes for systems of equations. The way that we solved the problem was to consider flux splitting schemes (Section 6.2.1), where the equation was written in a form that allowed forward differencing and backward differencing, whichever is appropriate for each component of the vector. We never discussed it, but there is no need fix the differencing for all time and space. If you review the method, you will see that we could adapt the scheme to allow for the evaluation of the eigenvalues at each grid point (the eigenvalues might depend on $(k\Delta x, n\Delta t)$ or even u_{jk}^n) and difference at that grid point appropriately depending on the sign of the eigenvalue. It would be obvious to ask: is this a lot of work and are the results worth the extra work? The answer is that it is a lot of extra work and only time will tell whether our results are worth the extra work. As a part of the answer to these questions, we implement the above idea in a relatively easy setting.

Specifically, let us consider the viscous Burgers' equation and consider using centered differences on the diffusion term and one sided differences on the convective term, vv_x. For example, we can use the explicit scheme

$$\frac{u_k^{n+1} - u_k^n}{\Delta t} + u_k^n \Delta_x^n u_k^n = \nu \frac{\delta_x^2}{\Delta x^2} u_k^n \qquad (6.9.1)$$

where

$$\Delta_x^n u_k^n = \begin{cases} \frac{u_{k+1}^n - u_k^n}{\Delta x}, & \text{if } u_k^n < 0; \\ \frac{u_k^n - u_{k-1}^n}{\Delta x}, & \text{if } u_k^n \geq 0 \end{cases} \qquad (6.9.2)$$

and the n superscript on the Δ refers to the fact that the decision is being made based on u at the nth time step. We see that the above scheme uses forward differencing when the coefficient $u_k^n < 0$ and backward differencing when the coefficient $u_k^n \geq 0$. This approach should be tried on the problem in HW0.0.1.

Of course, if we have an explicit one sided scheme, we also have an implicit one sided scheme. Consider the nonlinear implicit difference scheme

$$\frac{u_k^{n+1} - u_k^n}{\Delta t} + u_k^{n+1} \Delta_x^{n+1} u_k^{n+1} = \nu \frac{\delta_x^2}{\Delta x^2} u_k^{n+1}, \qquad (6.9.3)$$

where Δ_x is as in (6.9.2) (except that now the $n+1$ superscript refers to the fact that the decision is being made with respect to u_k^{n+1} being positive or negative, rather with respect to u_k^n). If you consider this equation carefully, you decide that it is difficult to linearize by the methods we used in Section 3.4.2. Our approach is to first lag the superscript on the Δ in equation (6.9.3). Hence, we replace difference equation (6.9.3) by the following difference equation.

$$\frac{u_k^{n+1} - u_k^n}{\Delta t} + u_k^{n+1} \Delta_x^n u_k^{n+1} = \nu \frac{\delta_x^2}{\Delta x^2} u_k^{n+1} \tag{6.9.4}$$

Any of the methods discussed earlier in Section 3.4.2 can be used to linearize equation (6.9.4) and, hence, approximately solve equation (6.9.4). Specifically, if we linearize equation (6.9.4) by lagging part of the nonlinear term, we solve

$$\frac{u_k^{n+1} - u_k^n}{\Delta t} + u_k^n \Delta_x^n u_k^{n+1} = \nu \frac{\delta_x^2}{\Delta x^2} u_k^{n+1}, \tag{6.9.5}$$

as an approximation to solving equation (6.9.4).

To approximate the solution of equation (6.9.4) by linearizing about the previous time step, we must solve the following difference equation.

$$\frac{u_k^{n+1} - u_k^n}{\Delta t} + u_k^n \Delta_x^n u_k^{n+1} + (\Delta_x^n u_k^n) u_k^{n+1} - \frac{\delta_x^2}{\Delta x^2} u_k^{n+1} = u_k^n \Delta_x^n u_k^n \tag{6.9.6}$$

And finally, if we were to solve equation (6.9.4) iteratively using Newton's method (and remember, equation (6.9.6) is just one Newton step using u_k^n as the initial guess), we arrive at the following solution scheme.

- Let $u_k^{n+1,0} = u_k^n$, $k = 0, \cdots, M$.

- For $j = 1 \cdots$, solve

$$\frac{u_k^{n+1,j} - u_k^n}{\Delta t} + u_k^{n+1,j-1} \Delta_x^{n+1,j-1} u_k^{n+1,j} + (\Delta_x^{n+1,j-1} u_k^{n+1,j-1}) u_k^{n+1,j}$$
$$- \frac{\delta_x^2}{\Delta x^2} u_k^{n+1,j} = u_k^{n+1,j-1} \Delta_x^{n+1,j-1} u_k^{n+1,j-1} \tag{6.9.7}$$

where $\Delta_x^{n+1,j-1}$ is as in equation (6.9.2) except that $u_k^{n+1,j-1}$ is used to decide which choice to make instead of u_k^n.

And, of course, the iteration is stopped when for some j,

$$u_k^{n+1,j} - u_k^{n+1,j-1}$$

is sufficiently small. The measure of this last quantity can be either in the sup-norm, the l_2 norm or any other appropriate norm.

Of course, there are many other discretizations that we should consider. One very logical possibility (some of the reasons that it is such a logical choice will be discussed in Chapter 7) is to use the Crank-Nicolson scheme. Hence, the nonlinear system of equations to be solved is

$$\frac{u_k^{n+1} - u_k^n}{\Delta t} + u_k^{n+1}\frac{\delta_0}{4\Delta x}(u_k^{n+1} + u_k^n) = \frac{\nu\delta_x^2}{2\Delta x^2}(u_k^{n+1} + u_k^n). \tag{6.9.8}$$

We can now proceed to linearize and solve equation (6.9.8) by any or all of the methods used above. The Crank-Nicolson scheme is a very popular scheme for nonlinear problems.

Remark: We emphasize again that the above schemes can easily be adjusted to solve the inviscid Burgers' equation by eliminating the viscous term.

6.9.2   HW0.0.3

Since we now have schemes for solving systems of equations, it is time to attack HW0.0.3. As presented, HW0.0.3 involves three, coupled, nonlinear partial differential equations with a given initial condition. Part of the difficulty with HW0.0.3 is that the problem is given as an initial–value problem. Part of the solution procedure will be to choose an interval over which to work and to monitor the solution to ensure us that the interval is sufficient large (which will all not turn out to be much of a problem).

Let us begin considering the system of partial differential equations (0.0.7)–(0.0.9) on the interval $(-2, 2)$. The initial condition given is

$$\rho(x,0) = \begin{cases} 2 & \text{if } -2 \le x < 0 \\ 1 & \text{if } 0 \le x \le 2 \end{cases} \qquad p(x,0) = \begin{cases} 2 & \text{if } -2 \le x < 0 \\ 1 & \text{if } 0 \le x \le 2 \end{cases}$$

and $v(x,0) = 0$, $x \in [-2,2]$. We note that the partial differential equations contain the variables $\rho = \rho(x,t)$, $v = v(x,t)$, $E = E(x,t)$ and $p = p(x,t)$ whereas we are only given initial conditions for ρ, p and v. The fact that we have only three initial conditions is logical and correct since we have only three partial differential equations. The inconsistency is including both E and p in the partial differential equations. Only one of these two variables is mathematically necessary. The reason that both of the variables are included is that E, the total energy, evolves nicely into the derivation of the equations (i.e. when you derive the equations, the variable E enters the equation with an E_t term, which is very appealing both numerically and mathematically) while the pressure, p, is a much more intuitive, physically important variable. Hence, the method is to include both of these variables, using each of them where they are most convenient. For example, we obtain the initial condition for E from the initial conditions on p, ρ and v, solve for E (along with ρ and v) and plot the results of p (along with ρ and v).

Since equations (0.0.7)–(0.0.9) are nonlinear, we must decide how and when to linearize them. To help make this job a little bit easier, we note that we can rewrite system (0.0.7)–(0.0.9) as

$$\mathbf{v}_t + A\mathbf{v}_x = \boldsymbol{0} \tag{6.9.9}$$

where

$$\mathbf{v} = \begin{bmatrix} \rho \\ \rho v \\ E \end{bmatrix}$$

and

$$A = \begin{pmatrix} 0 & 1 & 0 \\ \frac{\gamma-3}{2}v^2 & -(\gamma-3)v & (\gamma-1) \\ \frac{\gamma-1}{2}v^3 \div \frac{v}{\rho}(E+p) & \frac{1}{\rho}(E+p)-(\gamma-1)v^2 & \gamma v \end{pmatrix}. \tag{6.9.10}$$

The easiest way to see that equation (6.9.9) is equivalent to equations (0.0.7)–(0.0.9) is to expand each of the equations (0.0.7)–(0.0.9), write out each of the equations in (6.9.9) and compare the two expressions. We note that we have left the ρv term together in the $\mathbf{v}$ vector. We could claim that we have done this for convenience, but we admit that we have done so because it will make things more convenient later (and it is the right way to do it). Mathematically and numerically the variables ρ, ρv and E are better variables than ρ, v and E. In light of this, at times it might be best to write $\mathbf{v}$ as $\mathbf{v} = [\rho \quad m \quad E]^T$ where $m = \rho v$ and rewrite A as

$$A = \begin{pmatrix} 0 & 1 & 0 \\ \frac{\gamma-3}{2}\frac{m^2}{\rho^2} & -(\gamma-3)\frac{m}{\rho} & (\gamma-1) \\ -\frac{m}{\rho^2}\left[\gamma E - (\gamma-1)\frac{m^2}{\rho}\right] & \frac{1}{\rho}\left[\gamma E - \frac{3(\gamma-1)}{2}\frac{m^2}{\rho}\right] & \gamma\frac{m}{\rho} \end{pmatrix}. \tag{6.9.11}$$

Before we can think about solving either equations (0.0.7)–(0.0.9) or system (6.9.9) on the interval $(-2, 2)$, we must have some boundary conditions at $x = -2$ and/or $x = 2$. We do not know how to assign boundary conditions to nonlinear partial differential equations. We will assume that we can consider the system of partial differential equations (6.9.9) and assign boundary conditions based on the eigenvalues of matrix A as we did for linear systems of partial differential equations (and with this assumption, we will not be too far from being correct). The eigenvalues of A are v, $v+c$ and $v-c$ where c is the speed of sound and is given by

$$c = \sqrt{\frac{\gamma p}{\rho}}.$$

Since initially v will be zero at both $x = \pm 2$, the eigenvalues of A at the boundaries are 0 and $\pm c$. Thus, we get to assign one boundary condition at each boundary (we have one positive eigenvalue at $x = -2$ and one negative eigenvalue at $x = 2$). At the moment we will not try to determine which variable can or must be assigned a boundary condition. We will mention that when we try to solve this problem numerically, if we use some sort of centered scheme, we will have to provide two numerical boundary conditions at each end of the interval.

So finally we have stated the mathematical problem that we wish to solve reasonably well (we have not been very explicit about the boundary conditions). Of course, we do not really know how to numerically approximate the solutions to this problem. However, we do have several logical approaches to try. An approach that is probably the most obvious approach is to consider the system written as in equation (6.9.9), plan on lagging (in time) the terms in the matrix A and solve the equation by any one of the schemes we have introduced earlier in this chapter treating the equation as if it were linear. For example, since the Lax-Wendroff scheme was reasonably successful when applied in this manner to Burgers' equation, we might try the difference scheme

$$\mathbf{u}_k^{n+1} = \mathbf{u}_k^n - \frac{R}{2}A^n\delta_0\mathbf{u}_k^n + \frac{R^2}{2}(A^n)^2\delta^2\mathbf{u}_k^n$$

where $R = \Delta t/\Delta x$ and $A^n = A(\mathbf{u}^n)$. We shall refer to this scheme as the **linearized Lax-Wendroff scheme**.

Of course, to go along with such a scheme we need an initial condition (which we were given) and boundary conditions at both boundaries. As we discussed earlier, we get to assign one boundary condition at each boundary and we must provide two numerical boundary conditions. We will return later to how we can choose which variable to which we assign a boundary condition and what type of numerical boundary conditions should be assigned. Instead, at this time we will cheat a bit. Since our solution is constant at and near the boundaries and it will take some time for any disturbance to reach the boundary, a boundary condition of the form $\phi_x = 0$ (where ϕ is either ρ, m or E) is a logical boundary condition to assign to whichever primitive variables we should assign boundary conditions ($\phi_x = 0$ implies no change in the x-direction). At the boundary $x = -2$, a first order approximation of $\phi_x = 0$ can be written as $\phi_0 = \phi_1$. Also, we saw in HW5.6.10 and HW5.6.11 that a numerical boundary condition of the form $\psi_0 = \psi_1$ (where again ψ will be ρ, m and/or E) is commonly a good numerical boundary condition. Hence, we will not stipulate to which variable we assign the zero Neumann boundary condition and to which variables we assign the numerical boundary condition. We will use

$$\rho_0 = \rho_1, \quad m_0 = m_1 \text{ and } E_0 = E_1$$

as the boundary conditions to our numerical problem at $x = -2$ and

$$\rho_M = \rho_{M-1}, \quad m_M = m_{M-1} \text{ and } E_M = E_{M-1}$$

as the boundary conditions to our numerical problem at $x = 2$. We should also note that since we refer to the second variable as m, we are suggesting that you use the form of A given in equation (6.9.11). Also, we are assuming that you will solve the equations using ρ and E (in addition to m) and not use p. The initial conditions on v, ρ and p will be enough to give initial conditions on m and E. However, we suggest that when it is time to plot the solutions, you provide plots of ρ, v and p (who knows what E should look like anyway?).

We suggest that it is now time to try to solve HW0.0.3. The information given above should provide you with a reasonable approach to obtaining an approximation of the solution. If the reader has an interest in trying one of the other schemes other than the linearized Lax-Wendroff scheme, the description above should work equally well for linearized versions of some other difference scheme.

PS Above we stated that the eigenvalues of A where v, $v + c$ and $v - c$. The eigenvalues of these coefficient matrices are very important. Most often, these eigenvalues are very hard to find. It would be nice if we could give you an easy way to find the eigenvalues of A above and/or for other coefficient matrices. However, there does not seem to be an easy way. At this time, the algebraic manipulators are not to the stage where they can find **and** simplify the eigenvalues. The "and" is boldfaced, because Maple can find the eigenvalues. However, Maple does not seem to be able to figure out that the complicated expressions it finds are equal to v or $v \pm c$. The easiest way at the moment to gain some confidence that the eigenvalues given above are correct is to use an algebraic manipulator to verify that the above values are the eigenvalues. The software can do this.

Also, when we have to be more careful about what kind of boundary conditions we give in our problem, we will have to know the eigenvectors of A also. The eigenvectors of A associated with $v - c$, v and $v + c$ are

$$
\mathbf{r}_1 = \begin{bmatrix} 1 \\ v - c \\ \frac{1}{2}v^2 - vc + \frac{1}{\gamma - 1}c^2 \end{bmatrix}, \quad \mathbf{r}_2 = \begin{bmatrix} 1 \\ v \\ \frac{1}{2}v^2 \end{bmatrix} \quad \mathbf{r}_3 = \begin{bmatrix} 1 \\ v + c \\ \frac{1}{2}v^2 + vc + \frac{1}{\gamma - 1}c^2 \end{bmatrix},
$$

respectively. Like the eigenvalues, the eigenvectors are difficult to find. Also, like the eigenvalues, an algebraic manipulator can be used to verify that the above vectors are in fact the eigenvectors of matrix A.

6.9.3 Parabolic Problems in Polar Coordinates

Clearly, this section is out of place. In Section 4.5 we claimed that it was too difficult at that time to approximate the solutions to parabolic equations given in polar coordinates by implicit schemes. Since that time, the Sherman-Morrison Algorithm was introduced in Section 5.6.1, was discussed as part of the solution scheme for approximating the solution of two dimensional hyperbolic equations with periodic boundary conditions in Section 5.8.5 and was extended to include systems of equations in Section

6.3.2.3. It is now assumed that we are ready to return to Section 4.5 and discuss the approximation of the solutions of parabolic partial equations in polar coordinates using implicit schemes.

6.9.3.1   Polar Coordinates Without the Origin

We begin with the easiest problem, that discussed in Remark 2 of Section 4.5, of solving a problem given in polar coordinates on a region that does not contain the origin. In this case, the matrix equation that must be solved is in the form

$$Q_1 \mathbf{u}^{n+1} = \mathbf{u}^n + \mathbf{G}^{n+1}$$

where

$$
Q_1 = \begin{pmatrix}
T_1 & -\gamma_1 I & \Theta & \cdots & & & \\
-\alpha_2 I & T_2 & -\gamma_2 I & \Theta & \cdots & & \\
\Theta & -\alpha_3 I & T_3 & -\gamma_3 I & \Theta & \cdots & \\
\vdots & \ddots & \ddots & & \ddots & & \\
\Theta & \cdots & \Theta & -\alpha_{M_r-2} I & T_{M_r-2} & -\gamma_{M_r-2} I \\
\Theta & \cdots & \cdots & \Theta & -\alpha_{M_r-1} I & T_{M_r-1}
\end{pmatrix}
$$

$$(6.9.12)$$

and T_j, α_j, etc. are as in Section 4.5 (except that in this case, r_1 corresponds to some radius $R_0 > 0$). This equation is difficult to solve both because the wide band due to the α and γ blocks and due to the fact that the diagonal matrices, T_j, are not tridiagonal matrices. However, this problem is very similar to the problem faced when we considered two dimensional, hyperbolic problems where we had a Dirichlet boundary condition in one direction (analogous to the r direction) and a periodic boundary condition in the other direction (due to the continuity condition in θ). The method we used to solve that problem in Section 5.8.5 was to use an ADI scheme and solve the resulting equations as many smaller matrix equations (and using the Sherman-Morrison Algorithm for solving the equations with periodic boundary conditions). We can proceed in the same way. If we reconsider difference scheme (4.5.27)–(4.5.28), we can approximately factor this difference equation to get

$$(1 - r_{\theta_j} \delta_\theta^2)(1 - r_{r_j} \Delta_r) u_{jk}^{n+1} = u_{jk}^n + \Delta t F_{jk}^{n+1},$$
$$j = 1, \cdots, M_r - 1, \ k = 1, \cdots, M_\theta - 1 \qquad (6.9.13)$$
$$(1 - r_{\theta_j} \delta_p^2)(1 - r_{r_j} \Delta_r) u_{j0}^{n+1} = u_{j0}^n + \Delta t F_{j0}^{n+1},$$
$$j = 1, \cdots, M_r - 1 \qquad (6.9.14)$$

where

$$\Delta_r u_{jk}^{n+1} = r_{j+1/2}(u_{j+1\,k}^{n+1} - u_{jk}^{n+1}) - r_{j-1/2}(u_{jk}^{n+1} - u_{j-1\,k}^{n+1}),$$

$$\delta_p^2 u_{j0}^{n+1} = u_{j1}^{n+1} - 2u_{j0}^{n+1} + u_{j M_\theta - 1}^{n+1},$$

$$r_{r_j} = \frac{\Delta t}{r_j \Delta r^2} \text{ and } r_{\theta_j} = \frac{\Delta t}{r_j^2 \Delta \theta^2}.$$

We see in HW6.9.1 that difference scheme (6.9.13)–(6.9.14) is accurate $\mathcal{O}(\Delta t) + \mathcal{O}(\Delta r^2) + \mathcal{O}(\Delta \theta^2)$. We can then split the scheme into solving the equations

$$(1 - r_{\theta_j} \delta_\theta^2) u_{jk}^* = u_{jk}^n + \Delta t F_{jk}^{n+1},$$
$$k = 1, \cdots, M_\theta - 1, \ j = 1, \cdots, M_r - 1, \quad (6.9.15)$$

$$(1 - r_{\theta_j} \delta_p^2) u_{j0}^* = u_{j0}^n + \Delta t F_{j0}^{n+1},$$
$$j = 1, \cdots, M_r - 1, \quad (6.9.16)$$

$$(1 - r_{r_j} \Delta_r) u_{jk}^{n+1} = u_{jk}^*,$$
$$j = 1, \cdots, M_r - 1, \ k = 0, \cdots, M_\theta - 1. \quad (6.9.17)$$

The implementation of the scheme is then to solve equation (6.9.15)–(6.9.16) as $M_r - 1$ $M_\theta \times M_\theta$ systems with periodic boundary conditions and solve (6.9.17) as M_θ $M_r - 1 \times M_r - 1$ systems with Dirichlet boundary conditions. Any codes developed in Section 5.8.5 should be very useful here.

Remark: We must realize that it was important in which order we factored equation (4.5.27)–(4.5.28) to get equations (6.9.13)–(6.9.14). If we tried to factor the equation in the opposite order, because r_{θ_j} depends on r_j, the job would be much more difficult or impossible.

HW 6.9.1 Show that difference scheme (6.9.13)–(6.9.14) is consistent $\mathcal{O}(\Delta t) + \mathcal{O}(\Delta r^2) + \mathcal{O}(\Delta \theta^2)$.

6.9.3.2   Polar Coordinates With the Origin

We next proceed to consider solving the original problem presented in Section 4.5, the parabolic problem on a region which includes the origin. To solve this problem, we must solve matrix problem (4.5.30). As we shall see, the difficulty due to the periodicity in the θ direction will be solved just as we did the problem without the origin. Hence, we must find a method to handle the special first row and first column. We begin by rewriting matrix equation (4.5.30) as

$$Q_1 \mathbf{u} = \mathbf{f} \quad (6.9.18)$$

and note that Q_1 can be partitioned as

$$Q_1 = \begin{pmatrix} \alpha & \mathbf{R}^T \\ \mathbf{C} & Q_1' \end{pmatrix} \quad (6.9.19)$$

where $\mathbf{R}^T = [\mathbf{r}^T \, 0 \cdots 0]$, $\mathbf{C} = [\mathbf{c}^T \, 0 \cdots 0]^T$, Q_1' consists of the second through $M_\theta(M_r - 1) + 1$-st rows and columns of Q_1, and α, $\mathbf{r}$ and $\mathbf{c}$ are as in Section 4.5. If we then partition $\mathbf{u}$ and $\mathbf{f}$ as

$$\mathbf{u} = \left[\begin{array}{c} u_1 \\ \mathbf{u}_2 \end{array} \right] \text{ and } \mathbf{f} = \left[\begin{array}{c} f_1 \\ \mathbf{f}_2 \end{array} \right]$$

where u_1 and f_1 are scalars (1-vectors) and $\mathbf{u}_2$ and $\mathbf{f}_2$ are $M_\theta(M_r - 1)$-vectors, equation (6.9.18) can be rewritten as

$$\alpha u_1 + \mathbf{R}^T \mathbf{u}_2 = f_1 \tag{6.9.20}$$

$$u_1 \mathbf{C} + Q_1' \mathbf{u}_2 = \mathbf{f}_2. \tag{6.9.21}$$

If we solve equation (6.9.20) for u_1,

$$u_1 = \frac{1}{\alpha}(f_1 - \mathbf{R}^T \mathbf{u}_2), \tag{6.9.22}$$

and use this to eliminate u_1 from equation (6.9.21), equation (6.9.21) can be written as

$$Q_1' \mathbf{u}_2 - \frac{1}{\alpha}(\mathbf{R}^T \mathbf{u}_2)\mathbf{C} = \mathbf{f}_2 - \frac{f_1}{\alpha}\mathbf{C}. \tag{6.9.23}$$

Then noting that $(\mathbf{R}^T \mathbf{u}_2)\mathbf{C} = (\mathbf{C}\mathbf{R}^T)\mathbf{u}_2$, we rewrite equation (6.9.23) as

$$\left(Q_1' - \frac{1}{\alpha}(\mathbf{C}\mathbf{R}^T)\right)\mathbf{u}_2 = \mathbf{f}_2 - \frac{f_1}{\alpha}\mathbf{C}. \tag{6.9.24}$$

If we return to Section 5.6.1, we see that the equation given in (6.9.24) is the perfect equation to be solved by Sherman-Morrison Algorithm. We should have suspected this earlier from the form of Q_1. In the notation of Proposition 5.6.1, we set $A = Q_1' - (1/\alpha)\mathbf{C}\mathbf{R}^T$, $B = Q_1'$, $\mathbf{w} = (1/\alpha)\mathbf{C}$, $\mathbf{z} = \mathbf{R}$ and $\mathbf{b} = \mathbf{f}_2 - (f_1/\alpha)\mathbf{C}$. We then solve

$$Q_1' \mathbf{y}_1 = \mathbf{f}_2 - \frac{f_1}{\alpha}\mathbf{C} \tag{6.9.25}$$

$$Q_1' \mathbf{y}_2 = \frac{1}{\alpha}\mathbf{C}, \tag{6.9.26}$$

set

$$\beta = \frac{\mathbf{R}^T \mathbf{y}_1}{1 - \mathbf{R}^T \mathbf{y}_2},$$

and get

$$\mathbf{u}_2 = \mathbf{y}_1 + \beta \mathbf{y}_2.$$

We then use equation (6.9.22) to find u_1. Thus, the Sherman-Morrison Algorithm can be used nicely to solve equation (4.5.30) as long as we can perform the above three steps. Clearly, if we can find $\mathbf{y}_1$ and $\mathbf{y}_2$, we can perform the last two steps to find $\mathbf{u}_2$ and u_1. To find $\mathbf{y}_1$ and $\mathbf{y}_2$, we must solve equations (6.9.25) and (6.9.26). Though these equations are not easy to solve, we can find a sufficiently good approximation to their solutions using the approach developed in Section 6.9.3.1 to solve the problem without the origin. In other words, equations (6.9.25) and (6.9.26) must be solved using an ADI scheme with periodic boundary conditions in one direction (solved using the Sherman-Morrison Algorithm) and Dirichlet boundary conditions in the other direction.

Hence, the solution technique for solving equation (4.5.30) for one time step is to rewrite equations (6.9.25)–(6.9.26) as

$$L_\theta \mathbf{y}_1^* = \mathbf{f}_2 - \frac{f_1}{\alpha}\mathbf{C} \qquad (6.9.27)$$

$$L_\theta \mathbf{y}_2^* = \frac{1}{\alpha}\mathbf{C}, \qquad (6.9.28)$$

$$L_r \mathbf{y}_1 = \mathbf{y}_1^* \qquad (6.9.29)$$

$$L_r \mathbf{y}_2 = \mathbf{y}_2^* \qquad (6.9.30)$$

where L_θ and L_r are given by the left hand sides of equations (6.9.15)–(6.9.16) and (6.9.17), respectively, and then proceed to compute β, $\mathbf{u}_2$ and u_1. And, of course, to save work, equations (6.9.27) and (6.9.28) can be solved together (a tri-diagonal algorithm with two right hand sides and because each of these problems involves the Sherman-Morrison Algorithm for treating the periodicity in θ, the solution will involve the tridiagonal algorithm with four right hand sides) and equations (6.9.29) and (6.9.30) can be solved together.

The implementation of the above scheme should not be very different from the implementation of the scheme for polar coordinates without the origin, (6.9.15)–(6.9.17). However, there are several important differences. Solving equations (6.9.15)–(6.9.17) was routine. The right hand side of equations (6.9.15)–(6.9.16) contained $u_{jk}^n + \Delta t F_{jk}^{n+1}$ and equation (6.9.17) handles the boundary conditions (at both $j = 0$ and $j = M_r$) by reaching for them as a part of the difference operator (so as we have done for all ADI schemes, when the appropriate tri-diagonal matrix equation is solved, these boundary conditions will have to be included in the right hand side of the Trid solver (added to the u_{1k}^* and $u_{M_r-1k}^*$ terms). When we solve equation (6.9.28), we must be aware that $\mathbf{f}$ (and, hence $\mathbf{f}_2$ and f_1) contains both the $u_{jk}^n + \Delta t F_{jk}^{n+1}$ terms and the boundary conditions at $j = M_r$ terms. Also, we must be aware that when the L_r operator reaches to $j = 0$, since we are solving the matrix equations (6.9.25) and (6.9.26) where the Q_1' knows nothing about reaching to $j = 0$, we must assume that the values at $j = 0$ are zero (and, hence, add nothing for this contribution to the right hand side of the Trid solver). And, of course, when we solve equations (6.9.28)–(6.9.30), we must remember that the right hand sides are now given by $(1/\alpha)\mathbf{C}$, $\mathbf{y}_1^*$ and $\mathbf{y}_2^*$ instead of anything like the right hand sides of equations (6.9.15)–(6.9.17) (the right hand side of equation (6.9.27) is not very natural looking either). We warn you of some of these rather obvious differences because, after using difference equations such as (6.9.15)–(6.9.17) to approximately solve partial differential equations in a very logical, natural setting, it is easy to make some rather stupid mistakes when we apply similar difference schemes to solve some "unnatural" equations such as (6.9.27)–(6.9.30).

HW 6.9.2 Redo HW4.5.1 using the implicit scheme derived in this section.

HW 6.9.3 Redo HW4.5.3 using the implicit scheme derived in this section.

HW 6.9.4 (a) Derive a Crank-Nicolson scheme designed to approximate the solution to equation (4.5.1) (analogous to difference equations (4.5.27)–(4.5.28)). Verify that the Crank-Nicolson scheme is second order in both time and space.
(b) Approximately factor the difference equation found in part (a) to derive analogs to difference equations (6.9.15)–(6.9.17).
(c) Derive a treatment at the origin (analogous to equation (4.5.29)) consistent with the Crank-Nicolson scheme derived above.

HW 6.9.5 Redo HW4.5.1 using the Crank-Nicolson scheme derived in HW6.9.13.

HW 6.9.6 Redo HW4.5.3 using the Crank-Nicolson scheme derived in HW6.9.13.

HW 6.9.7 (a) Develop a Peaceman-Rachford scheme for solving equation (4.5.1) (again analogous to difference equations (4.5.27)–(4.5.28)). Verify that the Peaceman-Rachford scheme will be second order in both time and space.
(b) Using a second order accurate treatment at the origin, describe the procedure used to solve a problem in polar coordinates that includes the origin.

6.9.4   An Alternate Scheme for Polar Coordinates

In the last section we developed the scheme for approximating the solutions of two dimensional, parabolic equations in polar coordinates by an implicit scheme. As we can see, this is a very complex process. The difficulty of applying implicit schemes for solving problems in polar coordinates (and, of course, cylindrical and spherical coordinates are more difficult) is usually circumvented by using an explicit scheme instead of an implicit scheme. Sometimes it is impossible or not best to use an explicit scheme. In this section we shall introduce an alternative implicit method for solving equations given in polar coordinates. The method will be a mixed spectral-difference method. The idea of using spectral methods to find an approximate solution to partial differential equations is a very old idea. In recent years, the area has expanded rapidly. The basic technique is to consider an infinite series solution to the partial differential equation and to use a truncation of that series as an approximate solution. Since we refer to the method as a mixed spectral-difference method, we will be using difference methods to determine the coefficients of the truncated series approximation.

We begin by again considering partial differential equation (4.5.1) or

$$v_t = \frac{1}{r}(rv_r)_r + \frac{1}{r^2}v_{\theta\theta} + F(r,\theta,t)\ 0 \leq r < 1,\ 0 \leq \theta < 2\pi,\ t > 0.$$
(6.9.31)

The method that we wish to introduce in this section involves looking for a solution of the form

$$v(r,\theta,t) = \frac{1}{2}a_0 + \sum_{\ell=1}^{J}(a_\ell \cos \ell\theta + b_\ell \sin \ell\theta)$$
(6.9.32)

where $a_\ell = a_\ell(r,t)$, $\ell = 0, \cdots, J$ and $b_\ell = b_\ell(r,t)$, $\ell = 1, \cdots, J$. Anyone who has ever worked with Fourier series solutions of partial differential equations will realize that for sufficiently large J, the expansion form of v given in (6.9.32) is a logical solution form to consider. However, for the case being considered above, (6.9.32) is an especially nice solution form to consider. We recall that the speed of convergence of a Fourier series is proportional to the smoothness of the function. In most applications of Fourier series solutions to partial differential equations where the variable in the expansion ranges over the interval of definition of some variable, we only get continuity at the boundary. In this application, the boundary of the expansion variable (0 and 2π) is really not a boundary of the problem. At $\theta = 0$ and $\theta = 2\pi$, the solution is reasonably smooth (it has at least two derivatives when we have classical solutions) and sometimes it is very smooth. Hence, we should be able to obtain a good approximation to the solution using the expanded form of the solution given by (6.9.32) with reasonably small values of J.

To derive the basic equation that we must solve, we insert v given as in (6.9.32) into partial differential equation (6.9.31) to get

$$\frac{1}{2}a_{0t} + \sum_{\ell=1}^{J}(a_{\ell t} \cos \ell\theta + b_{\ell t} \sin \ell\theta)$$

$$= \frac{1}{2}\frac{1}{r}(ra_{0r})_r + \sum_{\ell=1}^{J}\left\{\left[\frac{1}{r}(ra_{\ell r})_r - \frac{\ell^2}{r^2}a_\ell\right]\cos \ell\theta\right.$$

$$\left. + \left[\frac{1}{r}(rb_{\ell r})_r - \frac{\ell^2}{r^2}b_\ell\right]\sin \ell\theta\right\}$$
(6.9.33)

where, though at times it is inconvenient, $a_{\ell r}$, etc. denotes the partial derivative of a_ℓ with respect to r, etc. If we multiply equation (6.9.33) first by $\cos k\theta$ and integrate from 0 to 2π, and then by $\sin k\theta$ and integrate from 0 to 2π, we obtain the following equations for determining a_0 and a_ℓ, b_ℓ, $\ell = 1, \cdots, J$.

$$a_{0t} = \frac{1}{r}(ra_{0r})_r \tag{6.9.34}$$

$$a_{\ell t} = \frac{1}{r}(ra_{\ell r})_r - \frac{\ell^2}{r^2}a_\ell, \ \ell = 1, \cdots, J \tag{6.9.35}$$

$$b_{\ell t} = \frac{1}{r}(rb_{\ell r})_r - \frac{\ell^2}{r^2}b_\ell, \ \ell = 1, \cdots, J \tag{6.9.36}$$

The plan is to solve equations (6.9.34)–(6.9.36) by implicit finite difference methods. (There may be instances where we might want to use an explicit scheme to determine a_0, a_ℓ, and b_ℓ. However, here we are presenting this scheme as an alternative scheme to using an implicit scheme for solving partial differential equation (6.9.31). Hence, we use an implicit scheme to solve the coefficient equations.) For example, we might use

$$a_{0_j}^{n+1} - \frac{\Delta t}{\Delta r^2 r_j}\left[r_{j+1/2}\left(a_{0_{j+1}}^{n+1} - a_{0_j}^{n+1}\right) - r_{j-1/2}\left(a_{0_j}^{n+1} - a_{0_{j-1}}^{n+1}\right)\right]$$
$$= a_{0_j}^n, \ j = 1, \cdots, M_r - 1 \tag{6.9.37}$$

$$a_{\ell_j}^{n+1} - \frac{\Delta t}{\Delta r^2 r_j}\left[r_{j+1/2}\left(a_{\ell_{j+1}}^{n+1} - a_{\ell_j}^{n+1}\right) - r_{j-1/2}\left(a_{\ell_j}^{n+1} - a_{\ell_{j-1}}^{n+1}\right)\right]$$
$$+ \frac{\ell^2}{r_j^2}a_{\ell_j}^{n+1} = a_{\ell_j}^n, \ j = 1, \cdots, M_r - 1, \ \ell = 1 \cdots, J \tag{6.9.38}$$

$$b_{\ell_j}^{n+1} - \frac{\Delta t}{\Delta r^2 r_j}\left[r_{j+1/2}\left(b_{\ell_{j+1}}^{n+1} - b_{\ell_j}^{n+1}\right) - r_{j-1/2}\left(b_{\ell_j}^{n+1} - b_{\ell_{j-1}}^{n+1}\right)\right]$$
$$+ \frac{\ell^2}{r_j^2}b_{\ell_j}^{n+1} = b_{\ell_j}^n, \ j = 1, \cdots, M_r - 1, \ \ell = 1 \cdots, J \tag{6.9.39}$$

where, though again sometimes inconvenient, $a_{\ell_j}^{n+1}$, etc. denotes an approximation to a_ℓ at the $(n + 1)$-st time step at $r = j\Delta r$.

The above approach works fine as long as we stay away from $r = 0$ (where the inside radius r_0 is some constant $R_0 > 0$). If we want to consider a domain that includes the origin $r = 0$, we must be more careful. We return to the grids considered in Section 4.5 and work carefully to derive a difference equation at the origin. Specifically, we consider the control volume that is pictured in Figure 4.5.3 and the integral form of the conservation law associated with that control volume given by (4.5.10), or without the nonhomogeneous term

$$\int_0^{2\pi}\int_0^{\Delta r/2}(v^{n+1} - v^n)r\,dr\,d\theta = \int_{t_n}^{t_{n+1}}\int_0^{2\pi}v_r(\Delta r/2, \theta, t)\frac{\Delta r}{2}d\theta dt. \tag{6.9.40}$$

If we insert series (6.9.32) into equation (6.9.40), we get

$$\int_0^{\Delta r/2} \frac{1}{2}(a_0^{n+1} - a_0^n)2\pi r dr = \frac{\Delta r}{2} \int_{t_n}^{t_{n+1}} \frac{1}{2}a_{0r}(\Delta r/2, t)2\pi t dt. \tag{6.9.41}$$

(The sine and cosine terms integrate out. Hence, the a_ℓ, b_ℓ, $\ell = 1, \cdots, J$ do not appear at the $j = 0$ grid point.) If we approximate equation (6.9.41) by evaluating the $a_0^{n+1} - a_0^n$ term in the first integral at $r = 0$ (and then integrate with respect to r) and approximate the second integral using the rectangular rule, we get

$$a_{0_0}^{n+1} - \frac{4\Delta t}{\Delta r^2}\left(a_{0_1}^{n+1} - a_{0_0}^{n+1}\right) = a_{0_0}^n. \tag{6.9.42}$$

We get no equations for a_{ℓ_0} and b_{ℓ_0}, $\ell = 1, \cdots, J$. The reason for this is that on the discrete grid, u at $r = 0$ does not depend on θ. Hence, the θ-coefficients, a_ℓ, b_ℓ, $\ell = 1, \cdots, J$, cannot depend on $k = 0$.

To summarize the procedure,

1. we solve equations (6.9.42), (6.9.37) to determine the approximation to a_0, the M_r-vector $\mathbf{a}_0^{n+1}$ associated with the $(n+1)$-st time step,

2. for $\ell = 1, \cdots, J$, we solve equation (6.9.38) to determine the approximation to a_ℓ, the $(M_r - 1)$-vector $\mathbf{a}_\ell^{n+1}$ associated with the $(n+1)$-st time step, and

3. for $\ell = 1, \cdots, J$, we solve equation (6.9.39) to determine the approximation to b_ℓ, the $(M_r - 1)$-vector $\mathbf{b}_\ell^{n+1}$ associated with the $(n+1)$-st time step.

For convenience, we set $a_{\ell_0}^n = b_{\ell_0}^n = 0$ for all n and $\ell = 1, \cdots, J$ which will make both $\mathbf{a}_\ell^{n+1}$ and $\mathbf{b}_\ell^{n+1}$ into M_r-vectors also.

Of course, to solve systems (6.9.42), (6.9.37); (6.9.38) and (6.9.39) we must have an initial condition and a boundary condition at $k = M_r$. As in problem (4.5.1)–(4.5.3), we assume that we are given an initial condition and a boundary condition at $r = 1$ of the form

$$v(r, \theta, 0) = f(r, \theta), \quad 0 \le r < 1, \ 0 \le \theta < 2\pi \tag{6.9.43}$$
$$v(1, \theta, t) = g(\theta, t), \quad 0 \le \theta < 2\pi, \ t > 0. \tag{6.9.44}$$

Since g is known, we use an FFT (either finite or fast Fourier transform) to determine $a_{0_{M_r}}^n$, $a_{\ell_{M_r}}^n$, $b_{\ell_{M_r}}^n$, $\ell = 1, \cdots, J$, i.e. we determine $a_{0_{M_r}}^n$, $a_{\ell_{M_r}}^n$, $b_{\ell_{M_r}}^n$, $\ell = 1, \cdots, J$ so that

$$g(k\Delta\theta, n\Delta t) = \frac{1}{2}a_{0_{M_r}}^n + \sum_{\ell=1}^J \left(a_{\ell_{M_r}}^n \cos \ell k\Delta\theta + b_{\ell_{M_r}}^n \sin \ell k\Delta\theta\right).$$

Likewise, we take the FFT of initial condition (6.9.43) to determine a_{0k}^0, $a_{\ell k}^0$, $b_{\ell k}^0$, $k = 0, \cdots, M_r$ and $\ell = 1, \cdots, J$.

Remark 1: We should note that each of the systems that we must solve are tridiagonal systems of equations. Each time step will involve solving $2J + 1$ tridiagonal systems of equations.

Remark 2: To make the above descirbed scheme computationally efficient, you should stay in "transform space" until a solution is desired. Then equation (6.9.32) can be used to compute a solution at the present time, at each $r = j\Delta r$, $j = 0, \cdots, M_r - 1$, which is a function of θ (which can be evaluated at $\theta = k\Delta\theta$ or at any other desired values of θ).

HW 6.9.8 Redo HW4.5.1 using the spectral-difference scheme derived in this section.

HW 6.9.9 Redo HW4.5.3 using the spectral-difference scheme derived in this section.

HW 6.9.10 Derive a spectral-difference scheme that uses a Crank-Nicolson solver for the equations obtained for the Fourier coefficients.

7
Dispersion and Dissipation

7.1 Introduction

In the previous chapters we have included a large number of problems and examples involving computational results. Generally, the problems involved were theoretically easy enough so that we could compare our results with the exact solutions. There have been times when we have made no mention of these comparisons and/or did not explain why we sometimes obtain bad results. In this chapter we will develop several techniques for analyzing numerical schemes. Sometimes the results will help us choose the correct parameters (Δx, Δt, etc.) for obtaining a satisfactory solution. Sometimes the results will only tell us why our results are bad. In either case, it is hoped that the material will fill a gap in our knowledge about difference scheme approximations to partial differential equations that will make us better practitioners. As we shall see later, the material presented in this chapter applies to both parabolic and hyperbolic partial differential equations. Since the results are more dramatic for hyperbolic equations, we begin our discussion with results from two previous computations involving the solution of hyperbolic equations.

7.1.1 HW5.6.3

If we return to our results found in HW5.6.3 concerning the approximate solution of problem (5.6.30)–(5.6.32), we see in Figure 7.1.1 that the results in part (a) for $M = 20$ are qualitative correct (the wave propagates to the left and reappears at the right side at approximately the correct time) but they are not very good. The results from HW5.6.3(b) shown in Figure 7.1.2 using $M = 100$, however, are better. If we were solving a problem for which we did not know the exact answer, we would probably think that these solutions are perfectly acceptable. However, we know that the wave at $t = 0.8$ should not be damped. This is error. If we next observe the solutions given us from HW5.6.3(c), Figure 7.1.3, we see that by the time the difference scheme gets to time $t = 20.0$, the solution has been damped to approximately 20% of its true value. And finally, Figure 7.1.4 shows that by the time the solution gets to time $t = 40.0$, the computed solution is nearing a constant (nonzero).

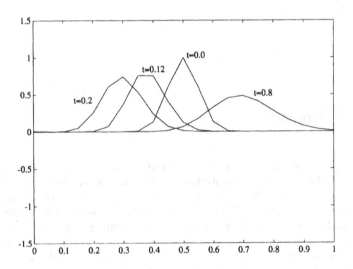

FIGURE 7.1.1. Approximate solutions to problem (5.6.30)–(5.6.32) at times $t = 0.0$, $t = 0.12$, $t = 0.2$ and $t = 0.8$ obtained by using difference scheme (5.3.1) with $\Delta x = 0.05$ and $\Delta t = 0.04$.

We might add that there is some good news. Note how accurately the phase speed of the waves have been calculated. All of the computed waves at times $t = 2.0$, 10.0, 20.0 and 40.0 are centered nicely about $x = 0$ as they should be. As we shall see, this is not always the case with other schemes.

Hence, we have used the FTFS difference scheme (5.3.1) to solve problem (5.6.30)–(5.6.32). If we had used the scheme to solve a more difficult problem, we might not know that the solutions are incorrect. We may have

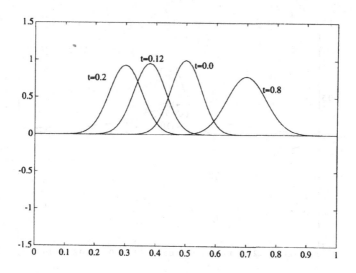

FIGURE 7.1.2. Approximate solutions to problem (5.6.30)–(5.6.32) at times $t = 0.0$, $t = 0.12$, $t = 0.2$ and $t = 0.8$ obtained by using difference scheme (5.3.1) with $\Delta x = 0.01$ and $\Delta t = 0.008$.

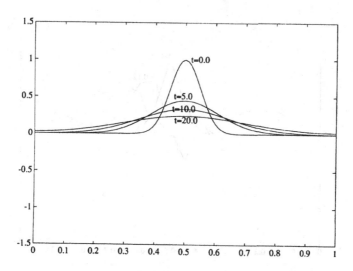

FIGURE 7.1.3. Approximate solutions to problem (5.6.30)–(5.6.32) at times $t = 0.0$, $t = 5.0$, $t = 10.0$ and $t = 20.0$ obtained by using difference scheme (5.3.1) with $\Delta x = 0.01$ and $\Delta t = 0.008$.

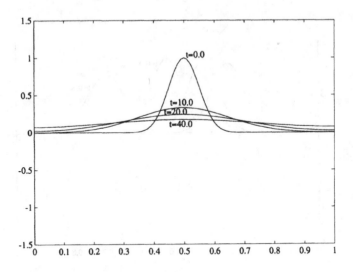

FIGURE 7.1.4. Approximate solutions to problem (5.6.30)–(5.6.32) at times $t = 0.0$, $t = 10.0$, $t = 20.0$ and $t = 40.0$ obtained by using difference scheme (5.3.1) with $\Delta x = 0.01$ and $\Delta t = 0.008$.

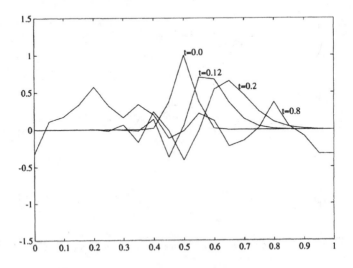

FIGURE 7.1.5. Crank-Nicolson solutions to the problem in HW5.6.5 at times $t = 0.0$, $t = 0.12$, $t = 0.2$ and $t = 0.8$ with $\Delta x = 0.05$ and $\Delta t = 0.04$.

thought that we were converging to some sort of steady state solution to our original problem. Obviously, the results are in some way due to the large time interval used (many time steps), but long time calculations will often be required when we do realistic applied problems. We must understand why a convergent scheme can give such bad results. We would like to know how to correct these results (if that is practically possible).

7.1.2   HW5.6.5

We next return to HW5.6.5 to consider results using the Crank-Nicolson difference scheme. In Figure 7.1.5 we again note that when $M = 20$ the solution is qualitatively correct but is very rough. In Figure 7.1.6 we note that when M is increased to 100, the solution is much smoother but we have the unwanted oscillation behind the computed wave at $t = 0.8$. And, if we observe Figure 7.1.7, we note that by the time we get out to $t = 10.0$, all of the character of the true solution is lost.

The errors for this solution appear to be both due to damping (note that the wave at $t = 0.8$ in Figure 7.1.6 has already lost 20% of its amplitude) and due to the apparent oscillation in Figure 7.1.6 and worse in Figure 7.1.7. If we cannot explain how to correct these results, we must at least explain why these results are as they appear. And more importantly, we must understand that if we are not careful, we can solve some difficult problem numerically and all of the interesting phenomena that appears *could* be due to numerical dissipation and/or dispersion.

In both case given above, we note that the results are not very good when we use $M = 20$. We saw that the results were worse yet when we used $M = 20$ for the long time calculations. Generally, $\Delta x = 0.05$ does not generate a sufficiently fine grid to adequately resolve the solutions to hyperbolic partial differential equations.

7.2   Dispersion and Dissipation for Partial Differential Equations

We see that when we solve partial differential equations analytically, we generally use a series term or a transform that makes the solution depend on a term of the form

$$v(x, t) = \hat{v}e^{i(\omega t + \beta x)} = \hat{v}e^{i\omega t}e^{i\beta x}. \tag{7.2.1}$$

The term given in (7.2.1) is the equation describing a wave in space and time. In the expression given in equation (7.2.1), ω is the **frequency** of the wave, β is the **wave number** and the **wave length**, λ is given by $\lambda = 2\pi/\beta$.

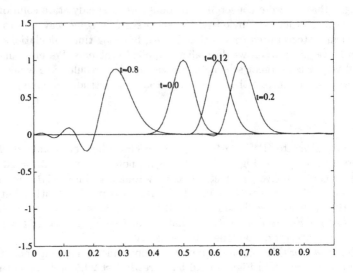

FIGURE 7.1.6. Crank-Nicolson solutions to the problem in HW5.6.5 at times $t = 0.0$, $t = 0.12$, $t = 0.2$ and $t = 0.8$ with $\Delta x = 0.01$ and $\Delta t = 0.008$.

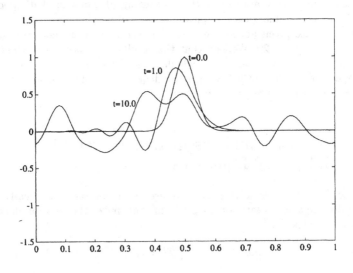

FIGURE 7.1.7. Crank-Nicolson solutions to the problem in HW5.6.5 at times $t = 0.0$, $t = 1.0$ and $t = 10.0$ with $\Delta x = 0.01$ and $\Delta t = 0.008$.

If we consider the expression (7.2.1) along with either the partial differential equation

$$v_t = \nu v_{xx} \tag{7.2.2}$$

or

$$v_t + a v_x = 0, \tag{7.2.3}$$

we see that it is clear that the function given in (7.2.1) can not satisfy these equations without a relationship between ω and β. We write this relationship as $\omega = \omega(\beta)$ and call this the **dispersion relation** (though we shall see later that it might be more logical to call it the dispersion-dissipation relation). For example, if we insert (7.2.1) into equation (7.2.2), we see that v is a solution if $\omega = i\nu\beta^2$. If we insert (7.2.1) into (7.2.3), we see that v is a solution if $\omega = -a\beta$. Very little thought reveals that there will generally have to be a dispersion relation between ω and β for the expression given in (7.2.1) to satisfy a partial differential equation.

We note in particular, that when we consider equation (7.2.2), the solution becomes

$$v(x,t) = \hat{v} e^{-\nu\beta^2 t} e^{i\beta x}. \tag{7.2.4}$$

We see that the wave does not move and decays with time. Whenever ω is purely imaginary, the wave will either grow or decay. We should also recall that this is what we generally see for parabolic equations, specifically in many of our parabolic calculations.

Likewise, when we consider equation (7.2.3), the solution becomes

$$v(x,t) = \hat{v} e^{i\beta(x-at)}. \tag{7.2.5}$$

Thus we see that in the general case when ω is real, the mode is propagated at with speed $-\omega/\beta$ and with no decay of the amplitude. We note that when ω is linear, the speed of propagation is independent of the frequency.

The decay and propagation of various Fourier modes is a very important part of the behavior of the solution of the partial differential equation. One of the pieces of analysis that we are able to perform on a difference equation that we are using to approximate the solution of a particular partial differential equation is to see how well the decay and propagation of various Fourier modes of the difference scheme matches with those of the partial differential equation. Also, often we can explain the behavior of the solutions (and why the solution is not what we want it to be) in terms of the decay and propagation of the Fourier modes.

We begin by emphasizing that the scheme will be unstable if some Fourier mode grows without bound. We *define* **dissipation** *of solutions of partial differential equations as when the Fourier modes do not grow with time and* at least one mode decays. We call *the partial differential equation* **nondissipative** *if the Fourier modes neither decay nor grow*. And we *define* **dispersion** *of solutions of partial differential equations as when Fourier modes of differing wave lengths (or wave numbers) propagate at different speeds.*

We see from our above discussion that the model parabolic equation (7.2.2) is dissipative. We note that if $\nu > 0$, modes associated with all wave numbers $\beta \neq 0$, dissipate. (If $\beta = 0$, the solution $v = \hat{v}$ is constant and is not determined by equation (7.2.2).) We also note that the solutions of the model hyperbolic equation, (7.2.3), are neither dissipative (the solutions are nondissipative) nor dispersive.

If we consider the equation

$$v_t + cv_{xxx} = 0, \tag{7.2.6}$$

the dispersion relation for the equation is

$$\omega = \beta^3 c. \tag{7.2.7}$$

The Fourier mode satisfying equation (7.2.6) is given by

$$v(x,t) = \hat{v}e^{i(\beta^3 ct + \beta x)} = \hat{v}e^{i\beta(x+\beta^2 ct)}. \tag{7.2.8}$$

We see that the wave propagates in the opposite direction of the propagating wave due to equation (7.2.3) and propagates with speed $\beta^2 c$. Clearly, modes associated with different wave numbers propagate with different velocities $(-\beta^2 c)$ so equation (7.2.6) is dispersive. We also note that equation (7.2.6) is not dissipative.

Remark: By this time it should be clear that partial differential equations containing only even ordered x derivatives will be dissipative (and not involve propagating waves), while partial differential equations containing only odd ordered x derivatives will be nondissipative and involve propagating waves (and be dispersive when the order is greater than one).

HW 7.2.1 Perform a dissipation-dispersion analysis for the model partial differential equations

1. $v_t + dv_{xxxx} = 0$

2. $v_t + ev_{xxxxx} = 0.$

HW 7.2.2 Perform a dissipation-dispersion analysis for the partial differential equation

$$v_t + av_x - \nu_1 v_{xx} + cv_{xxx} = 0.$$

7.3 Dispersion and Dissipation for Difference Equations

We now wish to see if the difference schemes that we use to approximate certain partial differential equations have the same dissipativity and dispersivity properties as the corresponding partial differential equations. Also, we might ask: do we always need them to be the same? We begin by considering the discrete analog of the Fourier mode (7.2.1)

$$u_k^n = \hat{u} e^{i(\omega n \Delta t + \beta k \Delta x)}. \tag{7.3.1}$$

As in the continuous case, we choose $\omega = \omega(\beta)$ so that the solution (7.3.1) will satisfy our difference equation. The function $\omega(\beta)$ will called the **discrete dispersion relation.**

We note that if we consider a difference scheme for an initial–value problem and the finite Fourier transform, the part of the term (7.3.1) given by $e^{i\beta k \Delta x}$ corresponds to the $e^{ik\xi}$ term in the finite Fourier transform. Hence, by considering $0 \leq \beta \Delta x \leq \pi$ we obtain information on all the modes present in the finite Fourier transform representation of the solution. Likewise, for an initial–boundary–value problem and a finite Fourier series, we see that the highest frequency present in expansion (3.2.6) is of the form $e^{2\pi i \frac{M}{2} k \Delta x}$. Again, considering $0 \leq \beta \Delta x \leq \pi$ will give all of the modes contained in the finite Fourier series expansion of our solutions. In either case, *to obtain dissipation and dispersion information for all modes associated with either finite Fourier transform or finite Fourier series solutions of a difference scheme, we consider $\beta \Delta x$ in the range $0 \leq \beta \Delta x \leq \pi$.*

Before we proceed, we claim that for difference equations the discrete dispersion relation, $\omega = \omega(\beta)$, will generally be complex (the same situation if we considered partial differential equations of the form $v_t + av_x = \nu v_{xx}$). Hence, we will find it convenient to set $\omega = \alpha + ib$ (where $\alpha = \alpha(\beta)$ and $b = b(\beta)$ are assumed to be real) where α and b are referred to as the **real discrete dispersion relation** and the **imaginary discrete dispersion relation**, respectively. If we insert $\omega = \alpha + ib$ into expression (7.3.1) we see that it has the form

$$u_k^n = \hat{u} e^{i[\alpha n \Delta t + ibn \Delta t + \beta k \Delta x]} = \hat{u} \left(e^{-b \Delta t} \right)^n e^{i\beta[k \Delta x - (-\alpha/\beta)n \Delta t]}. \tag{7.3.2}$$

Hence, we see that

- if $b > 0$ for some β, then the associated difference equation is dissipative,

- if $b < 0$ for some β, then solutions to the scheme will grow without bound (and the scheme will be unstable) and

- if $b = 0$ for all β, the scheme will be nondissipative.

Also,

- if $\alpha = 0$ for all β, there will be no wave propagation,

- if $\alpha \neq 0$ for some β, there will be wave propagation with velocity $-\alpha/\beta$ and

- if $-\alpha/\beta$ is a nontrivial function of β, the scheme will be dispersive.

In our first example we consider the common explicit scheme used to solve equation (7.2.2). This is a parabolic difference scheme designed to solve a parabolic partial differential equation. As we promised in Section 7.1, this case will be less interesting than the analyses for hyperbolic equations. For this reason, it will also be easier and, hopefully, a good place to introduce some of our techniques for analyzing dissipation and dispersion.

Example 7.3.1 Analyze the dissipation and dispersion for the difference scheme

$$u_k^{n+1} = u_k^n + r\delta^2 u_k^n \tag{7.3.3}$$

where $r = \nu\Delta t/\Delta x^2$.

Solution: Before we begin, we recall that in Example 3.1.1, we saw that difference scheme (7.3.3) is stable if and only if $r \leq 1/2$. For this reason, we would generally begin by assuming the condition $r \leq 1/2$. As we shall see in the work that follows, this is not necessary since the dispersion-dissipation analysis that we do will also give us the stability condition. We begin our analysis as we did for the continuous problems and insert expression (7.3.1) into the difference scheme (7.3.3) to get

$$\hat{u}e^{i[\omega(n+1)\Delta t + \beta k\Delta x]} = \hat{u}e^{i(\omega n\Delta t + \beta k\Delta x)} + r\left(\hat{u}e^{i[\omega n\Delta t + \beta(k+1)\Delta x]}\right.$$
$$\left. - 2\hat{u}e^{i(\omega n\Delta t + \beta k\Delta x)} + \hat{u}e^{i[\omega n\Delta t + \beta(k-1)\Delta x]}\right). \tag{7.3.4}$$

Dividing both sides of equation (7.3.4) by $e^{i(\omega n\Delta t + \beta k\Delta x)}$ yields

$$\begin{aligned}
e^{i\omega\Delta t} &= 1 + r\left(e^{i\beta\Delta x} - 2 + e^{-i\beta\Delta x}\right) \\
&= 1 + r(2\cos\beta\Delta x - 2) \\
&= 1 - 4r\sin^2\frac{\beta\Delta x}{2}. \tag{7.3.5}
\end{aligned}$$

Then since the last term is real, if we let $\omega = \alpha + ib$ and write

$$e^{i\omega\Delta t} = e^{-b\Delta t}e^{i\alpha\Delta t},$$

we see that $\alpha = 0$ and $\omega = ib$ where

$$b = -\frac{1}{\Delta t}\log\left|\left(1 - 4r\sin^2\frac{\beta\Delta x}{2}\right)\right|. \tag{7.3.6}$$

Hence, we see that the discrete dispersion relation is given by

$$\omega = -\frac{i}{\Delta t}\log\left|\left(1 - 4r\sin^2\frac{\beta\Delta x}{2}\right)\right|. \tag{7.3.7}$$

It is then easy to see that the general Fourier mode of wave number β assumes the form

$$u_k^n = \hat{u}e^{-bn\Delta t}e^{i\beta k\Delta x}. \tag{7.3.8}$$

In Example 3.1.1 we saw that when $r > 1/2$, $b < 0$ (because $|(1 - 4r\sin^2\beta\Delta x/2)| > 1$). Then the term

$$e^{-bn\Delta t} = \left(e^{-b\Delta t}\right)^n$$

will grow without bound (since when $b < 0$, $e^{-b\Delta t} > 1$) and the scheme will be unstable. Likewise we saw in Example 3.1.1 that when $r \leq 1/2$, $b \geq 0$ (because $|(1 - 4r\sin^2 \beta\Delta x/2)| \leq 1$. Hence, the term

$$e^{-bn\Delta t} = \left(e^{-b\Delta t}\right)^n$$

in equation (7.3.8) decays (or does not grow in the case when $b = 0$.). Thus the general wave of the form (7.3.1) (which in this case takes the form of (7.3.8)) does not move and generally decays (or dissipates). We should notice specifically that all modes except $\beta = 0$ decay. The $\beta = 0$ mode will persist and eventually dominate the solution.

Before we leave this example, consider a problem involving the heat equation on the interval $(0, 1)$, with zero Dirichlet boundary conditions. Using common separation of variables or Fourier series techniques, we know that the solution to such a problem can be written as

$$v(x,t) = \sum_{j=1}^{\infty} C_j e^{-\nu(j\pi)^2 t} \sin j\pi x.$$

If we choose $M = 100$ and $r = .9$, and return to Example 3.2.1 (specifically finite Fourier expansion (3.2.23)), we see that our numerical solution to the problem can be written as

$$u_k^n = -2\sum_{j=1}^{M} c_j^0 \left(1 - 4r\sin^2 \frac{j\pi\Delta x}{2}\right)^n \sin \pi jk\Delta x$$

$$= -2\sum_{j=1}^{M} c_j^0 e^{-b(j\pi\Delta x)n\Delta t} \sin \pi jk\Delta x.$$

Clearly, to compare the natural dissipation in the solution to the partial differential equation with the dissipation in the solution to the difference equation, we must compare $e^{-\nu(j\pi)^2 t}$ with $e^{-b(j\pi\Delta x)n\Delta t}$. Below we include the values of $e^{-\nu(j\pi)^2 t}$ and $e^{-b(j\pi\Delta x)n\Delta t}$ for $j = 1$, 3 and 5, $\nu = 1.0$ and 0.001, and $t = n\Delta t = 0.09$, 0.9 and 9.0.

- for the $j = 1$ mode (the slowest damping mode) and $\nu = 1.0$

    1. if $t = n\Delta t = 0.09$, $e^{-bn\Delta t} = 0.41123$ and $e^{-\nu\pi^2 t} = 0.411369$

    2. if $t = n\Delta t = 0.9$, $e^{-bn\Delta t} = 1.38331 \cdot 10^{-4}$ and $e^{-\nu\pi^2 t} = 1.38776 \cdot 10^{-4}$

    3. if $t = n\Delta t = 9.0$, $e^{-bn\Delta t} = 0.0$ and $e^{-\nu\pi^2 t} = 0.0$

- for the $j = 1$ mode and $\nu = 0.001$

    1. if $t = n\Delta t = 0.09$, $e^{-bn\Delta t} = 0.999111$ and $e^{-\nu\pi^2 t} = 0.999112$

    2. if $t = n\Delta t = 0.9$, $e^{-bn\Delta t} = 0.991153$ and $e^{-\nu\pi^2 t} = 0.991156$

    3. if $t = n\Delta t = 9.0$, $e^{-bn\Delta t} = .914974$ and $e^{-\nu\pi^2 t} = 0.915004$

- for the $j = 3$ mode and $\nu = 1.0$

    1. if $t = n\Delta t = 0.09$, $e^{-bn\Delta t} = 3.28641 \cdot 10^{-4}$ and $e^{-\nu\pi^2 t} = 3.37353 \cdot 10^{-4}$

    2. if $t = n\Delta t = 0.9$, $e^{-bn\Delta t} = 0.0$ and $e^{-\nu\pi^2 t} = 0.0$

    3. if $t = n\Delta t = 9.0$, $e^{-bn\Delta t} = 0.0$ and $e^{-\nu\pi^2 t} = 0.0$

- for the $j = 3$ mode and $\nu = 0.001$

    1. if $t = n\Delta t = 0.09$, $e^{-bn\Delta t} = 0.992011$ and $e^{-\nu\pi^2 t} = 0.992037$

    2. if $t = n\Delta t = 0.9$, $e^{-bn\Delta t} = 0.922926$ and $e^{-\nu\pi^2 t} = 0.923168$

3. if $t = n\Delta t = 9.0$, $e^{-bn\Delta t} = 0.448406$ and $e^{-\nu\pi^2 t} = 0.449581$

- for the $j = 5$ mode and $\nu = 1.0$ all values are essentially 0.0
- for the $j = 5$ mode and $\nu = 0.001$

    1. if $t = n\Delta t = 0.09$, $e^{-bn\Delta t} = 0.977839$ and $e^{-\nu\pi^2 t} = 0.978038$
    2. if $t = n\Delta t = 0.9$, $e^{-bn\Delta t} = 0.799233$ and $e^{-\nu\pi^2 t} = 0.800862$
    3. if $t = n\Delta t = 9.0$, $e^{-bn\Delta t} = 0.106349$ and $e^{-\nu\pi^2 t} = 0.108537$

We see above that though the dissipation in the numerical scheme is an approximate dissipation, it is a very accurate approximation. All of the results above are acceptable.

We now return to the first set of computations discussed in Section 7.1.1. Hopefully, we now have the tools necessary to understand the results given in Figures 7.1.1–7.1.4.

Example 7.3.2 (a) Analyze the dissipative and dispersive qualities of the FTFS difference scheme

$$u_k^{n+1} = u_k^n - R(u_{k+1}^n - u_k^n) \tag{7.3.9}$$

where $R = a\Delta t / \Delta x$.

(b) Use the dissipation results found in part (a) to explain the computational results described in Section 7.1.1.

(c) Use the dispersion results found in part (a) to explain the computational results described in Section 7.1.1.

Solution: (a) **Dissipation** We began by noting that for stability consideration we may as well assume that $a < 0$ and that $|R| \leq 1$. We next proceed as we did in Example 7.3.1, insert the expression for a general discrete Fourier mode, (7.3.1), into the difference scheme (7.3.9), divide by $\hat{u}e^{i(\omega n \Delta t + \beta k \Delta x)}$ and get the discrete dispersion relation

$$
\begin{aligned}
e^{i\omega\Delta t} &= e^{i\alpha\Delta t} e^{-b\Delta t} \\
&= 1 - R\{e^{i\beta\Delta x} - 1\} \\
&= 1 + R - R\cos\beta\Delta x - iR\sin\beta\Delta x.
\end{aligned} \tag{7.3.10}
$$

Thus

$$
\begin{aligned}
e^{-b\Delta t} &= |1 + R - R\cos\beta\Delta x - iR\sin\beta\Delta x| \\
&= \sqrt{(1+R)^2 - 2R(1+R)\cos\beta\Delta x + R^2}.
\end{aligned} \tag{7.3.11}
$$

Using the stability calculation done in Example 3.1.2, $b > 0$ (and, hence, $e^{-b\Delta t} < 1$) for some β whenever $a < 0$ and $|R| < 1$. We note that all of the modes $\beta \neq 0$ decay and the $\beta = 0$ neither grows nor decays. Hence, the scheme is dissipative. We should make a special note that when $R = -1$, $b = 0$ for all β and the scheme is nondissipative. (We should also recall then when we choose $R = -1$ in this algorithm, we get the exact solution to the partial differential equation.)

Dispersion If we then divide both sides of the expression (7.3.10) by the magnitude of the right hand side, we get that

$$e^{i\alpha\Delta t} = \cos\alpha\Delta t + i\sin\alpha\Delta t \tag{7.3.12}$$

$$= \frac{1 + R - R\cos\beta\Delta x - iR\sin\beta\Delta x}{|1 + R - R\cos\beta\Delta x - iR\sin\beta\Delta x|} \tag{7.3.13}$$

or

$$\tan\alpha\Delta t = \frac{-R\sin\beta\Delta x}{1 + R - R\cos\beta\Delta x}.$$

Thus the real discrete dispersion relation can be written as

$$\alpha = -\frac{1}{\Delta t}\tan^{-1}\left\{\frac{R\sin\beta\Delta x}{1+R-R\cos\beta\Delta x}\right\}$$

$$= -\frac{1}{\Delta t}\tan^{-1}\left\{\frac{R\sin\beta\Delta x}{1+2R\sin^2\frac{\beta\Delta x}{2}}\right\}. \qquad (7.3.14)$$

Since α is not linear ($\frac{d^2\alpha}{d\beta^2}\neq 0$), *the difference scheme is dispersive.*

We next would like to interpret the information concerning wave propagation and dispersion that we get from the real discrete dispersion relation (7.3.14). One approach is to carefully examine equation (7.3.14) using asymptotic methods. For instance, we start by noting that when $\beta\Delta x$ is near π (high frequency waves, β large, λ small), then

$$\alpha \approx \frac{-1}{\Delta t}\tan^{-1}(0).$$

We must be careful here in deciding which value to choose for $\tan^{-1}(0)$. If we choose $\beta\Delta x$ near π and consider equation (7.3.12)–(7.3.13), we see that $\sin\alpha\Delta t$ is near zero (the sign determined by $-R\sin(\beta\Delta x)$) and $\cos\alpha\Delta t$ is near

$$\frac{1+2R}{|1+2R|}.$$

When $R < -1/2$, the sine is positive and the cosine is negative so we are in the second quadrant and $\tan^{-1}(0) = \pi$. When $0 > R > -1/2$, the sine is positive and the cosine is positive so we are in the first quadrant and $\tan^{-1}(0) = 0$.

Thus for $\beta\Delta x$ near π, we see that the speed of propagation of high frequency waves is given by

$$-\alpha/\beta = \left\{\begin{array}{ll} 0 & \text{if } R > -1/2 \\ a/R & \text{if } R < -1/2. \end{array}\right.$$

(This result will be verified graphically later in Figure 7.3.8 and we will also see that the speed of propagation of the high frequency modes is not especially relevant.)

For the low frequency waves ($\beta\Delta x$ small, β small, λ large), we can use Taylor series expansions of $\sin\theta$, $\sin^2\frac{\theta}{2}$, $1/(1+z)$, and $\tan^{-1}z$ to show that

$$\alpha = -\frac{1}{\Delta t}\left\{R\beta\Delta x - \frac{R}{6}[1+3R+2R^2](\beta\Delta x)^3 + \mathcal{O}\big((\beta\Delta x)^4\big)\right\}, \qquad (7.3.15)$$

or

$$\alpha \approx -a\beta\left\{1 - \frac{1}{6}(1+2R)(1+R)(\beta\Delta x)^2\right\}. \qquad (7.3.16)$$

Thus the wave form can be written approximately as

$$\begin{aligned} u_k^n &= \hat{u}e^{i(\omega n\Delta t+\beta k\Delta x)} \\ &= \hat{u}e^{-bn\Delta t}e^{i(\alpha n\Delta t+\beta k\Delta x)} \\ &\approx \hat{u}e^{-bn\Delta t}e^{i\left[-a\beta n\Delta t\{1-\frac{1}{6}(1+2R)(1+R)(\beta\Delta x)^2\}+\beta k\Delta x\right]} \\ &= \hat{u}e^{-bn\Delta t}e^{i\beta\left[k\Delta x-\{a-\frac{a}{6}(1+2R)(1+R)(\beta\Delta x)^2\}n\Delta t\right]} \end{aligned} \qquad (7.3.17)$$

We see that if $R = -1/2$ or $R = -1$, the speed of the wave is a, just as in the analytic case. If $0 \geq R > -1/2$, then the speed is

$$a - \frac{a}{6}(1+2R)(1+R)(\beta\Delta x)^2 > a. \qquad (7.3.18)$$

And if $-1/2 > R \geq -1$, the speed of the wave is

$$a - \frac{a}{6}(1+2R)(1+R)(\beta\Delta x)^2 < a. \qquad (7.3.19)$$

Thus we see that if $|R| < 1/2$, the computed low frequency waves go slower than the analytically determined high frequency waves (remember $a < 0$). And if $1/2 < |R| \leq 1$, the computed low frequency waves go faster than their analytic counterpart.

(b) We now return to the computational example used in Section 7.1.1. It is easy to see that we have dissipation. In Fact, in Figure 7.1.4 we see that by the time we get to $t = 40.0$, the solution has almost damped to a constant solution (nonzero). If we return to the expression (7.3.11) (and the discussion following that derivation), we see that all of the modes except the $\beta = 0$ mode are dissipative. The $\beta = 0$ mode will not dissipate. Hence, the nonzero constant that we see in Figure 7.1.4 is the contribution of the $\beta = 0$ mode to the solution (and will determine the coefficient of the 0-mode in the finite Fourier series solution to the discrete problem).

If we again consider expression (7.3.11), we note that the slowest damping mode (least damping except for $\beta = 0$) is the $\beta \Delta x = 2\pi \Delta x$ mode. (We should realize that using the results of Chapter 3 and Theorem 3.2.2 we get the $\beta \Delta x = 2\pi \Delta x$ mode. We do not get the $\beta \Delta x = \pi \Delta x$ mode as in Example 3.2.1 because we are not able to reflect about either of the boundary points as we did in Example 3.2.1.) If we consider our case with $R = -.8$ and $\Delta x = 0.01$, we see that

$$e^{-b\Delta t} = \sqrt{0.68 + 0.32\cos(0.02\pi)} = 0.999684226.$$

Hence, it is not surprising that when we go 5000 time steps, we get

$$(e^{-b\Delta t})^{5000} = 0.206157288$$

(and the rest, except for the zeroth mode, dissipate faster) so that by the time we get to $t = 40.0$, only 20% of the $\beta \Delta x = 2\pi \Delta x$ mode remains in the solution.

We might next ask how we can correct these dissipation errors that have swamped our calculation. Of course, unless we want to change difference schemes, we cannot eliminate the dissipation. For this case, we can at least analyze (and maybe implement—depending on what we are willing to pay in computation time) what is necessary.

One approach is to claim that we want 90% of each mode in our solution to be present at time $t = 40.0$. The approach is then to determine M so that

$$(e^{-b\Delta t})^n = e^{-40b} \geq 0.9$$

for all β satisfying $\beta \Delta x = 2\pi k \Delta x$, $k = 0, \cdots, M/2$(when $M - 1$ is odd). Since $\beta = M\pi$ represents that fastest decaying mode in our calculation and $\beta \Delta x = M\pi \Delta x = \pi$, we must determine an n (which is equivalent to choosing a $\Delta t = 40/n$) so that

$$\left[\sqrt{0.68 + 0.32(-1)} \right]^n = 0.9.$$

It is not hard to see that *there is no solution to this problem*. In fact, it is not possible to determine M so that 90% of half of the modes are present at time $t = 40.0$.

In Figure 7.3.1 we plot $e^{-b\Delta t}$ for $R = -0.8$. The modes involved in the numerical solution of the problem are uniformly distributed across the interval $[0, \pi]$. Hence, the damping factor on the "middle mode" (assuming that M is divisible by four),

$$\left[e^{-b\Delta t} \right]_{k=M/4} = \sqrt{0.68},$$

damps at that given rate no matter which Δx is chosen. *Working with a smaller Δx will not help retain a larger percentage of the modes accurately in the solution.*

To understand why *making Δx smaller will give a more accurate solution and reduce the dissipation*, we return to the finite Fourier series discussed in Chapter 3, and specifically, to Theorem 3.2.2. If we calculate the finite Fourier transform (FFT, also used to refer to the fast Fourier transform which for our purposes is even better) of our initial function (in our case $u_k^0 = \sin^{40}(\pi k \Delta x)$), we obtain the coefficients for each term

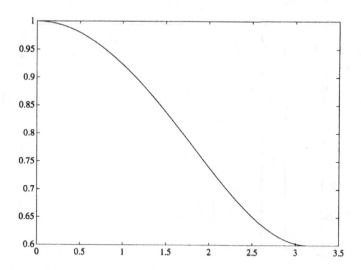

FIGURE 7.3.1. Plot of $e^{-b\Delta t}$ associated with difference scheme (7.3.9) with $R = -0.8$.

in the finite Fourier expansion of u_k^0. Using expression (7.3.2), we know that if the finite Fourier expansion of u^0 is given by

$$u_k^0 = \sum_{j=-(M-2)/2}^{M/2} c_j e^{2\pi i j k \Delta x} \qquad (7.3.20)$$

(when M is even), then the solution to difference scheme (7.3.9) with periodic boundary conditions at time $t = n\Delta t$ is given by

$$u_k^n = \sum_{j=-(M-2)/2}^{M/2} \left(e^{-b(2\pi j \Delta x)\Delta t}\right)^n c_j e^{2\pi i j [k\Delta x - (-\alpha(2\pi j \Delta x)/(2\pi j))n\Delta t]}. \qquad (7.3.21)$$

Hence, the j-th mode $j = 0, \cdots, M/2$ is involved in the solution at time $t = n\Delta t$ with coefficient $\left(e^{-b(2\pi j \Delta x)\Delta t}\right)^n c_j$. For any given mode, it is of interest to know whether that mode is significant in the initial condition (c_j significantly large) and how much the term $\left(e^{-b(2\pi j \Delta x)\Delta t}\right)^n c_j$ has damped the coefficient.

Let U denote the FFT of u_k^0, $k = 1, \cdots, M - 1$ (the Matlab function fft was used to calculate the FFT's presented here, but any FFT package or coding of equation (3.2.7) is sufficient). We then define the **power spectral density** of u^0 to be the vector $Pu = |U|/M$ where $|U|$ is the vector containing the magnitude of the elements of the vector U. We note that the power spectral density is a measurement of the energy at various frequencies and provides us with a measure of which modes are significantly present in the Fourier representation of u^0. In practice, due to the symmetries present in Pu, we plot only half of the vector (which contains information about all of the modes present).

In Figure 7.3.2 we see that the nontrivial part of the power spectrum associated with $\Delta x = 0.05$ spreads over more than half of the interval $[0, \pi]$ while in Figure 7.3.3, the

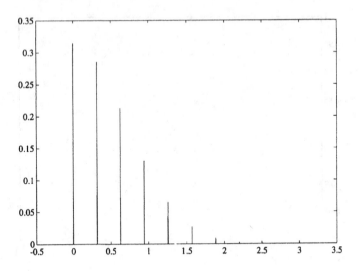

FIGURE 7.3.2. Plot of the power spectral density of u_k^0, $k = 1, \cdots, M - 1$ associated with $M = 20$, $\Delta x = 0.05$.

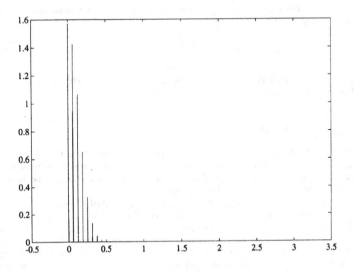

FIGURE 7.3.3. Plot of the power spectral density of u_k^0, $k = 1, \cdots, M - 1$ associated with $M = 100$, $\Delta x = 0.01$.

nontrivial part of the power spectrum associated with $\Delta x = 0.01$ is confined to the first 15% of the the interval $[0, \pi]$. Thus we see that *the reason that making Δx smaller improves the error due to dissipation is that by decreasing Δx, we move the significant modes in the computation (the modes with significantly large coefficients in the finite Fourier series representation) into the far left of the interval $[0, \pi]$ where the dissipation is much less.*

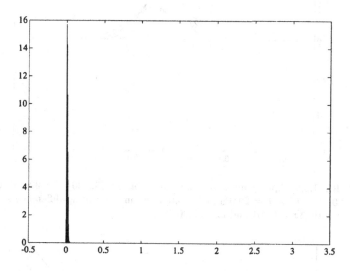

FIGURE 7.3.4. Plot of the power spectral density of u_k^0, $k = 1, \cdots, M - 1$ associated with $M = 1000$, $\Delta x = 0.001$.

For example, we see in Figure 7.3.4 that the power spectrum associated with $\Delta x = 0.001$ is contained in the far left side of the interval. In Figure 7.3.5 we see that though the results at $t = 40.0$ are not as good as we might like them to be, they are much better than those associated with $\Delta x = 0.01$. We should also be aware that we could make Δx smaller yet and further improve our results.

Hence we see that by reducing Δx we reduce the error due to dissipation in the calculation (this should not generally surprise us). Hopefully, we also understand the mechanism involved in dissipation error. At times, the above approach will let us predict the appropriate Δx necessary to give acceptable results (on either the problem of interest or on a model problem). At least, we should be aware of the fact that we do have error due to dissipation and what it looks like in our solution.

(c) We also saw in part (a) that there should be dispersion in the computed results shown in Figures 7.1.1–7.1.4 and Figure 7.3.5. At this time we do not know what dispersion should look like in computational results, but we should expect to see some waves moving away from the others (because they are going at the wrong speed). The first appearance of dispersion generally looks like the oscillation seen in Figure 7.1.6. Our calculations seem to indicate that there is none of this, that there is no dispersion. In fact, we pointed out in Section 7.1.1 that it was "good news" that our scheme approximated the wave speed well.

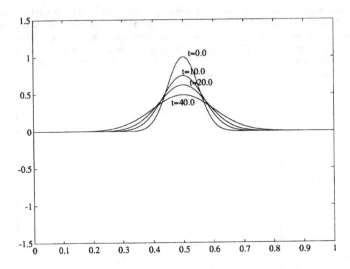

FIGURE 7.3.5. Approximate solutions to problem (5.6.30)–(5.6.32) at times $t = 0.0$, $t = 10.0$, $t = 20.0$ and $t = 40.0$ obtained by using difference scheme (5.3.1) with $\Delta x = 0.001$ and $\Delta t = 0.0008$.

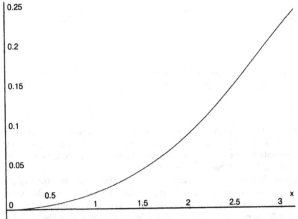

FIGURE 7.3.6. Plot of error in the propagation speed for the FTFS difference scheme (7.3.9) for $R = -0.8$.

If we return to our results in part (a), we see that when $R = -0.8$, the computed speed of the wave should be faster that the actual wave speed for low frequency waves (small $\beta\Delta x$). We also see that the wave speeds of the high frequency waves ($\beta\Delta x$ near π) are very inaccurate (since the high frequency waves are essentially not moving).

In Figure 7.3.1 we plotted $e^{-b\Delta t}$ associated with difference scheme (7.3.9) to show how the solutions dissipate. In Figure 7.3.6 we plot the error in the speed of propagation of the $\beta \Delta x$ mode

$$
\begin{aligned}
a - (-\alpha/\beta) &= -1.0 - \frac{1}{\beta \Delta t} \tan^{-1}\left\{ \frac{.8 \sin \beta \Delta x}{1 - 1.6 \sin^2 \frac{\beta \Delta x}{2}} \right\} \\
&= -1.0 + \frac{1}{.8 \beta \Delta x} \tan^{-1}\left\{ \frac{.8 \sin \beta \Delta x}{1 - 1.6 \sin^2 \frac{\beta \Delta x}{2}} \right\}
\end{aligned}
$$

for $0 \leq \beta \Delta x \leq \pi$. We note that when the error in the speed of a wave mode is significantly great, the dissipation of the mode is also significantly great. Hence, the modes that try to move at the wrong speed get damped out. It is commonly the case that *when we have dissipation in the problem or dissipation in the numerical scheme, the dissipation often helps to hide any dispersion that is present.*

And finally, another potential way to improve the dissipative and dispersive properties of our results is to change R. In Figures 7.3.7 and 7.3.8 we plot $e^{-b\Delta t}$ and the dispersive error for difference scheme (7.3.9) for various values of R. We see that there is no dispersion error when $R = -0.5$ and $R = -1.0$. We knew that this would be the case for $R = -1.0$ since we saw earlier that when $R = -1.0$, the difference scheme gives the exact solution to the model partial differential equation. (Because this occurs only for our "easy" model equation, we consider it "cheating" to use $R = -1.0$ for this problem.) Though for $R = -0.5$ we get no dispersive error, we see in Figure 7.3.7 that the difference scheme is very dissipative when $R = -0.5$. We note that the approach to reduce both the error due to dispersion and the dissipation is to choose R so that $| R |$ is as large as possible. Generally, *results are made better by choosing R as near to the stability limit as possible.*

Remark 1: In Example 7.3.2(a) above, we used an asymptotic analysis to investigate the error in the speed of propagation of various modes for small and large $\beta \Delta x$. In parts (b) and (c) of Example 7.3.2 we used graphics to aid us in analyzing both the dissipation and dispersion error of the scheme. The advantage of the asymptotic method is obviously that we obtain a formula that approximates the error. Two advantages of the graphical technique are that it is not necessary to make any assumptions concerning the size of $\beta \Delta x$ and it is easy to get a large amount of useful information with very little work.

Remark 2: Because it is necessary to plot the graph of a singular function, when we plot the errors in the speed of propagation, we have used Maple (any other symbolic manipulator should be sufficient). It is more difficult to accomplish this using a numerical package.

Remark 3: When we plotted the error in propagation speed (Figures 7.3.6 and 7.3.8), it was necessary to solve

$$
\tan \alpha \Delta t = \frac{-R \sin \beta \Delta x}{1 + 2R \sin^2 \beta \Delta x/2}.
$$

We must be careful about which branch of the *arctan* function we want. *We do not always want the principal value.* In this case, it is easy to return to expression (7.3.12)–(7.3.13). We see that

$$
\sin \alpha \Delta t = (-R \sin \beta \Delta x)/e^{-b\Delta t}
$$

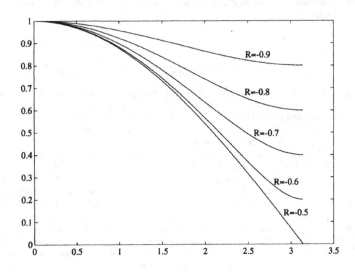

FIGURE 7.3.7. Plot of $e^{-b\Delta t}$ associated with the FTFS difference scheme (7.3.9) for $R = -0.5, \; -0.6, \; -0.7, \; -0.8, \; -0.9$.

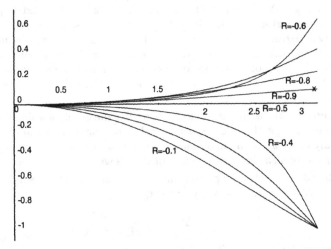

FIGURE 7.3.8. Plot of error in propagation speed associated with the FTFS difference scheme (7.3.9) for $R = -0.1, \; -0.2, \; -0.3, \; -0.4, \; -0.5, \; -0.6, \; -0.7, \; -0.8$ and -0.9.

and

$$\cos \alpha \Delta t = (1 + R - R\cos \beta \Delta x)/e^{-b\Delta t}$$
$$= \left(1 + 2R\sin^2 \frac{\beta \Delta x}{2}\right)/e^{-b\Delta t}.$$

Since for $R = -0.8$ (and any other negative R) $-R\sin \beta \Delta x \geq 0$, $\alpha \Delta t$ must be in the first or second quadrants. As $1 + 2R\sin^2 \beta \Delta x/2$ changes sign from positive to negative,

$$\tan^{-1}\left\{\frac{-R\sin \beta \Delta x}{1 + 2R\sin^2 \beta \Delta x/2}\right\}$$

will pass through $\pi/2$. For this example, the range of $\tan^{-1}$ is taken to be $[0, \pi]$.

Remark 4: In Figure 7.3.7 we have included dissipation information for $R = -0.5$, -0.6, -0.7, -0.8 and -0.9. It should be noted that the curves associated with $R = -0.1$, -0.2, -0.3 and -0.4 are the same as those for $R = -0.9$, -0.8, -0.7 and -0.6. Hence, we do not plot them twice. However, the curves representing the error in propagation speed are not the same.

HW 7.3.1 Use equation (7.3.14) to show that for $R = -1.0$ and $R = -0.5$, $(-\alpha/\beta) = -1.0$.

HW 7.3.2 Verify that the results given in Figure 7.3.8 agree with the asymptotic results derived in part (a) of Example 7.3.2.

Before we perform the dispersion analysis on the Crank-Nicolson scheme, we make several observations that make these analyses easier (some of which we already used in the last example). If we look back to how the discrete dispersion relation was obtained, it is clear that the discrete dispersion relation is given by

$$e^{i\omega \Delta t} = \rho(\beta \Delta x), \qquad (7.3.22)$$

where ρ is the symbol of the difference scheme. A quick look back at Chapter 3 shows that the discrete dispersion relations found in Examples 7.3.1 and 7.3.2 ((7.3.5) and (7.3.10)) could be found much easier using this relationship. Expression (7.3.22) above shows how ρ completely determines the dissipative and dispersive properties of the scheme. To discuss the dissipativity of the scheme, we compute

$$e^{-b\Delta t} = |\rho(\beta \Delta x)|$$

or

$$b = -\frac{1}{\Delta t}\log|\rho(\beta \Delta x)|.$$

Also, when we discuss the dispersivity of the scheme, we compute

$$\tan \alpha \Delta t = \frac{\mathcal{I}m\ \rho(\beta \Delta x)}{\mathcal{R}e\ \rho(\beta \Delta x)} \tag{7.3.23}$$

where $\mathcal{I}m(z)$ and $\mathcal{R}e(z)$ denote the imaginary and real part of the complex number z. Hence, all of the information we get from analyzing the real and imaginary discrete dispersion relations has always been available to us through the symbol of the difference scheme. This information must be extracted and analyzed.

We next consider the Crank-Nicolson scheme for solving hyperbolic partial differential equations of the form (7.2.3).

Example 7.3.3 (a) Analyze the dissipative and dispersive qualities of the Crank-Nicolson scheme

$$u_k^{n+1} + \frac{R}{4}(u_{k+1}^{n+1} - u_{k-1}^{n+1}) = u_k^n - \frac{R}{4}(u_{k+1}^n - u_{k-1}^n) \tag{7.3.24}$$

where $R = a\Delta t/\Delta x$.

(b) Use the dissipation and dispersion results found in part (a) to explain the computational results found in Section 7.1.2.

Solution: Dissipation We begin by returning to Section 5.4.4 to see that the symbol for difference scheme (7.3.24) is given by

$$\rho(\xi) = \frac{1 - \frac{iR}{2}\sin\xi}{1 + \frac{iR}{2}\sin\xi}. \tag{7.3.25}$$

It is then easy to see that

$$e^{-b\Delta t} = \mid \rho(\beta \Delta x) \mid = 1.$$

Hence, *the Crank-Nicolson scheme is nondissipative.*

Dispersion If we then apply (7.3.23), we see that the real discrete dispersion relation is given by

$$\tan \alpha \Delta t = \frac{-R\sin\beta\Delta x}{1 - \frac{R^2}{4}\sin^2\beta\Delta x} \tag{7.3.26}$$

and the error in the speed of propagation of the β mode is given by

$$a - (-\alpha/\beta) = a - \frac{-a}{R\beta\Delta x}\tan^{-1}\left\{\frac{-R\sin\beta\Delta x}{1 - \frac{R^2}{4}\sin^2\beta\Delta x}\right\}. \tag{7.3.27}$$

Of course, this error can be analyzed either using asymptotic expansions or using the graphical approach. In Figure 7.3.9 we include plots of the errors in the speed of propagation for $0 \leq \beta\Delta x \leq \pi$ and $R = 0.2, 0.4, 0.6, 0.8, 1.0, 1.5, 2.0, 5.0$ and 10.0. In Figure 7.3.9, it is easy to see that the errors in propagation speed of the various waves can be great. As is most often the case, the see that the high frequency waves tend to propagate with the most inaccurate speed.

(b) We now return to the computations done with the Crank-Nicolson scheme that were presented in Section 7.1.2. Though when we inspect Figure 7.1.6 it appears that there may be some dissipation error, from part (a) it is clear that *all of the inaccuracies in the results presented in Figures 7.1.5–7.1.7 are due to dispersion (and not dissipation).*

The oscillation trailing the main wave in Figure 7.1.6 is a classical dispersion error. Generally, those are the types of wiggles that are seen when a scheme is dispersive. We

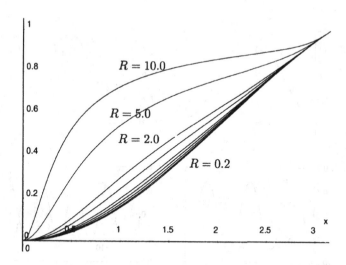

FIGURE 7.3.9. Plot of error in propagation speed associated with the Crank-Nicolson scheme (7.3.24) for $R = 0.2$, 0.4, 0.6, 0.8, 1.0, 1.5, 2.0, 5.0 and 10.0.

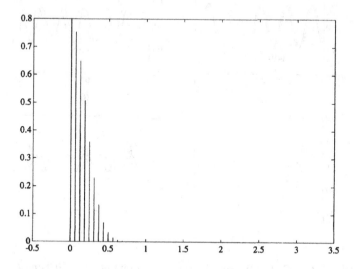

FIGURE 7.3.10. Plot of the power spectral density of the function $u_k^0 = \sin^{80}(\pi k \Delta x)$, $k = 0, \cdots, M - 1$ associated with $M = 100$, $\Delta x = 0.01$.

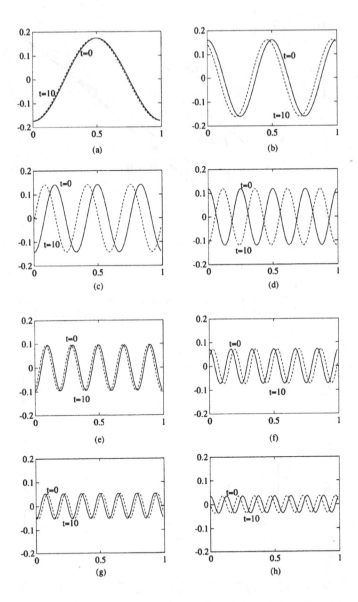

FIGURE 7.3.11. Plot of the contribution to the solution at $t = 0.0$ and $t = 10.0$ made by the modes associated with $j \pm 1, \cdots, \pm 8$. The contributions at $t = 0.0$ and $t = 10.0$ are represented by solid line — and a dashed line - - -, respectively.

see in Figure 7.1.7 that after 1250 time steps, enough dispersion has occurred so that the resulting approximate solution shows little resemblance to the exact solution.

If we return to Figure 7.3.9 and concentrate on the curve associated with $R = 0.8$, we see that there are clearly reasonably large errors in the propagation of many of the modes involved in the computation. Recall that if the finite Fourier series expansion of u^0, $k = 1, \cdots, M - 1$ is given by

$$u_k^0 = \sum_{j=-(M-2)/2}^{M/2} c_j e^{2\pi i j k \Delta x} \tag{7.3.28}$$

(for M even), then the solution at time $t = n\Delta t$ is given by

$$u_k^n = \sum_{j=-(M-2)/2}^{M/2} c_j e^{2\pi i j [k\Delta x - (-\alpha(2\pi j \Delta x)/(2\pi j))n\Delta t]} \tag{7.3.29}$$

(M odd). If we take the fast Fourier transform (FFT) of $u_k^0 = \sin^{80}(\pi k \Delta x)$, $k = 0, \cdots, M - 1$ and plot the power spectral density of u^0 (for $M = 100$), we see in Figure 7.3.10 that only the first ten modes are relevant to the computation. The speeds of propagation of these modes are 0.9991, 0.9965, 0.9922, 0.9863, 0.9787, 0.9695, 0.9589, 0.9468, 0.9334 and 0.9188, respectively. In Figure 7.3.11 we plot eight of these modes (which hoperfully are enough to make the point) as they appear in the solution at time $t = 0.0$ (which is also how they would also appear in the mathematical solution at $t = 10.0$) and how they appear in the approximate solution at time $t = 10.0$. Specifically, we are plotting

$$c_j e^{2\pi i j k \Delta x} + c_{-j} e^{-2\pi i j k \Delta x}$$

and

$$c_j e^{2\pi i j [k\Delta x - (-\alpha(2\pi j \Delta x)/2\pi j)10]} + c_{-j} e^{-2\pi i j [k\Delta x - (-\alpha(-2\pi j \Delta x)/(-2\pi j))10]}$$

for $j = 1, \cdots, 8$ and where the c_j's, $j = \pm 1, \cdots, \pm 8$ are determined by evaluating the FFT of the initial function u^0.

As can easily be seen in Figure 7.3.11, there are large phase errors in several of these modes and that the phase errors are different for each mode. These phase errors are due to the dispersion error in the scheme and are the cause of the bad results obtained in Figure 7.1.7. To verify that these modes *do* represent the solution of difference equation (7.3.24), in Figure 7.3.12 we plot the sum of the modes given in Figure 7.3.11 plus the 0-mode (which we did not plot because it is very uninteresting). We note that the results are the same as that given in Figure 7.1.7. *Figures 7.3.11 and 7.3.12 give a graphic illustration of the mechanism involved in dispersion errors in difference schemes.*

Now that we understand why we have such a bad solution at $t = 10.0$ in Figure 7.1.7, we must consider what we can do to obtain a better solution (if that is possible). The approach for improving the solution obtained by the Crank-Nicolson scheme (assuming that we still want $R = 0.8$) is to make Δx smaller. As was the case with the dissipation error in Example 7.3.2, making Δx smaller moves the relevant modes to the left in the interval $[0, \pi]$ where the error in the speed of propagation is smaller. In Figure 7.3.13 we plot the power spectral density of $u_k^0 = \sin^{80}(\pi k \Delta x)$, $k = 0, \cdots, M - 1$ associated with $M = 1000$ (and $\Delta x = 0.001$). In this case the errors in the speed of propagation of the first ten modes are 1.0000, 1.0000, 0.9999, 0.9999, 0.9999, 0.9997, 0.9996, 0.9994, 0.9993, 0.9991. Hence it is not surprising that the solutions given in Figure 7.3.14 are very accurate.

And finally, also as was the case with the dissipation error given in Example 7.3.2, the error in the Crank-Nicolson calculation can also be changed (and potentially improved) by choosing a different value of R. For example, we note from Figure 7.3.9, if we choose R smaller, the error in the speed of propagation will be smaller (though not much smaller). If we were to compute the solution of HW5.6.5 with $M = 100$ and $R = 0.2$

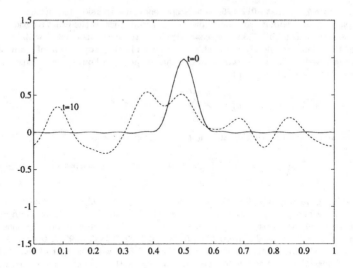

FIGURE 7.3.12. Plot of the sum of the modes presented in Figure 7.3.11 plus the zeroth mode and the ninth and tenth associated with $t = 0.0$ and $t = 10.0$.

($\Delta x = 0.01$ and $\Delta t = 0.002$), we would see that though the results are better than those presented in Figure 7.1.7, they are not much better. And though these results would still be unacceptable, they would take approximately four times as much computational time. See HW7.3.5.

Remark: The reason that we were able to correct the results obtain in Sections 7.1.1 and 7.1.2 was that there were very few relevant modes contained in the spectrum of both initial conditions and these modes fell far to the left in the $[0, \pi]$ interval. It is reasonable to argue that this would be a much more difficult task if we had solutions with a broad banded spectrum. It is generally true that to obtain good results from a difference scheme computation, the mesh size must be small enough so that the highest frequency modes are not important to the solution. It is nearly impossible to resolve all of the modes present in the calculation involving a broadly banded spectrum to a high degree of accuracy.

HW 7.3.3 Perform an asymptotic analysis on the propagation speed for the Crank-Nicolson difference scheme (as was done for difference scheme (7.3.9) in Example 7.3.2(a)).

HW 7.3.4 (a) Calculate the finite Fourier transform of the function $u_k^0 = \sin^{80}(2\pi j/M)$, $j = 0, \cdots, M - 1$ for $M = 1000$.
(b) Using the Fourier coefficients found in part (a), verify that the function can be represented in terms of the first ten modes.

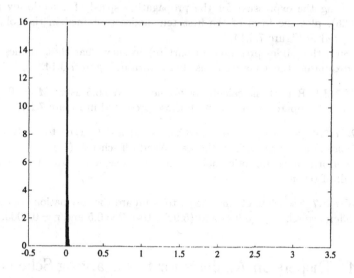

FIGURE 7.3.13. Plot of the power spectral density of the function $u_k^0 = \sin^{80}(\pi k \Delta x)$, $k = 0, \cdots, M - 1$ associated with $M = 1000$, $\Delta x = 0.001$.

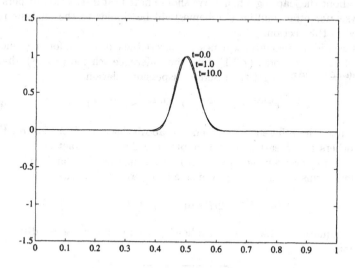

FIGURE 7.3.14. Crank-Nicolson solutions to the problem in HW5.6.5 at times $t = 0.0$, $t = 1.0$ and $t = 10.0$ with $\Delta x = 0.001$ and $\Delta t = 0.0008$.

(c) Using the expression for the propagation speed of a mode given in (7.3.26), plot the form of the first ten modes contained in the solution presented in Figure 7.3.14.

(d) Sum the modes presented in part (c) to show that these modes give approximately the same result as that shown in Figure 7.3.14.

HW 7.3.5 Repeat the calculations done in HW5.6.5 using $M = 100$ and $R = 0.2$. Compare the results with those presented in Figure 7.1.7.

HW 7.3.6 (a) Use plots of $| \rho(\beta \Delta x) |$ and $a - (-\alpha/\beta)$ to analyze the dissipation and dispersion of the Lax-Wendroff scheme, (5.3.8).

(b) Perform an asymptotic analysis on the propagation speed for the Lax-Wendroff scheme.

HW 7.3.7 Use plots of $| \rho(\beta \Delta x) |$ to compare the dissipation contained in difference schemes (5.9.4) and (5.9.5). Use $R = 0.5$ and $r = 0.0005$.

7.4   Dispersion Analysis for the Leapfrog Scheme

In this section we perform a dissipation-dispersion analysis for the leapfrog scheme for solving partial differential equation (7.2.2). As we shall see the results will be different from those of the two level schemes and will teach us a lot about the leapfrog scheme. We should note that dissipation-dispersion analyses for other multilevel schemes will be similar to the analysis presented in this section.

As we did for two level schemes, we insert the expression for the general discrete Fourier mode, (7.3.1), into the difference scheme (6.4.18), divide by $e^{i(\omega n \Delta t + \beta k \Delta x)}$ to get the discrete dispersion relation

$$e^{i\omega \Delta t} = e^{-i\omega \Delta t} - R(e^{i\beta \Delta x} - e^{-i\beta \Delta x}). \tag{7.4.1}$$

Of course, the differences between the above discrete dispersion relation and others found earlier in the chapter are due to the fact that we now have a three level scheme. If we multiply through by $e^{i\omega \Delta t}$ and replace the last two terms by the appropriate sine term, we are left with

$$\left(e^{i\omega \Delta t}\right)^2 + 2iR\sin(\beta \Delta x)e^{i\omega \Delta t} - 1 = 0. \tag{7.4.2}$$

Hence, equation (7.4.2) gives the following version of the discrete dispersion relation:

$$e^{i\omega \Delta t} = \pm\sqrt{1 - R^2 \sin^2 \beta \Delta x} - iR\sin \beta \Delta x. \tag{7.4.3}$$

We note that the expression on the right hand of equation (7.4.3) is the same as that for the eigenvalues of the amplification matrix associated with the leapfrog scheme, Example 6.4.3 (with ξ replaced by $\beta\Delta x$). We should also realize that though we could put it off until later (where we would still get that the scheme is unstable for $R \geq 1$), we make our lives a little easier by assuming that $R < 1$ so that $1 - R^2 \sin^2 \beta\Delta x \geq 0$ for all $\beta\Delta x$.

Before we proceed, we emphasize that because there are two solutions to equation (7.4.2), *the leapfrog scheme will have two waves associated with each wave number β.* We discuss these two waves more later, but for the moment we will proceed to study the dissipation and dispersion of each of these waves.

Since $| iR \sin \beta\Delta x \pm \sqrt{1 - R^2 \sin^2 \beta\Delta x} | = 1$ for all $\beta \in [0, \pi]$, $e^{-b\Delta t} = 1$ and *the leapfrog scheme is nondissipative.* Hence, like the Crank-Nicolson scheme, the leapfrog scheme is one of the few schemes that will propagate the wave forms without dissipation. We see that the real discrete dispersion relation is given by

$$e^{i\alpha\Delta t} = \pm\sqrt{1 - R^2 \sin^2 \beta\Delta x} - iR \sin \beta\Delta x \qquad (7.4.4)$$

or

$$\alpha_\pm = -\frac{1}{\Delta t} \tan^{-1} \frac{R \sin \beta\Delta x}{\pm\sqrt{1 - R^2 \sin^2 \beta\Delta x}}. \qquad (7.4.5)$$

An examination of equation (7.4.5) will show that the two β modes associated with the leapfrog solution of equation (7.2.3) have wave speeds that are the negative of each other. We know that the wave speed associated with partial differential equation (7.2.3) is a. Clearly, it will be impossible for $\alpha_+ = -\alpha_-$ to both approximate the wave speed a.

Using Taylor series expansions of $1/\sqrt{1 - z}$, $\sin z$ and $\tan^{-1} z$, we see that for small $\beta\Delta x$

$$-\alpha_\pm/\beta = \pm a + \mathcal{O}((\beta\Delta x)^2).$$

Hence, α_+, called the **principal root**, is the root that must approximate the propagation speed of our solution. For the leapfrog scheme to be useful (which it is), the other root, α_-, referred to as the **parasitic root**, must not contribute significantly to the solution. As we must do so often, we next try to explain when and why the leapfrog scheme will give good results. Specifically, we will try to explain why, even though the solution of the difference scheme has two modes moving in opposite directions, these do not pollute the computational results. We begin this discussion with the following example which includes periodic boundary conditions and the leapfrog scheme along with the explicit centered scheme as the initialization scheme.

Example 7.4.1 Use discrete separation of variables to solve the problem

$$u_k^{n+1} = u_k^{n-1} - R(u_{k+1}^n - u_{k-1}^n), \ k = 1, \cdots, M-1 \tag{7.4.6}$$

$$u_0^{n+1} = u_0^{n-1} - R(u_1^n - u_{M-1}^n), \tag{7.4.7}$$

$$u_M^{n+1} = u_0^{n+1} \tag{7.4.8}$$

$$u_k^0 = f_k, \ k = 0, \cdots, M \tag{7.4.9}$$

$$u_k^1 = f_k - (R/2)(f_{k+1} - f_{k-1}), \ k = 0, \cdots, M. \tag{7.4.10}$$

Solution: We begin as we did in Section 3.2, set $u_k^n = X_k T^n$, insert $u_k^n = X_k T^n$ into equations (7.4.6)–(7.4.8), separate the variables and get

$$\frac{T^{n+1} - T^{n-1}}{T^n} = -R\frac{X_{k+1} - X_{k-1}}{X_k} = \lambda \tag{7.4.11}$$

$$= -R\frac{X_1 - X_{M-1}}{X_0} \tag{7.4.12}$$

$$X_M = X_0 \tag{7.4.13}$$

where λ is a constant. Hence, we must solve

$$-R(X_{k+1} - X_{k-1}) = \lambda X_k, \ k = 1, \cdots, M-1 \tag{7.4.14}$$

$$-R(X_1 - X_{M-1}) = \lambda X_0, \tag{7.4.15}$$

$$X_0 = X_M \tag{7.4.16}$$

and

$$T^{n+1} - T^{n-1} = \lambda T^n. \tag{7.4.17}$$

It is not hard to see that the solution to (7.4.14)–(7.4.16) is the same as finding the eigenvalues and eigenvectors of the matrix

$$Q = \begin{pmatrix} 0 & -R & 0 & \cdots & 0 & R \\ R & 0 & -R & 0 & \cdots & \\ & & \ddots & \ddots & & \\ & \cdots & 0 & R & 0 & -R \\ -R & 0 & \cdots & 0 & R & 0 \end{pmatrix}. \tag{7.4.18}$$

It is easy to check that the eigenvalues and associated eigenvectors of Q are

$$\lambda_j = -2iR\sin(2\pi j/M), \ j = 0, \cdots, M-1 \tag{7.4.19}$$

and

$$\mathbf{u}_j = \begin{bmatrix} X_0^j \\ X_1^j \\ \vdots \\ X_{M-1}^j \end{bmatrix} = \begin{bmatrix} 1 \\ e^{2\pi ij/M} \\ e^{4\pi ij/M} \\ \vdots \\ e^{2\pi ij(M-1)/M} \end{bmatrix}, \ j = 0, \cdots, M-1. \tag{7.4.20}$$

(One way to find these eigenvalues and eigenvectors is to realize that the solution will involve a finite Fourier series and choose the basis functions accordingly.)

To solve equation (7.4.17), we first solve the characteristic equation

$$x^2 - \lambda_j x - 1 = 0$$

(and get $x_{\pm_j} = \pm\sqrt{1 - R^2 \sin^2(2\pi j/M)} - iR\sin(2\pi j/M))$ and set

$$T_j^n = A_j x_{+_j}^n + B_j x_{-_j}^n, \tag{7.4.21}$$

where A_j and B_j are arbitrary constants for $j = 0, \cdots, M - 1$.

Hence, the solution to equations (7.4.6)–(7.4.10) can be written as

$$u_k^n = \sum_{j=0}^{M-1} X_k^j (A_j x_{+_j}^n + B_j x_{-_j}^n). \tag{7.4.22}$$

To determine A_j and B_j we use the initial conditions (7.4.9) and (7.4.10). Setting $n = 0$ in equation (7.4.22), we get

$$u_k^0 = f_k = \sum_{j=0}^{M-1} X_k^j (A_j + B_j) \tag{7.4.23}$$

If we let $\hat{f}_j$ represent the j-th Fourier coefficient of $\mathbf{f}$, we see that

$$A_j + B_j = \hat{f}_j, \; j = 0, \cdots, M - 1. \tag{7.4.24}$$

If we insert equation (7.4.22) into initial condition (7.4.10), we get

$$u_k^1 = f_k - (R/2)(f_{k+1} - f_{k-1}) = \sum_{j=0}^{M-1} X_k^j (A_j x_{+_j} + B_j x_{-_j}). \tag{7.4.25}$$

If $\hat{f}_j$ represents the j-th Fourier coefficient of $\mathbf{f}$, then $(1 - iR\sin 2\pi j/M)\hat{f}_j$ is the j-th Fourier coefficient of $\{f_k - (R/2)(f_{k+1} - f_{k-1})\}$. Using this fact along with equation (7.4.25) gives us another equation for A_j and B_j,

$$A_j x_{+_j} + B_j x_{-_j} = (1 - iR\sin(2\pi j/M))\hat{f}_j. \tag{7.4.26}$$

Solving equations (7.4.24) and (7.4.26) for A_j and B_j yields

$$A_j = \frac{(1 - iR\sin(2\pi j/M))\hat{f}_j - \hat{f}_j x_{-_j}}{x_{+_j} - x_{-_j}} \tag{7.4.27}$$

$$B_j = \frac{\hat{f}_j x_{+_j} - (1 - iR\sin(2\pi j/M))\hat{f}_j}{x_{+_j} - x_{-_j}} \tag{7.4.28}$$

or

$$A_j = \left(\frac{1}{2} + \frac{1}{2\sqrt{1 - R^2 \sin^2 2\pi j/M}}\right)\hat{f}_j \tag{7.4.29}$$

$$B_j = \left(\frac{1}{2} - \frac{1}{2\sqrt{1 - R^2 \sin^2 2\pi j/M}}\right)\hat{f}_j. \tag{7.4.30}$$

Hence, equation (7.4.22) along with expressions (7.4.29) and (7.4.30) provide a solution to problem (7.4.6)–(7.4.10).

Remark: Comparing equations (7.4.4) and the form of $x_{\pm_j}$, we note that $x_{\pm_j} = e^{i\alpha_\pm(2\pi j/M)\Delta t}$. Hence, an alternate way of writing the solution (7.4.22) is

$$u_k^n = \sum_{j=0}^{M-1} X_k^j (A_j e^{i\alpha_+(2\pi j/M)n\Delta t} + B_j e^{i\alpha_-(2\pi j/M)n\Delta t}) \tag{7.4.31}$$

and we see that the $x_\pm$ terms correspond to the $\alpha_\pm$ waves discussed earlier.

We now return to our discussion of the principal and parasitic modes in the solutions to the leapfrog scheme. We expand the term

$$\frac{1}{2\sqrt{1 - R^2 \sin^2(2\pi j/M)}} = \frac{1}{2} + \frac{1}{4}R^2 \sin^2(2\pi j/M) + \cdots$$

$$= \frac{1}{2} + \mathcal{O}((2\pi j/M)^2). \tag{7.4.32}$$

Combining equations (7.4.29) and (7.4.30) with expansion (7.4.32) gives

$$A_j = \left(1 + \mathcal{O}((2\pi j/M)^2)\right)\hat{f}_j \tag{7.4.33}$$

$$B_j = \left(\mathcal{O}((2\pi j/M)^2)\right)\hat{f}_j. \tag{7.4.34}$$

Thus we see that the solution to (7.4.6)–(7.4.10) can be written as

$$\begin{aligned}
u_k^n &= \sum_{j=0}^{M-1} X_k^j\left(1 + \mathcal{O}((2\pi j/M)^2)\right)\hat{f}_j x_{+j}^n \\
&\quad + \sum_{j=0}^{M-1} X_k^j\left(\mathcal{O}((2\pi j/M)^2)\right)\hat{f}_j x_{-j}^n \tag{7.4.35} \\
&= \sum_{j=0}^{M-1} \left(1 + \mathcal{O}((2\pi j/M)^2)\right)\hat{f}_j e^{2\pi i j[k\Delta x - (-\alpha_+(2\pi j/M)/2\pi j)n\Delta t]} \\
&\quad + \sum_{j=0}^{M-1} \left(\mathcal{O}((2\pi j/M)^2)\right)\hat{f}_j e^{2\pi i j[k\Delta x - (-\alpha_-(2\pi j/M)/2\pi j)n\Delta t]}. \tag{7.4.36}
\end{aligned}$$

Clearly, the approximate solution to the one way wave equation with periodic boundary conditions using leapfrog scheme (7.4.6)–(7.4.10) consists of two distinct solutions. The first is associated with the principal root and the modes in the solution move with speed

$$(-\alpha_+/\beta) = a + \mathcal{O}((2\pi j/M)^2),$$

and the second is associated with the parasitic root, has a small coefficient $(\mathcal{O}((2\pi j/M)^2)$ (hence does not contribute much to the solution), and the modes in the solution move with speed

$$(-\alpha_-/\beta) = -a + \mathcal{O}((2\pi j/M)^2).$$

In Figures 7.4.1–7.4.2 we plot some of the results from HW6.4.3. Since $M = 20$ gives crudely correct results up to $t = 0.8$ and a mess at $t = 20.0$

we do not even bother to include those results. When we use $M = 100$, we get very nice looking results, correct at $t = 0.8$ and incorrect at $t = 20.0$ (though not as bad as the Crank-Nicolson results). As we have done before, in Figure 7.4.3 we include a plot at $t = 20.0$ computed using $M = 1000$ and $R = 0.8$. We note that we are able to obtain excellent results, even after $25,000$ time steps.

To explain the dispersion errors seen in Figures 7.4.1 and 7.4.2, in Figure 7.4.4 we plot the error in propagation speeds for the leapfrog scheme. Of course, we give the errors associated with the principal root α_+. We note that there is significant error in the wave speeds for larger β (and, as before, we obtained excellent results when we used $M = 1000$ because all of the relevant modes were associated with small values of $\beta\Delta x$). Also, we note that when we compare Figure 7.4.4 with Figure 7.3.9, we see that the dispersive errors for the leapfrog scheme are better than those for the Crank-Nicolson scheme.

Remark 1: We note that in the computations discussed in this section, we had $a = 1.0$, found that the principal root was α_+ and considered $R = 0.2, \cdots, 0.9$ in Figure 7.4.4. We should realize that if we considered an equation of the form of $v_t + av_x = 0$ with $a < 0$, say $a = -1$, then α_- would be the principal root and we would consider negative values of R. Other than these changes, all results would be analogous.

Remark 2: We saw in the computations above that the dispersion at time $t = 20.0$ can be corrected by taking a smaller Δx (and then for stability we are also required to take a smaller Δt). One of the other ways that we can try to correct the dispersive error is to consider a higher order scheme. See HW7.4.4.

HW 7.4.1 (a) Compare numerical results at $t = 20.0$, $M = 100$ and $R = 0.8$ between the Crank-Nicolson scheme and the leapfrog scheme. Explain the differences.
(b) Repeat part (a) with $M = 1000$.

HW 7.4.2 Plot the error in propagation speed for the leapfrog scheme associated with the parasitic root α_-.

HW 7.4.3 (a) Repeat HW5.6.3 using the leapfrog scheme.
(b) Analyze the dissipation and dispersion for the calculation done in part (a).

HW 7.4.4 (a) Use plots of $a - (-\alpha/\beta)$ to analyze the dispersion in the following difference scheme.

$$u_k^{n+1} = u_k^{n-1} - R\delta_0 u_k^n + \frac{R}{6}\delta^2\delta_0 u_k^n$$

(See HW2.3.3 and HW6.4.1.) Compare the dispersive properties of this

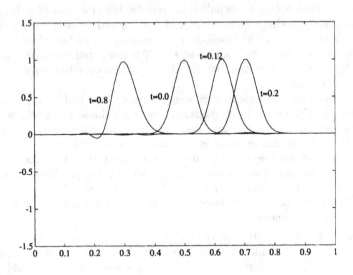

FIGURE 7.4.1. Leapfrog solutions to the problem in HW6.4.3 at times $t = 0.0$, $t = 0.12$, $t = 0.2$ and $t = 10.0$ with $\Delta x = 0.01$ and $\Delta t = 0.008$.

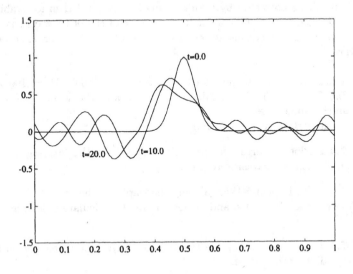

FIGURE 7.4.2. Leapfrog solutions to the problem in HW6.4.3 at times $t = 0.0$, $t = 10.0$ and $t = 20.0$ with $\Delta x = 0.01$ and $\Delta t = 0.008$.

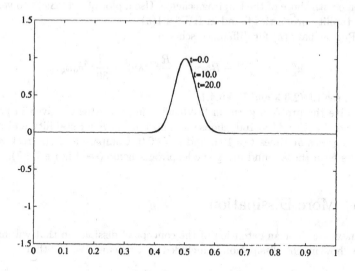

FIGURE 7.4.3. Leapfrog solutions to the problem in HW6.4.3 at times $t = 0.0$, $t = 10.0$ and $t = 20.0$ with $\Delta x = 0.001$ and $\Delta t = 0.0008$.

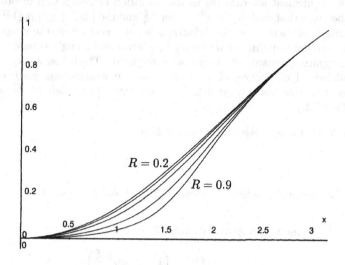

FIGURE 7.4.4. Plot of the error of propagation speed associated with the principal root of the leapfrog scheme, (6.4.18), for $R = 0.2$, 0.4, 0.6, 0.8 and 0.9.

scheme with those of the leapfrog scheme. Use a plot of $|\rho(\beta\Delta x)|$ to verify the stability properties found in HW6.4.1(a).

(b) Repeat part (a) for difference scheme

$$u_k^{n+1} = u_k^{n-1} - R\delta_0 u_k^n + \frac{R}{6}\delta^2\delta_0 u_k^n - \frac{R}{30}\delta^4\delta_0 u_k^n.$$

Again see HW2.3.3 and HW6.4.1.

(c) Solve the problem given in HW6.4.3 using the schemes given in parts (a) and (b). Use $M = 100$, choose an appropriate R for stability and plot the solutions at times $t = 10.0$ and $t = 20.0$. Compare and contrast your results with those found using the leapfrog scheme (see Figure 7.4.2).

7.5   More Dissipation

We next consider an extension of the concept of dissipation that will measure "how much" dissipation is included in a scheme. We say that

(i) a scheme is **dissipative of order** $2s$ if

$$|\rho(\xi)| \leq \left(1 - c\xi^{2s}\right)$$

for some $c > 0$ (independent of Δx, Δt and ξ) and all $\xi \in [-\pi, \pi]$.

Two equivalent alternatives to the definition of dissipation of order $2s$ can be given that replace the ξ^{2s} by $\sin^{2s}\frac{\xi}{2}$ and the $|\rho(\xi)|$ by $|\rho(\xi)|^2$. The advantage of these alternate definitions is that most often, the magnitude of the symbol contains the necessary sine terms and an ugly square root (if the magnitude squared of the symbol is not used). The following equivalent definitions of dissipative of order $2s$ make our calculations much easier. (The proof that these three definitions are equivalent is left to the reader in HW7.5.1.)

(i') A scheme is dissipative of order $2s$ if

$$|\rho(\xi)| \leq \left(1 - c'\sin^{2s}\frac{\xi}{2}\right)$$

for some $c' > 0$ (independent of Δx, Δt and ξ) for all $\xi \in [-\pi, \pi]$.

(i'') A scheme is dissipative of order $2s$ if

$$|\rho(\xi)|^2 \leq \left(1 - c''\sin^{2s}\frac{\xi}{2}\right)$$

for some $c'' > 0$ (independent of Δx, Δt and ξ) for all $\xi \in [-\pi, \pi]$.

It is then easy to see that

- for fixed r, $r \leq 1/2$, the FTCS difference scheme (7.3.3) (with symbol $\rho(\xi) = 1 - 4r \sin^2 \xi/2$) is dissipative of order 2,

- the Lax-Wendroff scheme (where the symbol ρ satisfies $|\rho(\xi)|^2 = 1 - 4R^2(1 - R^2) \sin^4 \xi/2$) is dissipative of order 4 for fixed R, $0 < R < 1$,

- the FTFS scheme, (5.3.1), (where the symbol ρ satisfies $|\rho(\xi)|^2 = 1 + 4R(1+R) \sin^2 \xi/2$) is dissipative of order 2 for fixed R, $-1 < R < 0$,

- the FTBS scheme, (5.3.4), is dissipative of order 2 for fixed R, $0 < R < 1$,

and

- the Lax-Friedrichs scheme, Table 5.1, is dissipative of order 2 for $|R| < 1$.

We note that for the calculations given above, the parameters r and R are assumed to be held constant (so they do not depend on Δt and Δx) during the computation.

Some of the information that we obtain from the order of dissipation of a scheme is how quickly the dissipation increases as $\beta \Delta x$ increases in the interval $[0, \pi]$. In Figure 7.3.1 we saw that as the plot of $e^{-b\Delta t}$ left the point $(0, 1)$, it dropped quickly. For that reason, it was necessary to choose Δx sufficiently small so that all of the relevant modes in the calculation were in the region where the dissipation was very small. A fourth order scheme would come out of the point $(0, 1)$ flatter, thus allowing the relevant modes to have larger wave numbers. Of course, we must always remember that a scheme that is order 4 that has a large c may not be much better than an order 2 scheme with a small c. But, generally, *for $p < s$, plots of $e^{-b\Delta t}$ for schemes dissipative of order $2s$ will leave the point $(0, 1)$ like $1 - c(\beta \Delta x)^{2s}$ which will be flatter than plots for schemes dissipative of order $2p$ which will leave $(0, 1)$ like $1 - c(\beta \Delta x)^{2p}$.*

Everything that we have done with dissipation has been for one dimensional scalar equations. It is clear that we also have and/or want dissipation in multi-dimensional schemes and schemes for solving systems of partial differential equations. To keep this discussion brief, we consider a general difference scheme for solving a system of partial differential equations in m independent variables. We assume that the scheme is in the form

$$\mathbf{u}^{n+1} = Q\mathbf{u}^n \tag{7.5.1}$$

and that $G = G(\boldsymbol{\xi})$ is the $K \times K$ amplification matrix for the scheme where $\boldsymbol{\xi}$ represents the vector $\boldsymbol{\xi} = [\xi_1, \cdots, \xi_m]^T$ (where for the two dimensional problems considered earlier, $\boldsymbol{\xi} = [\xi, \eta]^T$). We say that difference equation (7.5.1) is **nondissipative** if the eigenvalues of G, $\lambda_j = \lambda_j(\boldsymbol{\xi})$, $j = 1, \cdots, K$, satisfy $|\lambda_j| = 1$, $j = 1, \cdots, K$ for all $\xi_j \in [-\pi, \pi]$, $j = 1, \cdots, m$. The scheme is said to be **dissipative** if $|\lambda_j(\boldsymbol{\xi})| < 1$ for some j and some $\boldsymbol{\xi}$. Difference scheme (7.5.1) is said to be **dissipative of order** $2s$ if

$$|\lambda_j(\boldsymbol{\xi})| \leq 1 - c \parallel \boldsymbol{\xi} \parallel^{2s}$$

for some $c > 0$ (independent of Δx_j, $j = 1, \cdots, m$, Δt and $\boldsymbol{\xi}$) for all $\xi_j \in [-\pi, \pi]$, $j = 1, \cdots, m$. (We should realize that we will also have two equivalent alternatives to the above inequality, analogous to (i), (i'), and (i'').)

Hence, again we see that

- if $r_x + r_y \leq 1/2$, difference scheme (4.2.5) is dissipative order 2,

- difference scheme (4.2.30) is dissipative order 2,

- if $0 < R_x + R_y < 1$, difference scheme (5.8.9) is dissipative order 2,

- if $R_x^2 + R_y^2 < 1$, the two dimensional Lax-Friedrichs difference scheme (5.8.13) is dissipative order 2,

- the Beam-Warming difference scheme, (5.8.34)–(5.8.35), is nondissipative,

- if $|\mu_j \Delta t / \Delta x| < 1$, $j = 1, \cdots, K$, the Lax-Wendroff scheme, (6.2.45), is dissipative order 4,

- the two dimensional leapfrog difference scheme, (6.7.42), is nondissipative for $|R_x| + |R_y| < 1$ and

- the Beam-Warming difference scheme for systems, (6.7.29)–(6.7.30), is nondissipative.

And finally, before we leave this section, we note that there should be a connection between dissipation and stability. We state, without proof, the following theorem due to Kreiss, ref. [6].

Theorem 7.5.1 *Suppose that difference scheme (7.5.1) is accurate order $2s - 1$ or $2s - 2$ and dissipative of order $2s$. Then the scheme is stable.*

We note immediately that Theorem 7.5.1 applies to the FTFS scheme, FTBS scheme, the Lax-Wendroff scheme and the Lax-Friedrichs scheme.

HW 7.5.1 Prove that the three definitions of dissipative order $2s$ ((i), (i') and (i'')) are equivalent.

7.6 Artificial Dissipation

In the past we have approached several new schemes by taking a previous scheme that turned out to be unstable and trying to stabilize it. Both Dufort-Frankel and Lax-Wendroff were schemes of this sort. In this section we discuss altering schemes by adding dissipation. Artificial dissipation is used both to make an unstable scheme stable and to make a neutrally stable scheme ($\rho(\xi) \equiv 1$) more stable. The latter situation is faced often when we try to use a neutrally stable scheme to solve a nonlinear partial differential equation. Often, some extra dissipation is needed.

Let us begin by considering the unstable scheme

$$u_k^{n+1} = u_k^n - \frac{R}{2}\delta_0 u_k^n \tag{7.6.1}$$

for the partial differential equation

$$u_t + a u_x = 0. \tag{7.6.2}$$

We should recall that difference scheme (7.6.1) was also the starting point when we developed the Lax-Wendroff scheme. We alter difference scheme (7.6.1) by adding a term $\delta^2 u_k^n$ in an attempt to stabilize the scheme. We then have the difference scheme

$$u_k^{n+1} = u_k^n - \frac{R}{2}\delta_0 u_k^n + \epsilon\delta^2 u_k^n, \tag{7.6.3}$$

where ϵ is still free to be determined. An easy consistency calculation shows that in order for difference scheme (7.6.3) to be consistent with partial differential equation (7.6.2), we must have $\epsilon = \epsilon_1\Delta t$. In this case, we see that the scheme will be at least (we haven't determined ϵ_1 yet) order $\mathcal{O}(\Delta t) + \mathcal{O}(\Delta x^2)$. Of course, we know that if we choose $\epsilon = R^2/2$, we get the Lax-Wendroff scheme that is second order in both space and time.

To consider the stability of difference scheme (7.6.3), return to the stability result given in Example 3.1.3. Since we clearly want the hyperbolic part of the difference equation to dominate, we see that if we require that ϵ satisfy

$$\frac{R^2}{2} \le \epsilon \le \frac{1}{2},$$

then difference scheme (7.6.3) will be stable. We note that if we choose

- $\epsilon = 1/2$ we get the Lax-Friedrichs scheme, Table 5.1,

- $\epsilon = R^2/2$ we get the Lax-Wendroff scheme, (5.3.8) and

- $\epsilon = R/2$ we get the FTBS scheme, (5.3.4).

Of course, there are other difference schemes where it is sometimes helpful to add some artificial dissipation. For example, we have seen that both the Crank-Nicolson scheme, (7.3.24), and the leapfrog scheme, (6.4.18), are neutrally stable. There are times when we are so near instability, that round off, or in the case of the leapfrog scheme, parasitic modes can cause ripples in our solution that do not go away. One way to alleviate this problem is to add some artificial dissipation to the schemes. Obviously, we do not want to add so much that we either lower the order of the scheme or dissipate the waves enough so that the wave propagation properties of the schemes are lost. For example, we might consider the following two modifications of the Crank-Nicolson scheme

$$u_k^{n+1} + \frac{R}{4}(u_{k+1}^{n+1} - u_{k-1}^{n+1}) = u_k^n - \frac{R}{4}(u_{k+1}^n - u_{k-1}^n) + \epsilon\Delta t\delta^2 u_k^n, \tag{7.6.4}$$

$$u_k^{n+1} + \frac{R}{4}(u_{k+1}^{n+1} - u_{k-1}^{n+1}) = u_k^n - \frac{R}{4}(u_{k+1}^n - u_{k-1}^n) - \epsilon\Delta t\delta^4 u_k^n, \tag{7.6.5}$$

(where $\delta^4 u_k^n = \delta^2(\delta^2 u_k^n) = u_{k+2}^n - 4u_{k+1}^n + 6u_k^n - 4u_{k-1}^n + u_{k-2}^n$) and the following two modifications of the leapfrog scheme

$$u_k^{n+1} = u_k^{n-1} - R\delta_0 u_k^n + \epsilon\Delta t\delta^2 u_k^{n-1}, \tag{7.6.6}$$

$$u_k^{n+1} = u_k^{n-1} - R\delta_0 u_k^n - \epsilon\Delta t\delta^4 u_k^{n-1}. \tag{7.6.7}$$

It is not hard to see that both schemes are still $\mathcal{O}(\Delta t^2) + \mathcal{O}(\Delta x^2)$. Also, very little thought will show that *both modifications of the Crank-Nicolson scheme, (7.6.4) with $\epsilon\Delta t \leq 1/2$ and (7.6.5) with $\epsilon\Delta t \leq 1/8$, will still be unconditionally stable.* (See HW7.6.1.) In Example 7.6.1 below, we show that *the fourth order variation of the leapfrog difference scheme, (7.6.7), is stable if $R^2 \leq 1-16\epsilon\Delta t$.* In HW7.6.2, we see that *the second order variation of the leapfrog scheme is stable if $R^2 \leq 1 - 4\epsilon\Delta t$.*

Example 7.6.1 Discuss the stability of the fourth order variation of the leapfrog scheme, (7.6.7).

Solution: To analyze the stability of difference scheme (7.6.7) completely, it would be necessary to reproduce all of the work done in Example 6.4.3. Instead, we give a brief version, leaving the details and the parts omitted to the reader in HW7.6.3.

We begin by rewriting difference scheme (7.6.7) as

$$\mathbf{U}_k^{n+1} = \begin{pmatrix} -R\delta_0 & 1-\epsilon\Delta t\delta^4 \\ 1 & 0 \end{pmatrix} \mathbf{U}_k^n. \tag{7.6.8}$$

We then take the discrete Fourier transform of equation (7.6.8) and obtain the following amplification matrix

$$G(\xi) = \begin{pmatrix} -2iR\sin\xi & 1 - 16\epsilon\Delta t\sin^4\frac{\xi}{2} \\ 1 & 0 \end{pmatrix}. \tag{7.6.9}$$

The eigenvalues of the amplification matrix are given by

$$\lambda_{\pm} = \pm\sqrt{1 - R^2\sin^2\xi - 16\epsilon\Delta t\sin^4\frac{\xi}{2}} - iR\sin\xi. \qquad (7.6.10)$$

If we choose ϵ, Δt and R so that $R^2 \leq 1 - 16\epsilon\Delta t$, then $1 - R^2\sin^2\xi - 16\epsilon\Delta t\sin^4\frac{\xi}{2} \geq 0$. In this case we have $\mid \lambda_{\pm} \mid\leq 1$ and we can finish by mimicking the work done in Example 6.4.3. Hence, *if ϵ, Δt and R satisfy $R^2 \leq 1 - 16\epsilon\Delta t$, then difference scheme (7.6.7) is stable.*

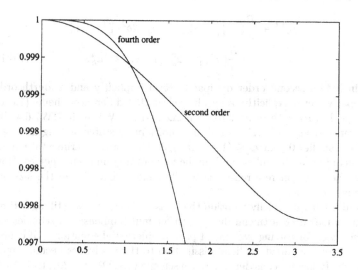

FIGURE 7.6.1. Plots of $e^{-b\Delta t}$ associated with both variations of the leapfrog scheme, (7.6.6) and (7.6.7), with $\epsilon\Delta t = 1/12$.

We note above that either a second order term or a fourth order term (or higher) can be included to stabilize the scheme. Generally, if it is important, the fourth order dissipation term requires a smaller ϵ (this is generally not important). In Figure 7.6.1 we plot $\mid \rho(\xi) \mid$ (or $e^{-b\Delta t}$) for $\epsilon\Delta t = 1/12$ for both variations of the leapfrog scheme. We note that the dissipation results for difference scheme (7.6.7)are much better than for difference scheme (7.6.6). This observation follows from the discussion of dissipative schemes order $2s$ given in Section 7.5. We note that difference schemes (7.6.4) and (7.6.6) are dissipative order 2 while difference schemes (7.6.5) and (7.6.7) are dissipative order 4. When we are considering artificial dissipation, we want the dissipation to stabilize the scheme, but we want as little added as possible. The argument given in Section 7.5 explains that the higher order dissipative schemes will affect fewer of the early modes in the expansion than will the lower order dissipative schemes. Hence, *for the purposes of adding artificial dissipation, we should generally add higher order dissipative terms.*

Another use of artificial dissipation that is sometimes useful is to include some dissipation on the implicit side of the Crank-Nicolson scheme. Of course, when we do this, so that we do not widen the band of the matrix, we include only a second order artificial dissipation (if a fourth order dissipation term were used implicitly, we would have to solve pentadiagonal matrices at each time step). For example, in

$$u_k^{n+1} + \frac{R}{4}(u_{k+1}^{n+1} - u_{k-1}^{n+1}) - \epsilon_i \Delta t \delta^2 u_k^{n+1} = u_k^n$$

$$- \frac{R}{4}(u_{k+1}^n - u_{k-1}^n) - \epsilon_e \Delta t \delta^4 u_k^n, \qquad (7.6.11)$$

we include a second order dissipative term implicitly and a fourth order dissipative term explicitly. Again it is clear that difference scheme (7.6.11) is second order with respect to both Δx and Δt. We see in HW7.6.4 that difference scheme (7.6.11) is still unconditionally stable as long as $\epsilon_i \geq 0$ and ϵ_e satisfies $0 \leq \Delta t \epsilon_e \leq (1 + 2\Delta t \epsilon_i)/8$. There are experimental reasons for including both explicit and implicit smoothing on such schemes. However, the only concrete reason is to allow more flexibility in the size of ϵ that is allowable.

And finally, we briefly mention that it is obvious that we still sometimes want and/or need artificial dissipation for multi-dimensional schemes and for schemes for solving systems of partial differential equations. Of course, analysis of the added artificial dissipation to these schemes becomes more difficult. Below we consider difference scheme (6.7.29)–(6.7.30), the two dimensional Beam-Warming scheme for solving the system of partial differential equations (6.7.2) (with $C_0 = 0$). As we did with the one dimensional, scalar Crank-Nicolson scheme, we include both second order implicit dissipation and and fourth order explicit dissipation to get

$$\left(I - \frac{R_x}{4}A_1\delta_{x0} - \Delta t \epsilon_i I \delta_x^2\right)\mathbf{u}_{jk}^* = \left(I + \frac{R_x}{4}A_1\delta_{x0} - \Delta t \epsilon_e I \delta_x^4\right) \times$$

$$\left(I + \frac{R_y}{4}A_2\delta_{y0} - \Delta t \epsilon_e \delta_y^4\right)\mathbf{u}_{jk}^n \quad (7.6.12)$$

$$\left(I - \frac{R_y}{4}A_2\delta_{y0} - \Delta t \epsilon_i I \delta_y^2\right)\mathbf{u}_{jk}^{n+1} = \mathbf{u}_{jk}^* \qquad (7.6.13)$$

A motivation for difference scheme (7.6.12)–(7.6.13) is to note that difference scheme

$$\left(I - \frac{R_x}{4}A_1\delta_{x0} - \Delta t \epsilon_i I \delta_x^2\right)\left(I - \frac{R_y}{4}A_2\delta_{y0} - \Delta t \epsilon_i I \delta_y^2\right)\mathbf{u}_{jk}^{n+1} =$$

$$\left(I + \frac{R_x}{4}A_1\delta_{x0} - \Delta t \epsilon_e I \delta_x^4\right)\left(I + \frac{R_y}{4}A_2\delta_{y0} - \Delta t \epsilon_e \delta_y^4\right)\mathbf{u}_{jk}^n$$

(equivalent to difference scheme (7.6.12)–(7.6.13)) is an approximate factorization of the Crank-Nicolson scheme (admittedly a strange approximate factorization),

$$\left(I - \frac{R_x}{4}A_1\delta_{x0} - \frac{R_y}{4}A_2\delta_{y0}\right)\mathbf{u}_{jk}^{n+1} = \left(I + \frac{R_x}{4}A_1\delta_{x0} + \frac{R_y}{4}A_2\delta_{y0}\right)\mathbf{u}_{jk}^{n},$$

i.e. all of the dissipation terms added are higher order. Hence, difference scheme (7.6.12)–(7.6.13) is $\mathcal{O}(\Delta t^2) + \mathcal{O}(\Delta x^2) + \left(\Delta y^2\right)$.

As usual, to be able to obtain a stability result for difference scheme (7.6.12)–(7.6.13), we assume that the matrices are simultaneously diagonalizable. We then take the discrete Fourier transform of equations (7.6.12)–(7.6.13) and diagonalize the the amplification matrix. This procedure reduces the stability to considering the eigenvalues of the amplification matrix,

$$\lambda_j = \frac{(1 + i\frac{R_x}{2}\mu_j\sin\xi - 16\epsilon_e\Delta t\sin^4\frac{\xi}{2})(1 + i\frac{R_y}{2}\nu_j\sin\eta - 16\epsilon_e\Delta t\sin^4\frac{\eta}{2})}{(1 + 4\epsilon_i\Delta t\sin^2\frac{\xi}{2} - i\frac{R_x}{2}\mu_j\sin\xi)(1 + 4\epsilon_i\Delta t\sin^2\frac{\eta}{2} - i\frac{R_y}{2}\nu_j\sin\eta)}.$$

We can then resort to the one dimensional result given for difference scheme (7.6.11) to see that $\epsilon_i \geq 0$ and $0 \leq \Delta t\epsilon_e \leq (1 + 2\Delta t\epsilon_i)/8$ is sufficient to imply that difference scheme (7.6.12)–(7.6.13) is unconditionally stable.

HW 7.6.1 (a) Show that if $\epsilon\Delta t \leq 1/2$, then difference scheme (7.6.4) is stable.
(b) Show that if $\epsilon\Delta t \leq 1/8$, then difference scheme (7.6.5) is stable.

HW 7.6.2 Show that if ϵ, Δt and R satisfy $R^2 \leq 1 - 4\epsilon\Delta t$, then difference scheme (7.6.6) is stable.

HW 7.6.3 Complete the analysis done in Example 7.6.1.

HW 7.6.4 Show that if $\epsilon_i \geq 0$ and $0 \leq \epsilon_e\Delta t \leq (1 + 2\Delta t\epsilon_i)/8$, then difference scheme (7.6.11) is stable.

HW 7.6.5 Repeat the calculation done in HW6.4.3 (and in Section 7.4) ($M = 100$ and $R = 0.8$ outputting the solution at $t = 10.0$ and $t = 20.0$) using both variations of the leapfrog scheme, (7.6.6) and (7.6.7). For both schemes try $\Delta t = 0.005$ and $\epsilon = 1.0$ and $\epsilon = 10.0$. Compare your results with each other and with those computed previously.

HW 7.6.6 Repeat the calculation done in HW7.6.5 using both variations of the Crank-Nicolson scheme, (7.6.4) and (7.6.5). Try $\Delta t = 0.01$ and $\epsilon = 1.0$ and $\epsilon = 100.0$.

HW 7.6.7 Repeat the calculations done in HW7.4.4 using a small amount of fourth order dissipation added to each of the schemes.

7.7 Modified Partial Differential Equation

In the preceding sections, we have been discussing the dispersion and dissipation of the schemes. In this section we introduce another way to view the dissipation and dispersion, including why the particular scheme is dissipative or dispersive and an approximation of how much dissipation or dispersion is contained in the scheme. The approach is to find the **modified partial differential equation** and we will introduce the concept by example.

We begin by considering the equation

$$v_t + av_x = 0 \tag{7.7.1}$$

and the FTFS difference scheme for approximating the solution of partial differential equation (7.7.1),

$$u_k^{n+1} = u_k^n - R\delta_+ u_k^n. \tag{7.7.2}$$

(At the moment we make no assumption on the sign of a). We perform the necessary calculations in the following example.

Example 7.7.1 Find the modified partial differential equation associated with difference scheme (7.7.2).

Solution: We begin by considering a function $u = u(x, t)$ that, when evaluated at the appropriate lattice points, is a solution to difference equation (7.7.2). We expand u in the appropriate Taylor series about the point $(k\Delta x, n\Delta t)$ and insert these series into difference equation (7.7.2) to get

$$
\begin{aligned}
0 &= \frac{u_k^{n+1} - u_k^n}{\Delta t} + a\frac{u_{k+1}^n - u_k^n}{\Delta x} \\
&= (u_t)_k^n + \frac{\Delta t}{2}(u_{tt})_k^n + \frac{\Delta t^2}{6}(u_{ttt})_k^n + \cdots \\
&\quad + a(u_x)_k^n + \frac{a\Delta x}{2}(u_{xx})_k^n + \frac{a\Delta x^2}{6}(u_{xxx})_k^n + \cdots.
\end{aligned} \tag{7.7.3}
$$

Instead of satisfying partial differential equation (7.7.1), $u = u(x, t)$ satisfies partial differential equation (7.7.3). Equation (7.7.3) represents an "infinite order" partial differential equation that is equivalent in some way to our difference equation. Of course, we will truncate equation (7.7.3) so that we are left with a partial differential equation that we understand.

If we truncated the equation and then tried the same approach for analyzing the dissipation and dispersion that we used in Section 7.2, we see that the u_{tt}, u_{ttt}, etc. terms cause problems because it makes the dispersion relation into a quadratic, cubic, etc. equation. To make it easier to determine how the u_{tt} and u_{ttt} terms affect the dissipativity and dispersion, we replace these terms by derivatives with respect to x. We might be tempted to return to equation (7.7.1) and claim that

$$u_{tt} = (-au_x)_t = -a(u_t)_x = -a(-au_x)_x = a^2 u_{xx} \tag{7.7.4}$$

and

$$u_{ttt} = -a^3 u_{xxx}. \tag{7.7.5}$$

This is not the case. Expressions (7.7.4) and (7.7.5) are only true for solutions to equation (7.7.1). The function u *is not a solution of equation (7.7.1), u is a solution of equation (7.7.3)*. For this reason we must work much harder to get expressions for u_{tt} and u_{ttt} (and, as we shall see, the expressions that we use are not the same as equations (7.7.4) and (7.7.5)).

We saw earlier, in Section 7.2, that the u_{xx} term was a dissipative term and the u_{xxx} was a dispersive term. In reality, all of the even order derivatives are dissipative (or they blow up) and all of the odd order derivatives greater than one are dispersive. However, what is generally done in building the modified partial differential equation is to include only the lowest even order term that is greater than any term contained in the original equation and the lowest order odd term greater than or equal to three that is also greater than any odd order term contained in the original equation. The calculation done below will ignore all derivatives of order four or above. It is not obvious that the higher order derivatives will have a negligible affect on the solution. This calculation must be done carefully, making the decision whether or not the higher order derivatives would affect the solution significantly.

To find expressions for u_{tt} and u_{ttt}, we begin by differentiating equation (7.7.3) with respect to t and tt to get

$$0 = (u_{tt})_k^n + \frac{\Delta t}{2}(u_{ttt})_k^n + \frac{\Delta t^2}{6}(u_{tttt})_k^n + \cdots$$
$$+ a(u_{xt})_k^n + \frac{a\Delta x}{2}(u_{xxt})_k^n + \frac{a\Delta x^2}{6}(u_{xxxt})_k^n + \cdots \qquad (7.7.6)$$

and

$$0 = (u_{ttt})_k^n + \frac{\Delta t}{2}(u_{tttt})_k^n + \frac{\Delta t^2}{6}(u_{ttttt})_k^n + \cdots$$
$$+ a(u_{xtt})_k^n + \frac{a\Delta x}{2}(u_{xxtt})_k^n + \frac{a\Delta x^2}{6}(u_{xxxtt})_k^n + \cdots . \qquad (7.7.7)$$

If we were to solve equations (7.7.6) and (7.7.7) for u_{tt} and u_{ttt}, respectively, and use these expressions to eliminate u_{tt} and u_{ttt} from equation (7.7.3), we see that we would have introduced u_{xt}, u_{xxt} and u_{xtt} terms (only up to including third order derivatives). We obtain expressions for u_{xt}, u_{xxt} and u_{xtt} by differentiating equation (7.7.3) with respect to x, xx and xt, and get

$$0 = (u_{tx})_k^n + \frac{\Delta t}{2}(u_{ttx})_k^n + \frac{\Delta t^2}{6}(u_{tttx})_k^n + \cdots$$
$$+ a(u_{xx})_k^n + \frac{a\Delta x}{2}(u_{xxx})_k^n + \frac{a\Delta x^2}{6}(u_{xxxx})_k^n + \cdots , \qquad (7.7.8)$$

$$0 = (u_{txx})_k^n + \frac{\Delta t}{2}(u_{ttxx})_k^n + \frac{\Delta t^2}{6}(u_{tttxx})_k^n + \cdots$$
$$+ a(u_{xxx})_k^n + \frac{a\Delta x}{2}(u_{xxxx})_k^n + \frac{a\Delta x^2}{6}(u_{xxxxx})_k^n + \cdots , \qquad (7.7.9)$$

and

$$0 = (u_{txt})_k^n + \frac{\Delta t}{2}(u_{ttxt})_k^n + \frac{\Delta t^2}{6}(u_{tttxt})_k^n + \cdots$$
$$+ a(u_{xxt})_k^n + \frac{a\Delta x}{2}(u_{xxxt})_k^n + \frac{a\Delta x^2}{6}(u_{xxxxt})_k^n + \cdots . \qquad (7.7.10)$$

If equations (7.7.6)–(7.7.10) are used systematically to eliminate the t derivatives (of order greater than one and less than or equal to three) from equation (7.7.3), we obtain the modifed partial differential equation associated with difference scheme (7.7.2),

$$0 = (u_t)_k^n + a(u_x)_k^n + \frac{a\Delta x}{2}(1+R)(u_{xx})_k^n$$
$$+ \frac{a\Delta x^2}{6}(2R+1)(R+1)(u_{xxx})_k^n + \cdots . \qquad (7.7.11)$$

By eliminating the reference to the point $(k\Delta x, n\Delta t)$ and truncating to include up to the third order derivations, the modified partial differential equation associated with difference scheme (7.7.2) can be written as

$$0 = u_t + au_x + \frac{a\Delta x}{2}(1+R)u_{xx} + \frac{a\Delta x^2}{6}(2R+1)(R+1)u_{xxx}. \qquad (7.7.12)$$

Remark: Though we are trying to solve partial differential equation (7.7.1), we see that if we use difference scheme (7.7.2), we have an approximate solution that comes closer to satisfying partial differential equation (7.7.12). It is clear, while partial differential equation (7.7.1) is neither dissipative nor dispersive, partial differential equation (7.7.12) is both dissipative and dispersive. As we shall see, the behavior of the solution to difference equation (7.7.2) is very similar to the behavior of the solution to partial differential (7.7.12). We consider the following example.

Example 7.7.2 Analyze the dissipative and dispersive properties of partial differential equation (7.7.12).

Solution: If we return to Section 7.2 and consider a partial differential equation of the form

$$u_t + au_x - \nu_1 u_{xx} + cu_{xxx} = 0, \qquad (7.7.13)$$

we see that the dispersion relation is given by

$$\omega = -a\beta + c\beta^3 + i\nu_1\beta^2$$

and β-mode solution will be of the form

$$u(x,t) = \hat{u}e^{-\nu_1\beta^2 t}e^{i\beta[x-(a-c\beta^2)t]}. \qquad (7.7.14)$$

The dissipative term in the solution (7.7.14) is

$$e^{-\nu_1\beta^2 t} = (e^{-\nu_1\beta^2})^t. \qquad (7.7.15)$$

For later comparisons, it is better to let $t = n\Delta t$ and write (7.7.15) as

$$e^{-\nu_1\beta^2 t} = (e^{-\nu_1\beta^2 \Delta t})^n. \qquad (7.7.16)$$

The propagating term in solution (7.7.14) is given by

$$e^{i\beta[x-(a-c\beta^2)t]}.$$

The partial differential equation is dispersive and the speed of propagation of the β-mode is $a - c\beta^2$.

If we then apply this result to partial differential equation (7.7.12), $\nu_1 = -\frac{a\Delta x}{2}(1+R)$ and $c = \frac{a\Delta x^2}{6}(2R+1)(R+1)$, we see that the dissipation is determined by

$$e^{-\nu_1\beta^2 \Delta t} = e^{\frac{a\Delta x}{2}(1+R)\beta^2 \Delta t} \qquad (7.7.17)$$

and the propagating term is given by

$$e^{i\beta[x-(a-\frac{a\Delta x^2}{6}(2R+1)(R+1)\beta^2)t]}.\tag{7.7.18}$$

We first note that if $a > 0$ (and, hence $R > 0$), (7.7.17) raised to the nth power will "blow up" with n and be unstable. Thus we must assume that $a < 0$ and $-1 \le R < 0$ (so that $1 + R$ will be greater than zero). Of course, we realize that we already knew that for difference scheme (7.7.2) to be stable, we must have $a < 0$ and $-1 \le R < 0$. In addition, we note that there is no dissipation for $\beta = 0$, the dissipation depends strongly on Δx and the dissipation will be greater for larger values of β.

When we did a dissipation analysis for difference scheme (7.7.2) in Section 7.3, we saw that the dissipation was determined by $e^{-b\Delta t}$. Thus, the comparison is between b and $\nu_1\beta^2 = -a\Delta x(1 + R)\beta^2/2$. If we return to equation (7.3.11) and use the Taylor series expansions for $\log(1 - z)$ and $\cos z$, we see that

$$\begin{aligned}
b &= -\frac{1}{2\Delta t}\log\left(1 - [-2R(1 + R) + 2R(1 + R)\cos\beta\Delta x]\right)\\
&= -\frac{1}{2\Delta t}\left\{[2R(1 + R) - 2R(1 + R)\cos\beta\Delta x] - \cdots\right\}\\
&= -\frac{1}{2\Delta t}\left\{\left[2R(1 + R) - 2R(1 + R)\left(1 - \frac{(\beta\Delta x)^2}{2} + \cdots\right)\right] - \cdots\right\}\\
&= -\frac{a\Delta x}{2}(1 + R)\beta^2 + \cdots.
\end{aligned}$$

Thus we see that the dissipative term we get via the modified equation approach, $\nu_1\beta^2$, is the first term in the Taylor series expansion of b from the exact dissipation analysis.

And, finally, if we return to expression (7.7.18), we notice that the speed of propagation of the β-mode is

$$a - \frac{a\Delta x^2}{6}(2R + 1)(R + 1)\beta^2,$$

which is the same as that obtained when we used the asymptotic expansion to approximate $-\alpha/\beta$ in Example 7.3.2(see (7.3.17)).

Thus we see that the modified partial differential equation (7.7.12) is both dissipative and dispersive. This dissipation and dispersion approximates the dissipation and dispersion of finite difference scheme (7.7.2). We also see that the dissipation and dispersion associated with modified partial differential equation (7.7.12) is the same as the asymptotic approximation of the exact dissipation and dispersion of finite difference scheme (7.7.2).

Remark: Clearly, we get no new dissipation/dispersion information using the modified equation approach compared to using the exact approach described in Section 7.3. The earlier analysis provided an exact dissipation/dispersion analysis while the modified partial differential equation approach provides a truncated approximate approach. How to perform these analyses is up to the person that must do the work. It is our opinion that the easiest approach is to perform the exact dissipation/dispersion analysis as was done in Section 7.3 and obtain the "results" by graphing $|\rho(\beta\Delta x)|$ and $a - (-\alpha/\beta)$ verses $\beta\Delta x$ for the appropriate R and r values. However, the modified equation approach gives a much clearer illustration of the mechanisms involved in the dissipation and dispersion that we see in the graphical approach. Both approaches have their merits.

The calculation to obtain the modified equation performed in Example 7.7.1 was not very nice. We did the first example that way because the approach contains all of the basic concepts of the method. However, there are easier ways to obtain the modified equation. In the next two examples we show two better ways for obtaining the modified partial differential equation. Both approaches are clearly equivalent to the approach used in Example 7.7.1, but both are easier. We begin by calculating the modified partial differential equation associated with the Crank-Nicolson scheme.

Example 7.7.3 Calculate the modified partial differential equation associated with the Crank-Nicolson scheme, (7.3.24).

Solution: We begin as we did in Example 7.7.1, expand u in the appropriate Taylor series about the point $(k\Delta x, (n + 1/2)\Delta t)$ (the $n + 1/2$ is because that is the point about which we expanded when we did the consistency analysis for the Crank-Nicolson scheme), plug these series into the Crank-Nicolson scheme and get

$$
\begin{aligned}
0 &= \frac{u_k^{n+1} - u_k^n}{\Delta t} + \frac{a}{2}\frac{u_{k+1}^{n+1} - u_{k-1}^{n+1}}{2\Delta x} + \frac{a}{2}\frac{u_{k+1}^n - u_{k-1}^n}{2\Delta x} \\
&= \left(u_t\right)_k^{n+1/2} + \frac{\Delta t^2}{24}\left(u_{ttt}\right)_k^{n+1/2} + a\left(u_x\right)_k^{n+1/2} + \frac{\Delta t^2}{8}\left(u_{xtt}\right)_k^{n+1/2} \\
&\quad + \frac{a\Delta x^2}{6}\left(u_{xxx}\right)_k^{n+1/2} + \mathcal{O}(5)
\end{aligned}
\tag{7.7.19}
$$

where $\mathcal{O}(5)$ *denotes the usual $\mathcal{O}$ notation with all potential products of Δt and Δx, up to a total of five of them.* We should make special note that the $\mathcal{O}(5)$ resulted from ignoring the terms containing fifth derivatives and higher. When the expansion is done, no terms remain that contain second or fourth order derivatives.

The calculation of the modified partial differential equation is now done in Table 7.7.1. We see that Row (a) contains all of the possible terms, up to and including the fourth order (order in terms of derivatives) terms. Row (b) represents the truncated version of equation (7.7.19) by listing the coefficients of each of the terms in equation (7.7.19) in the appropriate columns. As before, we want our result to contain a u_t term and terms involving x-derivatives. We do not want any other t-derivatives. It is easy then to look at Row (b) and see that we need expressions for u_{ttt} and u_{xtt}. Row (c) represents two derivatives of Row (b) with respect to t (so that this derivative acting on the u_t term in Row (b) will produce a u_{ttt} term) multiplied by the appropriate constant so that when Rows (b) and (c) are added together, the u_{ttt} term will add out.

We note that the introduction of Row (c) introduces another u_{xtt} term. Rows (d) and (e) represent two derivatives of Row (b) with respect to xt (so this derivative on the u_t term of Row (b) will produce a u_{xtt} term). Constants are chosen so that when Rows (d) and (e) are added together with Rows (b) and (c), the u_{xtt} terms will add out. Of course, Rows (d) and (e) could have been consolidated into one row. It was done this way for convenience and to lessen the chance for making mistakes. We should notice that Rows (d) and (e) are necessary to eliminate the u_{xtt} terms, but they introduce u_{xxt} terms.

So, finally, Rows (f) and (g) represent the appropriate constants multiplied by two derivatives of Row (b) with respect to x (so with the u_t term, we produce a u_{txx} term) necessary to eliminate the u_{xxt} terms.

Because Row (b) represents the expansion of the difference scheme and we assume that u is a solution to the difference scheme, we can consider all of these rows as equaling zero (the others are just derivatives of an expression that is equal to zero). Hence, we can add Rows (b)–(g) and get the modified partial differential equation associated with the Crank-Nicolson scheme

$$
T_4 : \ 0 = u_t + au_x + \frac{a\Delta x^2}{12}[2 + R^2]u_{xxx}.
\tag{7.7.20}
$$

		u_t	u_x	u_{tt}	u_{xt}	u_{xx}	u_{ttt}	u_{xtt}	u_{xxt}	u_{xxx}	u_{tttt}	u_{xttt}	u_{xxtt}	u_{xxxt}	u_{xxxx}
(a)															
(b)		1	a				$\frac{\Delta t^2}{24}$	$\frac{a\Delta t^2}{24}$		$\frac{a\Delta x^2}{6}$					
(c)	$-\frac{\Delta t^2}{24}\frac{\partial^2}{\partial t^2}(b)$						$-\frac{\Delta t^2}{24}$	$-\frac{a\Delta t^2}{24}$							
(d)	$-\frac{a\Delta t^2}{8}\frac{\partial^2}{\partial x\partial t}(b)$							$-\frac{a\Delta t^2}{8}$	$-\frac{a^2\Delta t^2}{8}$						
(e)	$\frac{a\Delta t^2}{24}\frac{\partial^2}{\partial x\partial t}(b)$							$\frac{a\Delta t^2}{24}$	$\frac{a^2\Delta t^2}{24}$						
(f)	$\frac{a^2\Delta t^2}{8}\frac{\partial^2}{\partial x^2}(b)$								$\frac{a^2\Delta t^2}{8}$	$\frac{a^3\Delta t^2}{8}$					
(g)	$-\frac{a^2\Delta t^2}{24}\frac{\partial^2}{\partial x^2}(b)$								$-\frac{a^2\Delta t^2}{24}$	$-\frac{a^3\Delta t^2}{24}$					

TABLE 7.7.1. Table for computing the modified partial differential equation associated with the Crank-Nicolson scheme.

We should make special note that the fourth order terms were included and do not appear in equation (7.7.20). This, of course, is much stronger than considering an expansion that only includes the third order terms. The T_4 notation included with modified partial differential equation (7.7.20) implies that the equation represents the modified equation that includes derivatives up to and including the fourth order derivatives (so that when none are present, we know that the appropriate coefficient is zero).

And, finally, we emphasize that the approach used in this example is no different from that used in Example 7.7.1. The approach using the table is merely a very systematic way to perform the steps done in Example 7.7.1.

Remark: We note that we can now use equation (7.7.20) to approximately analyze the dissipation and dispersion of the Crank-Nicolson scheme. Immediately, we note that since neither the second order nor the fourth order terms are present, there will be no dissipation through the fourth order derivative terms (and, a careful analysis of the method used above will convince us that there will be no even order terms, hence, no dissipation at all). And, as was the case in Example 7.7.2, we note that the propagating term from equation (7.7.20),

$$e^{i\beta[x-(a-c\beta^2)]} = e^{i\beta[x-(a-\frac{a\Delta x^2}{12}(2+R^2)\beta^2]},$$

will produce exactly the same error in the speed of propagation given by the asymptotic analysis done in HW7.3.3.

The final approach for computing the modified partial differential equation is clearly the best approach. In Example 7.7.4 below, we indicate the approach that can be used to determine the modified partial differential equation using one of the available symbolic manipulators. Once the user becomes familiar with these tools, this is clearly the easiest (and mistake free) approach.

Example 7.7.4 Describe how a symbolic calculation package can be used to determine the modified partial differential equation associated with the Lax-Wendroff scheme, (5.3.8).

Solution: We begin by writing the Lax-Wendroff difference scheme as

$$0 = \frac{u(x, t + dt) - u(x, t)}{dt} + a\frac{u(x + dx, t) - u(x - dx, t)}{2dx}$$
$$- \frac{a^2 dt^2}{2dx^2}(u(x + dx, t) - 2u(x, t) + u(x - dx, t)). \qquad (7.7.21)$$

We should note that this rather strange form is especially useful because it will be necessary to expand the scheme in a Taylor series with respect to dx and dt and differentiate the results with respect to x and t. We now include a pseudo code (which should translate equally well to any of the available symbolic manipulators) for symbolically computing the modified partial differential equation associated with equation (7.7.21).

1. Define LW : Define the function LW to be the right hand side of equation (7.7.21).

2. $LWT = Mtaylor(LW, [dx, dt])$: Defines LWT to be the Taylor series expansion of LW with respect to dx and dt.

3. $LWTT = Diff(LWT, t)$: Defines $LWTT$ to be the derivative of LWT with respect to t.

4. $LWTX = Diff(LWT, x)$

5. $UTT = Solve(LWTT = 0, D[22](u)(x, t))$: Solves $LWTT = 0$ for u_{tt}.

6. $UXT = Solve(LWTX = 0, D[12](u)(x, t))$: Solves $LWTX = 0$ for u_{tx}.

7. $LWTP1 = Substitute(LWT, D[22](u)(x, t) = UTT)$: Substitute the value found above for UTT into LWT.

8. $LWTP2 = Substitute(LWTP1, D[12](u)(x, t) = UXT)$: Substitute the value found above for UXT into $LWTP1$.

9. $LWTTT = Diff(LWT, tt)$: Defines $LWTTT$ to be the derivative of LWT with respect to tt.

10. $LWTTX = Diff(LWT, tx)$: Defines $LWTTX$ to be the derivative of LWT with respect to tx.

11. $LWTXX = Diff(LWT, xx)$: Defines $LWTXX$ to be the derivative of LWT with respect to xx.

12. $UTTT = Solve(LWTTT = 0, D[222](u)(x, t))$: Solves $LWTTT$ for u_{ttt}.

13. $UTTX = Solve(LWTTX = 0, D[221](u)(x, t))$: Solves $LWTTX$ for u_{ttx}.

14. $UTXX = Solve(LWTXX = 0, D[211](u)(x, t))$: Solves $LWTXX$ for u_{txx}.

15. $LWTP3 = Substitute(LWTP2, D[222](u)(x, t) = UTTT)$: Substitute the value found above for $UTTT$ into $LWTP2$ for u_{ttt}.

16. $LWTP4 = Substitute(LWTP3, D[221](u)(x, t) = UTTX)$: Substitute the value found above for $UTTX$ into $LWTP3$ for u_{ttx}.

17. $LWTP5 = Substitute(LWTP4, D[211](u)(x, t) = UTXX)$: Substitute the value found above for $UTXX$ into $LWTP4$ for u_{txx}.

If we run our code, we will have eliminated all of the t derivatives up to the third order (except, of course, for u_t which we want) and obtain

$$T_3 :\ 0 = u_t + au_x + \frac{a\Delta x^2}{6}(1 - R^2)u_{xxx} \qquad (7.7.22)$$

as a candidate for the modified partial differential equation associated with the Lax-Wendroff scheme. We see that the second order derivative does not appear. One might be tempted to think that this scheme is the same as was the case with the Crank-Nicolson scheme. It would even be easier to believe this when we allow the machine to do all of the work and do not have much of a feeling of the process that was followed. To be a careful experimenter and especially because we do not have to do the work, it is best to include the following piece of code that will check to see whether the fourth order terms would appear.

1. $LWTTTT = Diff(LWT, ttt)$: Defines $LWTTTT$ to be the derivative of LWT with respect to ttt.

2. $LWTTTX = Diff(LWT, ttx)$: Defines $LWTTTX$ to be the derivative of LWT with respect to ttx.

3. $LWTTXX = Diff(LWT, txx)$: Defines $LWTTXX$ to be the derivative of LWT with respect to txx.

4. $LWTXXX = Diff(LWT, xxx)$: Defines $LWTXXX$ to be the derivative of LWT with respect to xxx.

5. $UTTTT = Solve(LWTTTT = 0, D[2222](u)(x, t))$: Solves $LWTTTT$ for u_{tttt}.

6. $UTTTX = Solve(LWTTTX = 0, D[2221](u)(x, t))$: Solves $LWTTTX$ for u_{tttx}.

7. $UTTXX = Solve(LWTTXX = 0, D[2211](u)(x,t))$: Solves $LWTTXX$ for u_{ttxx}.

8. $UTXXX = Solve(LWTXXX = 0, D[2111](u)(x,t))$: Solves $LWTXXX$ for u_{txxx}.

9. $LWTP6 = Substitute(LWTP5, D[2222](u)(x,t) = UTTTT)$: Substitute the value found above for $UTTTT$ into $LWTP5$ for u_{tttt}.

10. $LWTP7 = Substitute(LWTP6, D[2221](u)(x,t) = UTTTX)$: Substitute the value found above for $UTTTX$ into $LWTP6$ for u_{tttx}.

11. $LWTP8 = Substitute(LWTP7, D[2211](u)(x,t) = UTTXX)$: Substitute the value found above for $UTTXX$ into $LWTP7$ for u_{ttxx}.

12. $LWTP9 = Substitute(LWTP8, D[2111](u)(x,t) = UTXXX)$: Substitute the value found above for $UTXXX$ into $LWTP8$ for u_{txxx}.

If we include the above piece of code with the first code we gave, we will then extend our modified partial differential equation up to fourth order terms and we get

$$T_4 : \quad 0 = u_t + au_x + \frac{a\Delta x^2}{6}(1 - R^2)u_{xxx} + \frac{a\Delta x^3}{8}R(1 - R^2)u_{xxxx}. \qquad (7.7.23)$$

Hence, we see that the Lax-Wendroff scheme is dissipative (we knew that) and the potential modified partial differential equation given in equation (7.7.22) gives a false impression of the dissipativity of the Lax-Wendroff scheme.

Remark 1: Example 7.7.4 shows that we must be careful how and when we decide that the series can be truncated. Of course, this is easier to do when the computer is doing all of the work for us.

Remark 2: The scheme described in Example 7.7.4 is an non-interactive scheme. It is easy to make this procedure into a function that is easy to apply. At times, depending on the qualities of the symbolic manipulator being used, some methods will have to be used to help simplify the expressions. One of the tools available is a "coefficient" function that will pick off the appropriate coefficients of some very ugly expansions.

Remark 3: Another approach for using symbolic manipulators to find the modified partial differential equation is to follow the approach used in Example 7.7.3, but writing the table on the symbolic manipulator and letting the symbolic manipulator do all of the computations.

Remark 4: Equation (7.7.23) makes an approximate dissipation-dispersion analysis easy (remember the modified equation will always provide only an approximation the the dissipation and dispersion). Clearly, the dispersive term is dominant over the dissipative term, so we should guess that the solution will appear dispersive and not dissipative (or, more dispersive than dissipative). However, when comparing the dispersivity of the Lax-Wendroff scheme to other dispersive schemes (Crank-Nicolson, leapfrog), we should realize that the Lax-Wendroff scheme will have an advantage. Though the dissipation is higher order, there is some dissipation available to dampen the high frequency terms that are moving with the most inaccurate speeds. In fact, the Lax-Wendroff scheme should behave much like either the Crank-Nicolson or the leapfrog scheme with the correct amount of artificial dissipation of order 4 (as defined in Section 7.5). Comparing modified equation (7.7.23) with the equations found in HW7.7.1(c) and Example 7.7.3 (equation (7.7.20)), we see that the speed of propagation of a β-wave for the Lax-Wendroff scheme is the same as that for the leapfrog scheme and is approximately the same as that for the Crank-Nicolson scheme.

HW 7.7.1 (a) Show that the modified partial differential equation associated with the Lax-Friedrichs scheme, Table 5.3.1, is given by

$$T_3 : \quad 0 = u_t + au_x - \frac{a\Delta x}{2R}(1 - R^2)u_{xx} - \frac{a\Delta x^2}{3}(1 - R^2)u_{xxx}.$$

(b) Show that the modified partial differential equation associated with the BTCS scheme, 5.4.6, is given by

$$T_3 : \quad 0 = u_t + au_x - \frac{a\Delta x}{2}Ru_{xx} + \frac{a\Delta x^2}{6}(1 + 2R^2)u_{xxx}.$$

(c) Show that the modified partial differential equation associated with the leapfrog scheme, 6.4.18, is given

$$T_4 : \quad 0 = u_t + au_x + \frac{a\Delta x^2}{6}(1 - R^2)u_{xxx}.$$

HW 7.7.2 Find the modified partial differential equations associated with difference schemes (5.9.4) and (5.9.5). Use the modified partial differential equations to compare the dissipation contained in these two schemes.

7.8   Discontinuous Solutions

Computational problems for hyperbolic partial differential equations often involve trying to compute discontinuous solutions. It should not be hard to believe that this is generally a difficult problem. Because the dissipation and the dispersivity of a scheme strongly affects what the schemes do to jumps in the solution, this seems to be an appropriate time and place to discuss dissipation and dispersion related to trying to resolve discontinuous solutions.

In HW5.6.7 we considered a problem of the form

$$v_t - v_x = 0 \; x \in (0, 1) \tag{7.8.1}$$
$$v(x, 0) = f(x) \; x \in [0, 1] \tag{7.8.2}$$
$$v(0, t) = v(1, t), t \geq 0 \tag{7.8.3}$$

where f was defined as

$$f(x) = \begin{cases} 1 & \text{if } 0.4 \leq x \leq 0.6 \\ 0 & \text{otherwise.} \end{cases} \tag{7.8.4}$$

In the solution, we found distinctively different results for each of the three schemes. When we used the FTFS scheme, we saw that the solution was damped some (that should not surprise us) and smeared. When we used the Lax-Wendroff scheme, the solution was oscillatory, with a small oscillation following the step. And, finally, when we used the Lax-Friedrichs scheme, we had a damped solution that looked a lot like that from the FTFS scheme, with a highly oscillatory wave superimposed on it. See Figures 7.8.1–7.8.3.

It should be clear that the results in Figure 7.8.1 are due to dissipation that we have described earlier in this chapter, the results in Figure 7.8.2 are due to dispersion that we found in Example 7.7.4 and the results in Figure 7.8.3 are due to dissipation plus "something else." Because, at the moment we only have one "something else," we might guess that the highly oscillatory wave superimposed on the damped wave is due to dispersion.

What we will try to show in the remainder of this section is how and why the results described above are related to dissipation and dispersion. Solutions containing jump discontinuities are very important in applied mathematics. We must be able to compute such solutions, and we must understand why and when we will have problems.

Example 7.8.1 Explain the solution behavior for the solution found by using the FTFS scheme, (5.3.1), illustrated in Figure 7.8.1.

Solution: If we return to the results shown in Figure 7.1.2, we should be surprised. The amplitude of the result given in Figure 7.8.1 has damped very little compared to the result given in Figure 7.1.2 (and the result given in Figure 7.1.2 has only run to $t = 0.8$). If we return to the plot of $e^{-b\Delta t}$ associated with the FTFS scheme (which is the same plot that is relevant here), we see that it must be the case that the modes determining the amplitude of the solution given in Figure 7.8.1 must be very near the $\beta\Delta x = 0$ axis. In Figure 7.8.4 we show the power spectral density of the function f. If we compare the power spectrum in Figure 7.8.4 with that given in Figure 7.3.3 (the power spectrum associated with the computational results given in Figure 7.1.2), we see that values plotted in Figure 7.8.4 are originally greater than those in Figure 7.3.3, get small faster than those in Figure 7.3.3 and spread out more than those in Figure 7.3.3. *The reason that the amplitude in Figure 7.8.1 is not damped as much as that in Figure 7.1.2 is because more (more than in Figure 7.1.2) of this amplitude is carried by the first few modes which do not damp as much as the later modes.*

The fact that the jumps in the step function are badly smeared are due to the fact that the spectral density of the step function, Figure 7.8.4, is spread out more than that in Figure 7.3.3. If we were to compute the finite Fourier series for the step function f and include only K of the modes (compute the series as we did in equation (7.3.20) and then consider

$$\sum_{j=-K}^{K} c_j e^{2\pi i j k \Delta x}),$$

we would see that the truncated series does a poor job at refining the corners of the step. *The jumps are smeared in Figure 7.8.1 because the sharpness in the jumps are due to*

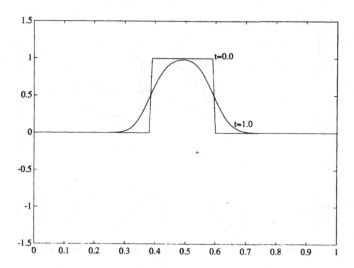

FIGURE 7.8.1. Approximate solution to problem (7.8.1)–(7.8.3) at time $t = 1.0$ found by using the FTFS scheme, (5.3.1).

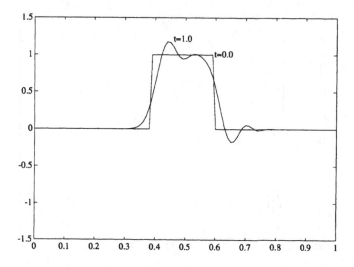

FIGURE 7.8.2. Approximate solution to problem (7.8.1)–(7.8.3) at time $t = 1.0$ found by using the Lax-Wendroff scheme, (5.3.8).

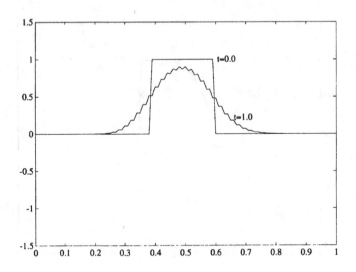

FIGURE 7.8.3. Approximate solution to problem (7.8.1)–(7.8.3) at time $t = 1.0$ found by using the Lax-Friedrichs scheme, (5.3.1).

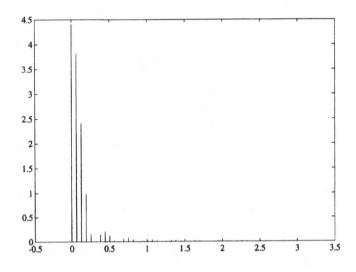

FIGURE 7.8.4. Plot of the power spectral density of the function $u_k^0 = f(k\Delta x)$, $k = 1, \cdots, M - 1$ associated with $M = 100$, $\Delta x = 0.01$.

high frequency modes and all of the high frequency modes are damped out. In fact, if we truncated the exact finite Fourier series solution of finite difference problem (7.3.21) to include only nine terms, we would get the solution shown in Figure 7.8.1. If we truncated the finite Fourier series of our initial function f to nine terms, there would be no jumps in this new "step function." Hence we see that, *the dissipation in the scheme causes the smearing and damping (less than we might have expected). If we are intent on using such a highly dissipative scheme, the only way to lessen the smearing of the jumps is to use a finer grid (and move more of the relevant modes in the finite Fourier series expansion of f nearer to $\beta\Delta x = 0$ where there is less damping.)*

Example 7.8.2 Explain the solution behavior for the Lax-Wendroff scheme, (5.3.8), illustrated in Figure 7.8.2.

Solution: In HW7.3.6 we did a dissipation-dispersion analysis of the Lax-Wendroff scheme. In Figures 7.8.5 and Figures 7.8.6, we plot the dissipation $| \rho(\beta\Delta x) |= e^{-b\Delta t}$ and the error in the propagation speed $a - (-\alpha(\beta\Delta x)/\beta)$ for the Lax-Wendroff scheme with $R = -0.8$. From these plots it is reasonably clear that the dispersion dominates the dissipation.(The dissipation curve comes out of the point $(0, 1)$ very flat (it is dissipative order 4) while the dispersion curve comes out much sharper.) The fact that the dispersion will generally (you are never guaranteed which modes are in your initial condition) dominate the dissipation can more easily be seen by the results of Example 7.7.4. In Example 7.7.4 we saw that the modified partial differential equation associated with the Lax-Wendroff scheme was

$$0 = u_t + au_x + \frac{a\Delta x^2}{6}(1 - R^2)u_{xxx} + \frac{a\Delta x^3}{8}R(1 - R^2)u_{xxxx}.$$

Clearly, the dispersive term is a lower order term than the dissipative term.

It is not trivial to look at Figures 7.8.5 and 7.8.6 or the modified partial differential equation and accurately predict the results seen in Figure 7.8.2. However, it is fairly clear that since we know that the corners (or sharp jumps) of the steps are defined by the high frequency modes, it is pretty clear that these modes will be dispersing badly and the steps will break up badly.

Remark: Note in Figure 7.8.6 that the error in propagation speed changes from negative (in most of the interval $[0, \pi]$) to positive near $\beta\Delta x = \pi$. Hence, we see that the longer wave length modes will be moving faster than they should move while the high frequency modes will be moving slower than they should move. This description is misleading and false. The fact is that the modes with a positive error in speed of propagation are those modes with such a high frequency that they are generally not important in the finite Fourier series expansion of the initial condition. As can be seen in Figure 7.8.4, the amplitudes of the modes in the very high frequency ranges are zero or negligible.

Since $a = -1.0$ in this example, the correct wave moves to the left with velocity -1.0. As can be seen in Figure 7.8.6, the longest modes either have no error or have very little error in their speed of propagation. The modes with $\beta\Delta x$ satisfying $0.5 \leq \beta\Delta x \leq 1.5$ have both significant error in propagation speeds and significant contributions to the initial condition. Since the error in the speed of propagation, $a-(-\alpha/\beta)$, is negative for these mid-range modes, the propagation speeds will satisfy $-1. < -\alpha/\beta < 0$.

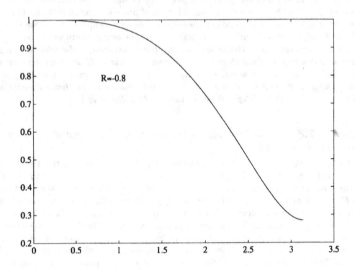

FIGURE 7.8.5. Plot of $e^{-b\Delta t}$ associated with the Lax-Wendroff difference scheme, (5.3.8).

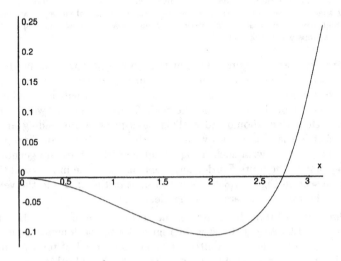

FIGURE 7.8.6. Plot of error in the propagation speed for the Lax-Wendroff scheme, (5.3.8), for $R = -0.8$.

Hence, these modes will lag behind the low frequency modes, which is what we see in Figure 7.8.2 (the "wiggles" are to the right of the main approximation of the step function).

Example 7.8.3 Explain the solution behavior for the Lax-Friedrichs scheme, Table (5.3.1), illustrated in Figure 7.8.3.

Solution: If we look at Figures 7.8.7 and 7.8.8, it is not difficult to believe that the scheme is generally dissipative as the plot given in Figure 7.8.3 shows. As was the case in Example 7.8.1, the step is smeared due to the fact that the high frequency modes necessary to define the sharp edges of the step are damped out.

To understand why there is a highly oscillatory wave superimposed on the solution, we look at Figure 7.8.7. We note that the $\beta \Delta x = \pi$ mode is not damped at all (and the highly oscillatory modes very near $\beta \Delta x = \pi$ are not damped very much). Hence, if the initial condition has such a mode or such a mode is introduced through the calculation, the mode will remain in the solution. The error in propagation speed plot shown in Figure 7.8.8 shows that these high frequency modes will move with a speed very different from the true speed of propagation.

A careful examination of Figures 7.8.8 and 7.8.4 make it unclear that the above explanation is adequate to completely explain the solution given in Figure 7.8.3. It is not clear from Figure 7.8.4 that the initial condition has a sufficiently large contribution from the $j = M/2$ ($\beta \Delta x = \pi$) mode to account for the results shown in Figure 7.8.3. Though the period of the oscillation that appears to be superposed on the smooth curve appears to correspond to the $j = M/2$ mode, the fact that the amplitude of this superposed oscillation is not constant makes it clear that the oscillation is not due to a pure $M/2$ mode.

After some thought (and a few computations to back up our claims), the explanation is clear. In Figure 7.8.9, we plot the power spectrum of u_k^0 for $\beta \Delta x$ from $\pi/2$ to π. Because the plot is now not dominated by the modes near $j = 0$, we see that there are significant modes near $j = M/2$. However, if we take an FFT of u_k^0, we see that the amplitude of the $j = M/2$ mode is not as great as the amplitude of the oscillation in Figure 7.8.3 (and, we have not yet explained the variation in the amplitude of the oscillation).

Solving problem (7.8.1)–(7.8.4) by discrete separation of variables (or using a discrete Fourier series), we see that the solution to problem (7.8.1)–(7.8.4) can be written as

$$u_k^n = \sum_{j=-(M-2)/2}^{M/2} c_j \lambda_j^n e^{2\pi i j k/M}$$

where $\lambda_j = exp[-b(2\pi j \Delta x)]exp[i\alpha(2\pi j \Delta x)]$ and c_j is the j-th coefficient of the discrete Fourier series of u_k^0. Here we have taken the easy way and defined λ_j as in equation (7.3.2). We could have just as well written λ_j as the eigenvalues of the appropriate matrix through the separation of variables process. Because of the dissipativity of the Lax-Friedrichs scheme, all of the mid range modes will be damped out quickly. The solution for early times will be due to the first few modes and the last few modes (where the modes we refer to here are trigonometric modes, not the exponential modes). In Figure 7.8.10, we plot the contribution of the last three modes to the discrete Fourier series expansion of u_k^0, i.e.

$$c_{-48}e^{-2\pi i 48k/M} + c_{-49}e^{-2\pi i 49k/M} + c_{50}e^{2\pi i 50k/M} + c_{49}e^{2\pi i 49k/M} + c_{48}e^{2\pi i 48k/M}.$$

(Since c_{-48}, c_{48} and c_{-49}, c_{49} are complex conjugates of each other, they add together to produce the $j = 48$ and $j = 49$ modes.) Comparing this plot with the plot of the solution (Figure 7.8.3), we see that the oscillation superposed on the solution appears to be due to these last three terms.

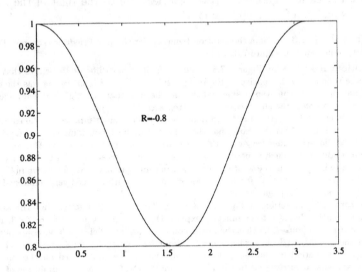

FIGURE 7.8.7. Plot of $e^{-b\Delta t}$ associated with the Lax-Friedrichs difference scheme, (5.3.1).

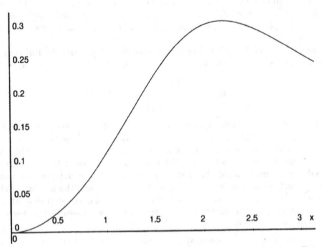

FIGURE 7.8.8. Plot of error in the propagation speed for the Lax-Friedrichs scheme, (5.3.1), for $R = -0.8$.

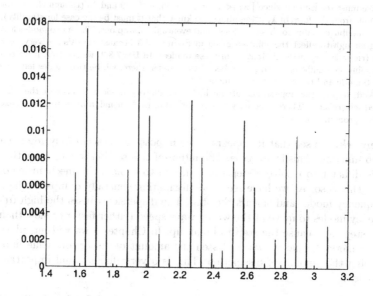

FIGURE 7.8.9. Plot of the modes of the power spectral density of $u_k^0 = f(k\Delta x)$, $k = 1, \cdots, M - 1$ associated with $M = 100$, $\Delta x = 0.01$ that lie between $\pi/2$ and π.

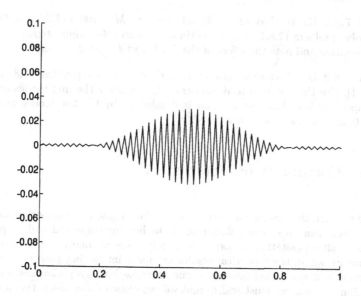

FIGURE 7.8.10. Plot of the contribution of the last three modes of the discrete Fourier series expansion of the step function (7.8.4)

We must realize that since $\mid \lambda_j \mid < 1$ for $j = -48, -49, 49$ and 48 (though this is not obvious from the form of λ_j given above, we know that it must be the case for stability), the contributions due to these modes should eventually damp out. These two modes last long enough to affect the solution given in Figure 7.8.3 because the λ's are very near one (there is very little dissipation in these modes). In HW7.8.2, we see that if we run the solution a sufficiently large number of time steps, there will be two modes left: the $j = 0$ mode and the $j = M/2 = 50$ mode.

And, finally, the appearance of the highly oscillatory mode is a part of the Lax-Friedrichs scheme. There is no easy way to get rid of it. As usual, a finer grid will make it less pronounced.

Remark: We see that it appears to be impossible to accurately resolve a step function. This is not true. The intent of this section was to show that it is difficult to resolve a step function and to explain the mechanisms of why this is so. As we have seen, the dissipation generally damps the high frequency modes and smears the step. The dispersion causes the high frequency modes to move at the wrong wave speed (either faster or slow than the step) and cause the step to break up. In Chapter 9 we will introduce some methods that we try to lessen the amount of dispersion while minimizing the amount of dissipation. This is a very difficult and important problem.

HW 7.8.1 Solve problem (7.8.1)–(7.8.3) where f is defined as in (7.8.4) using the leapfrog scheme, (6.4.18), the leapfrog scheme with dissipation, (7.6.6) and (7.6.7), the BTCS scheme, (5.4.6), and the forward, implicit scheme, (5.4.2). Compare your results with each other and the results found in HW5.6.7).

HW 7.8.2 Use the Lax-Friedrichs scheme with $M = 100$ and $\Delta t = 0.008$ to solve problem (7.8.1)–(7.8.4) to time $t = 40.0$ (5000 time steps). Plot the solution and note the affect of the $j = 0$ and $j = M/2$ modes.

HW 7.8.3 Use discrete separation of variables to solve problem (7.8.1)–(7.8.4). Use this solution to demonstrate how quickly the mid modes are damped out and how the solution is dominated by the few modes near $j = 0$ and the few modes near $j = M/2$.

7.9   Computational Interlude V

One reaction that we may have when we see how numerical dissipation and dispersion can affect our calculations is to become depressed and suspicious of all computational efforts. And surely, there are many results in the literature where the numerical results are more interesting than the real solution and the added interest is due to numerical dissipation and dispersion. Instead, we must realize that we can obtain these dissipation and dispersion errors and use the available analyses to try to eliminate them from our calculations. This will not always be easy or possible, but then we will at least know when our results are not good.

7.9.1   HW0.0.1

In Section 3.4.2 we suggesting using implicit schemes to solve HW0.0.1. We suggested several different combinations of generalizations of the BTCS scheme and the Crank-Nicolson scheme (linearization, linearization about the previous time step and using Newton's method to solve the nonlinear equation). When the results were not acceptable, we argued that we were unable to accurately resolve derivatives across the steep fronts that occur in the solution. That explanation is still plausible given the fact that if Δx and Δt are chosen extremely small (small enough so that only someone without enough to do would wait for the solution), the implicit methods will give good results for the viscous Burgers' problem. However, we would like to believe that some of the tools developed in this chapter should help explain why it is necessary to choose Δx and Δt to be ridiculously small.

We must always remember that when we are developing methods for non-linear problems from analogous linear methods for model linear problems, there is always the chance that we will have difficulties with the schemes that are truly nonlinear in origin. In that case, we will not be able to predict these difficulties based on analysis of linear model equations. However, it is usually smart to at least consider the analogous linear scheme on the linear model problem. If there are potential problems with the linear scheme on the linear model problem (as there was when we used the explicit scheme to solve HW0.0.1 for small values of ν), these will usually cause problems with the nonlinear scheme.

Thus, we are interested in studying the dissipation and dispersion of linear difference schemes (5.9.5) and (3.4.28). As we saw in HW7.7.2, the modified partial differential equation associated with difference scheme (5.9.5) is given by

$$T_4 : 0 = u_t + au_x - \left(\nu + \frac{1}{2}a^2\Delta t\right)u_{xx} + \left(a\nu\Delta t + \frac{1}{3}a^3\Delta t^2 + \frac{1}{6}a\Delta x^2\right)u_{xxx}$$

$$- \left(\frac{1}{2}\nu^2\Delta t + \nu a\Delta t^2 + \frac{1}{12}\nu\Delta x^2 + \frac{1}{4}a^4\Delta t^3 + \frac{1}{6}a^2\Delta t\Delta x^2\right)u_{xxxx}. \quad (7.9.1)$$

Using any of the approaches described in Section 7.7, we see that the modified partial differential equation associated with difference scheme (3.4.28) is

$$T_4 : \ 0 = u_t + au_x - \nu u_{xx} + \frac{a\Delta x^2}{12}(2 + R^2)u_{xxx} - \frac{\nu}{4}(2 + R^2)u_{xxxx}.$$

The consideration of the above modified partial differential equations gives us some information related to the behavior of their nonlinear analogs. Clearly, the Crank-Nicolson scheme, (3.4.28), is very dispersive. When ν is small (smaller than Δx^2), the dispersive term will dominate. Therefore, when the Crank-Nicolson scheme is used, it should not surprise us that waves are moving with many different speeds (and when that happens, the nonlinear effects (if there are to be some) may be more pronounced). However, when we consider the modified partial differential equations associated with difference scheme (5.9.5), as we should have expected, the scheme is dominated by dissipation. For a linear problem, it seems as if difference scheme (5.9.5) would damp out any errant traveling waves. To be completely sure (remembering that the modified partial differential equation gives the dissipation for small $\beta\Delta x$), one should consult the full dissipation analysis of difference scheme (5.9.5) provide for by the plot of $\mid \rho(\beta\Delta x) \mid$ found in HW7.3.7). One last piece of evidence to whether or not the effects seen in the implicit solution to HW0.0.1) are nonlinear effects is to apply the linear model schemes to a problem involving a discontinuity. With the dispersion that we know is present, it should not surprise us that the Crank-Nicolson scheme should have difficulties with such a problem. It should be interesting to see how the BTCS scheme performs on such a problem. See HW7.9.1.

Before we leave this section, we must discuss our last attempt to solve HW0.0.1. We last suggested to solve HW0.0.1 by using a scheme with one sided differences on the convective term so as to treat the equation as if it were a hyperbolic equation and not reach across the steep front. In Figure 7.9.1(a) we can see that it is possible to use an explicit scheme such as (6.9.1) and obtain good results. Clearly, we have accurately resolved the front. However, if we were not so clever as to stop at time $t = 0.21$ and instead ran the solution out to time $t = 0.7$, we see in Figure 7.9.1(b) that the solution gets damped badly. If we assume that the results should be similar to the linear model problem, we would think that the solution should not damp as it does in Figure 7.9.1(b) and that the damping was due to the dissipation of the numerical scheme. We may be especially suspicious of this damping due to the fact that we know that the one sided difference schemes are very dissipative. To investigate whether the solution given in Figure 7.9.1(b) might be the correct solution, we suggest that the above computation be repeated using $M = 1000$ and $\Delta t = 0.0007$. The results shown in Figure 7.9.1 are discussed further in Chapter 9, Part 2.

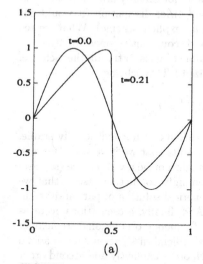
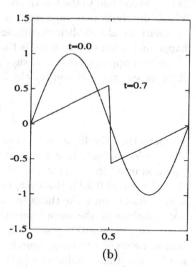

(a)   (b)

FIGURE 7.9.1. Plots of the solutions to Burgers' equation using $M = 100$, $\nu = 0.001$, $\Delta t = 0.007$ and difference scheme (6.9.1) at (a) $t = 0.21$ and (b) $t = 0.7$.

HW 7.9.1 Use difference schemes (5.9.5) and (3.4.28) to solve the following problem.

$$v_t - v_x = \nu v_{xx} \; x \in (0,1)$$
$$v(x,0) = f(x) \; x \in [0,1]$$
$$v(0,t) = v(1,t), t \geq 0$$

where $\nu = 0.00001$ and f was defined as

$$f(x) = \begin{cases} 1 & \text{if } 0.4 \leq x \leq 0.6 \\ 0 & \text{otherwise.} \end{cases}$$

7.9.2   HW0.0.3

In Section 6.9.2 we suggested using the linearized Lax-Wendroff scheme to solve HW0.0.3. The results show significant "wiggles" near each discontinuity that formed. After seeing the results in Example 7.7.4 or HW7.3.6 this should not have surprised us. The leading term in the error associated with the Lax-Wendroff scheme is a dispersive term.

There are several ways that we could try to correct these wiggles (assuming that such interesting wiggles are not a part of the correct solution). The easiest way to try is to add some artificial dissipation to the linearized Lax-Wendroff scheme. We could add either explicit fourth order dissipation (which would add to the fourth order dissipation already present in the Lax-Wendroff scheme) or some explicit second order dissipation (it would not be smart to add implicit dissipation to an explicit scheme). What we see happening when this is done, is that the discontinuities smear badly.

Another approach that is always available to us is to try another scheme. Here we suggest that you try the linearized BTCS scheme

$$\mathbf{u}_k^{n+1} - \frac{R}{2} A^n \delta_0 \mathbf{u}_k^{n+1} = \mathbf{u}_k^n.$$

We note that the linear version of this scheme is unconditionally stable. We should also note that based on the results that we have with Burgers' equation using this scheme, this choice might not be a wise choice (see Sections 3.4.2 and 6.9.1). However, the seemingly conflicting results that we sometimes obtain make the study of numerical solution of partial differential equations all the more interesting. And, finally, if oscillations occur as a part of your solution (especially high frequency oscillations), remember that some second order or fourth order artificial dissipation can be added to eliminate these oscillations (add fourth order explicitly and second order either explicitly or implicitly).

References

[1] Paul DuChateau and David Zachmann. *Applied Partial Differential Equations*. Harper & Row, Publishers, New York, 1989.

[2] Bertil Gustafsson. On the convergence rate for difference approximations to a mixed initial boundary value problem. Technical Report 33, Department of Computer Science, Uppsala University, Uppsala, Sweden, 1971.

[3] Roger A. Horn and Charles R. Johnson. *Matrix Analysis*. Cambridge University Press, Cambridge, 1985.

[4] Eugene Isaacson and Herbert Bishop Keller. *Analysis of Numerical Methods*. Dover Publication, Inc., New York, 1994.

[5] Heinz-Otto Kreiss and Jens Lorenz. *Initial–Boundary Value Problems and the Navier Stokes Equations*. Academic Press, Inc., Boston, 1989.

[6] H.O. Kreiss. On different approximations of the dissipative type for hyperbolic differential equations. *Comm. Pure Appl. Math.*, 17:335, 1964.

[7] Erwin Kreyszig. *Introductory Functional Analysis with Applications*. John Wiley & Sons, New York, 1978.

[8] Jerrold E. Marsden and Anthony J. Tromba. *Vector Calculus*. W.H. Freeman and Company, San Francisco, 1976.

[9] Robert D. Richtmyer and K.W. Morton. *Difference Methods for Initial–Value Problems.* Interscience Publishers, New York, 1967.

[10] Gary A. Sod. *Numerical Methods in Fluid Dynamics.* Cambridge University Press, Cambridge, 1985.

[11] John C. Strikwerda. *Finite Difference Schemes and Partial Differential Equations.* Wadsworth & Brooks/Cole Advanced Books & Software, Pacific Grove, California, 1989.

[12] Angus E. Taylor. *Introduction to Functional Analysis.* John Wiley & Sons, Inc., New York, 1957.

[13] J.W. Thomas. *Numerical Partial Differential Equations: Conservation Laws and Elliptic Equations.* Springer-Verlag, New York, Accepted for publication.

Index